Spatial Analysis Methods and Practice

This is an introductory textbook on spatial analy_
GIS. Each chapter presents methods and metrics, explains ro...   t
results and provides worked examples. Topics include:

- Describing and mapping data through exploratory spatial data analysis
- Analyzing geographic distributions and point patterns
- Spatial autocorrelation
- Cluster analysis and multivariate data
- Geographically weighted regression and linear regression
- Spatial econometrics

The worked examples link theory to practice through a single real-world case study, with software and illustrated guidance.

- Exercises are solved twice: first through ArcGIS and then GeoDa.
- Through a simple methodological framework the book describes the dataset, explores spatial relations and associations and builds models.
- Results are critically interpreted, and the advantages and pitfalls of using various spatial analysis methods are discussed.

This is a valuable resource for graduate students and researchers analyzing geospatial data through a spatial analysis lens – including those using GIS in the environmental sciences, geography and social sciences.

After completing his postdoctoral studies in the United States, George Grekousis now teaches geography-related courses as an associate professor in China. His interdisciplinary research focuses on spatial analysis, geodemographics, and artificial intelligence. Dr. Grekousis has been awarded several grants from well-known international bodies, and his research has been published in several leading journals, including *Computers, Environment and Urban Systems*, *PLOS One*, and *Applied Geography*.

Spatial Analysis Methods and Practice

Describe–Explore–Explain through GIS

GEORGE GREKOUSIS

Sun Yat-Sen University (SYSU)

With solved examples in ArcGIS, GeoDa and GeoDa Space

CAMBRIDGE
UNIVERSITY PRESS

University Printing House, Cambridge CB2 8BS, United Kingdom

One Liberty Plaza, 20th Floor, New York, NY 10006, USA

477 Williamstown Road, Port Melbourne, VIC 3207, Australia

314–321, 3rd Floor, Plot 3, Splendor Forum, Jasola District Centre, New Delhi – 110025, India

79 Anson Road, #06–04/06, Singapore 079906

Cambridge University Press is part of the University of Cambridge.

It furthers the University's mission by disseminating knowledge in the pursuit of education, learning, and research at the highest international levels of excellence.

www.cambridge.org
Information on this title: www.cambridge.org/9781108498982
DOI: 10.1017/9781108614528

© George Grekousis 2020

First published 2020

Printed in the United Kingdom by TJ International, Padstow Cornwall

A catalogue record for this publication is available from the British Library.

Library of Congress Cataloging-in-Publication Data
Names: Grekousis, George, author.
Title: Spatial analysis methods and practice : describe - explore - explain through GIS / George Grekousis, Sun Yat-Sen University (SYSU), China.
Description: First edition. | New York, NY : Cambridge University Press, 2020. | Includes bibliographical references and index.
Identifiers: LCCN 2019038257 (print) | LCCN 2019038258 (ebook) | ISBN 9781108498982 (hardback) | ISBN 9781108712934 (paperback) | ISBN 9781108614528 (epub)
Subjects: LCSH: Spatial analysis (Statistics) | Geographic information systems.
Classification: LCC QA278.2 .G737 2020 (print) | LCC QA278.2 (ebook) | DDC 910.285–dc23
LC record available at https://lccn.loc.gov/2019038257
LC ebook record available at https://lccn.loc.gov/2019038258

ISBN 978-1-108-49898-2 Hardback
ISBN 978-1-108-71293-4 Paperback

Contents

Preface

As spatial data are more and more widely available, and as location-based services, from smartphone applications to smart cities monitoring, are becoming standard to everyday's humans' interaction and communcation, a growing number of researchers, scientists and professionals, far crossing the typical boundaries of geography discipline, realize the need for in-depth analysis of georeferenced data. Although geographical information systems map and link attributes to locations, spatial data hide much more treasure than a glossy mapping representation. To unlock this information, spatial analysis is necessary, as it provides the methods and tools to transform spatial data into knowledge, assisting in enhanced decision making and better planning. As such, a tremendous demand for accurately analyzing georeferenced data (including big data) across a wide range of disciplines exists.

To respond to this demand, *Spatial Analysis Methods and Practice* is an introductory book in spatial analysis and statistics through GIS. The book presents spatial data analysis methods and geoinformation analysis techniques to solve various geographical problems, following a "Describe–Explore–Explain" approach. Each chapter focuses on a single major topic, introduces the related theory, explains how to interpret metrics' outputs in a meaningful way and, finally, provides worked examples.

The topics covered include:

- Chapter 1: Think Spatially: Basic Concepts of Spatial Analysis and Space Conceptualization (Exercises solved with ArcGIS, GeoDa)
- Chapter 2: Exploratory Spatial Data Analysis Tools and Statistics (Exercises solved with ArcGIS, GeoDa)
- Chapter 3: Analyzing Geographic Distributions and Point Patterns (Exercises solved with ArcGIS)
- Chapter 4: Spatial Autocorrelation (Exercises solved with ArcGIS, GeoDa)
- Chapter 5: Multivariate Data in Geography: Data Reduction and Spatial Clustering (Exercises solved with ArcGIS, GeoDa, Matlab)
- Chapter 6: Modeling Relationships: Regression and Geographically Weighted Regression (Exercises solved with ArcGIS, Matlab)
- Chapter 7: Spatial Econometrics (Exercises solved with GeoDa space)

The book offers both a theoretical (*Theory*) and a practical (*Lab*) section for each chapter and adopts a "learn-by-doing" approach. *Theory* presents in detail concepts, methods and metrics, while *Lab* applies these metrics in solved step-by-step examples through ArcGIS and GeoDA. Matlab scripts are also offered for two labs.

Theory

Spatial analysis methods and techniques are described in a comprehensive and consistent way through the following subsections:

- *Definition:* Each subsection begins with the definitions of the methods to be presented. This allows for easy tracing of the definition of a new theory, concept or metric.
- *Why Use:* The "why use" statement follows. It offers an initial understanding of the importance of a method or metric and also presents the type of problems that these methods and metrics are more suitable to be applied.
- *Interpretation:* It is used to explain how we should interpret the outcomes of spatial analysis methods and metrics and goes one step further from just reporting numbers or maps without any further critical discussion.
- *Discussion and Practical Guidelines:* This section discusses the pros and cons of each method and metric. It also provides valuable tips on how to implement them from a practical perspective. For example, guidelines are offered to assist on how to select the appropriate parameters' values (of statistics/metrics/tools), thus avoiding accepting uncritically the default values offered by software. Experimenting through various parameters' values and settings allows for better insight on the impact of each parameter to the final outcome. Potential case studies are also presented.
- *Concluding Remarks:* A list of important remarks and guidelines is presented at the end of each chapter, summarizing the key topics of the theory.
- *Questions and Answers*: A set of 10 questions and answers is presented for self-evaluation.

Lab

Lab focuses on gaining hands-on practical experience through well-designed solved examples. All the main metrics included in *Theory* are presented in the *Lab* of each chapter. This allows readers to gain knowledge on how to perform spatial analysis and report results through step-by-step ArcGIS or GeoDa commands. This section also highly emphasizes how to critically interpret results so that spatial analysis leads to knowledge extraction assisting in enhanced decision making and spatial planning.

A single worked example runs through the whole book. By working on a single case study, readers can delve deeper into the different approaches applied in spatial analysis. Chapter-by-chapter readers will gain a better understanding of the study region and thus interpretation of results would be easier and more meaningful.

The general structure of each lab is as follows:

- *Overall Progress:* A workflow is presented at the beginning of each lab showing the progress of the entire project. The exercises that each lab is consisted of along with the tools to be used and the expected outcomes are also presented graphically.
- *Scope of Analysis:* The problem to be solved is described.
- *Actions:* Step-by-step software guidance is provided to describe how to solve the problem and report results.
- *Interpreting Results:* Results are interpreted from the spatial analysis perspective and in relation to the problem at hand.

The book is a valuable resource to a wide audience and is not strictly addressed to geographers. Analysts, teachers, instructors, students of various majors and researchers from interdisciplinary fields who are eager to analyze geospatial data can benefit from this book. It provides the necessary concepts, methods, metrics and the technical skills through geospatial analysis tools (ArcGIS, GeoDa, and GeoDa Space) to study a variety of real-world problems pertaining to socioeconomic issues, locational analysis and planning, human and urban analysis, and efficiently assisting public policy and decision making. No previous knowledge of spatial analysis is required.

I am grateful for the help and advice of so many scholars, but as omissions and mistakes are inevitable, I would greatly appreciate messages pointing out corrections or suggestions, so that this book further improves. Errata will be published on the book's website.

1 Think Spatially
Basic Concepts of Spatial Analysis and Space Conceptualization

THEORY

Learning Objectives

This chapter

- Presents the basic concepts, terms and definitions pertaining to spatial analysis
- Introduces a spatial analysis workflow that follows a describe–explore–explain structure
- Presents in detail the reasons that spatial data are special – namely spatial autocorrelation, scale, the modifiable area unit problem, spatial heterogeneity, the edge effects and the ecological fallacy
- Explains why conceptualization of spatial relationships is extremely important in spatial analysis
- Presents the approaches used to conceptualize spatial relationships
- Explains how distance, contiguity/adjacency, neighborhood, proximity polygons and space–time window are used in space conceptualization
- Defines the spatial weights matrix, which is essential to almost every spatial statistic/technique
- Introduces the real-world project along with the related dataset to be worked throughout the book

After a thorough study of the theory and lab sections, you will be able to

- Implement a comprehensive workflow when you conduct spatial analysis
- Distinguish spatial from nonspatial data
- Understand why spatial data should be treated with new methods (e.g., spatial statistics)
- Understand the importance of applying conceptualization methods according to the problem at hand
- Understand essential concepts for conducting spatial analysis such as distance, contiguity/adjacency, neighborhood, proximity polygons and space–time
- Describe the spatial analysis process to be adopted for solving the real-world project of this book
- Presents the project's data with ArcGIS and GeoDa

1

1.1   Introduction: Spatial Analysis

"In God we trust. All others must bring data," said W. Edwards Deming (American statistician and professor, 1900–1993), as without *data*, there is little to be done. Counting objects or individuals and measuring their characteristics is the basis for almost every study. With the advent of *geographic information systems* (GIS), it is simple to link nonspatial data (e.g., income, unemployment, grades, sex) to spatial data (e.g., countries, cities, neighborhoods, houses) and create large *geodatabases*. In fact, when data are linked to *location*, then analysis becomes more intriguing, and *spatial analysis* and the science of geography take over, as raw data are of a little value. Analyzing data through spatial analysis methods and techniques allows us to add value by creating information and then knowledge. Within this context, spatial analysis can be defined in various ways:

- **Spatial analysis** is a collection of methods, statistics and techniques that integrates concepts such as location, area, distance and interaction to analyze, investigate and explain in a geographic context patterns, actions, or behaviors among spatially referenced observations that arise as a result of a process operating in space.
- **Spatial analysis** is the quantitative study of phenomena that manifest themselves in space (Anselin 1989 p. 2).
- **Spatial analysis** studies "how the physical environment and human activities vary across space – in other words, how these activities change with distance from reference locations or objects of interest" (Wang 2014 p. 27).
- **Spatial analysis** is "the process by which we turn raw data into useful information, in pursuit of scientific discovery, or more effective decision making" (Longley et al. 2011).
- **Spatial (data) analysis** is "a set of techniques designed to find pattern, detect anomalies, or test hypotheses and theories based on spatial data" (Goodchild 2008 p. 200).
- **Spatial analysis** is a broad term that includes (a) spatial data manipulation through geographical information systems (GIS), (b) spatial data analysis in a descriptive and exploratory way, (c) spatial statistics that employ statistical procedures to investigate if inferences can be made and (d) spatial modeling which involves the construction of models to identify relationships and predict outcomes in a spatial context (O' Sullivan & Unwin 2010 p. 2).

Why Conduct Spatial Analysis?

Spatial analysis concepts, methods, and theories make a valuable contribution to analysis and understanding of

- **Social Systems:** Spatial analysis methods can be used to study how people interact in social, economic and political contexts, as space is the underlying layer of all actions and *interconnections* among people.

- **Environment:** Spatial analysis methods can be applied in studies related to natural phenomena and climate change hazards, natural resources management, environmental protection and *sustainable* development.
- **Economy:** Spatial analysis methods can be used to analyze, map and model interrelations among humans and various economic dimensions of economic life.

The main advantage of spatial analysis is the ability to reveal patterns in data that had not previously been defined or even observed. For example, using spatial analysis techniques, one might identify the *clustering* of a disease occurrence and then develop mechanisms for preventing expansion or even eliminating it (Bivand et al. 2008). In this respect, spatial analysis leads to better *decision making* and *spatial planning* (Grekousis 2019).

In a broad sense, there are four types of spatial analysis:

- **Spatial point pattern analysis:** A set of data points is analyzed to trace if it exhibits one of three states: *clustered, dispersed, random*. Consider, for example, a spatial arrangement of stroke events in a study area. Are they clustered to a specific region or are strokes randomly distributed across space? Spatial analysis proceeds with a further investigation, such as to determine the driving factors that lead to this clustering (potentially the existence of nearby industrial zones and related pollution). Point patter analysis also includes centrographics, a set of spatial statistics utilized to measure the center, the spread and the directional trend of point patterns. In this type of analysis, data typically refer to the entire population and not to a sample.
- **Spatial analysis for areal data:** Data are aggregated into predefined zones (e.g., census tracts, postcodes, etc.), and analysis is based on how neighboring zones behave and whether relations and interactions exist among them (i.e., clusters of nonspatial data also form clusters in space). For example, do people with high or low income cluster around specific regions, or are they randomly allocated? *Spatial dependence, spatial heterogeneity, spatial autocorrelation, space conceptualization* (through spatial weights matrix) and *regionalization* (*spatial clustering*) are central notions in this type of analysis.
- **Geostatistical data analysis (continuous data):** Geostatistical analysis is the branch of statistics analyzing and modeling *continuous field* variables (O'Sullivan & Unwin 2010 p. 115). In this respect, geostatistical data comprise a collection of sample observations of a continuous phenomenon. Using various geostatistical approaches (e.g., interpolation), values can be calculated for the entire surface. Pollution, for instance, is monitored by a limited network of observation locations. To estimate pollution for every single point, we may apply interpolation techniques through geostatistical analysis. Geostatistical analysis is not covered in this book.

- **Spatial modeling:** Spatial modeling deals mainly with how spatial dependence, spatial autocorrelation and spatial heterogeneity can be modeled in order to produce reliable, predicted spatial outcomes. Spatial modeling can be used, for example, to model how the value of a house is related to its location. Spatial regression and spatial econometrics are key methods in spatial modeling.

Spatial Analysis Workflow

As spatial analysis is a wide discipline with a large variety of methods, approaches and techniques, guidance on how to conduct such analysis is necessary. This book introduces a new spatial analysis workflow that follows a describe–explore–explain structure in order to address what, where and why questions respectively (see Figure 1.1).

   Step A: Describe (What). This is the first step in the spatial analysis process. It describes the dataset through *descriptive statistics*. Descriptive statistics are used to summarize data characteristics and provide a useful understanding about the distribution of the values, their range and the presence of outliers.

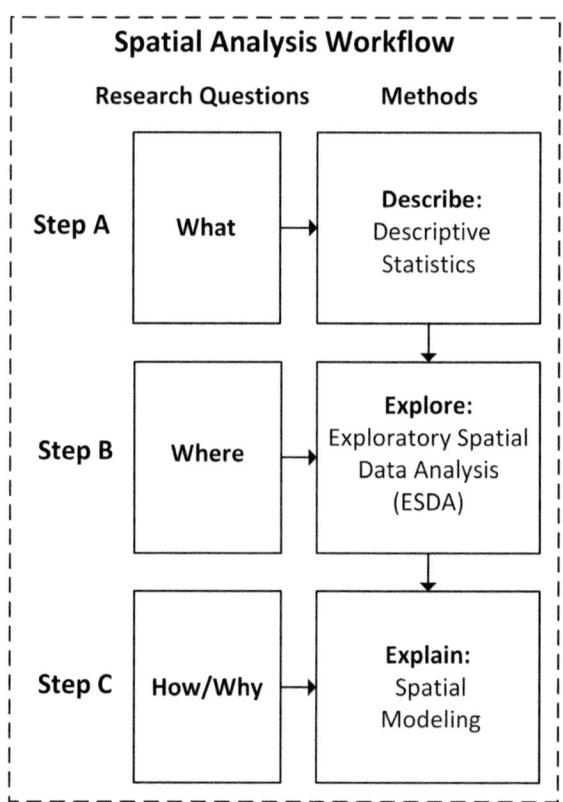

Figure 1.1 Spatial analysis workflow.

This step typically answers "what?" questions, such as what is the mean income of a neighborhood, or what is the population proportion living under the poverty level? This step offers an initial understanding of the dataset and its specific characteristics. Still, if the data have not been collected appropriately, then no analysis can lead to accurate and useful results. For this reason, any dataset should be checked for consistency and accuracy before any deeper analysis takes place. Datasets without detailed reports explaining the methods used and accuracies achieved should be avoided (always cite in your studies the link to the database used and the report that describes the methods used to collect the data along with the associated quality controls).

Step B: Explore (Where). In the second step, *exploratory spatial data analysis* (ESDA) is applied to explore and map data, locate outliers, test underlying assumptions or identify trends and associations among them, such as spatial autocorrelation presence or spatial clustering. In this step, we mostly answer "where?" questions, such as where are the areas with low/high values in income, is there any spatial clustering in the distribution of income per capita, where is it located, and where are the crime hot spots in a city?

Step C: Explain (Why/How). In the last step, explanatory statistical analysis through a spatial lens is applied to explain and understand causes and effects through models. In this step, we attempt to answer "why?/how?" questions. These methods do not just identify associations but also attempt to (a) unveil relations that explain why something happens and (b) trace the drivers behind a change. Typical questions in a geographic context include why do crime events cluster in a specific area? Is there any link to the specific socioeconomic characteristics of this area? Why is income per capita linked to location, how is income related to the size of a house? Which are the driving forces behind sea-level rise, and how does population increase drive urban land cover changes? In this type of analysis and in the context of this book, we treat variables as either independent or dependent. The dependent variable (effect) is the phenomenon/state/variable we attempt to explain. For example, if an analysis concludes that population increase (driver-independent variable) accounts for x% of urban land cover change (effect-dependent variable), then there is a linkage (a relation) established that explains the degree that the driver influences the effect. We have now built a model that explains why something happens which additionally can be used for predictions. This is a step beyond steps A and B, which mostly address "what happens" or "where something happens." Spatial regression and spatial econometrics will be described in this book concerning this stage of analysis. From the spatial analysis perspective, several additional questions could be also addressed at this stage: Can we learn something from this dataset and the applied methodology? Has new knowledge been created? What is the next step? How should future research proceed? When spatial analysis is completed, the knowledge created enhances decision making and spatial planning.

1.2   Basic Definitions

> **Box 1.1** More than 20 terms (in italics) related to spatial analysis, spatial statistics and *spatial thinking* (one more) have been mentioned in the preceding section. Some terms might be comprehensive, others entirely new and others quite vague. Let us start by building a common vocabulary and presenting some key definitions in this section. Definitions, terms and formulas typically vary among books, which confuses not only nonspecialists but scientists as well, causing much misunderstanding. This confusion has also hampered statistical, GIS and spatial analysis software, especially when referred to equations and formulas. This book presents the most commonly used names and symbols for terms and statistics.

Definitions

Spatial statistics employ statistical methods to analyze spatial data, quantify a spatial process, discover hidden patterns or unexpected trends and model these data in a geographic context. Spatial statistics can be considered part of ESDA, spatial econometrics and remote sensing analysis (Fischer & Getis 2010 p. 4). They are largely based on inferential statistics and hypothesis testing to analyze geographical patterns so that spatially varying phenomena can be better modeled (Fischer & Getis 2010 p. 4). Spatial statistics quantify and map what the human eye and mind intuitively see when reading a map depicting spatial arrangements, distributions or trends (Scott & Janikas 2010 p. 27; see also Chapter 2).

   Spatial modeling deals with the creation of models that explain or predict spatial outcomes (O'Sullivan & Unwin 2010 p. 3).

   Geospatial analysis is the collection of spatial analysis methods, techniques, and models that are integrated in geographic information systems (GIS; de Smith et al. 2018). Geospatial analysis is enriched with GIS capabilities and is used to design new models or integrate existing ones in a GIS environment. It is also used as an alternative term for "spatial analysis" (de Smith et al. 2018); strictly speaking, however, spatial analysis is part of geospatial analysis (see Box 1.2).

> **Box 1.2** It is not always easy to distinguish between the terms "geographic," "spatial" and "geospatial." These terms have been defined by many experts within different scientific contexts. The definitions provided here are not exhaustive; they serve as a basis for a common terminology. Even within the science of geography, terms can overlap, and distinctions can be vague. The term "geographic" refers to a location relative to the earth's surface combined with some type of representation. On the other

Box 1.2 (*cont.*)

hand, the term "spatial" does not refer solely to the earth's surface; its meaning is extended to a location combined with additional attribute data. The term "geospatial" is more computer oriented and refers to information based on both spatial data and models, combining geographical analysis with spatial analysis and modeling.

Spatial data refers to spatial entities with geometric parameters and spatial reference (coordinates and coordinate system) that also have other nonspatial attributes (see Figure 1.2; Bivand et al. 2008 p. 7). For example, we can describe a city by its population, unemployment rate, income per capita or the average monthly temperature. When these data are linked to a location through spatial objects (e.g., city postcodes), then we get spatial data. The range of attributes to be joined to the spatial objects depends on the problem being studied and the availability of datasets (e.g., census). Images that are georeferenced are also considered spatial data.

Conceptually, there are two ways to represent the geographic entities and as such to represent the world digitally: the object view of the world and the field view of the world (Haining 2010 p. 199).

Object view is a representation that describes the world with distinctive spatial objects that are georeferenced to a specific location using coordinates. In the object view, spatial objects are modeled as points, lines or polygons (also called features). This data model is called vector data model (O'Sullivan & Unwin 2010 p. 6). The object view of the world and the vector data model can be used to map, for example, demographic or socioeconomic data. Spatial objects might be represented differently when the scale of analysis changes.

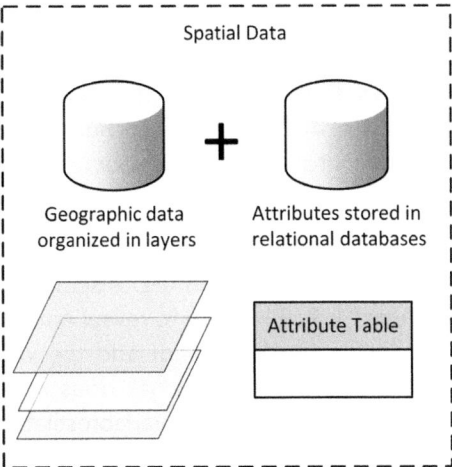

Figure 1.2 Spatial entities linked to attributes create spatial data.

For instance, a city might be represented as a point feature when examined on a national scale and as a polygon feature at a more local level.

Field view is a representation that describes the world as a surface of continuously varying properties (O'Sullivan & Unwin 2010 p. 7). The field view of the world is more appropriate for depicting a continuous phenomenon/ property (e.g., temperature, pollution, land cover type, height). A way to record a field is through the raster data model. In this model, rectangular cells (called pixels) organized in a regular lattice, depict the geographic variation of the property studied. Another way to model fields is by using triangulated irregular networks (TINs). We can convert spatial data from one model to the other (i.e., vector to raster conversion or vice versa) according to the study's needs.

Variable is any characteristic an object or individual might have. For example, age, height and weight are all variables characterizing humans, animals, or objects.

Attributes are information carried by spatial data. They are stored as columns in a GIS table. An attribute field is equivalent to a variable in classic statistics and has become the preferred term in GIS analysis, but these terms can be used interchangeably. An attribute might be the population of a postcode or the per-capita annual income in a census tract.

Data are produced when we measure the characteristics of objects or individuals.

Value is the result of a measurement (or response) of a characteristic. In statistics, the term **score** is also used to describe a variable's value.

Outlier is an unusual, very small or very large value compared to the rest of the values.

Dataset is a collection of variables of any kind of objects. Typically, a spatial dataset has a tabular format, whereby columns contain the attributes and rows contain the spatial entities.

Population is the entire collection of observations/objects /measurements about which information is sought.

Sample is a part of the entire population.

Level of measurement of a variable describes how its values are arranged in relation to each other (de Vaus 2002 p. 40). Variables/attributes are grouped at three levels of measurement: nominal, ordinal and interval or ratio (Haining 2010 p. 201; see Table 1.1).

- **Nominal** variables are variables with values that cannot be ordered. For example, race may be set as White = 1, Asian = 2, Hispanic = 3. This is a nominal variable, as the values "1, 2, 3" do not to reveal rank but are used as labels for the various categories. We cannot add the values of two different objects like, let's say, "1+3 = 4," as "4" does not reflect any meaningful value. Another example of nominal variables is the "Name of city" (e.g., Athens, Beijing, New York) or the "Land Cover Name" (e.g., Forest, Urban, Water). This type of attribute provides descriptive

Table 1.1 Level of measurement per data structure model (vector/raster) and examples per data type. Applicable logical and arithmetic operations are mentioned in parentheses. Many statistical procedures and techniques cannot be used at all levels of measurements, as different logical and arithmetic operations apply to different levels. For example, binary logistic regression is designed for dichotomous dependent variables and cannot be used for ratio variables. The level of measurement defines the pool of the statistical procedures to be used later in the analysis. From the statistical perspective, more techniques can be used to analyze ratio variables than can be used for nominal and ordinal variables; thus, ratio variables are preferred (de Vaus 2002 p. 43).

Level of measurement	Vector data model (object view)			Raster data model (field view)
	Point	Line	Polygon	Pixel
Nominal (=,$\neq$) Ordinal (=,$\neq$,>,<)	City name City most desirable to live (ranked)	Road name Road classification type: (Avenue, Highway)	Postcode ID Postcode classification according to education attainment	Land cover type Forest land cover subclasses
Interval (=,$\neq$,>,<,+,−)	Poverty level	Width of road	Poverty level for a postcode	Ground temperature
Ratio (=,$\neq$,>,<,+,−,x,/)	Population	Road freight	Postcode data: population, income per capita	Pollution PM2.5

information and can be used to label polygons on a map. The applicable operators are "equal" or "not equal" (=, $\neq$).

- **Ordinal** variables are variables whose categories can be ordered but whose numerical differences are not meaningful and cannot be calculated. For example, the variable "Student" might get the following values: "Exceptional" = 1, "Good" = 2, "Need to study harder" = 3. We can order categories from top to bottom (or vice versa), but there is no meaning in subtracting ("Exceptional" − "Good" = −1). We can apply the operators "equal," "not equal," "larger than" and "smaller than" (=, $\neq$, >, <). Spatial entity's attributes measured at nominal or ordinal levels are also called "categorical."

- **Interval and ratio variables** (also called "numerical") are variables for which each observation can be expressed in a numerically meaningful way. Numbers are not used only as labels but may be used to calculate statistics (e.g., the average). If the values of a numerical variable are limited to specific categories, then the variable is a discrete numerical, also called interval. The interval level is a class of ratio level. In interval-level measurement, categories are defined by fixed distances. Interval data allow for the operation of addition and subtraction (Haining 2010 p. 201). Still, interval variables do not preserve ratios (O'Sullivan & Unwin 2003 p. 13). Dichotomous variables (e.g., for the variable "sex," an individual might be Male = 1, Female = 0, or the inverse) can also be regarded as discrete interval-level variables. In this case, zero stands for the absence of something. If the set of possible values is not limited to some categories between low and high values, then the variable is a continuous numerical (also called "ratio"). Ratio variables have a meaningful zero. In ratio variables, we can use all operators (=, $\neq$, >, <, +, −, x, /).

To analyze variables/attributes, statistical methods can be employed. There are three major branches of classic statistics: **descriptive statistics, inferential statistics** and **explanatory statistics** (Linneman 2011 p. 20). We will deal with all of them in this book.

Descriptive statistics is a set of statistical procedures that summarize the basic characteristics of a given distribution. Descriptive statistics usually summarize a specific sample and are not appropriate for making inferences regarding the total population (unless we have the entire population at hand). As a result, they are not developed on the basis of probability theory as inferential statistics are. In this sense, the results of descriptive statistics apply only to the specific dataset they have been calculated for. Descriptive statistics make use of tables, graphs and simple statistical procedures (Linneman 2011 p. 21).

Inferential statistics is the branch of statistics that analyzes samples to draw conclusions for the entire population. Typical approaches for dealing with inferential statistics include tests of significance (hypothesis testing), confidence interval and Bayesian inference.

Explanatory statistics is the branch of statistics that uses methods and techniques to identify relations among variables and potentially "explain" causalities. In this type of statistics, variables are treated as dependent or independent (Linneman 2011 p. 21). The dependent variable is what we attempt to explain through a set of independent variables. Regression analysis is typically used in explanatory statistics.

1.3   Spatial Data: What Makes Them Special?

Consider that a realtor stores the contact numbers of his clientele in his cell phone. These contacts are data stored in his phone's memory in a casual type of database. If these data are linked to location (in terms of coordinates or addresses through geocoding), then they are transformed into spatial data. Each contact is now attached to a single spatial object (e.g., a point denoting the home address of each client that carries additional information – the attributes – such as phone number, name, date of birth or e-mail) that can be mapped. Transforming data to spatial data offers a lot more than a glossy visualization. It allows for in-depth geoprocessing analysis and advanced spatial querying. For example, which is the closest client to a specific point, where do the majority of clients live, where are clients who spend the most located, what is the best route to their homes, what percent of clients live within a zone of

1 km or 2 km from a predefined location, do clients cluster around specific neighborhoods, and what is the socioeconomic profile of these neighborhoods? Some questions can not be addressed in a timely manner before linking contacts to locations (imagine having thousands of clients) while others are impossible to be answered.

Analyzing spatial data through spatial analysis methods enriches our research by revealing hidden information. To unlock this information treasure, spatial data are analyzed using various descriptive, exploratory and explanatory statistical methods and techniques. The next chapters focus on these large classes of spatial techniques. Still, it may not be obvious if and why spatial data are different from nonspatial data and why we need to adopt new methods to analyze them. Let us see why spatial data are special (see also Box 1.3).

First, spatial analysis implies a focus on many geographically related parameters, such as location, distance, shape, area, neighborhood, adjacency and interaction. The inclusion of geographical parameters in spatial analysis differentiates data, methods and statistics from those of classical (nonspatial) data analysis (Anselin 1989 p. 2).

Second, many of the conventional statistical approaches used to analyze nonspatial data cannot be directly applied to spatial data because examining spatial data involves the following problems:

- The existence of **spatial autocorrelation/dependence**. Many statistical tests used for nonspatial data are based on the hypothesis that samples are randomly selected and observations are independent (O'Sullivan & Unwin 2010 p. 34). When we collect spatial data, however, this hypothesis is usually violated. This phenomenon is described as "spatial dependence." Tobler's first law of geography explains this basic property that rules spatial data, stating that "everything is related to everything else, but near things are more related than distant things" (Tobler 1979). Spatial dependence is closely related to spatial autocorrelation. Spatial autocorrelation is the degree of spatial dependency, association or correlation between the value of an observation of a spatial entity and the values of neighboring observations (Dall'erba 2009). The existence of spatial autocorrelation in a spatial dataset is not necessarily a problem. If there were no spatial dependence among objects, what would be the reason for spatial analysis? In other words, if space did not make any difference, then geographical analysis would be of no interest. Spatial autocorrelation exists in many geographical problems, so we have to adopt specific tools to handle it. Spatial autocorrelation will be analyzed in detail at Chapter 4.
- **Conceptualization of space.** Places are not isolated from each other. Social, economic and demographic interactions occur among adjacent or distant places. These interactions/relationships have a spatial dimension as they unfold over space, so location and distance matter. As

mentioned, according to Tobler's first law of geography, "near things are more related than distant things." To decode this rule, we have to define how "near" and "distant" an object should be from another object to be named as such and how an object should be depicted to calculate these distance metrics (e.g., point, line, polygon). When we apply methods to analyze spatial data, we have to determine mathematically how close a "close" object is, how "contiguity" is defined, what size a neighborhood is and how we can integrate "space–time" analysis. This means that we have to set a number of geographical parameters to define the spatial relationships among objects. This is called the **conceptualization of spatial relationships**, and it is a major difference between the methods applied to spatial and aspatial data. The next section of this chapter elaborates on this topic.

- The choice of the **geographical scale**. The appropriate geographical scale should be selected prior to any geographical/spatial analysis, as it directly affects the selection of the data model, the data to be collected, the methods to be used and the way conclusions will be drawn. For example, a city may be represented as a point at a national scale, as a polygon at the regional scale or as a set of polygons (e.g., postcodes) at a more local scale. For a more detailed analysis, we may go deeper at the census-track level (usually an area of 1,000 people).

- The choice of **scale of analysis** of data. The scale of analysis is closely related to the geographical scale, but it is not quite the same (although often it is used interchangeably). The scale of analysis defines the size and shape of the region that spatial statistics are calculated after the geographical scale has been set. It is essentially the level of understanding spatial phenomena and is closely related to the problem in question. The scale of analysis can be also considered as the geographical extent that spatial and temporal variability is analyzed through the construction of an appropriate spatial weights matrix. The geographical scale, on the other hand, refers to the scale of the data (i.e., national scale [1:100,000] or city scale [1:10,000]).

The scale of analysis might be different if we study a different variable within the same dataset. As such, although geographical scale typically remains the same across the dataset, the scale of analysis depends on the spatial distribution of the attributes' values. Generally, larger distances reflect broader trends (e.g., east to west) and smaller distances reflect more local trends (e.g., between neighborhoods). If we use a large scale of analysis, when we are looking at a local level, we might generalize and lose hidden spatial heterogeneity.

In a hypothetical example, unemployment clustering might be evident at 100 m and 1,000 m scale of analysis, reflecting patterns of clustering at both the census block level and the postcode level. In practice, the scale of analysis refers to the distance that the spatial features (e.g., postcodes) will be analyzed to calculate spatial statistics (see also Section 4.3).

- The **modifiable areal unit problem (MAUP).** Attribute data such as those from censuses or socioeconomic databases are often aggregated for privacy or simplicity reasons into predefined zones. The MAUP problem refers to the influence the zone design has on the outcomes of the analysis. A different designation would probably lead to different results. The main concern is that the definition of the zones (boundaries and extent) is arbitrary with respect to the specific geographical problem. For example, in many cities, postcodes are smaller in the center and become larger in the outskirts. How different would the statistical results be if postcodes were designed to have the same area? A typical and well-studied example of the MAUP is the 2000 US presidential election. Al Gore obtained more votes than George Bush but lost the election because of the way counties were designed inside each state (O'Sullivan & Unwin 2010 p. 38). A different designation of counties could have led to a different outcome. The MAUP relates to both the scale of the analysis and the aggregation of the data. In general, when we tend to have larger areal units, we tend to aggregate data at a higher level, such that generalization is more evident. Put simply, when generalization exists, aggregated values tend to be more similar to the overall mean (global mean), and deviations tend to be milder. In such a case, we may lose valuable information.
- **Space heterogeneity: Nonuniform space.** Space is not uniform. This is a major factor differentiating spatial data from nonspatial data. Our everyday life experience offers various examples of the non-smooth, noncontinuous and non-isotropic effects of space. For example, natural and planned breaks such as rivers, highways and parks alter space continuity. When space is not uniform, locations have different probabilities of a specific value/action or process. For example, the land value might be significantly higher in one side of the bank river compared to the other one. Spatial heterogeneity is the variation between the values of a set of observations inside the study area (Dall'erba 2009). When spatial heterogeneity exists, trends at specific directions (e.g., east to west) or large variations in neighboring observations might exist. For instance, socioeconomic differences are often very sharp in neighboring areas, something that reveals spatial heterogeneity (i.e., slums lie right next to high-income suburbs in many megacities).

When there is a nonuniform space or when spatial heterogeneity exists, collecting data might be problematic. Suppose a population is not distributed evenly across an urban area (population heterogeneity is high). If we collect and map stroke events across the city, we may find areas in which strokes are concentrated and form clusters. This does not necessarily mean that these areas have a higher risk of stroke events. The stroke-event clustering may be due to the fact that more people live in these areas (high population heterogeneity). Fewer strokes are expected in less-populated areas. In such a case, we have to take into account population distribution across space to better trace if there is a linkage between location and unexpectedly high rates of stroke. We might use measures of stroke density per capita for each subarea inside the city and adjust for population heterogeneity.

Spatial heterogeneity does not invoke the absence of spatial autocorrelation. Inside a study area, we may have areas with high spatial heterogeneity with negative spatial autocorrelation, and other areas with positive autocorrelation in a nonuniform space. In fact, spatial dependence is not easily distinguished from spatial heterogeneity. This is also referred to in the literature as the "inverse problem" (Anselin 2010). In spatial dependence, the correlation or covariance among variables at distinct locations is determined by the spatial arrangement of the objects in the geographic space (Anselin 2010). However, although clusters and patterns might be detected through various procedures such as spatial autocorrelation tests, we cannot determine if these clusters are due to structural change (heterogeneity) or to a true process that creates clusters irrespective of the space heterogeneity.

- The **edge effects problem.** In the edge effects problem, spatial units that lie in the center of the study area tend to have neighbors in all directions, whereas spatial units at the edges of the study area have neighbors only in some specific directions. Row standardization is typically used to account for this asymmetry in the count of neighbors (see the section on spatial weights later in this chapter).

- The **ecological fallacy.** This problem occurs when a relationship that is statistically significant at one level of analysis is assumed to hold true at a more detailed level as well. This is a typical mistake that occurs when we use aggregated data to describe the behavior of individuals. For example, if at the postcode level the variable "higher income" is strongly correlated to "higher education obtained," this does not necessarily mean that each person with higher education will have a high income. This is the fallacy problem – the belief that, if something stands true at an aggregation level, it is also true at a lower, more detailed level (e.g., the individual level). The correct interpretation is that postcodes linked to people with higher education tend to indicate higher incomes, not that each individual with higher education will have a high income. To reach a conclusion about the individual level and how education is linked to income, we should conduct research at this level of analysis (getting data at the individual level, not aggregated to some other level).

Box 1.3 Data that should have been treated as spatial are often analyzed without taking into account their spatial dimensions. This happens either because there is a lack of spatial thinking among the analysts or there is a

> **Box 1.3** *(cont.)*
>
> lack of knowledge of how to use spatial analysis tools. In this case, geographical space is withdrawn from the analysis, and data that should have been studied at various geographical and analysis scales are studied only at the scale for which data are available. As the spatial dimension is omitted, then one or more of the aforementioned reasons that make spatial data special emerge. For example, if spatial autocorrelation exists, then applying classical statistics often violates the assumptions essential for drawing statistically significant results, and the outcomes may be biased (Lee & Wong 2001). Thus, spatial data are special, and new methods and techniques that take into account spatial relationships and spatial conceptualization should be used for their analysis.

1.4 Conceptualization of Spatial Relationships

Definition
Conceptualization of spatial relationships is the modeling of the relationships and interactions between features across space. Put simply, it mathematically defines the terms *near*, *far*, *adjacent*, *contiguity*, *neighborhood*, *neighboring* and *distance* for a set of spatial objects by using specific values or functions.

Why Use
Referring once more to Tobler's first law of geography, objects that belong to the same neighborhood or are close to each other share common characteristics and are likely to interact more than those that are further away. Conceptualizing of spatial relationships is used to define what is to be regarded as close, far, adjacent or neighboring and is essential prior to any geographical analysis and spatial statistical tool implementation. Important decisions at this stage include

- The choice of the appropriate distance type to by applied (i.e., Euclidean, Manhattan, travel time or inverse distance)
- The contiguity definition (i.e., by polygons sharing common borders or by polygons in a predefined zone)
- The number of neighbors to be used for spatial statistics calculations (i.e., 10, 100 or just one)
- The method for selecting the nearest neighbor (i.e., by distance or by contiguity; if by distance, is it Euclidean or Manhattan? If by contiguity, do we consider those that share a border or those that overlap?)

Setting different spatial relationships leads to different spatial statistics outcomes for the same dataset. For this reason, it is crucial to define spatial relationships after thorough investigation avoiding the default values that most GIS software packages apply.

Interpretation

The more precise the conceptualization of spatial relationships for a set of spatial objects, the more accurate the outcomes of the statistical tests and models will be. If the applied conceptualization method fails to reflect the inherent structure of the spatial relationships of the spatial features, the outcomes of the analysis will be misleading. Suppose we want to define the catchment area (neighborhood) of a coffee shop so that we conduct later an analysis of the socioeconomic characteristics of the people living or working within this area. To model the spatial relationships between the people and the coffee shop, we make the following assumption: people within a 1 km radius of the coffee shop are more likely to interact with the coffee shop than are people outside this zone. This simple conceptualization of spatial relationships first defines the shape of the neighborhood (circle) and then its size (1 km). Different shapes (e.g., square or hexagon) and sizes (e.g., 2 km) could be used as well. Obviously, a catchment area defined by a 10 km radius would provide inaccurate results. It is more reasonable to expect that smaller distances would be more appropriate, as the main target group of this type of store is people who live or work within walkable distances. Different conceptualizations of the catchment area will lead to completely different results.

Discussion and Practical Guidelines

There are various methods of spatial relationships conceptualization, all of which attempt to better model the inherent structure of a spatial dataset in terms of spatial adjacency and neighborhood shaping. The following methods can be used:

A. **Distance.** Define distance threshold or distance function (see section 1.6):
    - Distance type (e.g., Euclidean, Manhattan, Minkowski, Network)
    - Fixed distance band (distance threshold)
    - Distance decay (distance function)

B. **Adjacency.** Define which objects are regarded as adjacent (see sections 1.6, 1.7):
    - Contiguity edges only (Rook's Case)
    - Contiguity edges corners (Queen's Case)
    - Higher-order contiguity
    - Interaction (this includes distance and adjacency)

C. **Neighborhood.** Define what makes a neighborhood (see section 1.8):
    - k-nearest neighbors
    - Proximity polygons
    - Delaunay triangulation

D. **Space–Time.** Define distance and time windows for spatiotemporal analysis (see section 1.8.2)

To decide which approach to use, we have to examine our problem from a conceptual perspective first and trace the potential spatial relationships. If we study traffic, it is more rational to use network or Manhattan distance instead of Euclidean distance. If we study population density in an urban agglomeration, it is more appropriate to use a distance decay function (e.g., inverse distance), as areas further away from the center are likely to be less densely populated.

Many spatial statistics tools require that a conceptualization method of spatial relationships be set prior to any analysis. Some of these statistics presented in detail in this book are as follows:

- Global spatial autocorrelation (Global Moran's *I*, General G-Statistic)
- Cluster and outlier analysis (Anselin Local Moran's *I*)
- Hot Spot Analysis (Getis-Ord Gi)
- Spatiotemporal Autocorrelations (Bivariate Moran's *I*, Differential Moran's *I*)
- Generate spatial weights matrix
- Geographically weighted regression

1.5 Distance Measure

Among the most common distance measures used in geographical analysis are the Euclidean distance, the Manhattan distance, the Minkowski distance, the Pearson's correlation distance, the Spearman correlation distance, the network distance, and the geodetic distance. In spatial statistics, Euclidean and Manhattan distance are those most widely used.

Definitions and Formulas

For two points A, B, where X_1, Y_1 are the coordinates of point A and X_2, Y_2 are the coordinates of point B, measured on a projected coordinate system (plane surface using Cartesian coordinates), distance can be defined as follows:

- **Euclidean** distance is the distance between two points A and B connected by a straight line calculated as (1.1):

$$S = \sqrt{(X_2 - X_1)^2 + (Y_2 - Y_1)^2} \qquad (1.1)$$

- **Manhattan** distance is the vertical plus horizontal difference (measured along the axes) between points A and B and is calculated as (1.2):

$$S = |X_2 - X_1| + |Y_2 - Y_1| \qquad (1.2)$$

Manhattan distance can be used to model the distance between points in urban environments when the street network is not available. Building blocks create barriers, and we usually cannot move directly from point A to point B in a straight line. Manhattan distance simulates this type of travel, which is why it should be used when we study attributes related to travel (e.g., access to schools, zone of influence of a shop) in an urban environment. When we analyze socioeconomic characteristics such as population, income or education, Euclidean distance is a more rational choice.

- **Minkowski distance** is a generalization of Euclidean and Manhattan distance (1.3):

$$S = ((X_2 - X_1)^p + (Y_2 - Y_1)^p)^{1/p} \tag{1.3}$$

For $p = 1$, we obtain the Manhattan distance and for $p = 2$, the Euclidean distance. Minkowski distance is used among others in Principal Component Analysis and Clustering Analysis (see Chapter 5).

Pearson's correlation distance considers two vectors to be close (similar) if they are highly correlated. In a multivariate dataset and for two variables with correlation r, Pearson's correlation distance is (1.4) (see Section 2.3.4 for correlation coefficient r definition and equation):

$$S = 1 - r \tag{1.4}$$

When we conceptualize spatial relationships by distance, in addition to choosing which of the preceding distance metrics to use, we should choose the range of distance within which spatial objects are regarded as having spatial relationships. The range of distance is closely related to the scale of analysis. Two widely used methods to define the range of distance are the fixed distance band and the distance decay.

1.5.1   Fixed Distance Band (Sphere of Influence)

Definition
Fixed distance band is a distance value expressing the size of a sphere of influence around a spatial object. All spatial entities inside the sphere of influence are weighted equally, while spatial entities outside the zone are assigned zero weight.

Why Use
Fixed distance band is used to shape neighborhoods containing equally treated spatial entities.

Interpretation
We define a sphere of influence by setting a fixed distance (a threshold distance value). Within this zone, all spatial objects are weighted equally, while objects outside the sphere are not accounted in calculations (for weight matrix calculation or spatial statistics), as they are not supposed to interact with objects inside the zone (i.e., their weight is zero).

Choosing an appropriate distance band value is crucial, as it sets the scale of the analysis and determines how spatial statistics will respond. For example, most of the spatial statistics metrics require at least one neighbor for each spatial feature. When the number of neighbors is small, the statistical results (e.g., z-scores) are unreliable. A small distance band might lead to objects with no neighbors. A large value might create excessively large neighborhoods and thus undesirable aggregations. In addition, with large distance bands, computational costs increase, especially for features with thousands of neighbors. We should use a distance band ensuring that each spatial object has at least one neighbor on the one hand and not too many on the other. When distributions are skewed, around eight neighbors should be included to yield reliable results. Various tools and procedures can be used to define the appropriate distance band for a specific dataset or study area, as explained next.

Tools for Determining Distance Band

There is no optimal distance band for a specific problem. Given heterogeneity in space, various scales of analysis can be applied. It depends on the problem at hand and the scopes of the analysis for determining the appropriate range of the distance band. Analyzing the distribution of schools in the core of a city in relation to the underlying population would require a relatively small scale of analysis ranging from some hundred meters to 1–2 km, as larger distances would not be reasonable for the daily transfer of children to schools. On the other hand, analyzing hospitals distribution for the same case study would require a larger distance band as hospitals have a larger catchment area. Common approaches to selecting the distance band include the following:

- Select a distance based on previous experience. If the researcher has solid knowledge of the spatial phenomenon studied and the case study area, or if there is a rich literature supporting a specific range of values to deal with a problem, then a distance band can be set accordingly.
- Use a distance band that ensures each object has at least one neighbor. Still, in the presence of locational outliers (see Chapter 3), this approach is inappropriate, as the distance band will be large enough. Although it will ensure at least one neighbor for some objects, it may create neighborhoods of hundreds of objects for others. Thus, the scale of analysis would be misleading. To overcome this problem, we should exclude locational outliers and calculate the distance at which each object has at least one neighbor. We should also set a minimum number of neighbors for those objects that do not have any neighbor inside the distance band selected. Using this approach, we define a more realistic distance band, ensuring that outliers are also included by calculating their weights based not on distance but on the count of nearby neighbors (for more details, see Section 4.3).
- Select a distance at which spatial autocorrelation is more pronounced (see Chapter 4). When spatial clustering exists, spatial autocorrelation techniques can be used to determine the appropriate distance band for the analysis. Such techniques include incremental spatial autocorrelation, optimized hot spot analysis and Ripley's k-function.

Discussion and Practical Guidelines

Before we define any appropriate distance band, we have to trace locational outliers (see Chapter 3). If the data are skewed, then the distance band should be neither so small that it includes only one or two neighbors nor so large that neighborhoods are as large as the study area because the z-scores computed in the various spatial statistics tests would be unreliable. In case of skewed data, z-scores are reliable when neighbors are approximately eight.

Most software packages calculate a default band whereby each object has at least one neighbor. Still, this is not always a suitable choice, especially when locational outliers exist (see the preceding section, "Tools for Determining Distance Band"). A fixed distance band is appropriate when we have a good theoretical background for the problem or good knowledge of the case study area. A fixed distance band also works well for point data and polygon data in hot spot analysis or other local spatial autocorrelation techniques.

1.5.2   Distance Decay

Definition

Distance decay is any function that implies a continuous, smooth and attenuating effect of distance on the attribute values of neighboring spatial entities (Longley et al. 2011 p. 108; see Figure 1.3).

Why Use

Distance decay is used to express the spatial interaction among various locations by applying weights that change relatively by distance so that closer objects have larger weights (stronger interaction) than those further away. It is used as the practical implementation of Tobler's first law of geography.

Interpretation

The distance decay concept considers that the physical or socioeconomic interaction between two points declines over space in a systematic way relative to distance. Various types of functions can be used, including reciprocal function (or inverse distance; see Figure 1.3A), negative power (or inverse distance squared for a power of 2; see Figure 1.3B), negative exponential or linear with a negative slope (which is uncommon). Inverse distance has a milder slope than inverse distance squared. A typical distance decay relationship (inverse distance or inverse distance squared) could be used to describe the height of the buildings in a city: the further away from the center, the lower the buildings become. Similarly, for population density, the further away from the city center, the lower the density.

Discussion and Practical Guidelines

It is generally recommended to use a fixed distance band in conjunction with a distance decay function (either as a cutoff point or a zone of indifference):

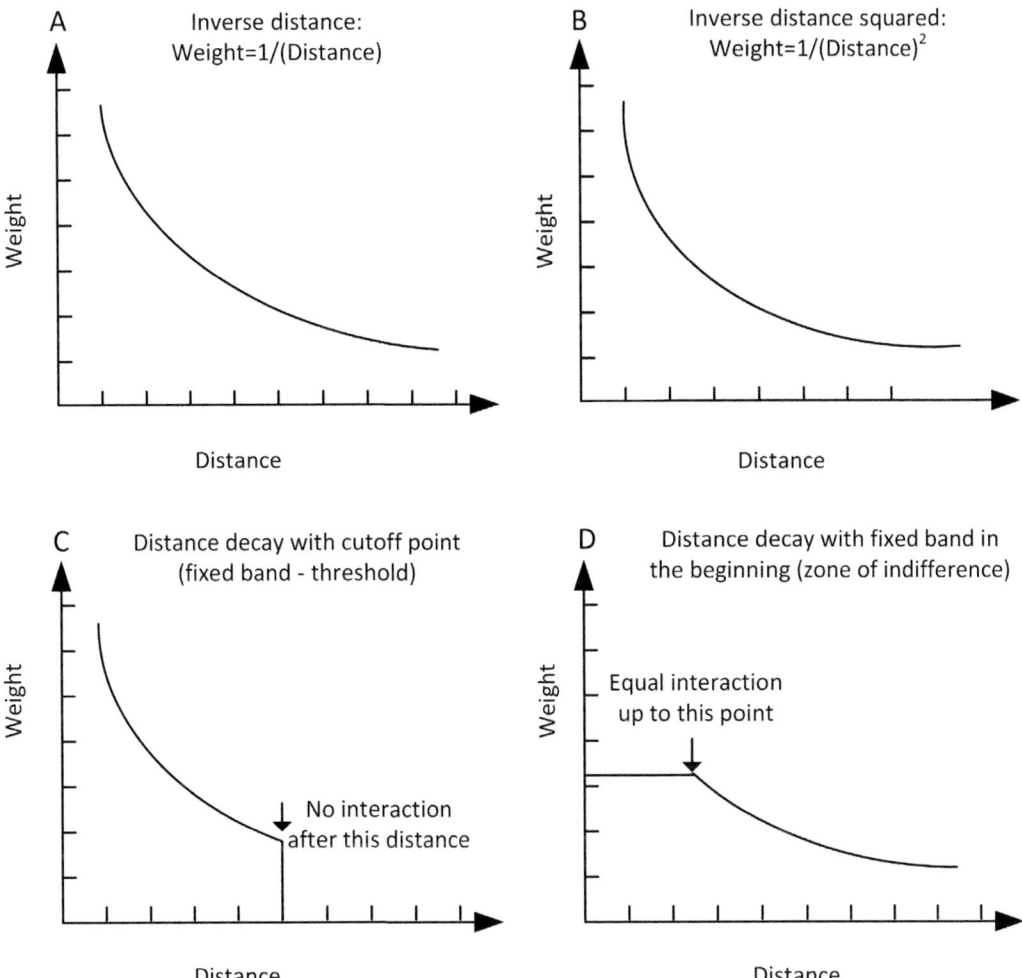

Figure 1.3 Distance decay graph. The larger the distance between two points, the less impact on each other. (A) Inverse distance has a similar but milder slope than Inverse distance squared. (B) We may apply inverse distance squared when we have evidence that distance influences objects to a higher degree than in inverse distance. (C) A threshold value (or fixed distance band) can be used to terminate a distance decay function. (D) A fixed distance band can be used prior to a distance decay function. This distance is also called the "zone of indifference," as all objects are treated equally inside it. After this distance, objects are weighted according to the distance decay.

- Fixed distance band applied as cut off point (see Figure 1.3C). Beyond this point, the distance decay is not calculated as no interaction is likely to be evident; any analysis using distances larger than this value would not provide valuable information.
- Fixed distance band applied prior to the distance function (zone of indifference) (see Figure 1.3D). In this case, we define a distance where

all spatial objects are treated equally. Beyond this distance, a decay function is applied, and weights are calculated accordingly. This approach is suitable when there is a zone where all objects have the same interaction. Instead of imposing sharp boundaries after this zone, we apply a distance function, so that there is a smooth effect of distance on the values of the neighboring entities.

Combining distance decay functions and a fixed distance band reduces computational time and reflects reality better. As objects at very large distances are likely to exhibit small interactions, the weights are almost zero, so instead of calculating them, we terminate the process at a predefined distance value. Distance decay functions may be used for continuous data. For instance, inverse Euclidean distance may be used to model temperature or pollution. On the other hand, inverse Manhattan distance might be more appropriate for point analysis inside urban environments (e.g., customer analysis) when the road network is not available. In hot spot analysis, distance decay inverse distance should be avoided, as it tends to produce small and isolated hot or cold spots (see Chapter 4).

Example

Suppose you want to select a coffee shop for your daily coffee consumption, and distance from your home is one of the most important criteria (in real life, additional parameters determine this choice, but we will use only distance; see Figure 1.4). It does not make a significant difference if the distance from your home to the coffee shop is 200 m or 250 m. In fact, if there are two coffee shops – one 200 m and the other 250 m from your location – you will most likely walk the extra 50 m if the second one is better. If the distance is 1 km, however, you may think twice about walking that far. Thus, we can use a fixed distance band for, say, 300 m around your home in which all available coffee shops are

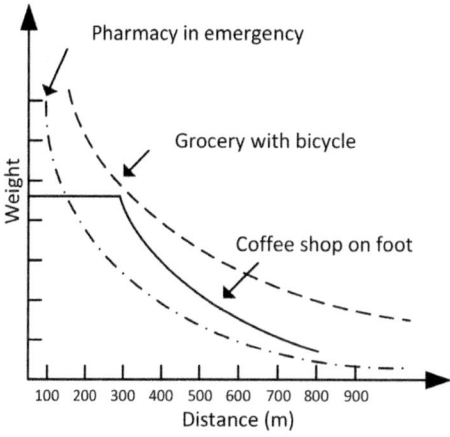

Figure 1.4 Examples of different distance decay used per analysis.

treated equally by you in terms of willingness to walk to reach them. After this distance, we may apply a distance decay function; for example, going to a coffee shop just 30 m further away (after your cutoff point of 300 m) is not irrational, whereas, outside this cutoff distance, all coffee shops receive proportionally lower weights in relation to distance and are not as likely to be visited as those inside the zone of influence. On the other hand, if you want to go to a pharmacy as soon as possible, then you do care which is the closest. An inverse distance squared would be an appropriate conceptualization method in this case, as you want to add more weight/importance to those closest to you. Finally, if you change mode and you aim to travel by bicycle, then you could elect to buy some food from your local grocery at a greater distance. An inverse distance would then be appropriate.

1.6   Contiguity: Adjacency Matrix

1.6.1   Polygons Contiguity

Definition
Contiguity is a spatial property that describes whether a target object and one or more other objects are in close proximity. In practice, contiguity refers to which polygons are assigned as neighbors for a single target object. The most common contiguity conceptualization methods for polygon features are:

- **Contiguity edges only (Rook's Case).** Only those polygons that share a common border (edge) are regarded as neighboring and are included in calculations for the target polygon (see Figure 1.5A).
- **Contiguity edges corners (Queen's Case).** In this case, polygons that share borders and also have common corners (nodes) are considered neighbors and are included in calculations for the target polygon (see Figure 1.5B).

Why Use
Contiguity is used to define neighborhoods used in the calculation of the spatial weights matrix and various spatial statistics.

Discussion and Practical Guidelines
The preceding methods of contiguity conceptualization are appropriate when we model data or contiguous processes represented by polygons in order to define neighborhoods or some type of interaction. The order of contiguity is also important in defining a neighborhood. First-order contiguity is when interaction and neighborhood are considered only for the immediately contiguous features, while a second-order contiguity reflects that the target object is affected from the neighbors of the neighbors (Anselin 2016). Something that should be considered

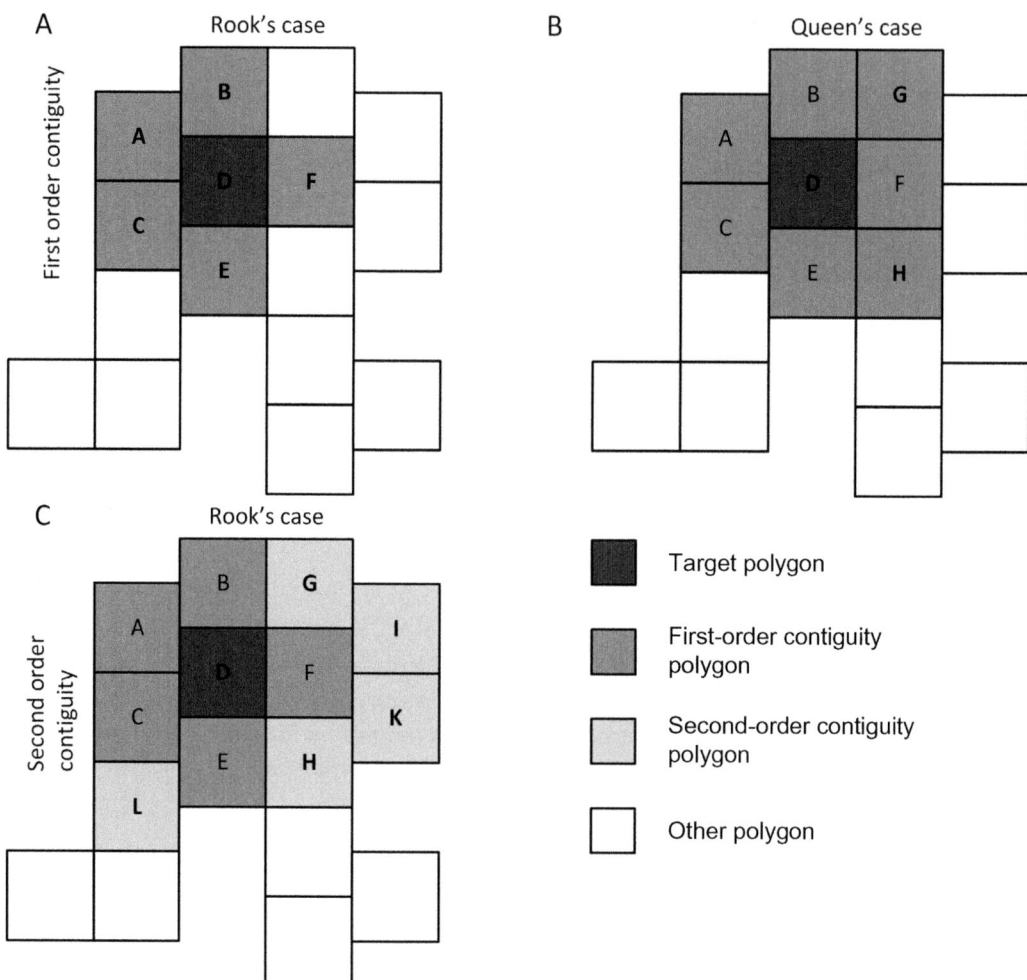

Figure 1.5 (A) Rook's (share border) definition of contiguity. The polygons immediately adjacent to target feature D are A, B, C, F and E (first-order contiguity). (B) Queen's case (share border and corners). G and H polygons are additionally included in the neighborhood of D, which is now composed of A, B, C, E, F, G and H (first-order contiguity). (C) Exclusive (pure) second-order contiguity for Rook's Case for D includes G, H, I, K and L as these are the neighbors of the first-order contiguity neighbors (A, B, C, E and F) of the target feature. In case of inclusive second-order contiguity the neighborhood of D is composed of both the neighbors and the neighbors of neighbors (A, B, C, E, F, G, H, I and K).

while calculating higher-order contiguity weights is whether or not the lower-order neighbors should be included (Anselin 2018; see Figure 1.5B). When higher order does not include lower order, it can be referred as exclusive (or pure) higher-order contiguity (Anselin 2018). When higher-order contiguity additionally includes lower order, it is named inclusive higher-order contiguity (Anselin 2018). For example, second-order contiguity that includes first order as well is named inclusive second-order contiguity. A pure higher-order contiguity (which does not include lower-order neighbors) is used when studying the effects of spatial lags on spatial autocorrelation and on spatial autoregressive models, where redundant and circular paths of polygon features are not taken into account (Anselin 2018).

Rook's or Queen's contiguity conceptualizations behave well when we consider that spatial interaction increases if two polygons share an edge, a node or both. In general, if the polygons are of similar size, then polygon contiguity is an appropriate conceptualization method. For polygons of different sizes (e.g., small polygons in the city center and large polygons in the outskirts), polygon contiguity applies different scales of analysis, which is undesirable. For this reason, row standardization should be used with polygon contiguity for spatial weights matrix calculation (see Section 1.9).

1.6.2   Adjacency Matrix

Definition
In the context of spatial analysis and for polygon representation, the **adjacency matrix** is a square matrix, the elements of which indicate whether pairs of polygons are adjacent or not.

Why Use
An adjacency matrix is used to represent the various forms of contiguity and to define neighborhoods for further spatial analysis.

Interpretation
An adjacency matrix is a symmetrical matrix the off-diagonal elements of which take values of either 0 or 1 and the diagonal elements of which have no values. If two spatial entities are adjacent (either first-order or higher-order), then the corresponding matrix element is set to 1 and 0 otherwise. For a given target polygon in the matrix, the polygons with an adjacency value of 1 (in the same row or column) define its neighborhood (the set of features to be taken into account for calculating spatial statistic for the target feature).

Discussion and Practical Guidelines
For the shaded polygons depicted in Figure 1.5A and Rook's Case contiguity (only share edges), the adjacency matrix is (1.5):

$$Adj = \begin{bmatrix} & A & B & C & D & E & F & SUM \\ A & * & 1 & 1 & 1 & 0 & 0 & 3 \\ B & 1 & * & 0 & 1 & 0 & 0 & 2 \\ C & 1 & 0 & * & 1 & 1 & 0 & 3 \\ D & 1 & 1 & 1 & * & 1 & 1 & 5 \\ E & 0 & 0 & 1 & 1 & * & 0 & 2 \\ F & 0 & 0 & 0 & 1 & 0 & * & 1 \\ SUM & 3 & 2 & 3 & 5 & 2 & 1 & * \end{bmatrix} \quad (1.5)$$

We use an asterisk (*) for polygons' adjacency with their selves. The sum of each row or column equals the total number of adjacent polygons to each single polygon. If we define as the neighborhood for a target object (polygon) those that are directly adjacent (first-order), then a polygon is a neighbor to the target object if the value in the matrix is 1; otherwise, they are not neighbors. For example, polygon A has three first-order neighbors (B, C, D).

1.7   Interaction

Definition
Interaction is the degree of linkage between two locations, the origin and the destination. It is calculated as a combination of distance and adjacency. In its simplest form, its formula is (1.6):

$$Interaction = w_{i,j} \frac{P_i P_j}{d_{i,j}^b} \qquad (1.6)$$

where

> i, j are the locations denoting the origin and destination respectively
> w is some short of weight between i and j. For example, w might be 0 if objects are not adjacent or 1 if they are.
> P_i and P_j are their respective values for a certain variable (e.g., population)
> d_{ij} is any distance function between i,j
> b is an exponent to determine the rate of declining

Why Use
In spatial statistical analysis, calculating interaction can be used as a way to calculate weights in a spatial weights matrix (see Section 1.9).

Interpretation
The larger the value, the stronger the interaction between two locations.

Discussion and Practical Guidelines
Spatial interaction models are mainly used to study spatial flows, such as in migration, tourism, commuting, international trade and money flow. For a spatial interaction to occur, three independent conditions must exist:

- A supply set of locations (destination) and a demand (origin) set of locations (e.g., commuters [demand] traveling to their job [supply])
- Alternative locations for both points of origin or destination
- Origins and destinations that are linked

1.8   Neighborhood and Neighbors

Definition
Neighborhood in the spatial analysis context is a geographically localized area to which local spatial analysis and statistics are applied based on the hypothesis that objects within the neighborhood are likely to interact more than those outside it.

Why Use
Defining the appropriate neighborhood is necessary for the accurate perform-ance of spatial statistics. Most of these statistics require a neighborhood definition and the construction of a spatial weights matrix that reflects the intensity of the relationships among the spatial entities.

Interpretation
The size of a neighborhood determines how observations are aggregated. Neighborhoods with too few or too many objects will probably yield unreliable statistical results.

Discussion and Practical Guidelines
Defining a neighborhood is not a trivial task, as there are many ways to conceptualize its formation. As explained previously, the distance band (Section 1.5.1) and polygon adjacency (Section 1.6.2) methods can be used to define neighborhoods. Other conceptualization methods include *k*-nearest neighbors, space–time proximity, proximity polygons and Delaunay triangulation, described next.

1.8.1   *k*-Nearest Neighbors (*k*-NN)

Definition
In the context of spatial analysis, **_k_-NN** is a method used to define a neighbor-hood based on the *k*-nearest neighbors to the target object.

Why Use
This method ensures that each object will have at least a certain number of neighbors. More broadly, it is used to

- Define the neighborhood (region) in which spatial objects will be accounted in spatial statistics calculations
- Define the neighborhood in which spatial objects are likely to be more similar than objects at further distances
- Model spatial relationships

Interpretation

k-NN is based on the calculation of a distance matrix used to store any type of distance among all objects in the dataset.

For example, the distance matrix (based on Euclidean distance) for all the polygon pairs in Figure 1.6 is (1.7):

$$
Distance =
\begin{bmatrix}
 & A & B & C & D & E \\
A & * & 4.0 & 2.2 & 2.9 & 4.3 \\
B & 4.0 & * & 4.1 & 2.5 & 4.2 \\
C & 2.2 & 4.1 & * & 2.4 & 2.6 \\
D & 2.9 & 2.5 & 2.4 & * & 1.9 \\
E & 4.3 & 4.2 & 2.6 & 1.9 & *
\end{bmatrix}
\tag{1.7}
$$

For *k* = 2, the nearest neighbors matrix (*NN*) is (1.8):

$$
NN =
\begin{bmatrix}
 & A & B & C & D & E & SUM \\
A & * & 0 & 1 & 1 & 0 & 2 \\
B & 1 & * & 0 & 1 & 0 & 2 \\
C & 1 & 0 & * & 1 & 0 & 2 \\
D & 0 & 0 & 1 & * & 1 & 2 \\
E & 0 & 0 & 1 & 1 & * & 2 \\
SUM & 2 & 0 & 3 & 4 & 1 &
\end{bmatrix}
\tag{1.8}
$$

To build the *k* = 2 nearest neighbor matrix (*NN*), for each row of the original *Distance* matrix (1.7), we keep the two (*k*) smaller values and replace them with 1. All other values are replaced with 0. This matrix (*NN*) resembles an adjacency matrix, but it is not symmetrical. Each row sums to 2 (as many as *k*), but each column results in a different outcome. The columns' sum reveals how many times a specific object (the header of the respective column) is nearest (either first or second in our case) to other polygons.

For *k* = 2, each neighborhood consists of three spatial objects (two neighbors and the target polygon). For example, the nearest neighbors to polygon

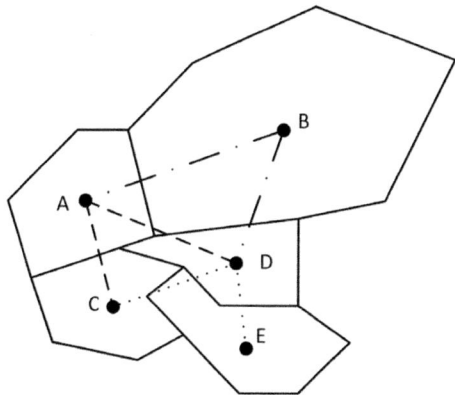

Figure 1.6 Example of *k*-nearest neighbors and distance matrix calculation.

B (check the row) are polygons A and D (see Figure 1.6). Polygon D is closer to B than to A and is thus called the "first nearest neighbor of B," while A is called the "second nearest neighbor of B" (see distance matrix). By inspecting rows A and D, we notice that polygon B is not nearest to any of these. This is not illogical, as other polygons are closer than B. Thus, the columns do not add up to the same number, but the rows should add up to two (k).

Discussion and Practical Guidelines

Setting the optimal number of k is not trivial. The selection is a trade-off between creating neighborhoods that are homogenous in characteristics and of similar aerial size. If spatial objects are distributed based on a competitive process – that is, if objects tend to lie from each other at similar distances throughout the entire study area – the k-nearest neighbors method produces reliable results when used in spatial statistics. Still, this method is vulnerable to polygon size or point density. When the distribution is not uniform or polygon sizes vary widely, using k-nearest neighbors might significantly change the scale of analysis. For example, postcodes in a city's business center are usually small in size, while postcodes in a city's outskirts are usually larger. If the same number of neighbors is set, the neighborhoods at the outskirts will be far larger than those in the center and will probably aggregate data, leading to valuable information loss. Thus, using a fixed k for the entire study area is not always desirable. It is advisable to use a distance decay function (with a distance band) and k-nearest neighbors only for those objects that do not have any neighbor at the fixed distance band. This will ensure that all objects will have at least one neighbor. When the distribution of values associated with the objects is skewed, a rule of thumb indicates that each object should have around eight neighbors.

1.8.2   Space–Time Window

Definition

Space–time window is a conceptualization method for defining a neighborhood based on both distance and time. An object is in the same neighborhood with the target object if it falls within the specified distance and also falls within the specified time interval.

Why Use

This method can be used to identify spatiotemporal hot spots or clusters created based on spatial and temporal proximity.

Interpretation

Spatial entities close to each other in both space and time are analyzed together, creating a spatio-temporal neighborhood. If a feature lies near to another in terms of distance but not in terms of space (or vice versa), it is not included in the spatial statistics calculations. In other words, two spatial entities close in distance might not be included in a spatiotemporal neighborhood if their temporal distance is large.

Discussion and Practical Guidelines

A distance band and a time window should be specified. If, for example, we provide a distance of 1 km and a time interval of three hours, all features found within 1 km from a target object that also have a date/time stamp within three hours of each other would be analyzed together, as they would form a spatio-temporal neighborhood. Selecting the appropriate distance band and time window is not trivial, and good knowledge of the region and the problem at hand is needed.

1.8.3   Proximity Polygons

Definition

Proximity polygons are polygons that divide space into regions so that the nearest centroid of each point in a region is the one in the polygon it lies in. Proximity polygons are also called Thiessen polygons or Voronoi polygons (O'Sullivan & Unwin 2003 p. 43).

Why Use

With proximity polygons, space is partitioned evenly regarding proximity. Thiessen and Voronoi polygons are tessellation methods that divide a plane into non-overlapping polygons and can be used for spatial interpolation to estimate catchment areas for public services or commercial businesses (Illian et al. 2008 p. 46).

Interpretation

For a given set of point features (e.g., post offices, police stations, coffee shops), proximity polygons can be used to create regions in which

- Each entity (e.g., post office) is the centroid of the polygon created
- Each point in space is included in only one polygon (no overlapping polygons)
- Each point in space is closest to the entity (e.g., post office) of the polygon centroid it belongs to
- Points at the boundaries lie at the same distance from the centroids of the respective polygons

Discussion and Practical Guidelines

By using proximity polygons, we create neighborhoods, or zones of influence, for a specific spatial entity. Taking post offices as an example, someone living within a specific proximity polygon is included in the neighborhood of the post office that lies in the centroid of this polygon. In environmental analysis, Thiessen polygons are used to estimate values of a continuous field (e.g., temperature, pollution). Any point inside the polygon will have the same value as the one at the centroid of the polygon. However, the sharp change in the neighboring points lying close to (i.e., in and out of) a shared boundary of two adjacent polygons can make this representation problematic. Delaunay triangulation is a smoother way to interpolate values, as is explained next.

1.8.4   Delaunay Triangulation and Triangular Irregular Networks (TIN)

Definition
Delaunay triangulation partitions space by creating triangles from point features or polygon centroids whose proximity polygons share an edge (O'Sullivan & Unwin 2010 p. 51).

Why Use
This method is suitable in cases where isolated polygons (e.g., islands) exist in the dataset or where the spatial distribution of objects is abnormal (e.g., when mixed large polygons with adjacent small polygons are scattered in the study area).

Interpretation
Points or centroids connected by triangle edges are regarded as neighbors. This ensures that each object will have at least one neighbor.

Discussion and Practical Guidelines
The edges of the triangles constructed using Delaunay triangulation are non-overlapping, creating a set of triangular facets that efficiently depict surfaces. These triangles create a network called triangular irregular network (TIN). Regarding space conceptualization, Delaunay triangulation creates neighborhoods consisting of quite regular triangles, as the minimum interior angle of all triangles is maximized, and long triangles are thus avoided. In a more general context, calculating TINS based on the height of points and combining them with vector data such as roads, streams or mountain peaks provide a more realistic view of the earth's surface.

1.9   Spatial Weights and Row Standardization

Definitions
Spatial weights are numbers that reflect some sort of distance, time or cost between a target spatial object and every other object in the dataset or specified neighborhood. Spatial weights quantify the spatial or spatiotemporal relationships among the spatial features of a neighborhood.
Spatial weights matrix is the matrix that stores the spatial weights.
Row standardization is the process of scaling the spatial weights to a range between 0 and 1. It is used to avoid biased data sampling or when data are aggregated from larger datasets.

Why Use

A spatial weights matrix is used to depict the degree of connection among the objects inside a specific neighborhood (Dall'erba 2009).

Interpretation

Any method of conceptualizing spatial relationships ends up creating a spatial weights matrix. This matrix quantifies the spatial relationships among the objects. It is then used as the fundamental matrix for most spatial analysis techniques. In its simplest form, it takes binary values (when used with fixed band distance, *k*-nearest neighbors and contiguity spatial relationships): 1 if two objects have a spatial relationship and 0 if two objects do not have a spatial relationship. For a distance decay function, interaction conceptualization method or user-defined weights, the matrix elements can receive any meaningful value. The larger the weight, the stronger the relationship.

Applying the inverse distance (1/(Distance)) conceptualization method for the polygons depicted in Figure 1.6, the elements of the spatial weights matrix (1.9) reflect the inverse of the distance of the corresponding set of polygons (see *Distance* matrix 1.7):

$$
Spatial\ Weights = \begin{bmatrix}
 & A & B & C & D & E & SUM \\
A & * & 0.244 & 0.455 & 0.345 & 0.233 & 1.276 \\
B & 0.244 & * & 0.250 & 0.400 & 0.238 & 1.132 \\
C & 0.455 & 0.250 & * & 0.417 & 0.385 & 1.506 \\
D & 0.345 & 0.400 & 0.417 & * & 0.526 & 1.688 \\
E & 0.233 & 0.238 & 0.385 & 0.526 & * & 1.382
\end{bmatrix} \qquad (1.9)
$$

A spatial weights matrix is almost always automatically generated by the software used when we apply spatial statistics by defining a conceptualization method.

Discussion and Practical Guidelines

Row standardization is a method for scaling weights to a range between 0 and 1. Each weight is divided by either the sum of all weights in its row or the sum of weights of the neighboring features (1.10). By doing so, results are adjusted so that the number of neighbors has no effect on the final results.

Standardized Spatial Weights

$$
= \begin{bmatrix}
 & A & B & C & D & E & SUM \\
A & * & 0.191 & 0.356 & 0.270 & 0.182 & 1 \\
B & 0.215 & * & 0.221 & 0.353 & 0.210 & 1 \\
C & 0.302 & 0.166 & * & 0.277 & 0.255 & 1 \\
D & 0.204 & 0.237 & 0.247 & * & 0.312 & 1 \\
E & 0.168 & 0.172 & 0.278 & 0.381 & * & 1 \\
SUM & 0.890 & 0.767 & 1.102 & 1.281 & 0.960 &
\end{bmatrix} \qquad (1.10)
$$

In the standardized matrix (1.10), every row adds up to 1. Column totals reflect how much interaction a spatial object has. For example, object D has the largest column sum, revealing that it has the strongest interaction with all the others. This also reveals a more central location. Object B has the lowest column total, indicating the least interaction with the remaining polygons. In the preceding example, we calculated the spatial matrix for all objects in the region. However, it is common to calculate spatial weights only for those that belong to the same neighborhood (according to the adjacency matrix).

Row standardization is recommended when there is a potential bias in the distribution of spatial objects and their attribute values due to poorly designed sampling procedures. For example, when we collect samples that are clustered in some parts of the study area (as a result of a poorly designed sampling method), some objects will have a higher probability of having more neighbors at close distances. As such, attribute values are likely to cluster together due to spatial autocorrelation. Row standardization mitigates these effects.

Row standardization should also be used when polygon features refer to administrative boundaries or any type of man-made zones. Census or socio-economic data are often aggregated from larger datasets to polygons of unequal sizes. Still, how polygons are designed might influence the outcomes of spatial analysis, something that is termed as the modifiable areal unit problem (see Section 1.3). For this reason, it is better to standardize weights as a percentage of the sum of their neighboring ones. This allows us to generalize our results at a higher level and mitigate zone design problems.

Row standardization is also recommended for binary weights matrix, fixed distance band and polygon contiguity conceptualization, to adjust for a different number of neighbors per observed object. Lastly, row standardization is not required for analyzing point features (i.e., traffic accidents, crime events) that have not resulted from aggregation or any sampling procedure.

1.10   Chapter Concluding Remarks

- The necessary steps of a spatial analysis workflow are describe, explore and explain.
- Spatial data might be points, lines, polygons or fields (i.e., pixels).
- Spatial data are special and thus must be treated differently.
- Spatial autocorrelation is a major reason why classic statistical approaches cannot be used with spatial data.
- The appropriate spatial relationship conceptualization method depends largely on the problem studied.
- A wrong selection of conceptualization method might yield unreliable statistical results.
- A very large or very small fixed distance might distort the results and lead to misinterpretation.

- Each of the conceptualization methods presented in this chapter assume that the effects of distance are continuous and uniform in every direction (isotropic). However, this is not always the case, as spatial heterogeneity is evident in reality.
- Having a non-isotropic or noncontinuous space does not prohibit the utilization of the preceding methods. It simply indicates that we should select the appropriate methods.
- The geographical scale determines the data representation, data collection, methods, research questions and outcomes. For this reason, setting the right scale is usually the first task in any geographical analysis.
- The selection of the appropriate scale of analysis defines the size and shape of the region that spatial statistics are calculated after the geographical scale has been set. It is essentially the level of understanding spatial phenomena and is closely related to the problem in question.
- It is quite common that smaller distances are more suitable for geographical analysis at the local scale.
- The scale of analysis can be also considered as the geographical extent that spatial and temporal variability is analyzed through the construction of an appropriate spatial weights matrix.
- We should not assume that a statistical relationship that holds at an aggregated level describes the behavior of individuals (the ecological fallacy).
- The influence of a spatial object on each neighboring one is not only a matter of distance. We thus have to include parameters other than distance effects. This will make the geographical analysis completer and more accurate.
- Row standardization is nearly always suggested in case of polygons.

Questions and Answers

The answers given here are brief. For more thorough answers, refer back to the relevant sections of this chapter.

Q1. What are spatial analysis and geospatial analysis? Is there any difference between these terms?

A1. Spatial analysis is a collection of methods, statistics and techniques, which integrates concepts such as location, area, distance, interaction to analyze, investigate and explain in a geographic context, patterns, actions or behaviors among spatially referenced observations that arise as a result of a process operating in space. Geospatial analysis is the collection of spatial analysis methods, techniques and models that are integrated into geographic information systems (GIS). Geospatial analysis is enriched with more advanced GIS capabilities and geoinformation applications than spatial analysis has and is used to design new models or integrate existing ones in a GIS environment. "Geospatial analysis" can be used as an alternative term for "spatial analysis"; strictly speaking, however, spatial analysis is part of geospatial analysis but focuses more on the methods than on the technologies.

Q2. What are the steps of a spatial analysis approach as presented in this book?

A2. Step A: Describe (What). This is the first step of a spatial analysis process and involves describing the dataset through descriptive statistics. Step B: Explore (Where). In the second step, exploratory spatial data analysis (ESDA) is used to explore and map data, locate outliers, test underlying assumptions and identify associations among them, such as spatial auto-correlation or spatial clustering. In this step, we mostly answer "where?" questions. Step C: Explain (Why/How). In the last step, explanatory statistical analysis through a spatial lens is applied to explain and under-stand causes and effects using models. In this step, we attempt to answer "why?/how?" questions

Q3. Based on the stages of a spatial analysis approach to solve a given problem, what are the key questions that spatial analysis attempts to address?

A3. Spatial analysis attempts to answer three basic sets of questions: **What, Where, and How/Why**. In the **"What"** set of questions, we study the status of specific variables. For example, what is the mean, maximum or minimum value of a variable? Along with exploratory spatial data analysis and related mapping techniques, it offers a first indication of how vari-ables are distributed in space. Then, we ask **"Where"** questions. Where are the areas with low/high values in income? Is there any spatial cluster-ing in the distribution of income per capita? Where is it located? Is there a crime hot spot? Where is it? These questions provide a solid analysis based on spatial statistics that quantify results as being significant or not and identify interesting spatial patterns that are not detectable other-wise. Finally, the analysis delves deeper by identifying and modeling relationships by answering **"How/Why"** questions. Models can be created that reveal how, for example, the value of a variable is related to its location.

Q4. What are the three levels of measurement on a variable/attribute? Give an example of each, and name their main differences.

A4. Variables/attributes are grouped at three levels of measurement: nom-inal, ordinal, and ratio. Nominal variables are variables with values that cannot be ordered. For example, race may be set as White = 1, Asian = 2 and Hispanic = 3. Nominal variables cannot be added or subtracted. Ordinal variables are variables whose categories can be ordered but whose numerical differences are not meaningful and cannot be calcu-lated. For example, the variable "Student" might get the following values: "Exceptional" = 1, "Good" = 2, "Need to study harder" = 3. This is an ordinal variable, as we can order categories from top to bottom (or vice versa), but there is no meaning in subtracting ("Exceptional" − "Good" = −1). Ratio are variables for which each observation can be expressed in a numerically meaningful way. For example, population is a ratio variable.

Q5. What is the conceptualization of spatial relationships? Why is it important?

A5. Conceptualization of spatial relationships is the modeling of the relationships and interactions between features across space. Put simply, is the mathematical definition of the terms *near, far, adjacent, contiguity, neighborhood, neighboring* and *distance* for a set of spatial objects by using specific values or functions. Spatial analysis techniques and calculations are based on how the spatial relationships among spatial objects have been modeled. Conceptualizing spatial relationships is important because the closer to reality a conceptualization of spatial relationships is, the more accurate the outcomes of the statistical tests or models will be. On the other hand, if the applied conceptualization method fails to reflect the inherent structure of the spatial relationships of the dataset, the analysis outcomes will be misleading.

Q6. What are the sphere of influence and the zone of indifference?

A6. A sphere of influence is a fixed distance value whereby all the spatial entities inside the zone it creates are weighted equally, while spatial entities outside the zone receive zero weight. It can be applied as a cutoff point at a distance decay function. A zone of indifference is a distance band used prior to a distance decay function. All objects are treated equally inside this zone. After this distance, objects are weighted according to the distance decay.

Q7. What is a neighborhood, and why is it important in spatial analysis?

A7. A neighborhood in the spatial analysis context is a geographically localized area to which local spatial analysis and statistics are applied based on the hypothesis that objects within the neighborhood are likely to interact more than those outside it. Defining the appropriate neighborhood is necessary for the accurate performance of spatial statistics. Most of these statistics require a neighborhood definition and the construction of a spatial weights matrix that reflects the intensity of the relationships among the spatial entities in this neighborhood.

Q8. What are Thiessen polygons and Delaunay triangulation? Why are they used?

A8. Thiessen polygons, also called "Voronoi polygons," are proximity polygons. Delaunay triangulation partitions space by creating triangles from point features or polygon centroids whose proximity polygons share an edge. Both methods are used to divide space into regions with specific attributes that can be used to define neighborhoods, calculate spatial interpolation or estimate catchment areas for public services or commercial businesses.

Q9. What are spatial weights and a spatial weights matrix? Why are they used?

A9. Spatial weights are numbers that reflect some sort of distance, time or cost between a target spatial object and every other object in the dataset or specified neighborhood. Spatial weights quantify the spatial or spatiotemporal relationships among the spatial features of a neighborhood. A spatial weights matrix stores the weights and is used to depict the degree of connection among the objects inside a specific neighborhood.

Q10. What is row standardization, and when should we apply it?

A10. Row standardization is the process of scaling the spatial weights to a range between 0 and 1. In row standardization, each weight for a spatial object is divided by either the sum of all weights in the row or the sum of weights of the neighboring features for this spatial object. By doing so, we adjust our results so that the number of neighbors has no effect on the final results. Row standardization is recommended mainly in two cases: (a) when there is a potential for bias in the distribution of spatial objects and their attribute values due to a poorly designed sampling procedure and (b) when data have been aggregated from larger datasets into polygons of different sizes.

LAB 1
THE PROJECT: SPATIAL ANALYSIS FOR REAL ESTATE MARKET INVESTMENTS

Each lab is consisted of two sections, namely Section A (ArcGIS) and Section B (GeoDa). Section A provides step-by-step instructions on how to solve the lab exercises using ArcGIS. Interpretation of the results and concluding remarks are also presented. Section B applies GeoDa functionalities to solve the same exercises. As such, readers may opt to solve the lab's exercises either using a leading commercial software or a well-established open-source freeware. The interpretation of the results as well as the conclusions related to the analysis are not repeated in Section B, as they are presented in the related Interpreting Results paragraphs of Section A. The reader should study these sections carefully, as they are independent of the software used. In addition, Overall Progress and Scope of Analysis sections precede Sections A and B to offer a better understanding of the spatial analysis process as well as the motivation for the analysis of each lab.

Overall Progress
Spatial Analysis/ Lab Workflow

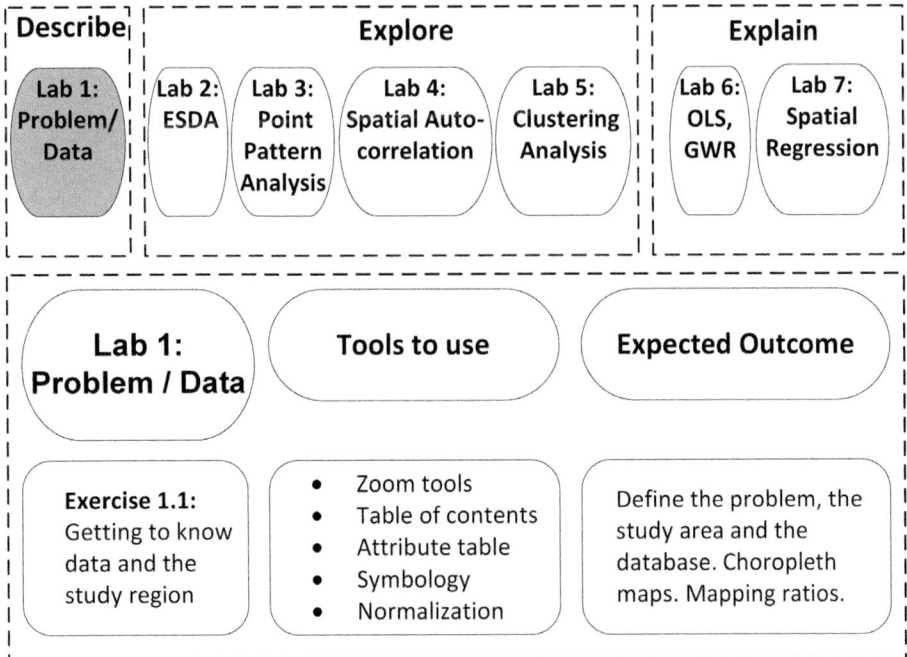

Figure 1.7 Lab 1 workflow and overall progress.

Scope of Analysis

A realtor wants to offer the best options to its clientele other than that of just listing its properties. As the company's moto is "location, location, location," it

Table 1.2 Project objectives and methods per lab.

No	Objectives	Methods	Lab
1	Location: High income	Mapping, Spatial autocorrelation	2,4,5
2	Location: Low Crime	Mapping, Centrographics, Spatial autocorrelation	3,4,5
3	Clustering: High spenders	Clustering	5
4	Modeling: Identify drivers	Regression, Spatial regression	2,6,7

provides geographical and socioeconomic data (for example, census data, income data, crime data) and offers location analytics through advanced spatial statistics and GIS. By conducting such an analysis, the company aims to become more competitive and also more reliable by answering tailored questions from each client regarding their unique investment needs. The more information available on location and surrounding areas, the higher the probability of successful investments. Spatial analysis offers the tools and mathematical background required to provide quantitative answers visualized through GIS maps that are usually better than personal opinions and general beliefs about a place.

This project deals with the following task. An investor seeks the best location to establish a successful new coffee shop and turns to the real estate company for advice and consultation.

Excluding rent and related costs (see Box 1.4), the investor is primarily interested in finding an appropriate neighborhood based on the following objectives (see Table 1.2):

1. As the service provided would be of premium quality, the coffee shop should be located in an area whose residents (i.e., potential clients/target group) have **high annual incomes**.
2. The coffee shop should be located in an area of **low crime**.
3. The target group should be likely to spend more money than average for expenses, including coffee-related products. **High spenders** should be identified.
4. The **socioeconomic drivers** behind people's monthly expenses (including those for coffee-related services) should be identified.

The fourth objective is not directly linked to finding an optimal location as described in the first three main objectives. It focuses on identifying the spatial relationships that can be used for modeling, market penetration and clientele analysis. Such an analysis should include detailed variables such as consumption preferences, everyday habits, type of job and the amount of money spent on coffee in coffee shops. For educational reasons and to keep the analysis brief, we will focus on primary socioeconomic variables.

Box 1.4 A complete study would also include such factors as the location of competitors, rent and related costs, access to public transport (i.e., subway), the budget of the investment, the number of daily passersby, the size of the

> **Box 1.4** (*cont.*)
>
> permanent population and the number of people working in nearby offices. It would be infeasible to answer these questions within a book and this project addresses only those dealing with space. The results can then be integrated into a market analysis conducted by marketers or business specialists to build a robust business and spatial plan. Spatial planning is key for success, not only in business but also in the implementation of national, regional and local policies on various issues, such as education, health, labor, emergencies and public administration.

In our case study, although the four objectives/questions seem simple, the analysis might prove endless (which is possible in spatial analysis). We will concentrate on a relatively large set of important questions and provide advanced modeling options. To address these questions, we will use many spatial analysis methods, such as exploratory spatial data analysis, spatial autocorrelation, data clustering, spatial clustering and spatial regression (see Tables 1.2 and 1.3).

The spatial analysis will go through three steps – Describe, Explore and Explain – while answering three basic questions: "What?" "Where?" and "How/Why?" (see Figures 1.1 and 1.7).

- For the **"What?"** set of questions, we will study the status of specific variables. For example, what is the mean income of the study area? What is the population with incomes higher than a specific value? Are there income outliers? This provides an initial understanding of the socioeconomic profile of the study area. Combined with exploratory spatial data analysis and related mapping techniques, it will offer a preliminary indication of how the variables are distributed in space.
- Then, we will ask **"Where?"** questions. Where are the areas with low/high income? Are there spatial clusters of areas with high income values? Is there a crime hot spot? These type of questions will provide a solid analysis based on spatial statistics that quantify results as being significant or not and also identify interesting spatial patterns that would not be detectable otherwise.
- Finally, the analysis will delve deeper by answering **"How/Why?"** questions. Several regression and econometric models will be created to model monthly expenditures (independent) based on a set of dependent variables (e.g., location, income).

Data

The study area is the city of Athens, Greece (referred to as the "city" hereafter). The spatial data refer to the postcodes of the city (polygons; see Table 1.4). The socioeconomic data refer to the 2011 census (see Table 1.5). Some of the census data are original, and some have been rescaled for reasons of confidentiality.

Tasks

The main analysis tasks are presented in Table 1.3.

Table 1.3 Project tasks per lab following the describe–explore–explain workflow.

Task	Lab	Tools to perform task	Why performing task
What (describe)			
Dataset and study area. Create and map ratios	1	Symbology	Define the problem, the study area and the database. To easily map ratios (e.g. population density).
Describe and map variables. Calculate and map z-scores	2	Choropleth maps, Histograms, Basic statistics (Skewness, Kurtosis, etc.), Normal QQ plot, Boxplots, Z-score rendering	To locate areas of high or low income. To find if distributions are skewed. To identify if distributions follow the normal distribution. To identify outliers.
Conduct correlation and pairwise correlation analysis in the dataset variables	2	Scatter plots, Scatter plot matrix	To identify if linear relationships exist among the variables. This will make modeling easier in a later step.
Where (explore)			
Analyze and measure the geographic distribution of crime events in the study area	3	Mean center, Median center, Standard distance, Standard deviational ellipse	To find if directional or temporal trends in crime locations exist. This will help assessing better the optimal location of the coffee shop.
Point pattern analysis	3	Average nearest neighbor, Ripley's k	To identify if the spatial pattern of crimes is random, dispersed or clustered. To find at which distance clustering or dispersion is more pronounced.
Create density maps	3	Kernel density estimation	To create a smooth map covering the study region depicting high or low densities in crime occurrences.
Locational outliers	3	Feature to point, Near	Identify if locational outliers exist and remove them to calculate spatial statistics.
Identify if spatial autocorrelation of income exists	4	Spatial weights matrix, Global Moran's I, Incremental Spatial Autocorrelation, Local Moran's I, Getis-Ord G*	To conceptualize space by creating the spatial weights matrix. To identify if high- or low-income values cluster in space. To locate hot or cold spots. To identify the scale of the analysis.

Identify if spatial autocorrelation of crime events exists	4	Optimized hot spot analysis	To identify in an optimized way if hot spots or cold spots of crime events exist. Hot spots of crime should be excluded from the potential locations for the coffee shop.
Multivariate data clustering	5	k-means clustering	Conduct geodemographical analysis based on a variety of socioeconomic variables.
Spatially constrained multivariate clustering	5	SCATTER	Spatial clustering (regionalization)
Similarity analysis	5	Similarity search (cosine similarity)	To identify similar postcodes to a target one as alternatively potential locations
Synthesis	5	Select by attributes/ location Export Reclassify	To identify the best location based on the evaluation criteria
How/Why (explain)			
Modeling relationships	6	Exploratory regression Ordinary least squares Geographically weighted regression	Model expenditures (independent) based on a set of dependent variables to identify the factors that increase expenditures
Modeling relationships	7	Spatial lag, spatial error, spatial regimes	Model expenditures (independent) based on a set of dependent variables utilizing spatial econometrics. Identify if expenditures are linked to spatial variables.

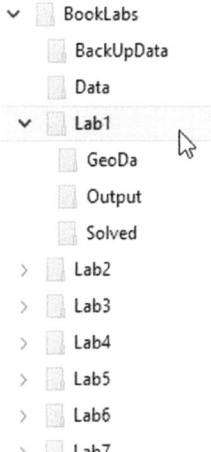

Figure 1.8 Dataset structure.

Dataset Structure

The structure of the datasets and related files under the folder BookLabs are as shown in Figure 1.8 (see also Box 1.5).

> **Box 1.5** Download Lab data from www.cambridge.org/9781108498982 and save them to I:\BookLabs\ (you can save data to other location if you prefer like C:\).

BackUpData folder stores the original data and serves as the backup of the data stored in the folder Data. In case of corrupted, accidentally deleted or wrongly edited data, you should copy the dataset from the BackUpData folder and paste it into the Data folder. Each subsequent folder (e.g., Lab1) stores the .mxd files for each specific lab and the Output and Solved folders. The Output folder is used to save the output files of your analysis, like shapefiles, graphs, pdf or images. This is the main folder of your data analysis. The Solved folder provides the solved exercise (e.g., Solved_Lab1_GettingToKnowDataSet.mxd) along with the final dataset after any tools applied and editing. Use this folder to compare with your results. An additional folder, GeoDa, is used only in exercises solved with GeoDa software (see Section B).

The spatial data (stored into the Data folder) used in this book are described in Table 1.4.

Table 1.4 Spatial data.

Files	Depicting
City.shp	90 postcodes (polygons) consisting the case study area
Downtown.shp	Outer polygon of the downtown area of the city
Assaults.shp	Point events of assaults crime
Burglaries.shp	Point events of burglaries crime
Crime.shp	Point events of crime (both assaults and burglaries)

The attribute data of the City shapefile are described in Table 1.5.

Table 1.5 Socioeconomic data refer to the 2011 census (rescaling has been applied for confidentiality). These variables are the attribute fields of `City.shp`.

Attributes	Description
Population	Total population (persons)
Density	Population density (persons per square meter)
Foreigners	Population per cent (%) of foreigners (other than Greek nationality)
Owners	Population percentage (%) owing a house (not paying rent)
SecondaryE	Population percentage (%) obtained secondary education or less
University	Population percentage (%) with bachelor degree
PhD_Master	Population percentage (%) obtained master or higher degree
Income	Average annual income per capita in euros
Insurance	Average monthly insurance cost per capita (in euros)
Rent	Average monthly rent (in euros)
Expenses	Average monthly per capita expenses for daily purchases (in euros – i.e., grocery, coffee)
Area	Area of post code in square meters
Postcode	Five-digit unique ID

Table 1.6 Basic symbols used in explaining interaction with software.

Symbols	Meaning
>	Next action
TOC	Table of contents
RC	Right-click
DC	Double-click
TAB =	Select TAB
=	Set value

Guidelines

This font and the term "ACTION:" indicate interactions with software. Folders, variables and file names will be also written in this font.
   The symbols used to explain actions are shown in Table 1.6.

Section A   ArcGIS

Exercise 1.1 Getting to Know the Data and Study Region

This exercise describes how a population can be mapped and how population density can be calculated and rendered.
   ArcGIS Tip: All mxd files have been created using ArcGIS 10.4 version. If the previous version is installed, open an empty mxd file and insert the shapefiles of each exercise from the Data folder.

Exercise 1.1 *(cont.)*

ArcGIS Tools to be used: Symbology, Zoom tools, Table of contents, Attribute table, Normalization

ACTION: Open dataset and map population

Navigate to the location you have stored the book dataset and double click Lab1_GettingToKnowDataSet.mxd

For example: I:\BookLabs\Lab1\Lab1_GettingToKnowDataSet.mxd

Tip: You can type this address directly to your windows explorer browser (just change the name of the drive-letter; if you have stored in C change I to C).

First, save the original file with a new name:

Main Menu > File > Save As > My_Lab1_GettingToKnowDataSet

In I:\BookLabs\Lab1\Output

TOC (Table of contents) > RC (Right-click) the City layer > Open Attribute Table (see Figure 1.9)

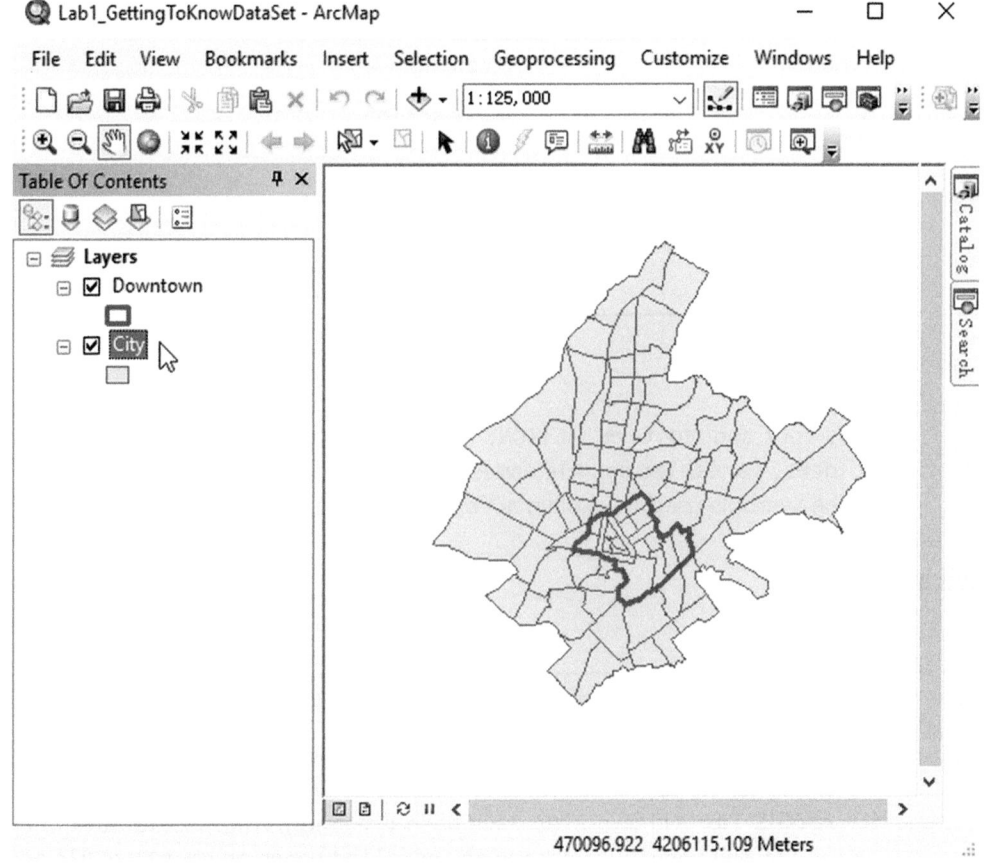

Figure 1.9 The case study area.

Exercise 1.1 *(cont.)*

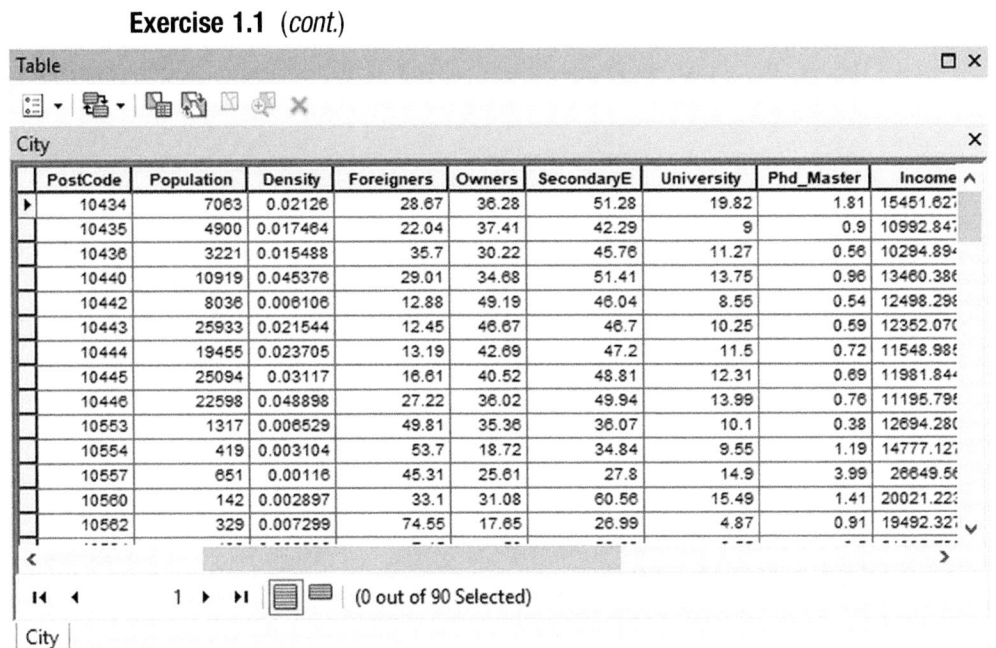

	PostCode	Population	Density	Foreigners	Owners	SecondaryE	University	Phd_Master	Income
▶	10434	7063	0.02126	28.67	36.28	51.28	19.82	1.81	15451.627
	10435	4900	0.017464	22.04	37.41	42.29	9	0.9	10992.847
	10436	3221	0.015488	35.7	30.22	45.76	11.27	0.56	10294.894
	10440	10919	0.045376	29.01	34.68	51.41	13.75	0.96	13460.386
	10442	8036	0.006106	12.88	49.19	46.04	8.55	0.54	12498.298
	10443	25933	0.021544	12.45	46.67	46.7	10.25	0.59	12352.070
	10444	19455	0.023705	13.19	42.69	47.2	11.5	0.72	11548.985
	10445	25094	0.03117	16.61	40.52	48.81	12.31	0.69	11981.844
	10446	22598	0.048898	27.22	36.02	49.94	13.99	0.76	11195.795
	10553	1317	0.006529	49.81	35.36	36.07	10.1	0.38	12694.280
	10554	419	0.003104	53.7	18.72	34.84	9.55	1.19	14777.127
	10557	651	0.00116	45.31	25.61	27.8	14.9	3.99	26649.56
	10560	142	0.002897	33.1	31.08	60.56	15.49	1.41	20021.223
	10562	329	0.007299	74.55	17.65	26.99	4.87	0.91	19492.321

I◀ ◀   1 ▶ ▶I   (0 out of 90 Selected)

City

Figure 1.10 A total of 90 postcodes with 17 fields are stored in the attribute table, of which 11 are socioeconomic variables (see Table 1.5).

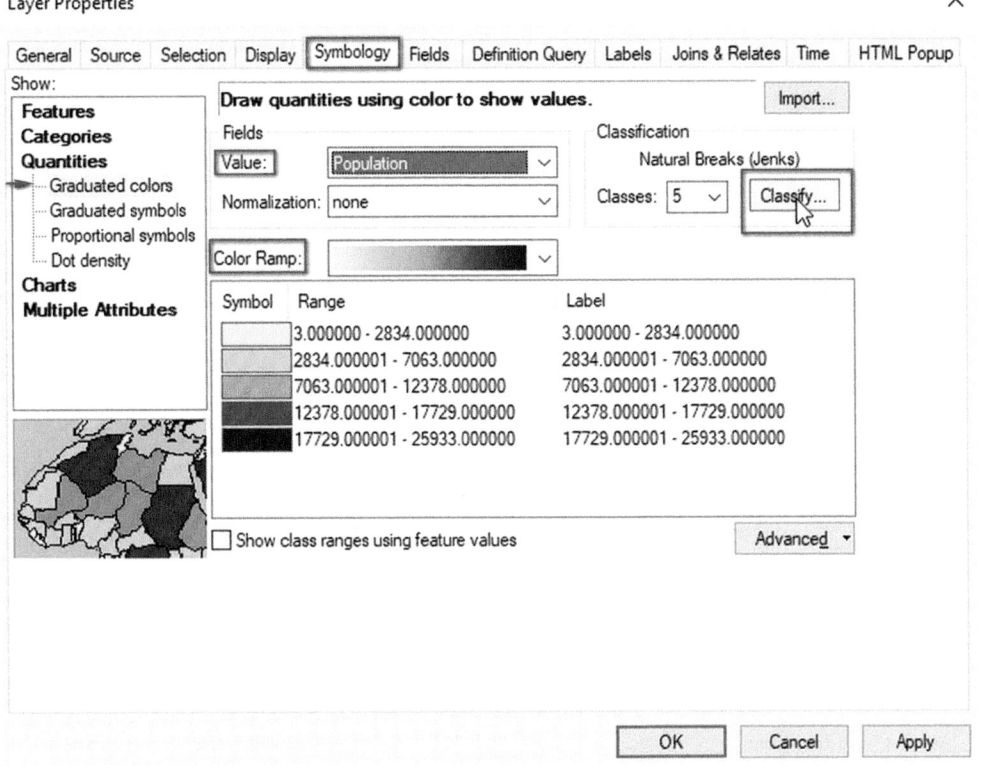

Figure 1.11 Layer properties dialog box for setting symbology.

Exercise 1.1 (*cont.*)

TOC > RC City > Properties > TAB = Symbology > Quantities > Graduated colors > Value = Population (see Figure 1.11)

Color Ramp = Yellow to Brown

Click Classify
Enter the following values in the Break Values window at the lower right (see Figure 1.12):

Break Values > 2000 > Enter > 5000 > Enter > 10000 > Enter > 15000 > Enter > 30000 > Enter > OK
RC Label > Format Labels > Select Numeric > Rounding > Number of decimal places = 2 > OK (see Figure 1.13)
Click Apply > OK

TOC > RC City > Save As Layer File > (see Figure 1.14)

Name = Population.lyr

In I:\BookLabs\Lab1\Output

Add the layer in the TOC.

Save

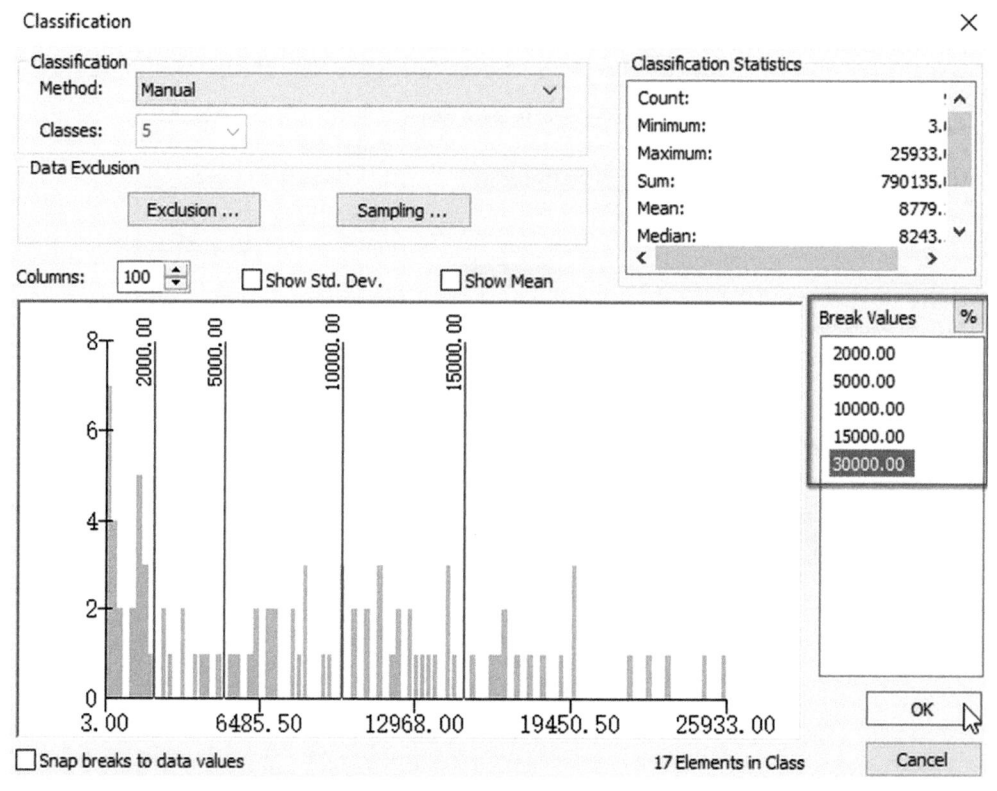

Figure 1.12 Setting categories range values.

Exercise 1.1 (*cont.*)

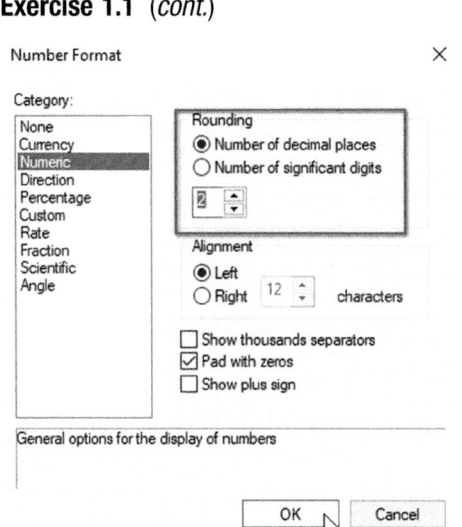

Number Format   ✕

Category:
| None |
| Currency |
| Numeric |
| Direction |
| Percentage |
| Custom |
| Rate |
| Fraction |
| Scientific |
| Angle |

Rounding
◉ Number of decimal places
○ Number of significant digits

[2] ⏶⏷

Alignment
◉ Left
○ Right   12 ⏶⏷   characters

☐ Show thousands separators
☑ Pad with zeros
☐ Show plus sign

General options for the display of numbers

OK   Cancel

Figure 1.13 Defining number format.

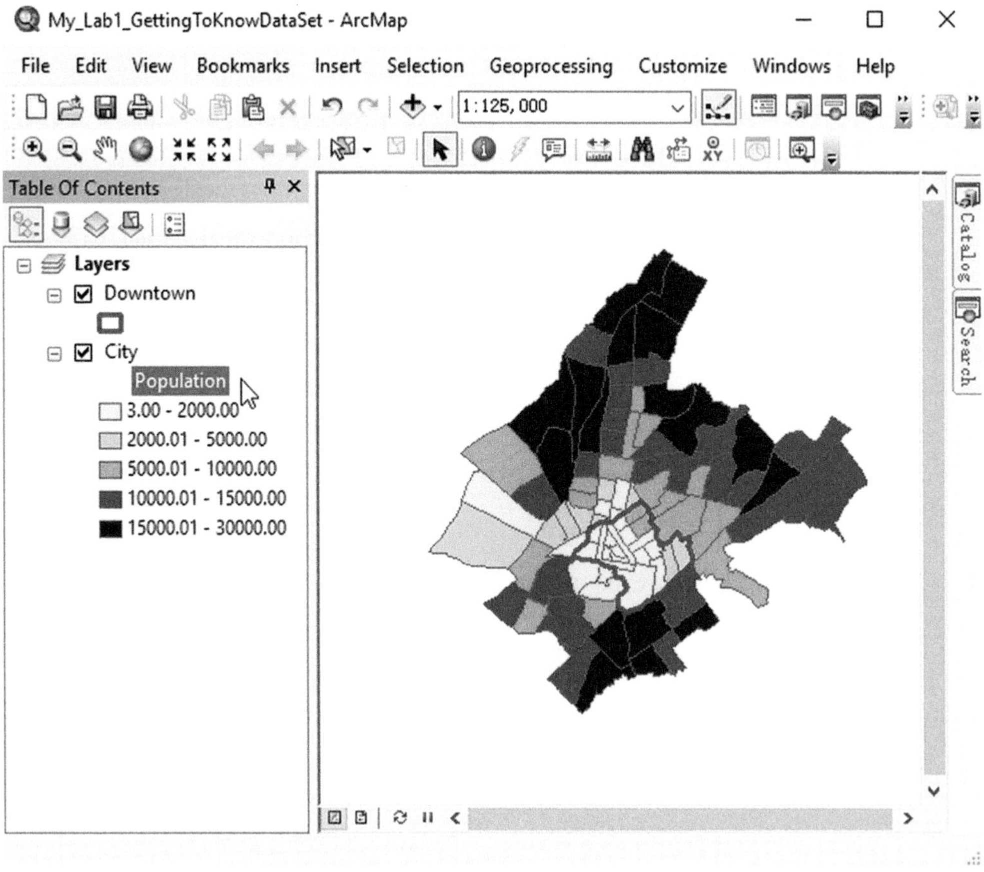

My_Lab1_GettingToKnowDataSet - ArcMap   —   ☐   ✕

File   Edit   View   Bookmarks   Insert   Selection   Geoprocessing   Customize   Windows   Help

1:125,000

Table Of Contents   ⛶ ✕

☐ 🗐 **Layers**
  ☐ ☑ **Downtown**
     ☐
  ☐ ☑ **City**
     Population
     ☐ 3.00 - 2000.00
     ☐ 2000.01 - 5000.00
     ☐ 5000.01 - 10000.00
     ■ 10000.01 - 15000.00
     ■ 15000.01 - 30000.00

Catalog   Search

Figure 1.14 Population choropleth map.

Exercise 1.1 (*cont.*)

Interpreting results: The case study area is consisted of 90 postcodes (spatial features; see Figure 1.9). By opening the attribute table of `City`, we inspect the variables stored and the corresponding values for each spatial feature (see Figure 1.10). Postcodes in downtown have a lower population than the postcodes in the outskirts (see Figure 1.14). As the central postcodes are smaller in size, it is advised to additionally depict population density, as it provides a better mapping of population distribution within a study area.

ACTION: Calculate and map population density

RC the City layer (not the Population.lyr) > Properties > TAB = Symbology > Quantities > Graduated colors

Value = Population

Normalization = Area

Color Ramp = Light Green to Dark Green

Classes = 4

Click Classify > Break Values > 0.0100 > Enter > 0.0200 > Enter > 0.0300 > Enter 1.000 > OK

Density and break values refer to population per square meter. In practice, 0.01 means that 0.01 people live within $1m^2$ or 1 person per $100m^2$.

RC Label > Format Labels > Numeric > Number of decimal places = 2 > OK > Apply > OK

TOC > RC City > Save As Layer File >

Name = PopDensity.lyr

In I:\BookLabs\Lab1\Output

Add the layer in the TOC.

Main Menu > File > Save

Tip: Saving `Population` normalized by area into a layer file (`.lyr`) saves the density representation. When you add a layer in the table of contents, it is given the name of the original shapefile created (`City` in this example) and not the name it was saved (i.e., `PopDensity.lyr`; see Figure 1.15).

Interpreting results: Choropleth map of population density depicts smaller densities (for most postcodes) in the city center (downtown – red polygon)

Exercise 1.1 (*cont.*)

that grow larger as we move outward (see Figure 1.15). We locate a cluster of densely populated postcodes at the northern part of the city. On the other hand, population density is lower in downtown area probably because of its business and historic character (with fewer permanent residents). Similarities with the population map (see Figure 1.14) can be identified, but overall, population density map offers a better insight on how population is distributed across the postcodes. For example, in the downtown area, postcodes are described in more detail with population density compared to population.

　　ArcGIS tip: The normalization procedure in ArcGIS is used to divide one variable with another. This offers the ability to calculate rates of change (population increase), percentages (e.g., land cover share), per capita

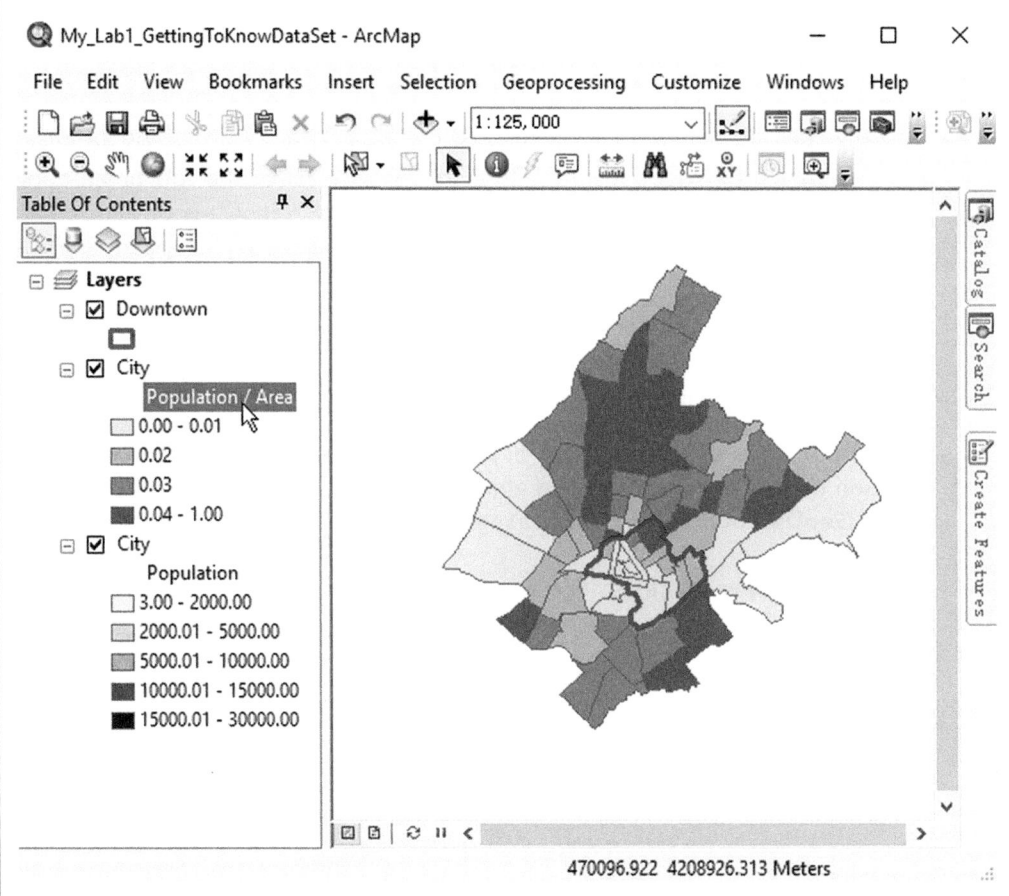

Figure 1.15 Population density.

Exercise 1.1 (*cont.*)

numbers (e.g., income per capita) and densities (e.g., population density). It should not be confused with the normalizing process that rescales data to a range [0,1] or [−1,1] (see Section 2.4). The normalization tool in ArcGIS is an adjustment that divides two variables. For example, if we have the aggregated income of all people living in each postcode, we can calculate and map the per capita income. A drawback of this tool is that we cannot obtain the values of the population density in a new field. We simply map the results. We can easily though produce this ratio using field calculator procedures. The normalization tool in ArcGIS is very useful when we need to test for various combinations of ratios. Once we decide which of the tested ratios to retain, we can calculate the values by using the field calculator.

Section B   GeoDa

Box 1.6 Download and install GeoDa free and open-source software through http://geodacenter.github.io. Browse also the documentation section where you can find a detailed workbook. GeoDa is developed by Dr. Luc Anselin and his team and their contribution to the spatial analysis field is paramount.

Exercise 1.1   Getting to Know the Data and Study Region

This exercise describes how a population can be mapped and how population density can be calculated and rendered.
   GeoDa Tools to be used: Category editor, Zoom tools, Table

ACTION: Open dataset and map population

Navigate to the location you have stored the book dataset and click Lab1_GettingToKnowDataSet_GeoDa.gda inside the GeoDa folder

For example:

I:\BookLabs\Lab1\GeoDa\Lab1_GettingToKnowDataSet_GeoDa.gda

Exercise 1.1 (*cont.*)

Tip: You can type this address directly to your windows explorer browser (just change the name of the drive letter; if you have stored in C, change I to C). Spatial data for GeoDa exercises are stored into GeoDa folder and not into Data folder. For spatial data and attribute values see Tables 1.4 and 1.5.

```
Main Menu > Click the Table icon (see Figures 1.16 and 1.17).
Click then on the Map-CityGeoDa window to activate it.
Main Menu > Map > Custom Breaks > Create New Custom Breaks >
(see Figure 1.18)
On the Variable Settings window (see Figure 1.19) select:

Population > OK
On the window 'New Custom Categories Title' type: Custom Breaks
(Population) and click OK
```

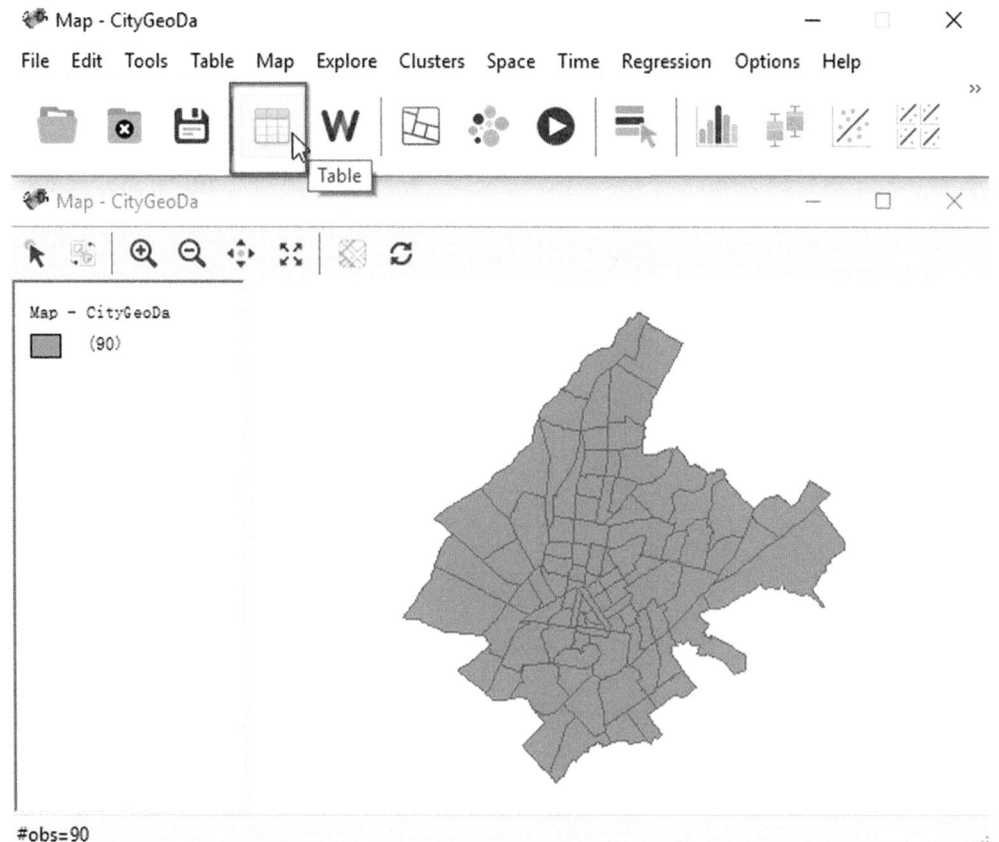

Figure 1.16 The case study area in GeoDa.

Exercise 1.1 (cont.)

Table - CityGeoDa

	Municipali	PostCode	Population	Secondaryf	University	Phd_Master	Income	Insurance	Rent	Expenses	Area	Regimes	Density	Foreigners	Owners
1	ATHENS	10434	7063	51.280000	19.820000	1.810000	15451.630000	212.980000	632.040000	178.583919	332220.136904	2	0.020000	28.670000	38.280000
2	ATHENS	10435	4900	42.290000	9.000000	0.900000	10992.850000	118.410000	623.700000	79.890622	285381.940898	2	0.020000	22.040000	37.410000
3	ATHENS	10436	3221	45.760000	11.270000	0.560000	10294.890000	108.400000	624.930000	72.294962	207968.061788	2	0.020000	35.700000	30.220000
4	ATHENS	10440	10919	51.410000	13.750000	0.860000	13460.380000	116.000000	610.380000	98.658519	240635.086409	2	0.050000	29.010000	34.680000
5	ATHENS	10442	8036	46.040000	8.550000	0.540000	12498.300000	146.160000	492.590000	99.116531	1315988.950180	2	0.010000	12.880000	49.190000
6	ATHENS	10443	25933	46.700000	10.250000	0.590000	12352.070000	141.770000	480.810000	106.902080	1203704.016840	2	0.020000	12.450000	46.670000
7	ATHENS	10444	19455	47.220000	11.500000	0.720000	11548.990000	133.530000	577.480000	92.826920	820720.175172	2	0.020000	13.190000	42.690000
8	ATHENS	10445	25094	48.810000	12.310000	0.690000	11981.840000	138.630000	568.090000	108.358586	809075.936223	2	0.030000	16.610000	40.520000
9	ATHENS	10446	22598	49.940000	13.990000	0.760000	11195.800000	108.430000	608.680000	117.530858	462145.820779	2	0.050000	27.220000	36.020000
10	ATHENS	10553	1317	40.070000	10.100000	0.380000	12694.280000	157.320000	529.170000	64.621173	201720.185958	2	0.010000	49.810000	35.360000
11	ATHENS	10554	419	34.840000	9.550000	1.190000	14777.130000	304.580000	655.830000	178.809314	134969.727457	1	0.000000	53.700000	18.720000
12	ATHENS	10557	651	27.820000	14.900000	3.690000	26649.570000	445.720000	855.960000	442.114011	561343.177923	1	0.000000	45.310000	25.610000
13	ATHENS	10560	142	60.560000	15.490000	1.410000	20021.220000	462.310000	623.960000	195.567287	49023.275437	1	0.000000	33.100000	31.880000
14	ATHENS	10562	329	26.990000	4.870000	0.910000	19492.330000	441.640000	916.820000	164.762596	45076.450886	1	0.010000	74.550000	17.650000
15	ATHENS	10564	488	23.690000	6.550000	0.200000	21802.800000	404.940000	565.430000	176.925739	124950.154386	1	0.000000	7.450000	50.000000
16	ATHENS	10672	1406	34.780000	35.850000	11.380000	28202.090000	403.190000	777.240000	301.895511	122351.422990	2	0.010000	9.530000	39.000000
17	ATHENS	10682	1992	26.810000	19.920000	3.260000	22312.880000	365.600000	679.350000	318.759713	184624.421286	2	0.010000	19.300000	38.180000
18	ATHENS	11362	13511	50.480000	19.920000	2.050000	14097.120000	178.540000	600.270000	177.494529	298200.543540	2	0.050000	14.370000	40.050000
19	ATHENS	10683	2834	45.730000	27.200000	3.740000	17602.320000	236.920000	775.570000	200.488897	97401.934682	2	0.030000	11.970000	32.560000
20	ATHENS	11141	19842	47.990000	21.580000	1.850000	15545.310000	219.410000	607.030000	185.823585	843011.636198	2	0.020000	5.830000	44.750000
21	ATHENS	11142	18378	47.440000	14.860000	0.800000	14860.840000	204.280000	538.930000	134.450185	870428.122102	2	0.020000	8.330000	52.440000
22	ATHENS	10680	6433	41.430000	24.150000	4.070000	28829.760000	302.830000	641.110000	182.958107	148636.577176	2	0.040000	23.390000	30.900000
23	ATHENS	10681	6433	41.430000	24.150000	4.070000	17452.160000	245.370000	749.290000	178.264871	117805.016379	1	0.050000	23.390000	30.900000
24	ATHENS	11143	16526	49.070000	15.760000	1.010000	14385.850000	191.910000	587.420000	144.023849	939605.087959	2	0.020000	18.100000	46.000000

#rows=90

Table - CityGeoDa

	PostCode	Population	Secondaryf	University	Phd_Master	Income
1	10434	7063	51.28	19.82	1.81	154
2	10435	4900	42.29	9.00	0.90	109
3	10436	3221	45.76	11.27	0.56	102
4	10440	10919	51.41	13.75	0.96	134
5	10442	8036	46.04	8.55	0.54	124
6	10443	25933	46.70	10.25	0.59	123
7	10444	19455	47.20	11.50	0.72	115
8	10445	25094	48.81	12.31	0.69	119
9	10446	22598	49.34	13.99	0.76	111

#rows=90

Figure 1.17 A total of 90 postcodes with 15 fields are stored in the attribute table of which 11 are socioeconomic variables (see Table 1.5).

Exercise 1.1 (*cont.*)

Map - CityGeoDa   —   □   ✕

File   Edit   Tools   Table   Map   Explore   Clusters   Space   Time   Regression   Options   Help

Map - CityGeoDa   —   □   ✕

Map - CityGeoDa
   ▢   (90)

Themeless Map
Quantile Map   >
Percentile Map
Box Map (Hinge=1.5)
Box Map (Hinge=3.0)
Standard Deviation Map
Unique Values Map
Co-location Map
Natural Breaks Map   >
Equal Intervals Map   >
Custom Breaks   >   Create New Custom
Rates-Calculated Map   >
Conditional Map
Cartogram
Map Movie

#obs=90

Figure 1.18 Creating a choropleth population map

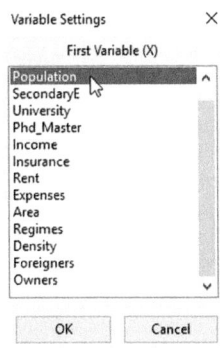

Variable Settings   ✕
        First Variable (X)
Population
SecondaryE
University
Phd_Master
Income
Insurance
Rent
Expenses
Area
Regimes
Density
Foreigners
Owners

        OK   Cancel

Figure 1.19 Variables selection dialog box.

Exercise 1.1 (*cont.*)

On the Category Editor, change only the following fields:Breaks = User Defined (see Figure 1.20)

Categories = 5

Type the following values directly in the break fields: break 1 = 2000 / break 2 = 5000 / break 3 = 10000 / break 4 = 15000 / break 5 = 20000 > Close the dialog box
The map is updated (see Figure 1.21)
Main Menu: Save

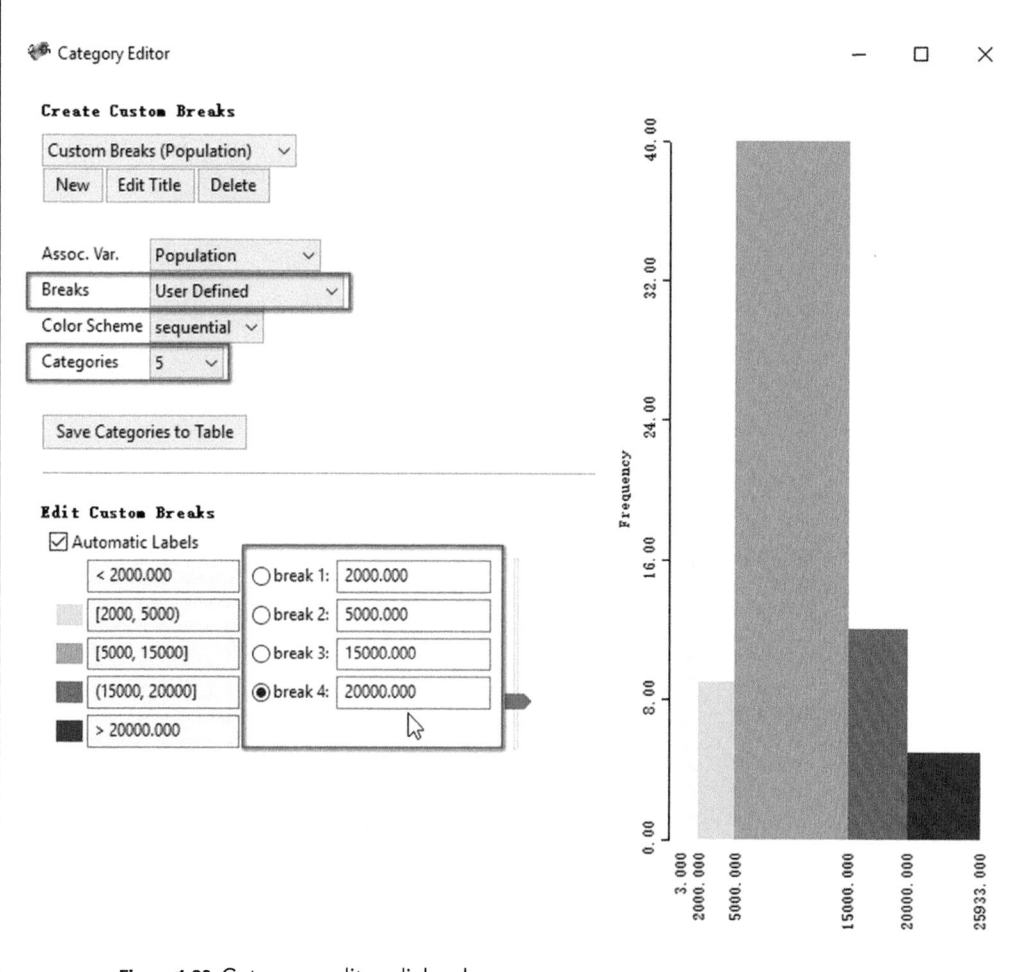

Figure 1.20 Category editor dialog box.

Exercise 1.1 (*cont.*)

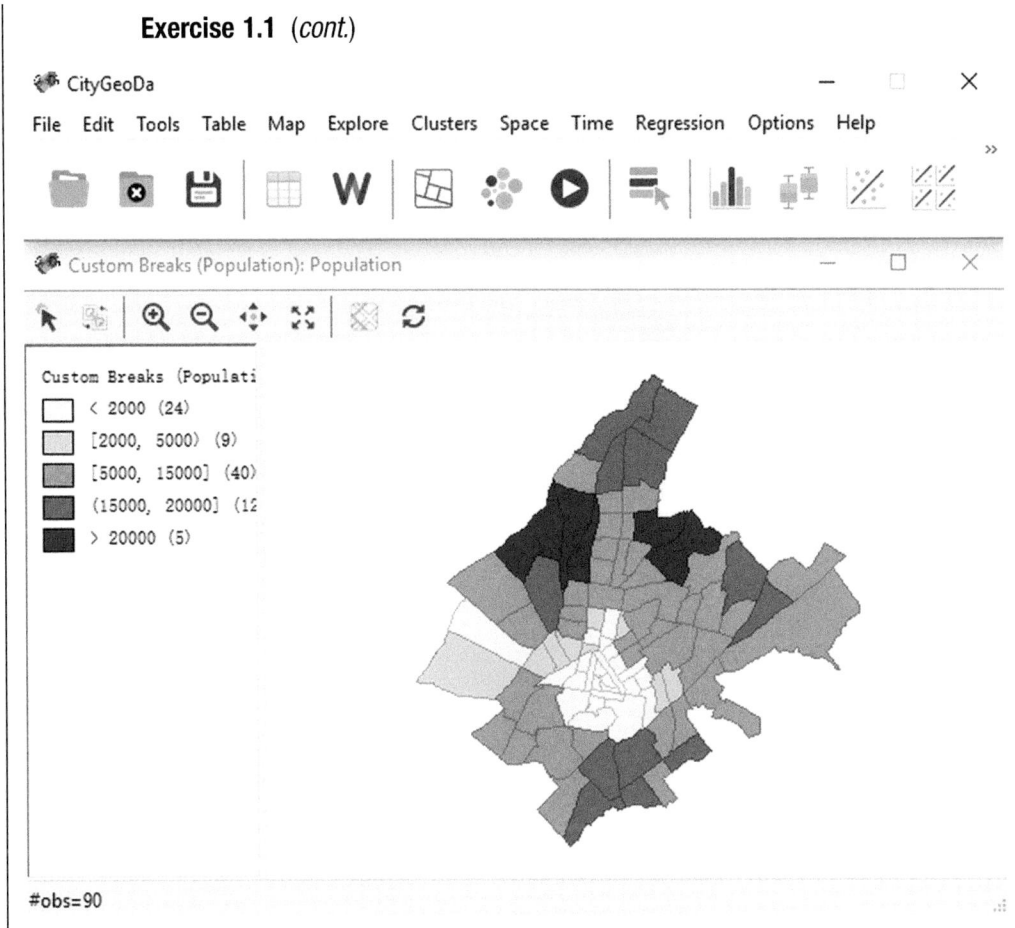

Figure 1.21 Population choropleth map.

Interpreting results: See Section A.

Tip: The interpretation of the results as well as the conclusions related to the analysis are not repeated here, as they are already presented in section A in the "Interpreting results" paragraphs of each exercise. The reader should study these sections carefully as they are independent of the software used.

ACTION: Calculate and map population density

Main Menu > Options > Rates > Raw Rate

Event Variable = Population (see Figure 1.22)

Base Variable = Area

Map Themes = Natural Breaks

Categories = 4 > OK (see Figure 1.23)

Save

Exercise 1.1 (*cont.*)

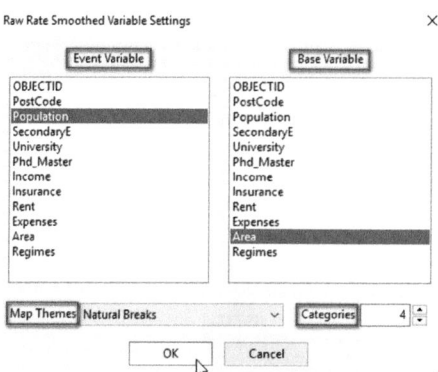

Raw Rate Smoothed Variable Settings   ×

Event Variable	Base Variable
OBJECTID	OBJECTID
PostCode	PostCode
Population	Population
SecondaryE	SecondaryE
University	University
Phd_Master	Phd_Master
Income	Income
Insurance	Insurance
Rent	Rent
Expenses	Expenses
Area	Area
Regimes	Regimes

Map Themes Natural Breaks   ∨   Categories   4 ⬍

OK   Cancel

Figure 1.22 Setting the population density.

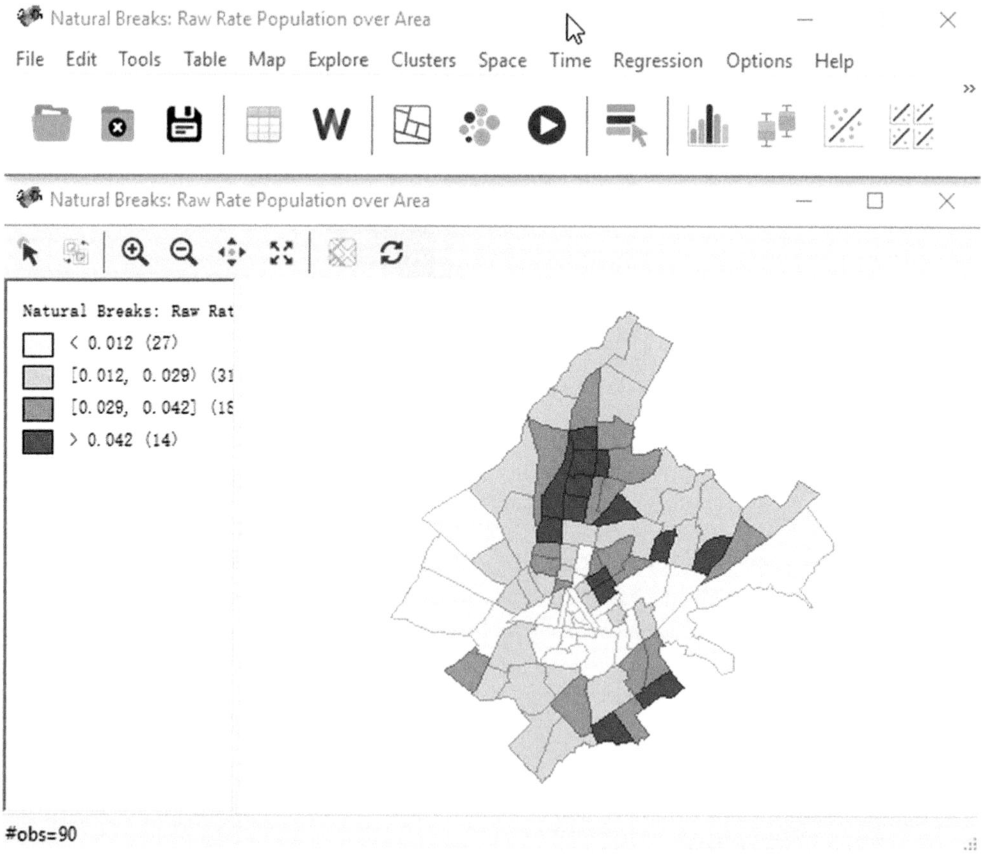

🦊 Natural Breaks: Raw Rate Population over Area   —   ×

File   Edit   Tools   Table   Map   Explore   Clusters   Space   Time   Regression   Options   Help
                                                                                        »

🦊 Natural Breaks: Raw Rate Population over Area   —   □   ×

Natural Breaks: Raw Rat
☐ < 0.012 (27)
☐ [0.012, 0.029) (31
▨ [0.029, 0.042] (18
■ > 0.042 (14)

#obs=90

Figure 1.23 Population density map.

Interpreting results: See Section A.

2 Exploratory Spatial Data Analysis Tools and Statistics

THEORY

Learning Objectives

This chapter deals with

- The notion of exploratory spatial data analysis
- The presentation of descriptive statistics
- Spatial statistics and their importance in analyzing spatial data
- Analyzing univariate data
- Simple exploratory spatial data analysis tools such as histograms, boxplots and other visual methods utilized for deeper insight of spatial datasets
- Bivariate analysis
- Correlation and pairwise correlation
- Normalization, rescaling and adjustments
- Introducing basic notions of statistical significant tests
- The importance of hypothesis setting in a spatial context
- The importance of normal distribution in classic statistics and how it is integrated into spatial analysis

After a thorough study of the theory and lab sections, you will be able to

- Have a solid knowledge of descriptive statistics
- Use descriptive statistics for univariate analysis
- Understand and use exploratory spatial data analysis techniques to map and analyze variables attached to spatial objects
- Create plots, link them to maps and identify interesting data patterns
- Conduct bivariate analysis and identify whether two variables are linearly related; use plots to further examine their relation
- Rescale data to make comparisons between variables easier and also allow for better data handling
- Apply ESDA tools through ArcGIS and GeoDa

2.1   Introduction in Exploratory Spatial Data Analysis, Descriptive Statistics, Inferential Statistics and Spatial Statistics

Definitions

Exploratory Spatial Data Analysis (ESDA) is a collection of visual and numerical methods used to analyze spatial data by

(a)   Applying classical nonspatial descriptive statistics that are dynamically linked to GIS maps and spatial objects

(b)   Identifying spatial interactions, relationships and patterns, through the use of a spatial weights matrix (defined by the appropriate conceptualization method), hypothesis testing and various metrics

ESDA methods and tools are used to

- Describe and summarize spatial data distributions
- Visualize spatial distributions
- Examine spatial autocorrelation (i.e., trace spatial relationships and associations)
- Detect spatial outliers
- Locate clusters
- Identify hot or cold spots

Descriptive statistics is a set of statistical procedures that summarize the essential characteristics of a distribution through calculating/plotting:

- Frequency distribution
- Center, spread and shape (mean, median and standard deviation)
- Standard error
- Percentiles and quartiles
- Outliers
- Boxplot graph
- Normal QQ plot

Inferential statistics is the branch of statistics that analyzes samples to draw conclusions for an entire population.

Spatial statistics employ statistical methods to analyze spatial data, quantify a spatial process, discover hidden patterns or unexpected trends and model these data in a geographic context. Spatial statistics are largely based on inferential statistics and hypothesis testing to analyze map patterns so that spatially varying phenomena can be better modeled (Fischer & Getis 2010 p. 4). Unlike nonspatial methods, spatial statistics use spatial properties such as location, distance, area, length and proximity directly in their mathematical formulas (Scott & Janikas 2010 p. 27). Spatial statistics quantify and further map what the human eye and mind intuitively see and do when reading a map that depicts spatial arrangements, distributions, processes or trends (Scott & Janikas 2010 p. 27).

Why Use Descriptive Statistics and ESDA

Describing a dataset is usually the first task in any analysis. This quickly provides an understanding of data variability and allows for the identification of possible errors (e.g., a value that is not acceptable, omissions [blank cells] or outliers [scores that differ excessively from the majority]). To describe a dataset, we use **descriptive statistics** (also called "summary statistics"). Typical questions that **descriptive** statistics may address in a geographic context include the following: What is the average income in a neighborhood? What is the percentage of people having graduated from the university in a postcode? How many customers of a specific coffee shop live within a distance of less than 10 minutes walking time? What is their purchasing power, and what is the standard deviation of their income?

Descriptive statistics are useful for calculating specific characteristics (e.g., average or standard deviation), thus providing insights for data distributions. However, they do not provide linkages among the results and the spatial objects arranged in a map. The main characteristic of ESDA tools is that they are dynamically linked to maps in a GIS environment. For example, when a point in a scatter plot is brushed (selected), a spatial object is also highlighted on the corresponding map. Likewise, brushing spatial objects on the map, the relevant points/areas/bars are highlighted in the graphs. The basis of exploratory spatial data analysis is the notion of spatial autocorrelation (see Chapter 4), whereby spatial objects that are closer tend to have similar values (in one or more attributes). As such, ESDA offers a more sophisticated analysis, as it discovers patterns in data through mapping and statistical hypothesis testing (see Chapter 3).

ESDA's strength rests on two major features (Dall'erba 2009; Haining et al. 2010 p. 209):

- ESDA extracts knowledge based on its data-mining capacity, as the information that the attribute values carry is relevant to the location of data. This is extremely useful when no prior theoretical framework exists – for example, in many interdisciplinary social science fields.
- ESDA utilizes a wide range of graphical methods combined with mapping, making the analysis more accessible to people who are not accustomed to model building.

Descriptive statistics are used in conjunction with ESDA tools. Sometimes, the boundaries between them are unclear, at least for simple tools. For this reason, many books include histograms, scatter plots or boxplots in descriptive statistics and others in ESDA. The distinction is not of major importance as long as one understands how each tool works. In essence, the only difference with simple ESDA tools (e.g., histograms, scatter plots, boxplots) is that they offer the ability to link graphs to spatial objects, which enhances their power when used in research analysis (Fischer & Getis 2010 p. 3). In this book, simple tools such as histograms, scatter plots and boxplots are presented from the spatial analysis perspective and are linked to GIS maps. More advanced ESDA topics that focus on both spatial and attribute association (e.g., point patterns analysis, spatial autocorrelation) are presented in Chapters 3 and 4. Broadly, simple ESDA tools can be used prior to the modeling phase, and advanced ESDA tools can act as model builders to identify spatial relationships and hidden patterns in spatial data (Fischer & Getis 2010 p. 3).

Why Use Spatial Statistics

Spatial statistics can be considered part of various spatial analysis methods such as ESDA, spatial point pattern analysis, spatial clustering and spatial econometrics. Spatial statistics are mainly used to

- Analyze geographic distributions through centrographic measures (see Chapter 3). In a way similar to descriptive statistics, geographic distributions can be measured to analyze their mean center and standard distance. Spatial statistics are calculated based on the location of each feature; this is a major difference from their homologous descriptive statistics, which refer solely to the nonspatial attributes of the spatial features. Although spatial statistics related to measuring geographic distributions can be weighted using an attribute value, the results refer to a spatial dimension. The spatial features used are typically points and polygons (centroids).
- Analyze spatial patterns. Spatial statistics can be used to analyze the pattern of a spatial arrangement. When this arrangement refers to point features, then the analysis is called point pattern analysis. Through such analysis, we determine whether a point pattern is random, clustered or dispersed (Chapter 3). The analysis of the spatial pattern that the attribute values (of spatial features) form in space is part of the spatial autocorrelation analysis examined in Chapter 4.
- Identify spatial autocorrelation, hot spots and outliers (see Chapter 4).
- Perform spatial clustering (see Chapter 5).
- Model spatial relationships. Spatial statistics can also be used to identify the associations and relationships between attributes and space; examples include spatial regression methods and spatial econometric models (analyzed in Chapters 6 and 7).
- Analyze spatially continuous variables such as temperature, pollution, soils, etc. In general, the type of spatial statistical analysis dealing with continuous field variables is named "geostatistics" (O'Sullivan & Unwin 2010 p. 115). Geostatistics focus on the description of the spatial variation in a set of observed values and on their prediction at unsampled locations (Sankey et al. 2008 p. 1135).

Spatial statistics are built upon statistical concepts, but they incorporate location in terms of geographic coordinates, distance and area. They extend classic statistical measures and procedures and offer advanced insights in analyses of data. In geographical analysis, spatial statistics are not used separately from statistics but are complementary. However, there is a fundamental difference between classical and spatial statistics. In classical statistics, we make a basic assumption regarding the sample: it is a collection of independent observations that follow a specific, usually normal, distribution. Contrariwise, in spatial statistics, because of the inherent spatial dependence and the fact that spatial autocorrelation exists (usually), the focus is on adopting techniques for detecting and describing these correlations. In other words, in classical statistics, observation independence should exist while, in spatial statistics, spatial dependence usually exists. Classical statistics should be modified accordingly to adapt to this condition.

2.2 Simple ESDA Tools and Descriptive Statistics for Visualizing Spatial Data (Univariate Data)

This section presents the most common ESDA techniques and descriptive statistics for analyzing univariate data (only one variable of the dataset is analyzed each time; bivariate data analysis is examined in the next section). These include

- Choropleth maps
- Frequency distributions and histograms
- Measures of the center, spread and shape of a distribution
- Percentiles and quartiles
- Outlier detection
- Boxplots
- Normal QQ plot

2.2.1 Choropleth Maps

Definition
Choropleth maps are thematic maps in which areas are rendered according to the values of the variable displayed (Longley et al. 2011 p. 110).

Why Use
Cloropleth maps are used to obtain a graphical perspective of the spatial distribution of the values of a specific variable across the study area.

Interpretation
The first task when spatial data are joined to nonspatial data (i.e., attributes from a census) is to map them, creating "choropleth maps." For example, population, population density and income per capita can be rendered in a choropleth map. There are two main categories of variables displayed in choropleth maps: (a) spatially extensive variables and (b) spatially intensive

variables. In spatially extensive variables, each polygon is rendered based on a measured value that holds for the entire polygon – for example, total population, total households or total number of children. In the spatially intensive category, the values of the variable are adjusted for the area or some other variable. For example, population density, income per capita and rate of unemployment are spatially intensive variables because they take the form of a density, ratio or proportion. Some argue that the first category is not always appropriate and that variables should be mapped using the second category (Longley et al. 2015 p. 111) because, as each polygon has a different size, mapping real values directly might be misleading. For example, a very small and very large polygon with identical populations will be rendered in the same color if we map population in absolute values. However, adjusting population to polygon area, thus depicting population density, would lead to rendering these two polygons in different colors, as their density values are very different. Thus, the type of variable used to create a choropleth map clearly depends on the problem and on the message one wants to communicate through the mapping.

Through choropleth maps, we visually locate where values cluster or whether they exhibit similar spatial patterns. We may describe such formations using expressions such as "In the western part of the study area, variable X has low scores, while, in the northern part, scores are higher," or "High scores of variable X are clustered in the city center." This is a descriptive way of reading a map and the related symbology. There are no statistics yet, but it communicates a great deal. It may even be better than many statistical analyses, since maps often speak for themselves, provided that the maps and symbols are accurate. Nevertheless, scientific analysis must always be accompanied by statistical analysis in order to prove the findings in a statistically sound way. The next step in mapping variables through a choropleth map is to apply descriptive statistics to summarize the data and use exploratory spatial data analysis methods to visualize the values associated with locations.

Discussion and Practical Guidelines

One might present findings through maps and graphs in an inappropriate way and give the wrong impression or even a misleading message (Tufte 2001). Sometimes this happens through ignorance, and sometimes it is done deliberately to mislead. For example, the choice of colors, scale, map projection or even map title might be misleading (see Box 2.1). As professionals, we have to create accurate maps and graphs and always use solid statistics to back up our findings.

Box 2.1 The Mercator projection was invented by Mercator in 1569 to help explorers navigate to the sea. Since the earth's shape is approximated

Box 2.1 (*cont.*)

by an ellipsoid, every two-dimensional map induces distortions in either area, length or angle direction. Mercator created a projection that keeps the angle right whether in two-dimensional or real-earth terms. By drawing a line on this map between two points and calculating the angle from north, ships could go directly to the destination with no divergence. However, this projection does not preserve area. In this projection, Brazil seems to have the same area as Alaska. In fact, Brazil is five times the size of Alaska. This does not mean that the map is wrong. It is just used wrongly, as its purpose is to map angles correctly. To compare areas, we have to use other projections. To avoid misleading interpretations, a map should be used for the purposes it was created for.

2.2.2   Frequency Distribution and Histograms

Definitions

Frequency distribution table is a table that stores the categories (also called "bins"), the frequency, the relative frequency and the cumulative relative frequency of a single continuous interval variable (de Vaus 2002 p. 207; see Table 2.1).

The **frequency** for a particular category or value (also called "observation") of a variable is the number of times the category or the value appears in the dataset.

Relative frequency is the proportion (%) of the observations that belong to a category. It is used to understand how a sample or population is distributed across bins (calculated as *relative frequency = frequency/n*)

Table 2.1 Frequency distribution table. Example for $n = 15$ postcodes and their population. Five (frequency) postcodes have population between 800 and 899 (bin) people, which is 33.33% (relative frequency = 5/15) of the total postcodes. Overall, 66.67% (cumulative relative frequency = 13.33% + 20.00% + 33.33%) of the postcodes have a population of at least 899 people.

Population range/bins	Frequency	Relative frequency %	Cumulative relative frequency %
600–699	2	13.33	13.33
700–799	3	20.00	33.33
800–899	5	33.33	66.67
900–999	3	20.00	86.67
1000–1199	2	13.33	100.00
$n =$	15	100.00	

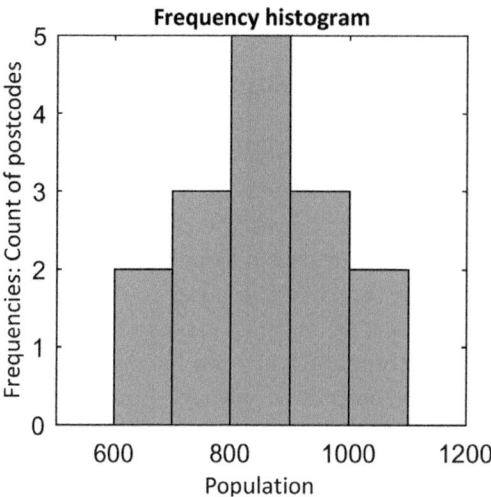

Figure 2.1 Frequency distribution histogram for the population variable in Table 2.1. Each bar depicts the number of postcodes (frequencies in the y-axis) for each population bin (in the x-axis).

The **cumulative relative frequency** of each row is the addition of the relative frequency of this row and above. It tells us what percent of a population (observations) ranges up to this bin. The final row should be 100%.

A **frequency distribution histogram** is a histogram that presents in the x-axis the bins and in the y-axis the frequencies (or the relative frequencies) of a single continuous interval variable (de Vaus 2002 p. 207; see Figure 2.1).

A **probability density histogram** is defined so that

(a)   The area of each box equals the relative frequency (probability) of the corresponding bin
(b)   The total area of the histogram equals 1

Why Use

Frequency distribution tables and histograms are used to analyze how the values of the studied variable are distributed across the various categories. The histogram can also be used to determine if the distribution is normal or not. Additionally, it can be used to display the shape of a distribution and examine the distribution's statistical properties (e.g., mean value, skewness, kurtosis). Interesting questions can then be answered that may assist spatial analysis or spatial planning (e.g., about how many postcodes have a population of less than a specific value; see Table 2.1). Histograms should not be confused with bar charts, which are used mainly to display nominal or ordinal data (de Vaus 2002 p. 205).

Discussion and Practical Guidelines

To calculate frequencies, we first set the number of bins in which the values will be grouped by dividing the entire range of values into sequential intervals (bins). The choice of the appropriate number of bins as well as their range depends on the project at hand and the scope of the analysis; it should be meaningful. A trial-and-test method is an appropriate approach for choosing how many bins to use. Using many bins is suitable if there is high variance and a large population, but using too many bins makes interpretation difficult. On the other hand, a relatively small number of bins might conceal data variability (for a small number of bins, we can use other graphs, such as pie graphs). As a rule of thumb, a value that falls on the boundary of two bins should be placed on the upper bin. For example, if the intervals for the variable "age" are set for every five years (e.g., 0–5, 5–10 and so on), then a five-year-old child should be grouped with the 5–10 bin. More mathematically, bins should be set as 0 to $<$ 5, 5 to $<$ 10 and so on.

After defining the bins and ranges, we count how many values (observations) lie within each bin. This count is the frequency. If we add all frequencies, we should obtain a total that equals the sample (n). A frequency distribution table also includes the relative frequency (percentage) and the cumulative relative frequency (cumulative percentage; see Table 2.1). All relative frequencies have to add up to 100%. For large frequencies, the plots are not well presented; some bins might be much larger than others. By using relative frequency, we change our scale on the y-axis to 0–1 (0%–100%), but all essential characteristics of the frequency distribution – such as location, spread and shape – are unchanged (see next section; Peck et al. 2012 p. 28). Likewise, the final cumulative relative frequency should be 100%.

Based on the frequency distribution table, a frequency distribution histogram can also be plotted (see Figure 2.1). In this histogram, each frequency is centered over the corresponding bin and is represented by a rectangle. For same-width bins, the area of the rectangle is proportional to the corresponding frequency. Histograms can also be created for the relative frequency. Each bar is centered on the same values in the x-axis as in the frequency distribution histogram, and the height equals the relative frequency, with the sum of the heights of all bins equaling 1. The relative frequency can be considered as the probability of a value occurrence.

Another option is to normalize (divide) the relative frequency by the width of each bin. In this case, we have on the y-axis the relative frequency per unit of the variable on the x-axis. This type of histogram is called probability density histogram. In this case, the bins are not of the same width. In fact, the area of each bin equals the probability of occurrence of a specific value or range of values and the area of all bins should equal 1. As the sample size increases, creating more bins for the same range of values, the density histogram can be fitted by a continuous function called a "probability density function" (PDF) (see Figure 2.2). The PDF is used to find the probability that a

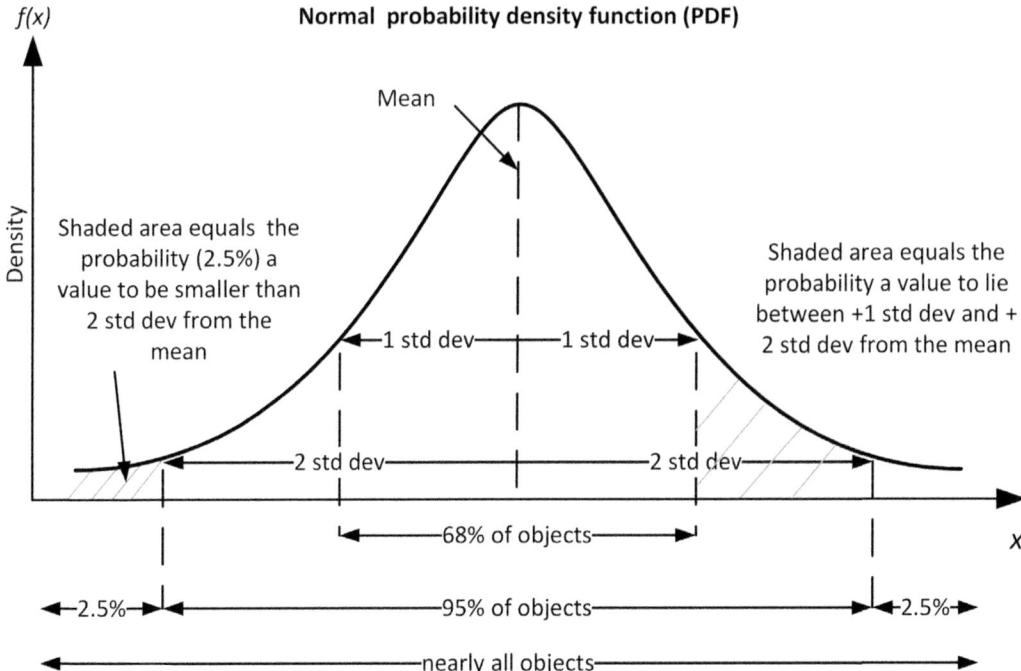

Figure 2.2 Standard deviation of a normal distribution; 68% of objects lie within one standard deviation from the mean (34% in each direction).

value X falls within some interval. A PDF is widely used in statistics and spatial statistics. Most of the times the units on the x-axis is the standard deviation of the variable.

For a normal distribution, the normal PDF is (Illian et al. 2008 p. 53, see Figure 2.2)

$$f(x|\mu,\sigma) = \frac{1}{\sigma\sqrt{2\pi}} e^{-\frac{1}{2}\left(\frac{x-\mu}{\sigma}\right)^2} \tag{2.1}$$

where

μ is the mean of the population
σ is the population standard deviation
σ^2 is the population variance
x is the value of the variable

The probability distribution function of normal distribution takes two parameters: the population mean and the population standard deviation. In such a distribution, we expect that 68% of the values lie one standard deviation from the mean and that 95% of the values lie inside two standard deviations from the mean (in both directions; see Figure 2.2). The area between the intervals and the curve equals the probability by which we anticipate a value to appear in our dataset. For example, the probability that a value will range between +1 standard deviation and +2 standard deviations in a normal distribution is 13.5% (the shaded area on the right). Values larger than two standard deviations from the mean are expected at less than 2.5% (right tail). In other words, the probability of obtaining a value larger than two standard deviations from the mean is less than 2.5%. Likewise, values smaller than two standard deviations from the mean, are expected at less than 2.5%. Some of the most commonly used significance tests in spatial statistics (as explained later) make use of the normal PDF to test if a hypothesis is true by calculating the probability under a specific interval (significance level).

Normal distribution (also called Gaussian distribution) is the most commonly used distribution in statistics, as many physical phenomena are normally distributed (e.g., human weight and height). In a normal distribution, the values of a variable are more likely to be closer to the mean, while larger or smaller scores have low probabilities of occurring. Normal distribution is used in many statistical tests to draw conclusions regarding the distribution studied. It has a zero mean, one unit standard deviation, symmetrical histogram and a bell-shaped shape (see Figure 2.3A). Not all bell-shaped histograms reveal a normal distribution, as a normal distribution decreases from the top to the tails in a certain way (see Figure 2.3G; for more on this, see Section 2.6).

In general, any distribution can be described by three important essential features: center, spread and shape. Analyzing these features provides information for (a) center, (b) extent, (c) general shape, (d) location and number of peaks and (e) the presence of gaps and outliers, as discussed next.

2.2.3   Measures of Center

Definitions

Measures of central tendency provide information about where the center of a distribution is located. The most commonly used measures of center for numerical data are the mean and the median (mode is another measure of center and is the value that occurs most often in a sample).

The **mean** is the simple arithmetic average: the sum of the values of a variable divided by the number of observations (calculated for interval data), as in (2.2):

$$\bar{x} = \frac{\sum_{i=1}^{n} x_i}{n} \tag{2.2}$$

where

　　n is the total number of observations
　　x_i is the score of the ith observation
　　Σ is the symbol of summation (pronounced sigma)
　　$\bar{x}$ is the sample mean value

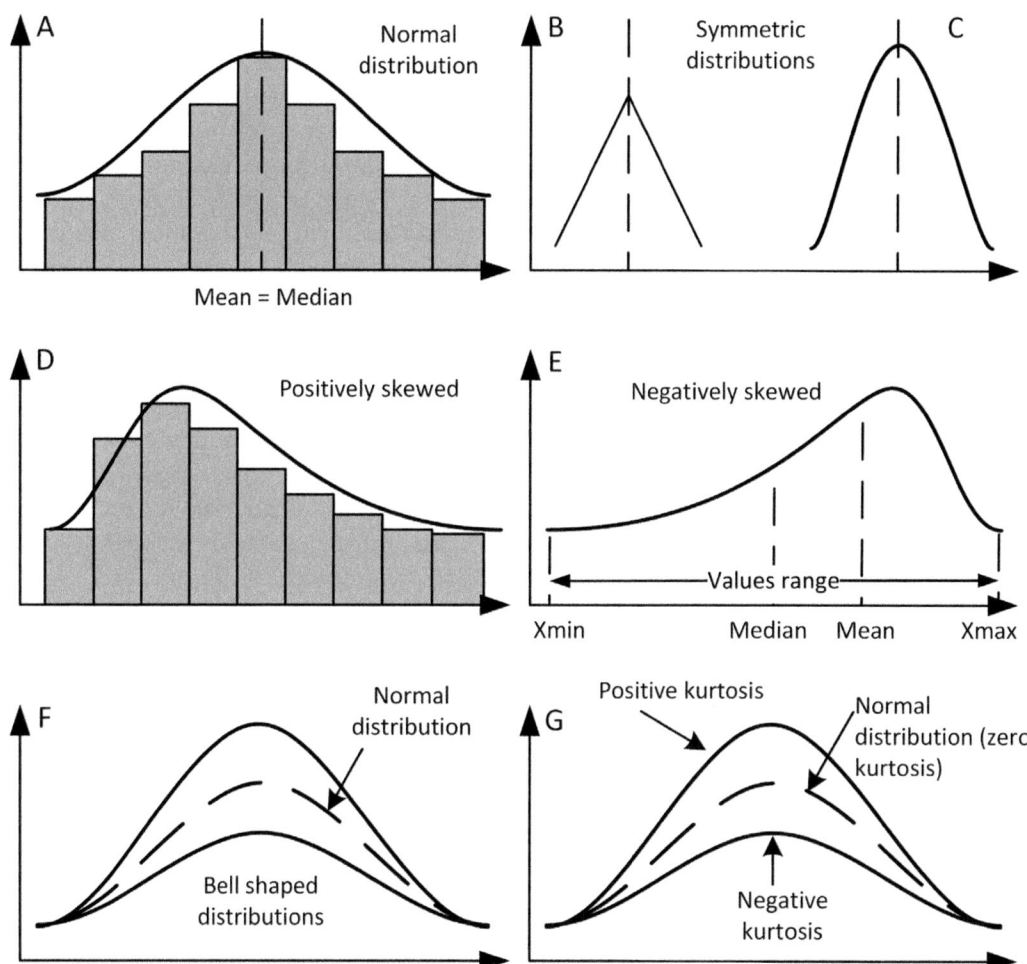

Figure 2.3 (A),(B),(C) Symmetric histograms. (D) Positively skewed. (E) Negatively skewed. (F) Different types of bell shaped distributions. (G) Curves with various kurtosis values.

The **median** is the value that divides the sorted scores form smaller to larger in half. It is a measure of center.

Why Use

The mean is used to describe the center of a distribution, while the median is used to split the frequency distribution histogram into two equal area parts.

Interpretation

If we list scores from smallest to largest, the middle score is the median. It cuts scores in two equal parts. Fifty percent of the objects have values larger than the median, and 50% of the objects have values less than the median. In addition, the median splits a frequency distribution histogram in two equal area parts, while the mean is the balance point of the distribution histogram (the sum of the values at the left of the mean equals the sum of the values at the right).

Discussion and Practical Guidelines

We should be cautious when we interpret the mean because (a) the same mean might be the result of completely different distributions and (b) extreme values or outliers change the mean value and inflate the skewness of a distribution significantly. When we exclude outliers or extreme values, the mean is likely to change significantly. In a normal distribution, the mean is zero (calculated for standard deviations) and is located in the center of the distribution (see Figure 2.3A). The median overcomes the outlier problem, as it is based on ranked positions and not on real values. When *n* is odd, there is a single median. When *n* is even, there are two "middle values," and we take their average to obtain the median. The median is located in the center of a normal distribution and coincides with the mean (see Figure 2.3A). In other types of distributions, the median tends to deviate from the mean. Typically, but not always (depending on the values), the median lies at the left of the mean in a negatively skewed distribution (see Figure 2.3E) and at its right in a positively skewed one. The median can be calculated for both ordinal and interval data.

2.2.4   Measures of Shape

Definitions

Measures of shape describe how values (e.g., frequencies) are distributed across the intervals (bins) and are measured by skewness and kurtosis.

   Shape of the distribution is the curved line (sometimes a straight line) that approximates the middle top of each bin in a continuous way. The x-axis closes the shape, and the area can be calculated. If a shape is symmetrical around a vertical line, the part of the histogram to the right is the mirror of the left part (see Figure 2.3A–C).

   Skewness is the measure of the asymmetry of a distribution around the mean.

   Kurtosis, from the graphical inspection perspective, is the degree of the peakedness or flatness of a distribution.

Why Use

Skewness is used to identify how values are distributed around the mean, while kurtosis reveals how stretched a distribution is on the y-axis compared to the normal distribution (see Figure 2.3G). Peakedness and flatness are actually based on the size of the tails of a distribution. For this reason, kurtosis is prone to outliers, as outliers tend to stretch the distribution tails and significantly change the mean. Thus, the upper hill of a curve may move upward or downward according to the strength and location of an outlier.

Interpretation

If a histogram is not symmetric, it is skewed. If the right tail (part) of the histogram tends to stretch considerably more than the left tail, this histogram is named "positively skewed" or "right skewed" (see Figure 2.3D). If the left tail is stretched, we call the histogram "negatively skewed" or "left skewed" (see Figure 2.3E). Skewness of greater than 1 or less than −1 typically indicates a nonsymmetrical distribution (de Vaus 2002 p. 225). Values higher than 1.5 or less than −1.5 indicate large skewness, meaning that the data tend to stretch away from the mean in some direction. For example, suppose that Figure 2.3D depicts the frequency distribution (y-axis) of annual per capita income (x-axis). In this case, the distribution of income is positively skewed. Only a few people have very high income (lie in the right tail further from the mean) while the majority has income less than the mean (which is left of the median). Income is unequally distributed: More have less and less have more.

A zero kurtosis indicates a near-normal distribution peakedness. A negative kurtosis indicates a more flat distribution (lower than normal), while a positive kurtosis reveals a distribution with a higher peak than the normal distribution. Strictly speaking, the kurtosis for a normal distribution is 3 (de Vaus 2002 p. 227). Most statistical software subtract 3 from the final figure to adjust to the zero definition. This provides a quicker understanding, as the positive or negative values are directly interpreted as distributions over or under a normal distribution. Some popular software, such as Matlab and ArcGIS, do not follow the zero definition and regard a kurtosis of 3 as the normal distribution

2.2.5   Measures of Spread/Variability – Variation

Definitions

Measures of spread (also called measures of variability, variation, diversity or dispersion) of a dataset provide information of how much the values of a variable differ among themselves and in relation to the mean. The most common measures are as follows (de Smith 2018 p. 150):

- Range (Peck et al. 2012 p. 185)
- Deviation from the mean
- Variance
- Standard deviation
- Standard distance (see Section 3.1.4)
- Percentiles and quartiles (see Section 2.2.6)

A **range** is the difference between the largest and smallest values of the variable studied, as in (2.3):

$$Range = x_{max} - x_{min} \tag{2.3}$$

where x_{max} is the maximum value of a variable, and x_{min} is the minimum value of the same variable (see Figure 2.3E).

   Deviation from the mean is the subtraction of the mean from each score, as in (2.4):

$$Deviation = (x_i - \bar{x}) \tag{2.4}$$

where

   x_i is the score of the ith object
   $\bar{x}$ is the sample mean value

The sum of all deviations is zero (sometimes, due to rounding up, the sum is very close to zero), as in (2.5):

$$\sum_{i=1}^{n} (x_i - \bar{x}) = 0 \tag{2.5}$$

Sample Variance is the sum of the squared deviations from the mean divided by $n - 1$ (sample variance) as in (2.6) (see Table 2.2):

$$s^2 = \frac{\sum_{i=1}^{n} (x_i - \bar{x})^2}{n - 1} \text{ (sample variance)} \tag{2.6}$$

Squared values are used to turn negative deviations to positive. To calculate the variance for the entire population, denoted by σ, we simply divide by n, as in 2.7:

$$\sigma^2 = \frac{\sum_{i=1}^{n} (x_i - \bar{x})^2}{n} \text{ (population variance)} \tag{2.7}$$

Standard deviation is the square root of variance (2.8, 2.9) (see Table 2.2).

$$s = \sqrt{s^2} \text{ (sample standard deviation)} \tag{2.8}$$

$$\sigma = \sqrt{\sigma^2} \text{ (population standard deviation)} \tag{2.9}$$

Table 2.2 Sample and population statistical symbols. Sample statistics are denoted by Latin letters and population parameters by Greek letters.

Measure	Sample statistic symbol	Population parameter symbol
Mean	$\bar{x}$ pronounced: ex bar	μ: pronounced: mu (miu)
Variance	s^2 pronounced: es squared	σ^2 pronounced: sigma squared
Standard deviation	s_x pronounced: es of ex	σ pronounced: sigma

Why Use

Range is used to assess the variation of values in a variable, while deviation from the mean is used to calculate how far away a score lies from the mean. Variance is used to measure the spread of values in a variable. Standard deviation indicates the size of a typical deviation from the mean. In essence, variance and standard deviation reflect the average distance of the observations from the mean (de Vaus 2002 p. 224). Standard deviation is easier to interpret than variance, as it is measured at the same unit of the variable studied.

Interpretation

The greater the range, the more variation in the variable's values, which might also reveal potential outliers. Large values of s^2 (variance) reveal a great variation in the data, indicating that many observations have scores further away from the mean. If the variation is large, we may cut off the top and bottom 5% or 10% of the dataset to produce a more compact distribution. This typically happens in satellite image analysis for color enhancement.

A positive standard deviation value indicates the number of standard deviations above the mean, and a negative value indicates the number of standard deviations below the mean. Standard deviation is used to estimate how many objects in the sample lie further away from the mean in reference to the z-score (e.g., 1 or 2) (see Section 2.5.5). In any normal distribution and for a specific variable:

- Approximately 68% of all values fall within one standard deviation of the mean (z-score = 1).
  [$(\bar{x} - 1^*$ standard deviation) up to $(\bar{x} + 1^*$ standard deviation)]
- Approximately 95% of all values fall within two standard deviations of the mean (z-score = 2).
  [$(\bar{x} - 2^*$ standard deviation) up to $(\bar{x} + 2^*$ standard deviation)]
- Nearly all values fall within three standard deviations of the mean (z-score = 3).
  [$(\bar{x} - 3^*$ standard deviation) up to $(\bar{x} + 3^*$ standard deviation)]

Discussion and Practical Guidelines

It is important to note that variation is not the same as variance (a synonym for variation is variability). Variation and variability are not some specific quantities. They are typically used as general terms expressing fluctuations in values. These fluctuations are calculated through the measures of spread.

Sample variance is the total amount of the squared deviation from the mean divided by the sample (n), but not exactly; it is divided by the sample (n) minus 1. Why minus 1? If it was just n, it would be the average of the total amount of the squared deviation, which makes more sense. In fact, minus 1 is necessary. It has been observed that variation tends to be underestimated when we use samples. Overestimating variance is better than underestimating it (Linneman 2011 p. 90). In more advanced statistics, the term $n - 1$ in this formula reveals the degrees of freedom (*df*). Degrees of freedom generally equal the sample (*n*) minus the number of parameters estimated. It is actually the number of objects (of the sample) that are free to vary when estimating statistical parameters. "Free to vary" means that these objects have the freedom to take any value (inside the set in which the function is defined), while others are constrained by restrictions. If we are interested in the standard deviation for the entire population, denoted by σ, we simply divide by *n* (2.7).

Selecting between the sample statistic and the population parameter formula (see Table 2.2) depends on the nature of our analysis and the available data. Suppose we want to calculate the standard deviation of income for a specific city. If we have data for the entire population of this city (through census), we should apply the population standard deviation formula. The results are not an estimate but a real population calculation. If we want to estimate the standard deviation of income for the entire country, based only on this city sample (we infer from the city to the country), then we should apply the sample standard deviation formula (see more on inferential statistics in Section 2.5).

Finally, by combining the standard deviation and z-scores (see Section 2.5.5), we can describe how objects (and their values) lie within a distribution. For example, suppose that the mean value of incomes in 30 postcodes of a city is 15,000 US dollars and the standard deviation is 2,000 US dollars. The standard deviation of 2,000 US dollars means that, on average, incomes vary away (in both directions) from the mean by 2,000 US dollars. If the distribution of income follows a normal distribution (in practice, it does not), approximately 68% of the postcodes (nearly 20) would have incomes in the range of [*15,000 – 1*standard deviation up to 15,000 + 1*standard deviation*], or between 13,000 and 17,000 US dollars.

Additional questions using the standard deviation and z-score can be asked. For example, how many postcodes are likely to have incomes higher than 19,000 US dollars? This is an important type of questions, especially when we focus on certain subpopulations. We first calculate the z-score: (*value-mean*)/ (*standard deviation*) = (*19,000– 15,000)/2,000 = 2* standard deviations away (see Eq. 2.21). This value means that a postcode with an income of 19,000 US dollars lies two standard deviations above the mean. As mentioned above, only 5% of objects lie more than two standard deviations from the mean in case of normally distributed variable. This is 2.5% in each direction. In the preceding example and to answer the original question, 2.5% of the postcodes (that is one postcode) have income larger than 19,000 US dollars.

2.2.6   Percentiles, Quartiles and Quantiles

Definition

A **percentile** is a value in a ranked data distribution below which a given percentage of observations falls. Every distribution has 100 percentiles.

The **quartiles** are the 25th, 50th and 75th percentiles, called "lower quartile" (Q1), "median" and "upper quartile" (Q3) respectively.

The **interquartile range (IQR)** is obtained by subtracting the lower quartile from the upper quartile as in 2.10:

$$IQR = Upper\ quartile - Lower\ quartile = Q3 - Q1 \qquad (2.10)$$

Quantiles are equal-sized, adjacent subgroups that divide a distribution.

Why Use

Percentiles are used to compare a value in relation to how many values, as a percentage of the total, have a smaller or larger value. The lower quartile (Q1), the upper quartile (Q3) and the median are commonly used to show how scores are distributed for every 25 percentiles. The interquartile range provides a measure of the variability of the 50% of the objects around the median. Quantiles are often used to divide probability distributions into areas of equal probabilities. In fact, percentiles are quantiles that divide a distribution to 100 subgroups.

Interpretation

If the 20th percentile of a distribution is 999, then 20% of the observations have values less than 999. If a student's grade lies in the 80th percentile, the student achieved better grades than did 80% of his/her classmates. The 50th percentile is the median. As mentioned, the median is the score that splits ranked scores in two; thus, 50% of the objects have higher scores, and 50% have lower ones.

Discussion and Practical Guidelines

Percentiles and quartiles are not prone to outliers, as they are based on ranks of objects. For example, the maximum score would lie in the last percentiles whether it is an outlier or not. Quartiles provide an effective way to categorize a large amount of data into a mere four categories. Finally, GIS software uses quantiles to color and to symbolize spatial entities when there are many different values.

2.2.7   Outliers

Definition

Outliers are the most extreme scores of a variable.

Why Use

They should be traced for three main reasons:

- Outliers might be wrong measurements
- Outliers tend to distort many statistical results
- Outliers might hide significant information worth being discovered and further analyzed

Interpretation (How to Trace Outliers)

For a univariate distribution, outliers can distort the mean and the standard deviation. In bivariate and multivariate analyses, many statistics – such as correlation coefficient, trend lines and regression analysis – will provide false results if outliers exist. The most common way of tracing outliers is by graphical representation through a histogram or a boxplot. In case of histograms, if there is an isolated bar in the far left or far right part of the histogram, it is a serious indication of having outliers in the data. If skewness is very large (positive or negative), it is also an indication of outliers' presence. Another approach is to regard outliers as those scores lying more than 2.5 standard deviations from the mean. In fact, it is not easy to set a specific number of standard deviations in order to identify an outlier. When we calculate how many standard deviations from the mean a potential outlier is, we have to consider that the outlier itself raises the standard deviation and also affects the value of the mean. A scatter plot (see Section 2.3.1) is another effective way to locate an outlier in bivariate analysis.

There is also a set of methods of tracing outliers by analyzing the residuals in regression analysis (e.g., standardized residuals), but we will not refer further to these methods in this book. We should be cautious about labeling an object as an outlier because removing it from the dataset leads to new values for the mean, standard deviation, and other statistics. Outliers should be eliminated only if we comprehend why they exist and whether it is likely that similar values reappear. Outliers often reveal valuable information. For example, an outlier value for a room's temperature indicates the potential for a fire, allowing preventive action. An outlier value for credit card use may reveal a different location than those commonly used (e.g., in a different country) and thus potential fraud (Grekousis & Fotis 2012). Thus, defining a value as an outlier depends on the broader context of the study, and the analyst should decide to eliminate or include it in the dataset carefully.

Discussion and Practical Guidelines
We can handle traced outliers based on the following guidelines:

- Scrutinize the original data (if available) to check whether the outliers' scores are due to human error (e.g., data entry). If scores are correct, attempt to explain such high or low value, as it is unlikely to be just a random phenomenon.
- Transform the variable. Still, data transformation does not guarantee outliers' elimination. In addition, it may not be desirable to transform the entire dataset for only a small number of outliers.
- Delete outlier from the dataset or change its score to be equal to the value of three standard deviations (de Vaus 2002 p. 94). The choice depends on the effect it will have on the results, but deletion is preferred. In either case, the deleted or changed score should be reported.
- Temporarily remove the outlier from the dataset and calculate the statistics. Then include the outliers again in the dataset for further analysis. For example, suppose we study the socioeconomic profiling of postcodes. Some postcodes might have extremely high incomes per capita relative to others, but they also carry additional socioeconomic information that might not include outlier values. If we completely remove the postcodes

with outlying incomes, we will lose valuable information (regarding the other variables). For this reason, it is wiser to temporarily remove the income outliers only for those statistics that have a distorting effect and include them again for further analysis later.

2.2.8   Boxplot

Definition
A **boxplot** is a graphical representation of the key descriptive statistics of a distribution.

Why Use
To depict the median, spread (regarding percentiles) and presence of outliers.

Interpretation
The characteristics of a boxplot are as follows (see Figure 2.4):

- The box is defined by using the lower quartile Q1 (25%; left vertical edge of the box) and the upper quartile Q3 (75%; right vertical edge of the box). The length of the box equals the interquartile range $IQR = Q3 - Q1$.
- The median is depicted by using a line inside the box. If the median is not centered, then skewness exists.
- To trace and depict outliers, we have to calculate the whiskers, which are the lines starting from the edges of the box and extending to the last object not considered an outlier.
- Objects lying further away than $1.5 \times IQR$ are considered outliers.
- Objects lying more than $3.0 \times IQR$ are considered extreme outliers, and those between ($1.5 \times IQR$ and $3.0 \times IQR$) are considered mild outliers. One may change the 1.5 or 3.0 coefficient to another value according to the study's needs, but most statistical programs use these values by default.
- Whiskers do not necessarily stretch up to $1.5 \times IQR$ but to the last object lying before this distance from the upper or lower quartiles.

If a distribution is positively skewed, the median tends to lie toward the lower quartile inside the box of a boxplot (see Figure 2.5A). A boxplot with the median line near to the center of the box and with symmetric whiskers slightly longer than the box length tends to represent a normal distribution (see Figure 2.5B). Negatively skewed distributions tend to look like graph C (see Figure 2.5C). Outliers might lie in any direction. They can also be traced as isolated bins (see Figure 2.5A and C).

Discussion and Practical Guidelines
Apart from describing a single distribution, boxplots can be used to compare distributions of the same variable but for different groups. To compare such distributions, we use parallel boxplots (see Figure 2.6). In this case, boxplots are

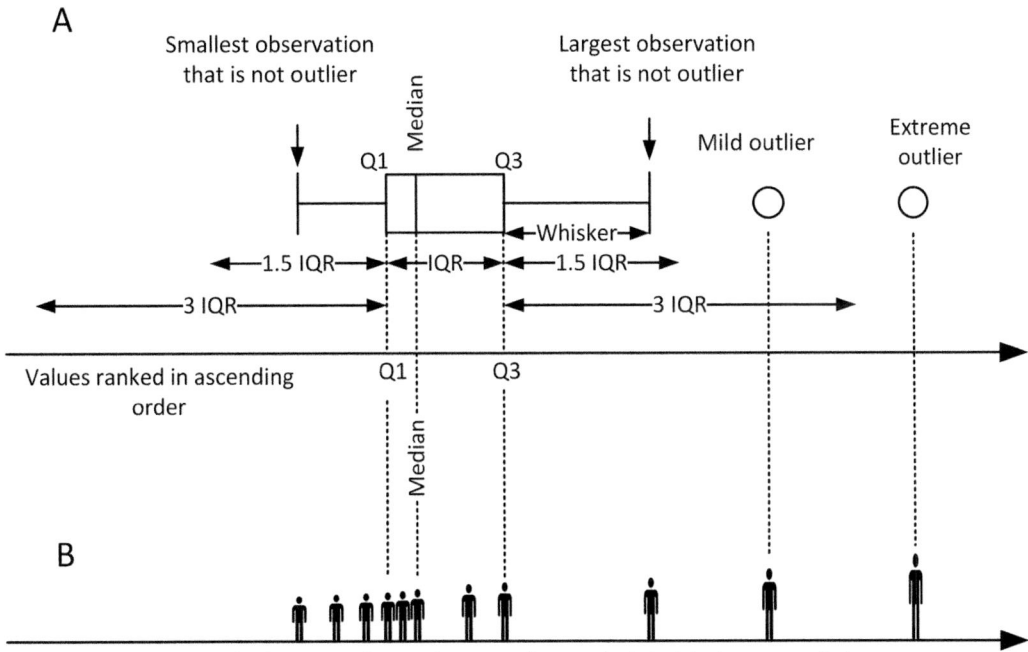

Figure 2.4 (A) Basic boxplot characteristics. In this graph, outliers exist only in the right part. Generally, outliers might exist in both parts concurrently. Whiskers stop at the largest or smallest observation that is not an outlier. In the left part, the minimum value of the variable lies less than $1.5 \times IQR$ away from the lower quartile (Q1), so there is no outlier. Whisker lengths are not necessarily the same in the two parts. (B) Eleven people ranked in ascending order according to their height. The far-right person is a basketball player, and he is considerably taller than the rest (outlier).

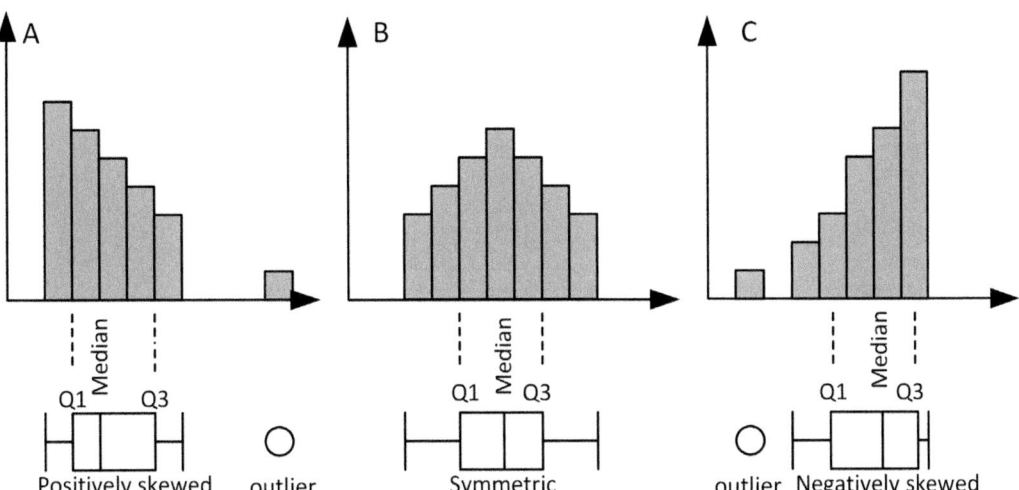

Figure 2.5 Boxplot general look for different type of distributions.

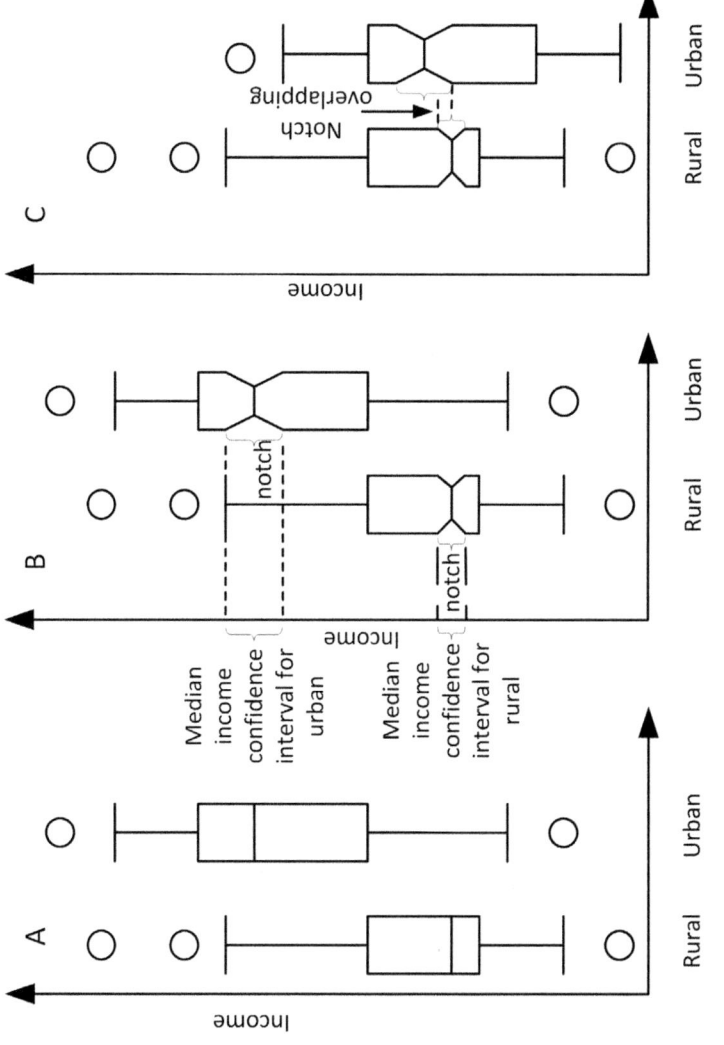

Figure 2.6 Boxplots plotted side by side to compare distributions and median. (A) A simple graphical inspection shows that the median income in urban areas is larger than the median income in rural areas. (B) Notched boxplot. (C) Notch overlapping indicates not statistical difference between the median values.

plotted side by side in a vertical representation, which is more common than the horizontal representation. We plot each group on the x-axis and the values of the common variable on the y-axis. A graphical examination is the first step in comparing these distributions – for example, to see if their medians are different. In Figure 2.6, we use parallel boxplots to describe two groups, urban and rural populations (groups on the x-axis), in relation to annual income (variable on the y-axis). Statistical tests such as a Mann–Whitney U-test should be used to check if any of the observed difference between the medians are statistically significant.

A particular type of boxplot, the notched boxplot, is used to provide a more accurate graphical representation for comparison purposes (Chambers et al. 1983; see Figure 2.6B). For each group, it provides the 95% confidence interval of the median (see Section 2.5.3 for confidence intervals). If these intervals do not overlap when we compare the two distributions (we inspect the y-axis), there is strong evidence at the 95% confidence level that the medians differ (see Figure 2.6B).

The confidence interval is calculated using the following formula (2.11):

$$median \pm 1.57 \frac{IQR}{\sqrt{n}} \tag{2.11}$$

In other words, the height of the notches equals 3.14 times the height of the main box divided by the square root of the sample size. This interval depicts the values in the same units as those of the variable studied. If we compare the intervals of two parallel boxplots and find that there are no common values, we may conclude, at a 95% confidence level, that the true medians of the group do differ. In Figure 2.6C, although the medians look different, the notches overlap, and we cannot conclude that there is a statistically significant difference in their medians. To better assess if there is indeed a statistically significant difference in the median between two groups, we should use statistical tests (e.g., a Mann–Whitney U test, as mentioned).

2.2.9 Normal QQ Plot

Definition
The **normal QQ plot** is a graphical technique that plots data against a theoretical normal distribution that forms a straight line.

Why Use
A normal QQ plot is used to identify if the data are normally distributed.

Interpretation
If data points deviate from the theoretical straight line, this is an indication of non-normality (see Figure 2.7). The line represents a normal distribution at a 45° slope. If the distribution of the variable is normal, then points will lie on this reference line. If data points deviate from the straight line and curves appear (especially in the beginning or at the end of the line), the normality assumption is violated. For instance, the plot in Figure 2.7 reveals non-normally distributed data.

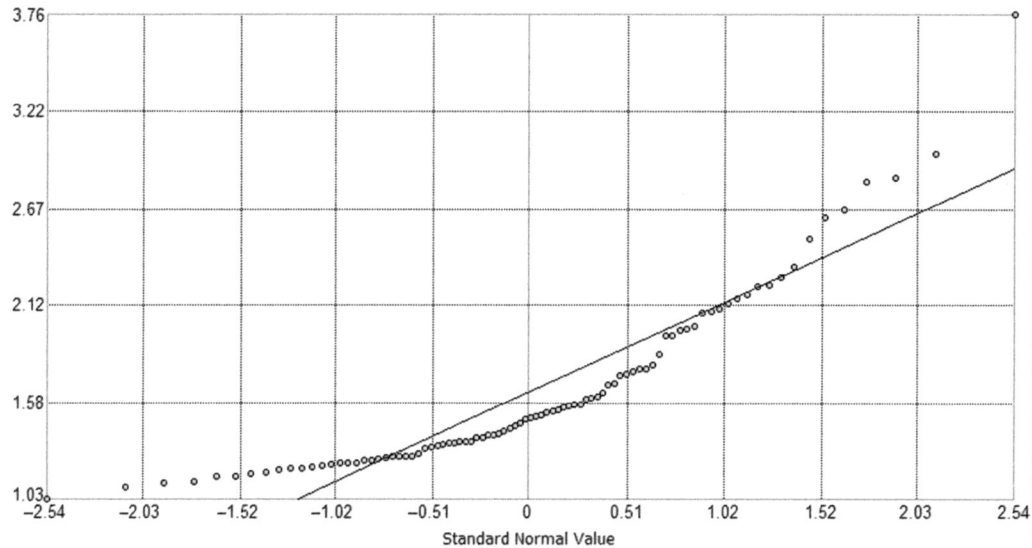

Figure 2.7 Normal QQ plot.

2.3   ESDA Tools and Descriptive Statistics for Analyzing Two or More Variables (Bivariate Analysis)

Spatial analysis often focuses on two different variables simultaneously. This type of analysis is called "bivariate," and the dataset used is called a "bivariate dataset." The study of more than two variables, as well as the dataset used, is called "multivariate." Multivariate methods will be presented in Chapter 5.

The most common ESDA techniques and descriptive statistics for analyzing bivariate data include

- Scatter plot
- Scatter plot matrix
- Covariance and variance–covariance matrix
- Correlation coefficient
- Pairwise correlation
- General QQ plot

2.3.1   Scatter Plot

Definition
A **scatter plot** displays the values of two variables as a set of point coordinates (see Figure 2.8).

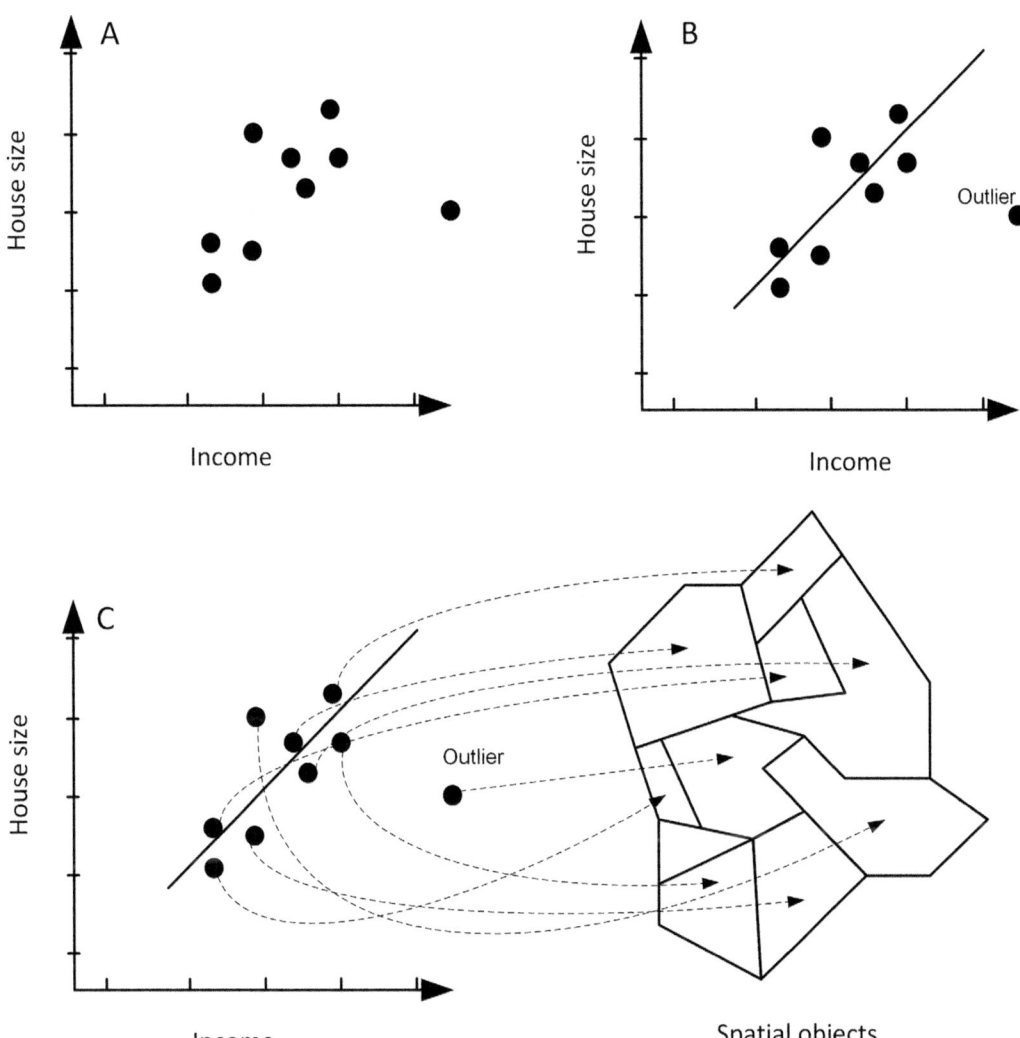

Figure 2.8 (A) Scatter plot for variables *Income* and *House size*. (B) A linear trend superimposed (with positive slope) reveals a positive linear association between the two variables. As income increases, so does house size. A value on the far right of the graph indicates a potential outlier. (C) In exploratory spatial data analysis, a one-to-one linkage exists, whereby each dot in the scatter plot stands for single spatial unit depicted in a single location on a map. Notice also that the outlier in the scatter plot is not a locational outlier, as the spatial entity does not lie far away from the rest of the entities (for locational outliers, see Section 3.1.6).

Why Use
A scatter plot is used to identify the relations between two variables and trace potential outliers.

Interpretation

Inspecting a scatter plot allows one to identify linear or other types of associations (see Figure 2.8A). If points tend to form a linear pattern, a linear relationship between variables is evident. If data points are scattered, the linear correlation is close to zero, and no association is observed between the two variables. Data points that lie further away on the x or y direction (or both) are potential outliers (see Figure 2.8B).

Discussion and Practical Guidelines

The first thing to inspect in any bivariate analysis is a scatter plot, which displays data as a collection of points (X, Y). In spatial data analysis, a scatter plot is a map with as many points as the spatial objects (rows in the dataset table; Figure 2.8C). For each data row in the database table, we create a set of coordinates. For example, for a single object, the value of variable A (*Income*) is the X coordinate, and the value of variable B (*House size*) is the Y coordinate (see Figure 2.8C). The X, Y coordinates can be switched (A to Y and B to X) with no significant change in the analysis. Each point in the scatter plot stands for a single spatial object in the map (polygons in Figure 2.8C). As the scatter plot belongs to the ESDA toolset, it offers the ability to highlight the spatial unit to which a point in the plot is linked in the map while brushing it. For instance, if we brush the outlier point, we directly locate which polygon corresponds to this value. Likewise, we can select one or more polygons in the map and identify their values in the scatter plot. We can also test if neighboring polygons in the map cluster in the scatter plot and identify if the object clustering in space creates attribute clusters as well.

2.3.2   Scatter plot matrix

Definition

A **scatter plot matrix** depicts the combinations of all possible pairs of scatter plots when more than two variables are available (see Figure 2.9).

Why Use

The visual inspection of all pair combinations facilitates (a) the locating of variables with high or no association, (b) the identification of relationship type (i.e., linear nonlinear) and (c) outlying points.

Interpretation

The closer the data points are to a linear pattern, the higher their linear correlation is to be. On the other hand, the more scattered a pattern is, the weaker the linear relationship between the two studied variables. The further away a data point lies from the main point cloud, the more likely it is to be an outlier.

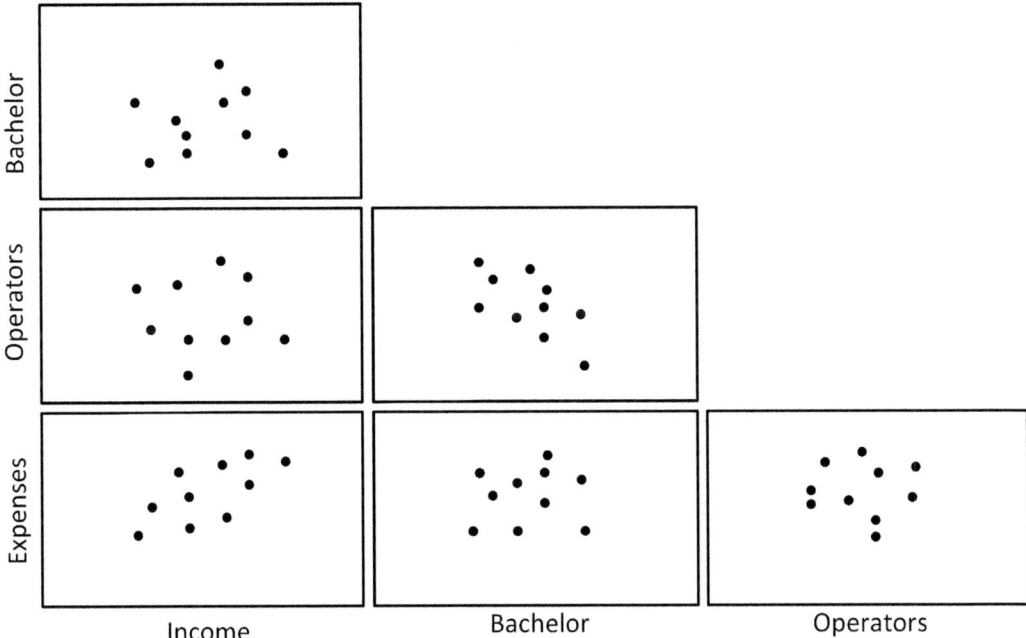

Figure 2.9 All combinations of scatter plot pairs for four census variables: Income, Expenses, Bachelor degree (education), Operators (occupation).

Discussion and Practical Guidelines

By inspecting a scatter plot matrix, one can quickly identify a linear or other type of association for multiple combinations of variables in a single graph. By identifying which variables exhibit high associations, we can proceed to further analysis, such as mapping them using choropleth maps or examining for potential bivariate spatial autocorrelation (see Chapter 4). A scatter plot matrix is a quick and efficient ESDA technique for identifying the association of all variable combinations, and is usually an efficient way to begin an analysis.

2.3.3   Covariance and Variance–Covariance Matrix

Definition

Covariance is a measure of the extent to which two variables vary together (i.e., change in the same linear direction). Covariance $Cov(X, Y)$ is calculated as (2.12) (Rogerson 2001 p. 87):

$$Cov(X, Y) = \frac{\sum_{i=1}^{n} (x_i - \bar{x})(y_i - \bar{y})}{n - 1} \qquad (2.12)$$

where

  x_i is the score of variable X of the i-th object
  y_i is the score of variable Y of the i-th object
  $\bar{x}$ is the mean value of variable X
  $\bar{y}$ is the mean value of variable Y

This formula is for sample covariance. For population covariance, we divide by n instead of $n - 1$.

Why Use
Covariance measures the extent to which two variables of a dataset change in the same or opposite linear direction.

Interpretation
For positive covariance, if variable X increases, then variable Y increases as well. If the covariance is negative, then the variables change in the opposite way (one increases, the other decreases). Zero covariance indicates no correlation between the variables.

Covariance can also be presented along with the variance of each variable in a variance–covariance matrix (2.13). In this matrix, the diagonal elements contain the variance of each variable (calculated based on the dataset matrix A [2.14]), and the off-diagonal elements contain the covariance of all pairs of combinations of these variables.

$$
\begin{aligned}
&\textit{Variance Covariance}\\
&=
\begin{bmatrix}
s^2_{1,1} & Cov(X_1 X_2) & Cov(X_1 X_3) & \cdots & Cov(X_1 X_p) \\
Cov(X_2 X_1) & s^2_{2,2} & \cdots & \cdots & \vdots \\
Cov(X_3 X_1) & \cdots & \ddots & \cdots & \vdots \\
\vdots & \cdots & \cdots & \ddots & \vdots \\
Cov(X_p X_1) & \cdots & \cdots & \cdots & s^2_{p,p}
\end{bmatrix}
\end{aligned}
\tag{2.13}
$$

$$
A =
\begin{bmatrix}
X_1 & X_2 & \cdots & X_p \\
a_{1,1} & \cdots & \cdots & a_{1,p} \\
\vdots & \cdots & \cdots & \cdots \\
a_{n,1} & \cdots & \cdots & a_{n,p}
\end{bmatrix}
\tag{2.14}
$$

Where p is the total number of variables (X) and n is the total number of observations (a) of the dataset A (2.14).

Discussion and Practical Guidelines

The variance–covariance matrix is applied in many statistical procedures to produce estimator parameters in a statistical model, such as the eigenvectors and eigenvalues used in principal component analysis (see Chapter 5). It is also used in the calculation of correlation coefficients. Covariance and variance–covariance are descriptive statistics and are widely used in many spatial statistical approaches.

2.3.4   Correlation Coefficient

Definition

Correlation coefficient $r_{(x,y)}$ analyzes how two variables (X, Y) are linearly related. Among the correlation coefficient metrics available, the most widely used is the Pearson's correlation coefficient (also called Pearson product-moment correlation), given by (2.15) (Rogerson 2001 p. 87):

$$r_{(x,y)} = \frac{Cov(X,Y)}{s_x s_y} \tag{2.15}$$

where

Cov(X, Y) is the sample covariance between the variables
s_x is the sample standard deviation of variable X
s_y is the sample standard deviation of variable Y

The population correlation coefficient is calculated using the population covariance and the population standard deviations.

Why Use

Correlation not only reveals if two variables are positively or negatively linearly related, but it also defines the degree (strength) of this relation on a scale of -1 to $+1$ by standardizing the covariance.

Interpretation

A positive correlation indicates that both variables either increase or decrease. A negative correlation indicates that a variable increases when the other decreases and vice versa. There are six main classes of correlation (see Figure 2.10). A strong positive correlation (for values larger than 0.8) indicates a strong linear relationship between the two variables; when variable X

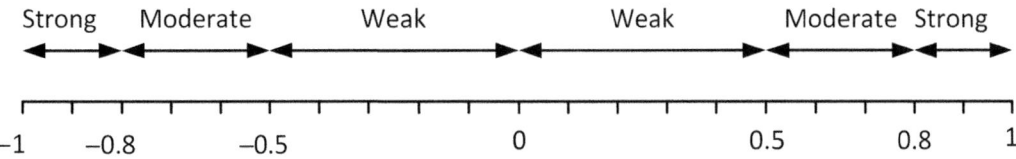

Figure 2.10 Labeling correlation.

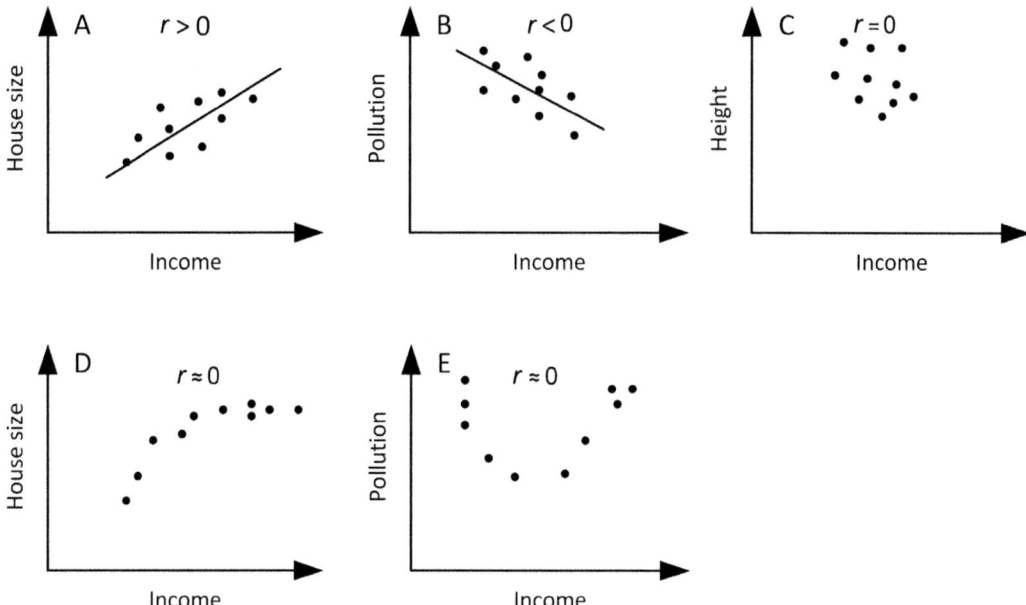

Figure 2.11 Linear correlation examples. (A) Strong positive correlation. A linear regression fit has been superimposed to the data to highlight the linear relationship. (B) Strong negative correlation. (C) No correlation (independent variables). (D) No linear correlation, but a curve pattern appears in the data. We can either use a nonlinear model or transform the data. (E) No linear correlation, but a pattern is observed in the data.

increases (or decreases), then variable Y also increases (or decreases) to a similar extent. A moderate positive correlation for values between 0.5 and 0.8 indicates that correlation exists but is not as intense as in a strong correlation. Observing a weak positive or weak negative correlation does not allow for reliable conclusions regarding correlation, especially when the values tend to zero. However, when the values lie between 0.3 and 0.5 (or between −0.5 and −0.3), and according to the problem studied, we may label correlation as "substantial." A moderate negative correlation (−0.8 to −0.5) means that correlation exists but is not very strong. Finally, a strong negative correlation (−1 to −0.8) indicates a strong linear relationship between the two variables (but with different directions: one decreasing and the other increasing or vice versa).

A Pearson's correlation close to zero indicates that there is no linear correlation, but this does not preclude the existence of other types of relation, as in plots D and E in Figure 2.11. Only if we use the scatter plot can we assess the potential for other types of relation.

Discussion and Practical Guidelines

Correlation coefficient is a statistical test, and its results have to be checked for statistical significance based on the null hypothesis that there is no

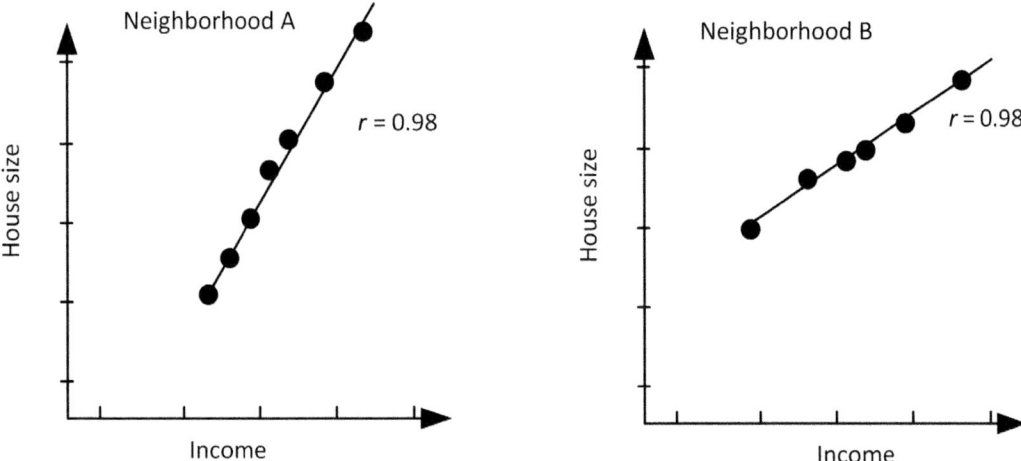

Figure 2.12 Correlation coefficient of income and house size for two different neighborhoods.

correlation between the two variables. A *p*-value is calculated expressing the probability of finding the observed value (correlation coefficient) if the null hypothesis is true (see Section 2.5.5 for a detailed analysis). A significance level should be set in advance (e.g., 0.05). If the *p*-value is smaller than the significance level (e.g., 0.001), then we reject the null hypothesis and conclude that the correlation observed is statistically significant at the 0.05 significance level. When reporting correlation values, the results should always be accompanied by their *p*-values and a related statement regarding their significance.

The slope of a regression line (the superimposed line in the data) is not the same as the correlation coefficient value unless the two variables used are in the same scale (e.g., through standardization; see Figure 2.12). The correlation provides us with a bounded measure $[-1, 1]$ of the association between the two variables. The closer to 1 or -1 it is, the closer it is to a perfect linear relationship. The slope in a regression line is not bounded by any limit and shows the estimated change in the expected value of Y for a one-unit change of X. This cannot be produced from the correlation itself. Although a positive slope is an indication of association and thus of positive correlation (i.e., the slope and correlation have the same sign), it cannot provide us with the measure of this association, as the correlation coefficient does. Nevertheless, when the variables are standardized, the slope equals the correlation coefficient.

Data in A and B depict the relation of income and house size for a set of households in two different neighborhoods (see Figure 2.12). Although the correlation is the same in both neighborhoods, the slope is different. Identical correlation means that in both neighborhoods, income is almost perfectly linearly related to housing size (all dots lie on a line), and the points have similar deviations from the line. A difference in the slopes would indicate that, in neighborhood A, the increase in house size for one additional unit of income is far larger than that in neighborhood B. Why this happens should be determined through additional data analysis. It might be because neighborhood A lies in suburbs, where more space is available, and that neighborhood B lies close to the city center, where houses in the same price range are smaller.

Finally, correlation is a measure of association **and not** of causation. Correlation is often used to prove that one action is the result of another. This assumption is wrong if made with no further analysis. Correlation establishes only that something is related to something else. Causation and relationship/ association are different. High correlation reveals a strong relation but not necessarily causation. Imagine a study on daily sales of ice cream along with the daily sales of cold drinks during the summer. If we calculate their correlation, we will probably identify a strong correlation since sales of ice cream and cold drinks are at their peak in the summer. Although this suggests an association or mathematical relationship (strong correlation), no functional relationship of causation is observed. It is not that high sales of ice cream drive (cause) the high sales of cold drinks (effect), nor is it the other way around. Another factor drives the relationship: temperature. High temperatures during summer drive (cause) the consumption (effect) of these products. Thus, the sales of ice cream and cold drinks have a functional relationship with temperature but do not have a relationship between them; this type of correlation is called "spurious." However, if we study the link between personal income and educational attainment, we will probably find a strong correlation, as people with higher income tend to have obtained at least a bachelor's degree. This is a sign of causation, as it is widely accepted that people with higher educational attainment are likely to get well paid jobs. This is merely an indication of causation and is not a definite cause-and-effect relationship. Determining whether such a link between these two variables exists would require additional scientific analysis and meticulously designed statistical experiments through explanatory analysis.

2.3.5   Pairwise Correlation

Definition
Pairwise correlation is the calculation of the correlation coefficients for all pairs of variables.

Why Use
When dealing with a large dataset, we can simultaneously calculate the correlations between all pairs of variables to identify potential linear relationships quickly.

Table 2.3 Pairwise correlation matrix for five variables. Example solved can be found in Chapter 6.

	Var1	Var2	Var3	Var4	Var5
Var1	1.0000	0.7630	0.3655	0.3560	0.4371
Var2	0.7630	1.0000	0.3281	0.2563	0.2609
Var3	0.3655	0.3281	1.0000	0.9372	0.7151
Var4	0.3560	0.2563	0.9372	1.0000	0.7399
Var5	0.4371	0.2609	0.7151	0.7399	1.0000

Interpretation
For n variables, the result is a square n-by-n matrix with the coefficient correlation values stored in the off-diagonal cells (see Table 2.3). Diagonal cells have a value of 1 to indicate the correlation of a variable with itself. The correlation is then interpreted as explained in Section 2.3.4.

Discussion and Practical Guidelines
We can also create a pairwise matrix plot, which displays on the off-diagonal cells (a) the scatter plots of variable pairs, (b) the correlation coefficients for each set of variables,and (c) the trend line. In the diagonal cells, the histogram of each variable is presented (see Figure 2.13). This matrix is more informative than the scatter plot matrix, which includes only the scatter plots (see Section 2.3.2).

2.3.6 General QQ plot

Definition
A **general QQ plot** depicts the quantiles of a variable against the quantiles of another variable.

Why Use
This plot can be used to assess similarities in the distributions of two variables (see Figure 2.14). The variables are ordered, and cumulative distributions are calculated.

Interpretation
If the two variables have identical distributions, then the points lie on the reference line at 45°; if they do not, then their distributions differ.

2.4 Rescaling Data

Definition
Rescaling is the mathematical process of changing the values of a variable to a new range.

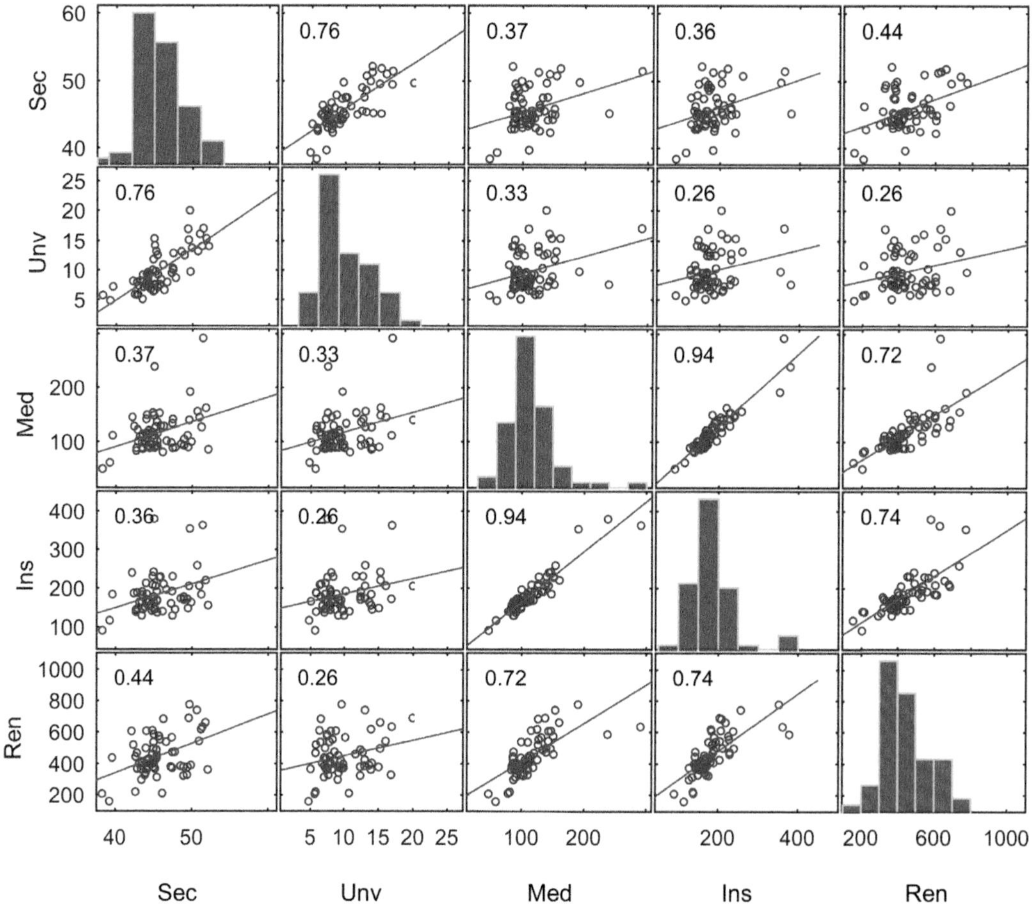

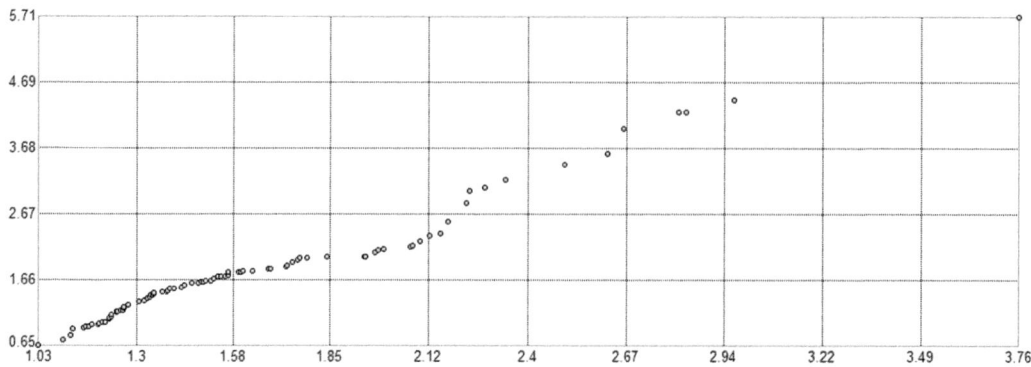

Figure 2.13 Correlation pairwise matrix plot.

Figure 2.14 General QQ plot of *Income* (x-axis) and *Expenses* (y-axis). Points deviate from a reference line at 45°, and their distributions cannot be regarded identical.

Why Use

When variables have very different scales, it is very hard to compare them directly. By rescaling data, the spread and the values of the data change, but the shape of the distribution and relative attributes of the curve remain unchanged. Large differences in value ranges and scales, infer mainly two problems (de Vaus 2002 p. 107):

- First, comparing their descriptive statistics – such as mean, median and standard deviation – is hard, and interpretation is not straightforward.
- Second, we cannot combine data with widely differing upper and lower scores and create an index or ratio as one variable might have a larger impact on the index formula, due solely to its scale. The resulting values are likely to be too large or too small, which will also be hard to interpret.

Rescaling data is also widely used in multivariate data analysis for data reduction and data clustering (see Chapter 5). For example, as clustering methods are based in the calculation of a dissimilarity matrix which contains the statistical distance (e.g., Euclidean distance) among data points, failing to rescale data leads to assigning disproportionally more importance in variables with significantly larger values with respect to the other ones.

Methods

To avoid such problems, three rescaling methods can be used:

- **Normalize:** The following formula is a typical method of creating common boundaries (2.16):

$$Xrescale = \left(\frac{X - Xmin}{Xmax - Xmin}\right) n \qquad (2.16)$$

where *Xrescale* is the rescaled value, *X* is the original value, *Xmin* is the minimum value, *Xmax* is the maximum value and *n* is the upper limit defined by user of the final variable. This rescaling method is prone to outliers, as they will scale data to a very small interval (the Xmax–Xmin range will be large).

For $n = 1$, the rescaled variable ranges from 0 to 1. This is called "normalization" and scales all numeric variables in the range [0, 1]. We can also normalize data to the [−1, 1] range using (2.17):

$$Xrescale = -1 + 2\left(\frac{X - Xmin}{Xmax - Xmin}\right) \qquad (2.17)$$

- **Adjust:** Another method of rescaling data is to divide a variable (or multiply it by assigning weights) by a specific value. For example, instead of using dollars to describe a person's income, we could use "income" divided by "average income of the country." This new value has several advantages. Due to inflation, income is not directly comparable across years. This ratio provides average income as an analogy of personal income, which is comparable across years. This is called "adjustment": we adjusted personal income to average income. Adjustment is the process of removing an effect (e.g., inflation, seasonality) to obtain more comparable data. Adjustments are also needed to compare incomes in different places, such as countries. We cannot directly compare the income of someone in a Western country to that of someone in a developing country because the income variable has wide differences within its range. Adjustments could be expressed in many other ways depending on the problem studied and the research question/hypothesis tested.

- **Standardize:** Calculate z-scores. A z-score is the number of standard deviations a score lies from the mean (see Section 2.5.5, Eq. 2.21). Put simply, a standardized variable expresses the original variable values in standard deviation units (de Vaus 2002 p. 108). A standardized variable always has a mean of 0 and a variance of 1 (de Smith 2018 p. 201).

Discussion and Practical Guidelines (Normalization vs. Standardization)

Normalization and standardization are widely used in statistics and subsequently in spatial analysis. But which method is more appropriate? There is a not a straightforward answer, but let us see some basic differences:

- As a standardized variable always has a zero mean and a unit variance, it provides little information when we want to compare the means between two distributions. When the mean values are important, normalization is more descriptive.
- With standardization, the new values are not bounded. On the contrary, with normalization, we bound our data between 0 and 1 (or −1 to 1). Having comparable upper and lower bounds might be preferable and more meaningful for some studies (e.g., marketing analysis) especially when the mean values are important.
- In the presence of outliers, normalizing the non-outlying values will scale them to a very small interval. This does not happen with standardization.
- Standardization is most useful for comparing subgroups of the same variable, such as comparing between rural and urban incomes (Grekousis et al. 2015a).
- Standardization is also used in multivariate data analysis (see Section 5.1) so that variables do not depend on the measurement scale and are comparable with each other (Wang 2014 p. 144). It is occasionally preferred to normalization as it better retains the importance of each variable due to the non-bounding limitation. For example, in case of outliers,

normalized data are squeezed at a small range, and as such, when dissimilarities (through statistical distances) are calculated, they contribute less to the final values.

• Many algorithms (especially artificial neural networks' learning algorithms) assume that data are centered at 0. In this case, standardization seems a more rational choice than normalization.

Keep in mind that rescaling is not always desirable. In case we have data of similar scales or proportions (e.g., percentages) or we want to assign weights to the variables with larger values, we might not consider normalizing, adjusting or standardizing. It depends on the problem in question and the available dataset to decide if and which rescaling type to apply.

2.5 Inferential Statistics and Their Importance in Spatial Statistics

In the previous sections, we discussed the use of descriptive statistics and related ESDA tools in summarizing the key characteristics of the distribution of an attribute. However, to delve deeper into the statistical analysis of a problem, we should apply more advanced methods. For example, though we can summarize a sample, we cannot make any inference related to the statistical population it refers to. Descriptive analysis is accurate only for the specific sample we are analyzing, and the results and conclusions cannot be expanded to the statistical population it was drawn from. Suppose we query 30 households in a neighborhood about income, the size of the house and the number of cars owned by each family. Using descriptive statistics, we calculate the average family income, the number of cars owned per family and the frequency distribution of their house size. This is a very good start, but it does not tell us much about the wider area. For example, what is the average income of the census tract this neighborhood belongs to? What is the average house size in the city? What is the relation between income and the number of cars in the city? These questions attempt to generate results at a larger scale but cannot be directly answered using the descriptive statistics calculated from the 30-household sample. Making inferences from a sample to a population requires inferential statistics.

Definition

Inferential statistics is the branch of statistics that analyzes samples to draw conclusions for the entire population. In other words, through inferential statistics, we infer from the sample data what are the characteristics of the population.

Why Use

Inferential statistics are used when we need to describe the entire population through samples. With inferential statistics, we analyze a sample, and the findings are generalized to a larger population. Descriptive statistics, by contrast, hold true only for the specific sample they were calculated for; this should be acknowledged in any analysis. Scholars often generalize and use descriptive statistics to summarize populations. A sample can describe a population if specific procedures (through inferential statistics) have been followed, but this should be clearly stated.

Importance to Spatial Statistics

Spatial statistics use inferential statistics to make inferences about a statistical population. For example, spatial autocorrelation tests, spatial regression models and spatial econometrics use inferential statistics methods. Being able to evaluate the results of spatial statistics and draw correct conclusions requires understanding of inferential statistics. Incorrect interpretations of advanced spatial statistics are common and usually stem from ignorance of inferential statistics theory. A firm knowledge of inferential statistics is required for anyone undertaking spatial analysis through spatial statistics, and the next sections cover the rudiments of the following inferential statistics topics:

- What are parametric and nonparametric methods and tests?
- What is a test of significance?
- What is the null hypothesis?
- What is a p-value?
- What is a z-score?
- What is the confidence interval?
- What is the standard error of the mean?
- What is so important about normal distribution?
- How can we identify if a distribution is normal?

2.5.1   Parametric Methods

Definitions

Parametric methods and tests are statistical methods using parameter estimates for statistical inferences (see Table 2.4; Alpaydin 2009 p. 61). They assume that the sample is drawn from some known distribution (not necessarily normal) that obeys some specific rules. They belong to inferential statistics.

Population parameters are values calculated from all objects in the population and describe the characteristics of the population as a whole. Population parameters are fixed values. Each population parameter has a corresponding sample statistic.

Sample statistics are characteristics of a sample. They can be used to provide estimates of the population parameters. Sample statistics do not have fixed values and are associated with a probability distribution called a sampling distribution. In practice, any value calculated from a given sample is called a statistic (Alpaydin 2009, p. 61).

Table 2.4 Parametric and nonparametric statistics according to the scope of analysis and measurement level. Parametric and nonparametric statistics can be found in many textbooks, studies and papers. A scientist should have a rough idea about these tests in order to comprehend why they are selected among so many different test options and what they intend to identify. The table focuses on statistics most commonly used to determine if differences or correlations among variables exist.

Identify	Parametric statistics (normal)	Nonparametric statistic (non- normal)	Level of measurement (for nonparametric)
Difference between two independent groups	t test (interval)	Mann–Whitney U-test Kolmogorov–Smirnov Z test Chi square	Ordinal/Interval Ordinal/Interval Nominal/Ordinal/ Interval
Difference between more than two independent groups	Analysis of variance and F test (interval)	Kruskall Wallis analysis of ranks Median test Chi square	Ordinal/Interval Interval Nominal/Ordinal/ Interval
Difference between two related groups	t test for dependent samples (interval)	Sign test Wilcoxon's test McNemar	Ordinal/Interval Ordinal/Interval Nominal/Ordinal/ Interval
Correlation between variables	Pearson's r (interval)	Spearman's Rho Kendall's tau Gamma	Ordinal Ordinal Ordinal

Parameter estimates are estimates of the values of the population parameters. Parameter estimates are estimated from the sample using sample statistics (e.g., sample mean, sample variance). As soon as parameters are estimated, they are plugged into the assumed distribution, and the final size and shape of the distribution for the specific dataset are determined (Alpaydin 2009 p. 61). The most commonly used methods of estimating population parameters are the maximum likelihood estimation and the Bayesian estimation.

How Parametric Methods Work

To better understand how inferential statistics and parametric methods work, we should take a look at the following basic flowchart (see Figure 2.15): The parametric statistical approach is composed of four basic steps:

A) **Population:** Define the population that the study refers to.
B) **Sampling:** Using a sampling procedure, extract data from the entire population. We use samples because we cannot practically measure each single object of a population.
C) **Sample:** Make assumptions about the distribution of the entire population – for example, that it follows the normal distribution. Next, estimate the population parameters using sample statistics.

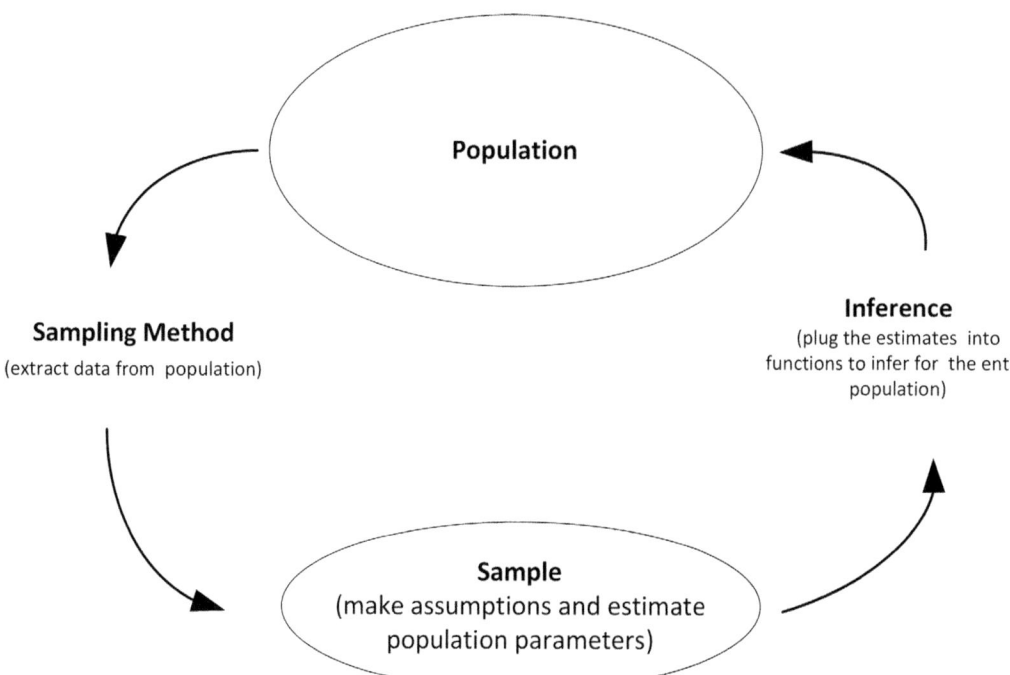

Figure 2.15 Parametric statistical approach.

D) **Inference:** Plug these parameters into functions describing the assumed distribution (e.g., probability density function), to obtain an estimation of the entire population.

Let us consider an example. Suppose we want to analyze the sales of a product according to the customers' age structure. The four basic steps are:

A) **Population:** Customers

B) **Sampling:** We first randomly select our sample (random sampling is not the only sampling method) from the total population. The sample might be *n* people who filled in questionnaires.

C) **Sample:** We then assume that the variable "sales of product" is normally distributed over the age of a potential customer (the age intervals on the *x*-axis and relative frequency of sales on the *y*-axis follow a normally shaped bell). We estimate two sample statistics: sample mean (as an estimate of the population mean) and the sample standard deviation (as an estimate of the population standard deviation) using the respective formulas.

D) **Inference:** Applying these two estimates in the normal probability density function (Eq. 2.1) allows for further analysis. For example, what is the probability that this product will be selected by customers aged 20 to 25? If the probability obtained from the probability distribution function is not desirable (e.g., lower than the company's target), we might decide to invest more in advertisements targeting this age group. We can also calculate confidence intervals, which is the range of values within which the above estimates are likely to lie at a 95% certainty. By conducting simple parametric statistical analysis, we achieved decision making.

Discussion and Practical Guidelines

Inferential statistics use sample statistics, so termed because they refer to a sample and not to the entire population. In general, the characteristics of samples are denoted by Latin letters and the characteristics of populations by Greek letters (see Table 2.2). Formulas for the same measure (e.g., standard deviation) might change when populations or samples are calculated. The term **"statistic" is used only for samples**. The term **"parameter" refers only to a population**. (Tip to remember: "population" starts with "P," as does "parameter"; "statistics" starts with "S," as does "sample.") **Parameters** are descriptive measures of the entire population and are fixed values. Parameter estimates are calculated based on sample statistics and are not predictable. They are associated with probability distributions and margins of errors.

To sum up, the main goal of inferential statistics is to **estimate the original population parameters** and measure the **amount of error** of these estimates by depending on the sample data available. In other words, as we cannot directly measure these parameters in the entire population, we use the sample to estimate the true parameters.

For example, the normal distribution needs only two parameters to be defined, the mean and the standard deviation. Once we estimate these parameters (by using the sample data), we can plug them into a probability distribution function (see Section 2.2.2) to generate the distribution curve. The objective of describing the population is now achieved. In inferential statistics, each distribution is defined entirely by only a small number of parameters (usually one to three).

The parameters' values are called "estimates," as we do not calculate a value but produce an estimation with an associated error. The correct expression is "parameter estimate," not "parameter calculation." There are two basic approaches to evaluating a parameter estimation:

(a) Using a confidence interval (see Section 2.5.3)
(b) Hypothesis testing (see Section 2.5.5)

Parametric methods and tests are more accurate and have higher statistical power if the assumption of the distribution adopted (e.g., normal) is true relative to nonparametric methods (see next section). Another advantage is that the problem is reduced to the estimation of a small number of parameters (e.g., mean and variance). Inferences are valid only if the assumptions made in the parametric approach hold true. When the assumptions (e.g., randomly

selected and independent samples) fail, they have a greater chance of producing inaccurate results. In the spatial context, we have to assess if the assumptions hold before we use a parametric test. Randomness and thus complete
spatial randomness are rare in space due to spatial autocorrelation (as a result
of spatial dependence in most of the geographical problems – see Chapter 4).
Spatial statistics that overcome these problems should be created. However,
the new tests share similar terminology with classic statistics and are also based
on significance tests, hypothesis testing and confidence intervals.

2.5.2   Nonparametric Methods

Definition

Statistical methods used when normal distribution or other types of
probability distributions are not assumed are called "**nonparametric**" (Alpaydin 2009 p. 163). In nonparametric methods, we do not make assumptions
about the distribution of the data and the parameters of the population we
study. The distribution and the number of parameters are no longer fixed.

Why Use

If assumptions are violated or if we cannot be certain whether they hold true,
we may turn to nonparametric statistics/methods/models. Nonparametric
models are not based on assumptions regarding either the sample (data) or
the population drawn from (e.g., linear relationship or normal distribution).
Nonparametric methods are based on the data, and their complexity depends
on the size of the training dataset. The only assumption made is that similar
inputs lead to similar outputs (Alpaydin 2009 p. 163). In nonparametric
methods, the parameters' set is not fixed and can increase or decrease
according to the available data.

Discussion and Practical Guidelines

Nonparametric methods do not imply the absence of parameters. They imply
the non-predefined nature of the analysis and suggest that parameters are
flexible and adapt according to the data (for example, the nonparametric kernel
density estimation method [Section 3.2.4] has the smoothing parameter h).
Nonparametric tests are widely used when populations can be ordered in
a ranked form. Ordinal data are thus well structured for nonparametric tests.
One can use an index to rank interval data and nonparametric tests if needed.
Nonparametric statistics make fewer assumptions and are simpler in structure;
for this reason, their applicability is wide. Nonparametric methods include
Mann–Whitney U test (also known as Wilcoxon test), Kolmogorov–Smirnov
test, Kruskall–Wallis one-way analysis of ranks, Kendall's tau, Spearman's rank
correlation coefficient, kernel density estimation and nonparametric regression
(de Vaus 2002 p. 77; Table 2.4).

2.5.3   Confidence Interval

Definition
Confidence interval is an interval estimate of a population parameter. In other words, a confidence interval is a range of values that is likely to contain the true population parameter value. A confidence interval is calculated once a confidence level is defined.

    Confidence level for a confidence interval reflects the probability that the confidence interval contains the true parameter value. It is usually set to 95% or 99%. It should not be confused with the significance level (see Section 2.5.5). A confidence level of 95% reflects a significance level of 5%.

Why Use
Confidence interval is used to estimate a range of values in which a population parameter lies for a certain confidence level (probability).

Interpretation
How accurately a statistic estimates the true population parameter is always an issue. The confidence interval of a statistic (e.g., the mean) estimates the interval within which the population parameter (e.g., mean) ranges. The confidence interval is expressed in the same unit used for the variable.

    Confidence intervals are constructed based on a confidence level X% defined by the user, such as 95% or 99%. Confidence levels indicate that, if we conducted the sampling procedure several times, the confidence interval would include the estimated population parameter X out of 100 times. Have in mind that the confidence level, (for example, 95%), does not indicate that for a given interval there is a 95% probability that the parameter lies within this interval. It indicates that 95% of the experiments will include the true mean, but 5% will not. Based on the definition of confidence interval by Neyman (Neyman 1937), once an interval is defined, the parameter either is included or not. As such, the probability does not refer to whether the population lies inside the interval but on the reliability of the estimation process of getting an interval that includes the true population parameter.

Discussion and Practical Guidelines
We will not detail how confidence intervals are calculated, as computing them directly is rare. It is more important to understand their use. Suppose we have selected a sample of households from a city and we want to estimate the mean household income of the households for the entire city (population). If the sample's mean income is 15,000 US dollars and the margin of error for the 95% confidence level is ±500 US dollars, this typically means that we can be 95% confident that the mean income of the households in this city ranges between 14,500 and 15,500 (confidence interval). There is still a 5% chance that the mean income lies in another range. To reduce the range of the interval, we can

use a larger sample size. This will reduce the standard error and thus the interval produced.

Confidence intervals are different from significant tests (explained in Section 2.5.5). A confidence interval provides a more complete view of a variable. Instead of deciding whether or not to reject the sample estimate, a confidence interval estimates the margin of error of the sample estimate (de Vaus 2002 p. 187). The margin of error is the value to be added or subtracted from the statistic – e.g., sample mean – which reflects the interval length. This reminds us that there is no absolute precision in any estimate.

Confidence intervals (and standard error of the mean discussed in the next section) are common in reports and papers related to geographical analysis, and one should be able to interpret these statistics in the context provided.

2.5.4   Standard Error, Standard Error of the Mean, Standard Error of Proportion and Sampling Distribution

Definitions

The **standard error** of a statistic is the standard deviation of its sampling distribution (Linneman 2011 p. 540). The standard error reveals how far the sample statistic deviates from the actual population statistic.

Standard error of the mean is the standard deviation of the sampling distribution of the mean.

A **sampling distribution** is the distribution of a sample statistic for every possible sample of a given size drawn from a population.

The standard error of the mean refers to the change in mean in each different sample. This procedure is more straightforward than it seems. The standard error of the mean is calculated by the following formula (2.18): (O'Sullivan & Unwin 2003 p. 403):

$$\sigma_{\bar{x}} = \frac{s}{\sqrt{n}} \tag{2.18}$$

s is the sample standard deviation of the distribution studied.
n is the sample (number of objects).

In case that attribute values (scores) are expressed as percentages (P), the standard error is calculated by the following formula (2.19) (Linneman 2011 p. 177):

$$\sigma_{\bar{x}} = \sqrt{\frac{P(1 - P)}{n}} \tag{2.19}$$

For example, if for $n = 30$ (sample size of students), 6 out of 30 (20%) smoke, then the standard error of this value would be (2.20) (de Vaus 2002 p. 193):

$$\sigma_{\bar{x}} = \sqrt{\frac{0,20(1 - 0,20)}{30}} = 7.3\% \qquad (2.20)$$

Why Use

The standard error of a statistic shows how accurately this statistic estimates the true population parameter. The standard error of the mean, for example, is used to estimate how precisely the mean of a sample has been calculated. It measures how close the sample mean is to the real population mean. The standard error is also used to calculate the confidence interval based on the z-score and the confidence level (de Vaus 2002 p. 188).

Interpretation

Low values of the standard error of the mean indicate more precise estimates of the population mean. The larger the sample is, the smaller the standard error calculated. This is rational, as the more objects we have, the closer to the real values our approximation will be. According to two rules from probability theory:

- There is 68% probability that the population parameter is included in a confidence interval of ±1 standard error from the sample estimate.
- There is 95% probability that the population parameter is included in a confidence interval of ±1.96 standard errors from the sample estimate (de Vaus 2002 p. 188).

Discussion and Practical Guidelines

When we estimate a population parameter (e.g., the mean), we obtain a single value of the statistic, also called the "point estimate." As this point estimate has been calculated based on a sample, it is subject to a sampling error. It is crucial to identify how much error this point estimate is likely to contain. In other words, we cannot be confident that one sample represents the population accurately. We have to select more than one sample, calculate the statistic for each one and then analyze the formed distribution of the statistic produced. As such, estimating population parameters usually requires the use of a sampling distribution and the calculation of the standard error, which reflects how well the statistic estimates the true population parameter.

For example, to calculate the standard error of the mean, we should take many samples and then calculate the mean for each one. The resulting distribution is a sampling distribution of the mean. According to central limit theorem for a sample size larger than 30 objects/observations (O'Sullivan & Unwin 2003 p. 403):

- The distribution of the sample means will be approximately normal (as the sample size increases), regardless of the population distribution shape.
- The mean of the sampling distribution of the mean is approximating the true population mean.

As it is unfeasible to draw hundreds of samples to create a sampling distribution, we use a single sample drawn from a population that hypothetically belongs to a sampling distribution and then estimate the value of the standard error of this sampling distribution (for example, for the standard error of the mean, we use Equation [2.16] [Linneman 2011 p. 166]).

2.5.5   Significance Tests, Hypothesis, p-Value and z-Score

Definition

A **test of significance** is the process of rejecting or not rejecting a hypothesis based on sample data. A test of significance indicates the probability that the results of the test are either due to sampling error or reflect a real pattern in the population the sample was drawn from (de Vaus 2002 p. 166). A test of significance is used to determine the probability that a given hypothesis is true.

The **p-value** is the probability of finding the observed (or more extreme) results of a sample statistic (test statistic) if we assume that the null hypothesis is true. It is a measure of how unlikely the observed value or pattern is to be the result of the process described by the null hypothesis. It is calculated based on the z-score.

The **z-score** (also called z-value) expresses distance as the number of standard deviations between an observation (for hypothesis testing calculated by a specific formula for a statistical test) and the mean. It is calculated (for samples) by the following formula (2.21):

$$z\ score = \frac{x - \bar{x}}{s} \tag{2.21}$$

where

   x_i is the score of the ith object
   $\bar{x}$ is the sample mean value
   s is the sample standard deviation

The z-score is widely used in standardization (see Section 2.4), in determining confidence intervals and in statistical significance assessments (O'Sullivan & Unwin 2003 p. 389).

Significance level α is a cutoff value used to reject or not reject the null hypothesis. Significance level α is a probability and is user-defined, usually taking values such as $\alpha = 0.05$, 0.01 or 0.001, which stand for 5%, 1% and

0.1% probability levels. The smaller the p-value the more statistically significant the results. A significance level of 5% reflects a confidence level of 95%.

Interpretation

In general, significance tests use samples to decide between two opposite statements (the null hypothesis and the alternative hypothesis). For example, a null hypothesis (H_0) can be that the sample observations result purely from a random process. The alternative hypothesis (H_1) states the opposite: that the sample observations are influenced by a nonrandom cause. This type of hypothesis is the most common in statistical testing. Another null hypothesis could be that there are no differences between two samples drawn from different distributions. The alternative hypothesis states that there are differences between the samples. The statement of the null hypothesis can be set according to the problem, but the alternative is the opposite. Statistical tests reject or do not reject the null hypothesis. In this respect, they should be designed carefully to reflect the problem at hand.

- *Rejecting the Null Hypothesis ($p \leq \alpha$)*
  A low p-value means that the probability that the observed results are the outcome of the null hypothesis under consideration is low. If the p-value is smaller than α, then we reject the null hypothesis.

  Rejecting the null hypothesis means that there is a (100% − α) probability that the alternative hypothesis H_1 is correct.

  More analytically, we accept H_1 as true but, in relation to a probability, not a certainty. This means that there is always a chance that H_1 will be accepted as true when it is not (Type I error; de Vaus 2002 p. 172). For example, we might reject the null hypothesis with a probability of 95%, but there still is a 5% (significance level) chance that it is true.

- *Not Rejecting the Null Hypothesis ($p > \alpha$)*
  When the p-value is larger than significance value α, then we cannot reject the null hypothesis. In such a case, we have to be very careful in interpreting the results. We do not use the word "accept" for the null hypothesis. When we do not reject the null hypothesis, this does not mean that we accept it; it simply means that **we fail to reject** it, as there is insufficient evidence to do so.

  Not rejecting the null hypothesis means that we do not have enough evidence to reject it but we cannot accept it without further analysis.

  For example, in case we test whether a distribution is normal or not (Ho: the distribution is normal) and $p > \alpha$, we cannot straightforwardly accept the distribution is normal. We have to examine other characteristics, such as plots or descriptive statistics, to conclude if we will ultimately accept the null hypothesis. In correlation analysis using Pearson's r statistic, failing to reject the null hypothesis (Ho: there is no linear correlation between two variables) does not necessarily mean that there is no correlation between the variables. If we create a scatter plot, we might trace a nonlinear correlation. An efficient way to avoid non-rejecting null hypothesis vagueness is to switch hypotheses. When deciding which alternative hypothesis to use, we have to consider our research objective. Some texts may say that not rejecting the null hypothesis means that you can accept it, but this should be done with caution.

Two types of error can result from a significance test (de Vaus 2002 p. 172):

- Type I error: when we reject the null hypothesis when it is true. This is determined by the significance level, which reflects the probability of Type I error.
- Type II error: when we do not reject the null hypothesis when we should.

In the spatial context, these errors reflect the multiple comparison problem and spatial dependence (see Section 4.6 on how to deal with these problems).

Discussion and Practical Guidelines

A typical workflow for significance testing is as follows (provided that a sample is collected):

1.   Make the statement for the null hypothesis (e.g., that the sample is drawn from a population that follows a normal distribution)
2.   Make the opposite statement for the H_1 hypothesis (e.g., that the sample is drawn from a population that is not normally distributed)
3.   Specify the significance level α.
4.   Select a test statistic (some formula) to calculate the observed value (e.g., sample mean).
5.   Compute the p-value, which is the probability of finding the observed (or more extreme) results of our sample if we assume that the null hypothesis is true. Put otherwise, the p-value is the probability of obtaining a result equal to (or more extreme than) the one observed in our sample if the null hypothesis is true.
6.   Compare the p-value to the significance value α. If $p \leq \alpha$, then we can reject the null hypothesis and state that the alternative is true and that the observed pattern, value, effect or state is statistically significant. If $p > \alpha$, then we cannot reject the null hypothesis, but we cannot accept it either.

Significance tests are used to assess if the probability that the null hypothesis is true (e.g., correlation or differences among samples) is due to sampling error or the existence of patterns in the population. One of the problems of using tests of significance is that they produce a binary yes/no (rejected/not rejected) answer, which does not entirely apply to the real world. In addition, significance tests do not provide a sense of the magnitude of any effect that the hypothesis is testing. On the contrary, the confidence interval approach, presented in the previous section, is more sufficient on this aspect (see Box 2.2).

Box 2.2 Think of significance and hypothesis tests as a trial. If you are accused of tax fraud, the tax bureau charge you. The null hypothesis is "guilty," and the alternative hypothesis is "not guilty." Your layer will bring forward evidence to prove that you are not guilty. If the evidence is sufficient, then the null hypothesis will be rejected and you will be declared innocent. If the evidence is not sufficient, then you cannot prove that you are not guilty, and you will most likely pay a fine. Still, this does not necessarily mean that you are guilty. Maybe the lawyer's evidence was not strong enough. In a parallel statistical world, the null hypothesis (guilty) is rejected if there is strong evidence (sample data with small errors). If the evidence is not sufficient, you cannot reject the null hypothesis, but this does not mean that the null hypothesis is true either.

Example in a Geographical Context

Suppose we study the following geographical research question: do people in cities earn more money than people in rural areas? This is a typical comparison question and is very interesting from the social and geographical perspectives. The key element here is location. Our **research question** is whether the rural–urban distinction affects income. It is a binary question, so hypothesis testing can offer valuable insights. Suppose we collected our samples and found that people living in cities have a 50% higher mean income than people living in rural areas. The real question is then as follows:

> Is the difference in mean income between rural and urban areas real, or it is just a random result due to sampling error (caused by not asking the right people about their income)?

In other words, can we state in a statistically sound way that this reflects the entire population? Remember that, even if we have the perception that people usually tend to earn more money in cities, we have to prove it or give a statistical context that supports our belief. Otherwise, it is just an opinion.

We could use the following hypotheses (statements) for our research question:

H_0 = People in cities do not have a mean income different from that of people in rural areas (null hypothesis).
H_1 = People in cities have a mean income different (50% higher) from that of people in rural areas (alternative hypothesis).

The two samples here are "people living in cities" (e.g., 1,000 people asked about their income during the last year) and "people living in rural areas" (e.g., 1,000 people asked about their income during the last year). The population characteristic to be analyzed is "income."

To answer this question (or, more correctly, to attempt to come to a conclusion), we have to run a significance test. If the entire population could be measured, we would not need such tests. We could decide which statement is correct based on the entire population. As this is usually impossible, significance tests are necessary.

- Rejecting the null hypothesis
  If the null hypothesis is rejected, we reject the hypothesis that the two samples and their distributions are the same (*not different*). In other words, there is a "good chance" that there is a difference between the distributions, and this difference is not just a result of randomness. This means that there are some patterns (reasons) by which these distributions are different. The "good chance" is estimated with a probability value using the significance level.

  If we select $\alpha = 0.05$ as the significance level and the resulting p-value calculated by the test is $p = 0.0015$, then we can reject the null hypothesis because $p < \alpha$. This means that H_1 is true and that there are differences between urban and rural incomes. The research hypothesis that income in cities is 50% higher than that in villages (as calculated earlier) is now statistically backed up. This result is typically expressed in statistical language as follows:

  > People living in urban areas have an income statistically different (50% higher) than that of people living in rural areas at the 5% significance level.

  The chance of finding the observed differences (e.g., in mean income) if the null hypothesis is true (no differences) is only $p = 0.0015$, or 0.15%. This can be stated using the significance level as follows:

  > There is a less than 5% chance that this difference is the result of sampling error.

  In other words, we have a 95% (100% − α) probability (confidence) that our distributions are different.

  Again, the results of all statistical significance tests are based on probability distribution functions. They produce probabilities, not certainties. In the preceding example, the conclusion is not that the "distributions are different" but that the "distributions are likely to be different with a probability of 95%." There is always a 5% chance that the distributions are similar. As a result, the smaller the significance level, the higher the chance that the results are close to the real values. The 95% probability in our example means that, if we randomly selected 100 different samples from a population, people living in cities would have incomes 50% higher than people living in rural area in 95 of the cases. More generally, 95 of the cases would reject the null hypothesis. Still, there is always the chance that five samples would not display the difference hypothesized.

- Not rejecting the null hypothesis
  Suppose that, for the same $\alpha = 0.05$ significance level, the resulting p-value calculated by the test were $p = 0.065$. In this case, we cannot reject the null hypothesis because $p > \alpha$. If we cannot reject the null hypothesis, we have to state the following:

  > We have insufficient evidence to reject the null hypothesis that people in cities do not have a mean income different from that of people in rural areas.

  Not rejecting the null hypothesis does not mean that we accept it, nor that the observed difference among the distributions is wrong. As we cannot reach a solid conclusion, we have to carry out other experiments and use other methods to decide whether incomes differ between rural and urban areas.

2.6   Normal Distribution Use in Geographical Analysis

Importance to Spatial Analysis

Spatial analysis is commonly used to study either the distribution of the locations of events/polygons or the spatial arrangement of their attributes (i.e., socioeconomic variables). As geographical analysis is an interdisciplinary field, many reports and research papers use purely statistical methods to supplement the core spatial analysis. For example, suppose we want to analyze the spatial distribution of income in relation to educational attainment. It is reasonable to begin with classic statistical analysis like the calculation of the Pearson's correlation coefficient between income and educational attainment. Spatial statistics can then be applied such as bivariate spatial autocorrelation to determine if these two variables tend to spatially cluster together. In geographical analysis, spatial statistics go along with classical statistics.

Statistics often deal with a normal distribution because many well-defined statistical tests (e.g., Pearson's correlation coefficient, analysis of variance, t-test, regression, factor analysis) are based on the assumption that the examined distribution is normal. If the observed distribution does not resemble a normal distribution (e.g., is skewed), then many statistical procedures are not accurate and cannot be used or should be used with caution. For example, if a distribution is skewed, the probability of having small values (compared to large values) differs. If data are collected through random sampling, different proportions of values are highly likely to fall into specific intervals. This will lead to an overrepresentation or underrepresentation of specific values in the sample that should be taken into account. We should highlight here that most statistics are based on a normal underlying distribution for attribute values and on a Poisson probability distribution for point patterns (Anselin 1989 p. 4). Poisson probability distribution is used to assess the degree of randomness in point patterns (Oyana & Margai 2015 p. 75, Illian et al. 2008 p. 57).

How to Identify a Normal Distribution

There are three simple methods of determining if a distribution is normal or not:

1. Create a histogram and superimpose a normal curve. Plot inspection can enable a rough estimation of whether the distribution approximates the normal curve.
2. Calculate the skewness and kurtosis for the distribution. If the distribution is skewed and/or kurtosis is high/low, we have a clear indication that the distribution is not normal (see Section 2.2.4).
3. Create a normal QQ plot (see Section 2.2.9).

What to Do When Distribution Is Not Normal

If the distribution is not normal, we have three options (we assume that outliers have been removed):

Option 1. Use nonparametric statistics (see Table 2.4).

Option 2. Apply variable transformation. An efficient way to avoid a non-normal distribution is to transform it (if possible) to a normal distribution. Table 2.5 presents transformations that can be used to transform a variable according to its skewness value. This transformation will not necessarily lead to a normal distribution, but there is a good chance that it will.

Option 3. Check the sample size. According to the central limit theorem, if the sample is larger than 30–40, parametric statistics may be used without affecting the results' credibility. This theorem states that, given certain conditions, as the size of a random sample increases, its distribution approaches a normal distribution. In other words, even if our distribution is not normal, we can use parametric statistics if we have a large sample (de Vaus 2002 p. 78). Such a violation of the normality assumption does not cause major problems (Pallant 2013). It is not easy to define the ideal value by which a sample can be regarded as large. According to the literature, values of 30 to 40 are regarded as sufficient for a sample to be considered large and follow the central limit theory. In spatial analysis, this means that we need a sample of more than 30 spatial entities (e.g., postcodes, cities, countries) to use parametric statistics. When fewer spatial entities are involved, it is essential to check the variables for normality if we want to make inferences for a larger population.

Table 2.5 Transformations to reduce skewness and restore normality.

Skew	Transformation	Formula	Used When
High positive skew	Reciprocals or Negative reciprocal. Reciprocal of a ratio may often be interpreted as easily as the ratio. For example: population density (people per unit area) becomes area per persons.	$Y_n = 1/Y$ $Y_n = -1/Y$ Add 1 in values less than 1 up to 0	Reciprocals are useful when all values are positive. The negative formula is used to transform negative values. Reciprocal transformation reverses order among positive values: largest becomes smallest.
Moderate to low positive skew	Square root Logarithmic transformation	$Y_n = Y^{\wedge}0.5$ $Y_n = \log y$ or $Y_n = \ln y$	May have many zero's or very small values. May have a physical exponent (e.g., area). May have only positive values.
Low to moderate negative skew	Power square	$Y_n = Y^{\wedge}2$	May have a logarithmic trend (decay, survival, etc.).
High negative skew	Power cube	$Y_n = Y^{\wedge}3$	May have a logarithmic trend (decay, survival, etc.).

2.7 Chapter Concluding Remarks

- Exploratory spatial data analysis and related tools offer a comprehensive visual representation of statistics by linking graphs, scatter plots or histograms with maps.
- Data are just numbers stored in tables in a database management system. By analyzing data, we add value, creating information and then knowledge.
- Spatial statistics employ statistical methods to analyze spatial data, quantify a spatial process, discover hidden patterns or unexpected trends and model these data in a geographic context.
- Choropleth maps are typically the first thing created after a geodatabase is built.
- Designing and rendering choropleth maps is an art form. As professionals, we have to create accurate maps and graphs, always use solid statistics to back up our findings and avoid misleading messages.
- Inspecting the basic characteristics of a variable is essential prior to any sophisticated statistical or spatial analysis. Calculating the mean value, maximum value, minimum value and standard deviation of a variable provides an initial description of its distribution.
- Creating a frequency distribution histogram and inspecting for potential normality are also necessary.

- In general, calculating the common measures of center, shape and spread gives quick insights into the dataset and should be conducted prior to any other analysis.
- Boxplots are very helpful descriptive plots, as they depict measures of center, shape and spread and allow for comparison of two or more distributions.
- Scatter plot matrices and pairwise correlation matrices give a snapshot of the relationships among pairs of variables in a dataset. It is advisable to build such plots right from the beginning to gain quick insights into the data.
- Locating outliers is necessary, as statistical results might be distorted if outliers are not removed or properly handled.
- Rescaling variables with large differences in their scale and range, through normalization, adjustment or standardization is necessary when want to compare them or to use them in the same formula.
- While observations independence should exist in classical statistics, spatial dependence usually exists in spatial statistics, and classical statistics should be modified accordingly.
- A test of significance is the process of rejecting or not rejecting a hypothesis based on sample data. It is used to determine the probability that a given hypothesis is true.
- A p-value is the probability of finding the observed (or more extreme) results of a sample statistic (test statistic) if we assume that the null hypothesis is true.
- Rejecting the null hypothesis means that there is a probability (calculated as the difference: $100\% - \alpha$) that alternative hypothesis H_1 is correct (α is the significance level).
- Not rejecting the null hypothesis means that there is not sufficient evidence to reject the null hypothesis, but we cannot accept it either without further analysis.
- Statistics often deal with normal distribution because many well-defined statistical tests are based on the assumption that the examined distribution is normal.

Questions and Answers

The answers given here are brief. For more thorough answers, refer back to the relevant sections of this chapter.

Q1. Why are spatial statistics used?

A1. Spatial statistics employ statistical methods to analyze spatial data, quantify a spatial process and discover hidden patterns or unexpected trends in these data in a geographic context. Spatial statistics are built upon statistical concepts, but they incorporate location parameters such as coordinates, distance and area. They extend classic statistical measures and procedures and offer advanced insights for data analysis. In geographical analysis, spatial statistics are not used separately from statistics but in complementary ways.

Q2. What is the main difference between spatial statistics and descriptive statistics?

A2. There is a fundamental difference between classical and spatial statistics. In classical statistics, we make a basic assumption regarding the sample: it is a collection of independent observations that follow a specific, usually normal, distribution. Contrariwise, in spatial statistics, because of the inherent spatial dependence and the fact that spatial autocorrelation exists (usually), the focus is on adopting techniques for detecting and describing these correlations. In other words, in classical statistics, observations independence should exist while, in spatial statistics, spatial dependence usually exists. Classical statistics should be modified accordingly to adapt to this condition

Q3. What is a choropleth map, and why is it used?

A3. Choropleth maps are thematic maps in which areas are rendered according to the values of the variable displayed. Through choropleth maps, we visually locate if values cluster together or whether they exhibit similar spatial patterns. Rendering choropleth maps is usually the first task when spatial data are joined to nonspatial data (i.e., attributes from a census).

Q4. Which are the two main types of choropleth maps? Give examples.

A4. There are two main categories of variables displayed in choropleth maps: (a) spatially extensive variables and (b) spatially intensive variables. In spatially extensive variables, each polygon is rendered based on a measured value that holds for the entire polygon – for example, total population, total households or total number of children. In the spatially intensive category, the values of the variable are adjusted for the area or some other variable. For example, population density, income per capita and rate of unemployment are spatially intensive variables because they take the form of a density, ratio or proportion.

Q5. What are the measures of spread? Name a few.

A5. Measures of spread (also called measures of variability, variation, diversity or dispersion) of a dataset are measures that provide information of how much the values of a variable differ among themselves and in relation to the mean. The most common measures are range, deviation from the mean, variance and standard deviation.

Q6. What is an outlier? Why should we trace outliers?

A6. Outliers are the most extreme scores of a variable. They should be traced for three main reasons: (a) outliers might be wrong measurements, (b) outliers tend to distort many statistical results, and (c) outliers might hide significant information worth to be discovered and further analyzed.

Q7. How should we handle outliers?

A7. We can handle traced outliers based on the following guidelines:
- Scrutinize the original data (if available) to check whether the outliers' scores are due to human error (e.g., data entry). If scores are correct, attempt to explain such high or low values.
- Transform the variable. However, data transformation does not guarantee outliers' elimination. In addition, it may not be desirable to transform the entire dataset for only a small number of outliers.
- Delete the outlier from the dataset or change its score to be equal to the value of three standard deviations.
- Temporarily remove the outlier from the dataset and calculate the statistics. Then include the outliers again in the dataset for further analysis.

Q8. What is the Pearson's correlation coefficient? Does it reveal association or causation?

A8. A correlation coefficient $r_{(x, y)}$ analyzes how two variables (X, Y) are linearly related. Among the correlation coefficient metrics available, the most widely used is the Pearson's correlation coefficient (also called Pearson product-moment correlation). Correlation is a measure of association and not of causation. Causation and relationship/association are different. High correlation reveals a strong relation but not necessarily causation.

Q9. How can a Pearson's correlation coefficient be interpreted (how many classes of correlation exist)?

A9. There are six main classes of correlation. A strong positive correlation (for values larger than 0.8) indicates a strong linear relationship between the two variables; when variable X increases (or decreases), then variable Y also increases (or decreases) to a similar extent. A moderate positive correlation for values between 0.5 and 0.8 indicates that correlation exists but is not as intense as in a strong correlation. Observing a weak positive or weak negative correlation does not allow for reliable conclusions regarding correlation, especially when the values tend to zero. However, when the values lie between 0.3 and 0.5 (or between –0.5 and –0.3), and according to the problem studied, we may label correlation as "substantial." A moderate negative correlation between –0.8 and –0.5 means that correlation exists but is not very strong. Finally, a strong negative correlation between –1 and –0.8 indicates a strong linear relationship between the two variables (but with different directions: one decreasing and the other increasing or vice versa).

Q10. What is a typical workflow of a significance test?

A10. A typical workflow for significance testing is as follows (provided that a sample is collected):

1. Make the statement for the null hypothesis (e.g., that the sample is drawn from a population that follows a normal distribution)

2. Make the opposite statement for the H_1 hypothesis (e.g., that the sample is drawn from a population that is not normally distributed)

3. Specify the significance level α.

4. Select a test statistic (some formula) to calculate the observed value (e.g., sample mean).

5. Compute the p-value, which is the probability of finding the observed (or more extreme) results of our sample if we assume that the null hypothesis is true. Put otherwise, the p-value is the probability of obtaining a result equal to (or more extreme than) the one observed in our sample if the null hypothesis is true.

6. Compare the p-value to the significance value α. If $p \leq \alpha$, then we can reject the null hypothesis and state that H_1 is true and that the observed pattern, value, effect or state is statistically significant. If $p > \alpha$, then we cannot reject the null hypothesis, but we cannot accept it either.

LAB 2
EXPLORATORY SPATIAL DATA ANALYSIS (ESDA): ANALYZING AND MAPPING DATA

Overall Progress

Spatial Analysis/Lab Workflow

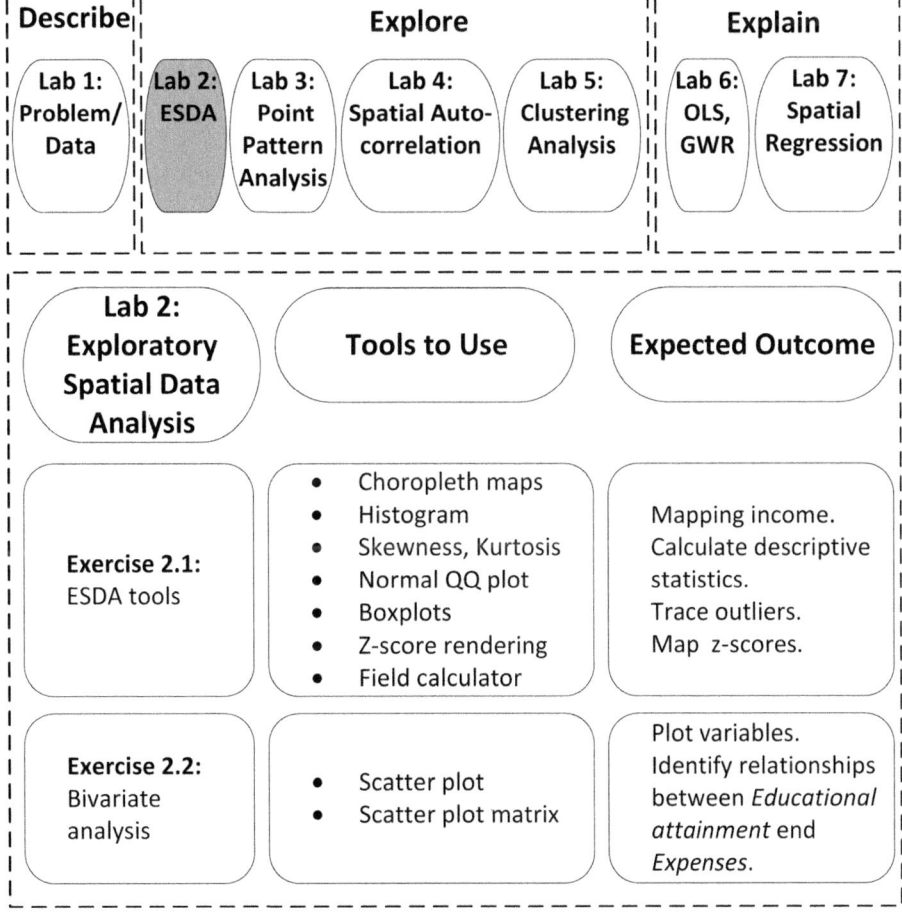

Figure 2.16 Lab 2 workflow and overall progress.

Scope of the Analysis: Income and Expenses

This Lab deals with

- **Objective 1**: Locating high income areas (see Table 1.2).
- **Objective 4:** Identify socioeconomic drivers of high monthly expenses (including those for coffee-related services)

As explained in Lab 1, the investor is interested in identifying, among others:

1. The areas where the residents have high incomes (i.e., the areas in which clients will have stronger purchasing power)
2. Whether there is any relation between specific socioeconomic variables (e.g., educational attainment) and monthly expenses (including coffee products and services). This will allow to better delineating the demographical profile of the target group.

To address the preceding questions, we use simple ESDA tools and descriptive statistics (see Figure 2.16). A more thorough analysis for Objective 1 is carried out in Lab 4, and a deeper analysis for Objective 4 is presented in Labs 6 and 7.

Section A   ArcGIS

Exercise 2.1 ESDA Tools: Mapping and Analyzing the Distribution of Income

In this exercise, we (a) map Income (yearly average income of the residents living in each postcode), (b) plot related graphs and (c) apply descriptive statistics to analyze income's spatial and statistical distribution.

   ArcGIS Tools to be used: Choropleth map, Histogram, Normal QQ plot, Boxplots, Z-score rendering

ACTION: Create choropleth map of income

Navigate to the location you have stored the book dataset and click the Lab2_SimpleESDA.mxd

Main Menu > File > Save As > My_Lab2_SimpleESSA.mxd

In I:\BookLabs\Lab2\Output

TOC > RC City > Properties > TAB = Symbology > Quantities > Graduated colors> (see Figure 2.17)

Value = Income

Color Ramp = Yellow to Brown

Classes = 4 > Click Classify > Break Values > 15000 > Enter > 20000 > Enter > 25000 > Enter 40000 > OK (See also Figures 1.12 and 1.13)

RC Label > Format Labels > Numeric > Rounding > Number of decimal places = 2 > OK > Apply > OK

TOC > RC City > Save As Layer File >

Name = Income.lyr

In I:\BookLabs\Lab1\Output

Exercise 2.1 *(cont.)*

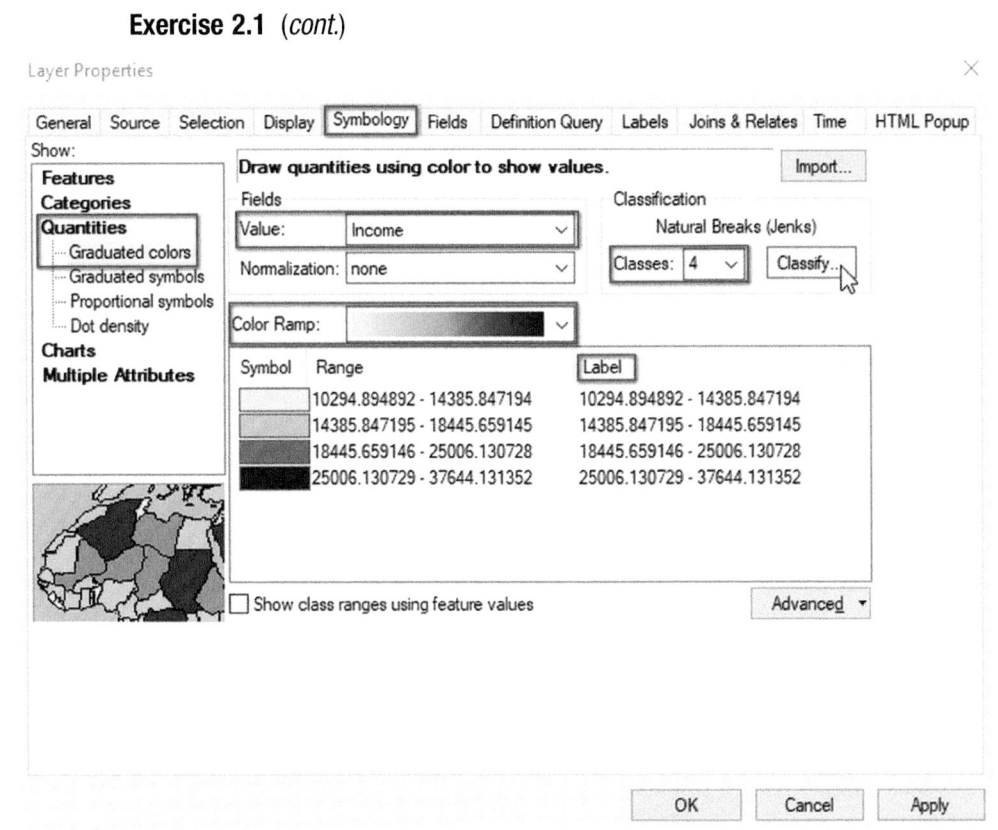

Figure 2.17 Layer properties dialog box for rendering Income.

Tip: Saving City into a layer allows us to save the income representation. Keep in mind that when you add the layer in the table of contents, it receives the name of the original shapefile created (i.e., City, in this example) and not the name it was saved as (i.e., CityIncome.lyr).

Interpreting results: Four postcodes groups of average annual income are created: (a) a group with income of less than 15,000 per year (low-income areas), (b) a group with income between 15,000 and 20,000 (average income areas), (c) a group with income between 20,000 and 25,000 (average to high-income areas) and (d) a group with income over 25,000 (high-income areas; see Figure 2.18). The map shows that high-income areas are centrally located (dark red). Most of the postcodes with average-to-high or high incomes lie inside the downtown area (red polygon). Lower-income areas can be found in the northern, western and southern postcodes. This depiction of the Income variable provides a first indication of how income is spatially distributed (spatial arrangement of values) across the postcodes of the city (see Box 2.3). To analyze how the values of Income are distributed in relation to the mean value (not spatially in this case), we can use the frequency distribution histogram.

Exercise 2.1 (*cont.*)

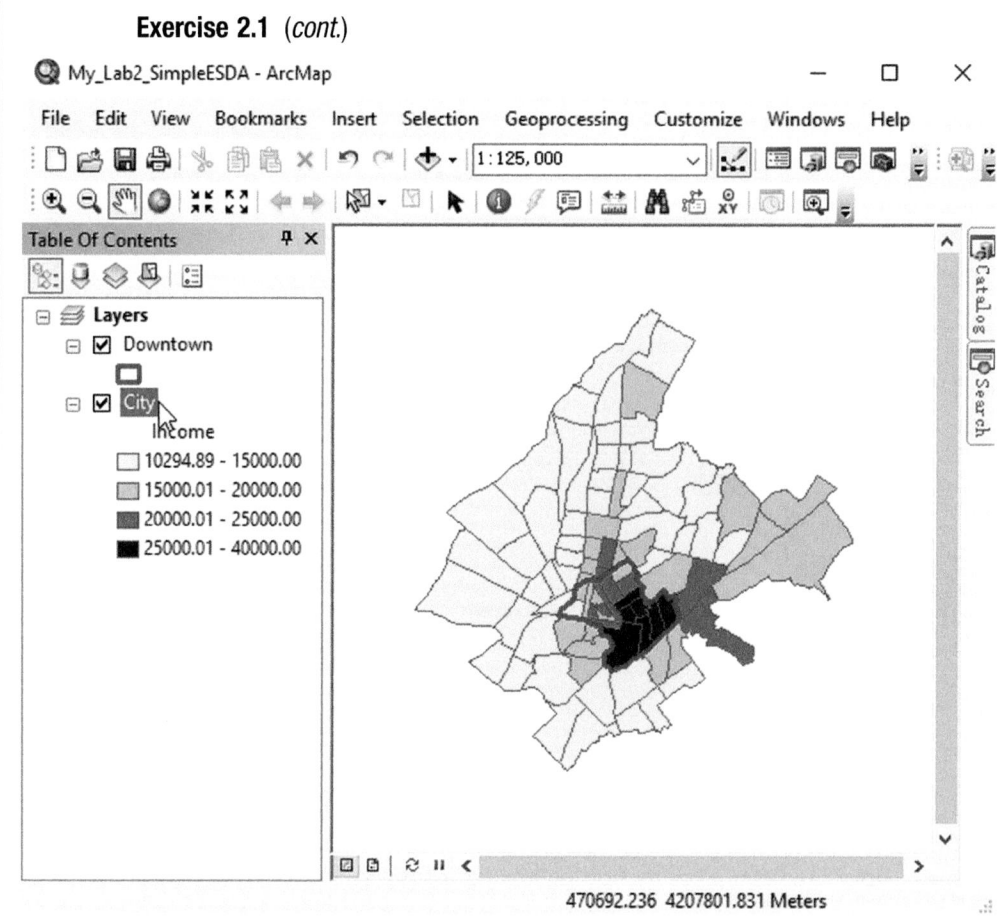

Figure 2.18 Income map classified into four groups.

Box 2.3 Analysis Criterion C1 to be used in synthesis Lab 5.4: The location of a coffee shop should lie within areas with: Income > 20,000 euros. [C1_HighIncome.shp]

Main Menu > Selection > Select By Attributes

Layer = City

SELECT * FROM City Where: "Income" >=20000

OK

TOC > RC City > Data > Export Data > Output feature class: I:
\BookLabs\Lab2\Output\C1_HighIncome.shp

Main Menu > Selection > Clear Selected Features

Add the shapefile to the data view just to inspect the output, but you can then remove it to proceed with the exercise.

Exercise 2.1 *(cont.)*

ACTION: Create histogram

Main Menu > Customize > Extensions > Check Geostatistical Analyst > Close (if checked already do not uncheck)

Main Menu > Customize > Toolbars > Check Geostatistical Analyst

Toolbar = Geostatistical Analyst > Geostatistical Analyst > (see Figure 2.19) Explore Data > Histogram

Select Layer = City (see Figure 2.20)

Attribute = Income

Geostatistical Analyst
Geostatistical Analyst

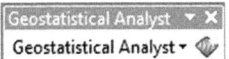

Figure 2.19 Geostatistical analyst toolbar.

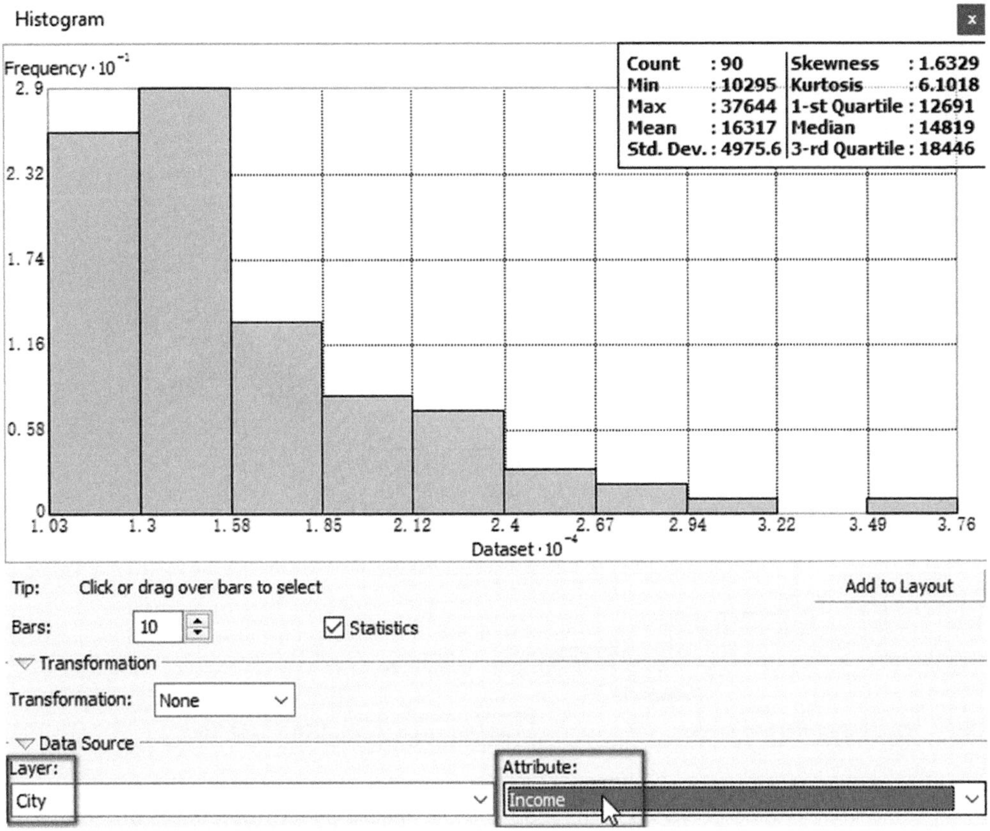
Histogram
Frequency · 10^-1
2.9
2.32
1.74
1.16
0.58
0
Count : 90
Min : 10295
Max : 37644
Mean : 16317
Std. Dev. : 4975.6
Skewness : 1.6329
Kurtosis : 6.1018
1-st Quartile : 12691
Median : 14819
3-rd Quartile : 18446
1.03
1.3
1.58
1.85
2.12
2.4
2.67
2.94
3.22
3.49
3.76
Dataset · 10^-4
Tip: Click or drag over bars to select
Add to Layout
Bars: 10
Statistics
Transformation
Transformation: None
Data Source
Layer:
City
Attribute:
Income

Figure 2.20 Histogram of income.

Exercise 2.1 (*cont.*)

Move the graph at the left corner of the map, as shown in Figure 2.21.

TOC > RC City > Open Attribute Table

Select a bin in the graph and examine the highlighted post codes in the map.

On the histogram, click Add to Layout > Move the graph at the lower left corner of the layout area > Get back to the Data View > Close the histogram plot

Main Menu > Selection > Clear Selected Features > Close Attribute Table of City >

Main Menu > File > Save

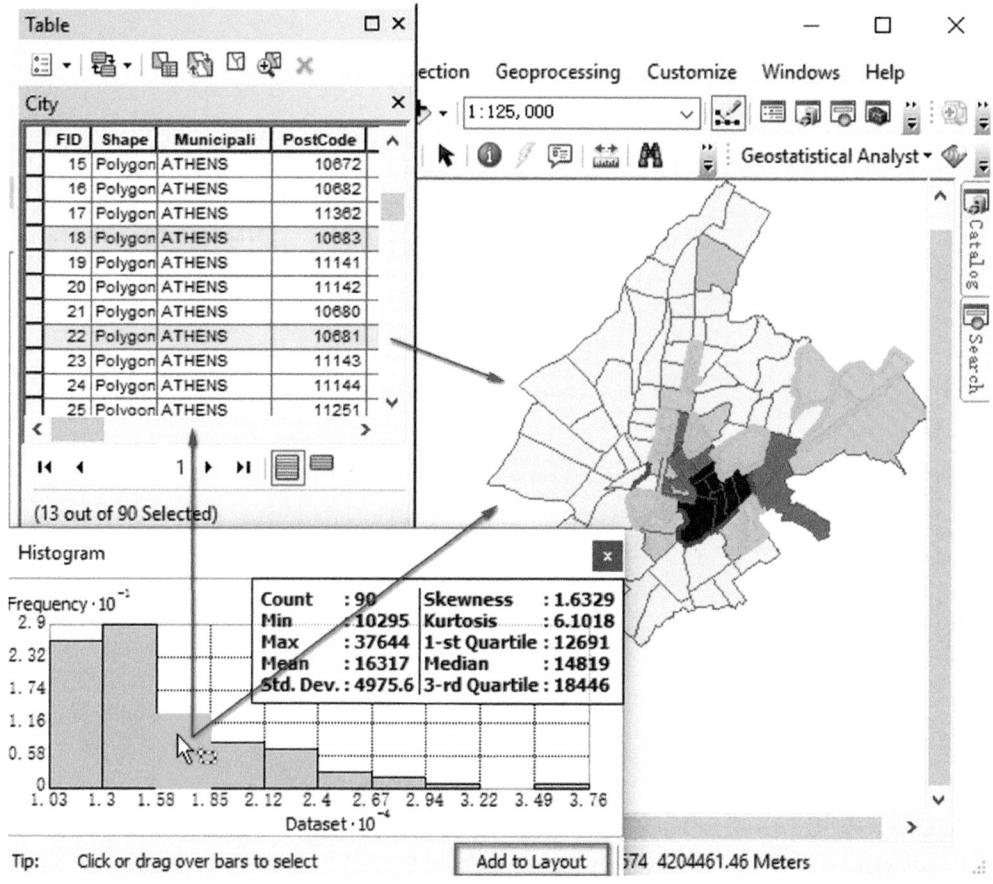

Figure 2.21 Brushing capabilities of ESDA tools. By selecting a bin in the graph, the corresponding polygons in the map and the lines in the attribute table of the City layer are highlighted.

Exercise 2.1 (*cont.*)

Interpreting results: The histogram depicts basic descriptive statistics of income (e.g., mean value, standard deviation, skewness, kurtosis, first and third quartiles; see Figure 2.20). Results show that the distribution of income is positively skewed and deviates significantly from a normal distribution. To better test if the distribution is normal, we should build a normal QQ plot (see Figure 2.23).

The histogram is linked to the shapefile's polygons and the layer's attribute table, which is one of the benefits of using ESDA tools in spatial analysis. By brushing a bin in the histogram, the related polygons are highlighted in the map and the relevant lines of the attribute table are selected (see Figure 2.21). Likewise, if we select a polygon or a row in the attribute table, the relevant bin in the histogram will be highlighted. A graphical inspection shows a bin in the far right of the histogram that should be further analyzed, as it might reveal the existence of outliers (postcodes with extremely large income values).

Postcodes with extreme values of income (spatial outliers) can be traced using standard deviation by defining an outlier to be an observation that lies at least 2.5 standard deviations from the mean (see Section 2.2.7). The standard deviation and the mean of Income are 4,975.6 and 16,317, respectively (see Figure 2.21). Thus, an income value is an outlier if it is larger than

```
Outlier > Mean + 2.5 x Standard Deviation = 16317 + 2.5 x 4975.6
= 28756.5
```

or if it is smaller than

```
Outlier < Mean - 2.5 x Standard Deviation = 3878
```

By sorting Income in the attribute table of the layer City, we check whether postcodes with incomes above or below these values exist. In fact, two postcodes are labeled as outliers, those having income values larger than 28,756.5.

An alternative way to test for outliers and also obtain a different representation of the Income distribution is by using boxplots and z-score rendering (see Figures 2.25 and 2.29).

ACTION: Create Normal QQ plot to identify if income follows a normal distribution

```
Toolbar = Geostatistical Analyst > Geostatistical Analyst >
Explore Data > Normal QQPlot
```

Exercise 2.1 (*cont.*)

Select Layer = City (see Figure 2.22)

Attribute = Income

Select the upper right dots on the plot and examine the highlighted polygons (see Figure 2.23).

On the Normal QQ plot, click Add to Layout > Move the graph next to the Histogram Plot > Get back to the Data View > Close the histogram plot

Main Menu > Selection > Clear Selected Features

Interpreting results: The normal QQ plot reveals that income values deviate from the straight line of a normal distribution (see Figure 2.22). We can thus argue that the distribution of the variable Income is not normally distributed. By brushing the points at the far right of the plot, we locate the postcodes that deviate significantly from the line (expected values of income if the distribution were normal; see Figure 2.23). These postcodes belong to the high-income group and are also clustered inside the downtown area, an interesting finding from the spatial analysis perspective.

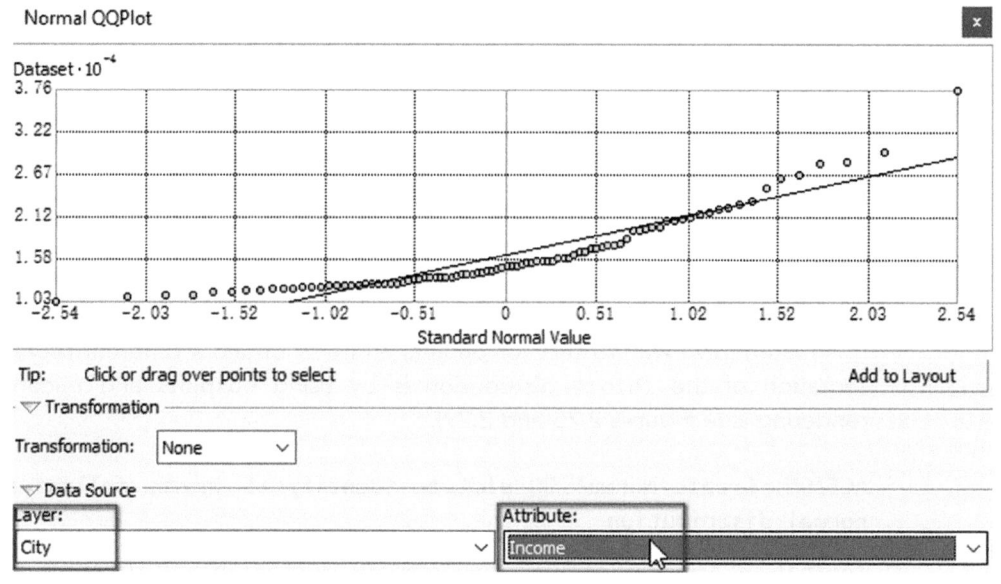

Figure 2.22 Normal QQ plot for income.

Exercise 2.1 *(cont.)*

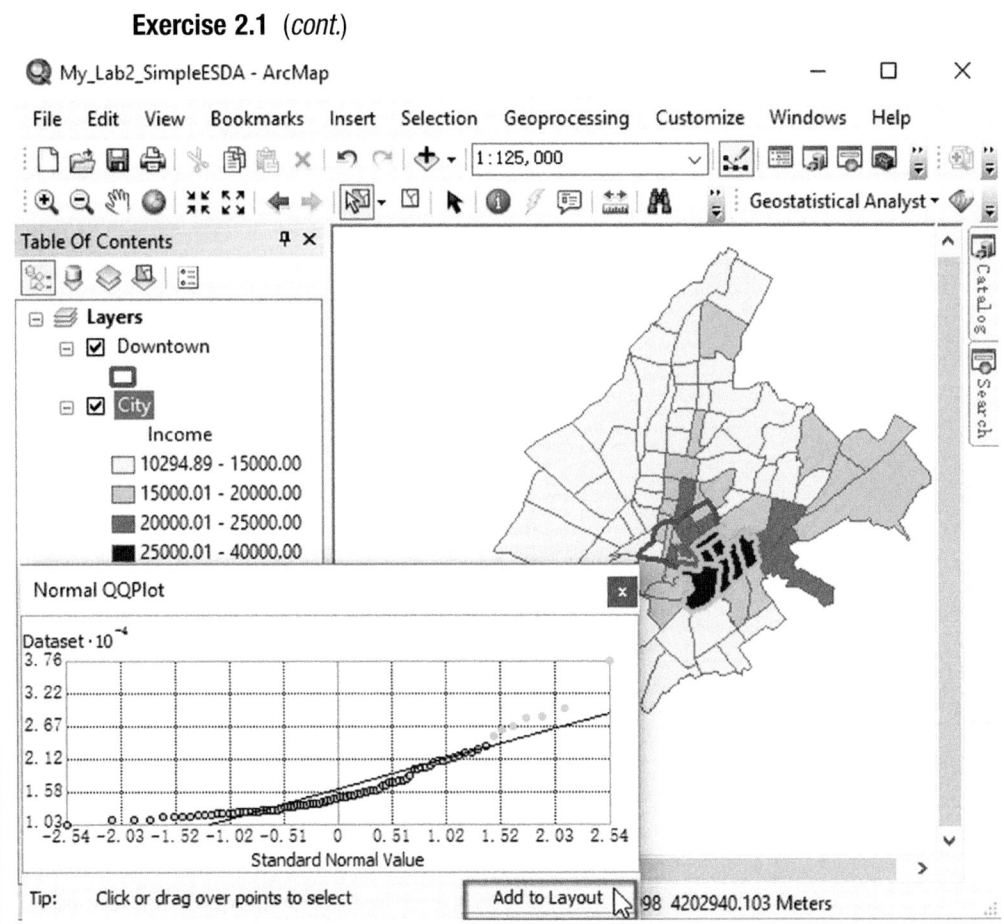

Figure 2.23 Locating postcodes with high income (far right of the QQ plot) that deviate significantly from the normal distribution line.

ACTION: Create box plot

Main Menu > View > Graphs > Create Graph >

Graph type = Box Plot (see Figure 2.24)

Layer/Table = City

Value field = Income > Next >

Title = Income > Finish

Brush the extreme outlier in the graph (marked with *) and locate the postcode in the map (see Figure 2.25).

Exercise 2.1 (*cont.*)

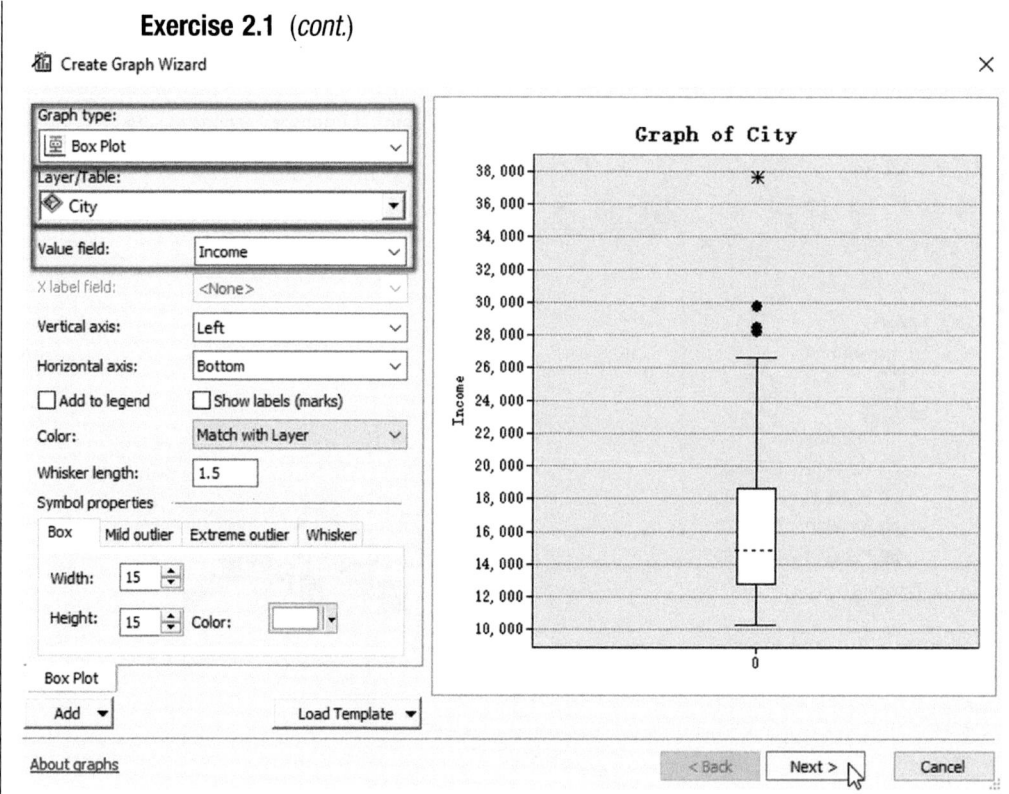

Figure 2.24 Boxplot dialog box.

RC on the Boxplot > Add to Layout > Move the boxplot to an empty space > Get back to Data View > Close graph >

Main Menu > Selection > Clear Selected Features

Interpreting results: The boxplot depicts three mild outliers (symbolized with a dot) and one extreme outlier (symbolized with an asterisk) (see Figure 2.24; see Section 2.2.8 for how mild and extreme outliers are defined based on the interquartile distance – not on the standard deviation). Mild and extreme outliers are located in the upper side of the box, refering to high-income values. Outliers of low-income values are not traced. By brushing the extreme outlier dot in the boxplot, we locate on the map the respective postcode in the downtown area (see Figure 2.25). To decide how to handle the extreme outlier identified earlier, we should go through the practical guidelines specified in Section 2.2.7. First, we check if this observation is an incorrect database entry. In this dataset, the value is correct. Second, we should assess if such a value is reasonable based on (a) our knowledge of the specific area (if any) and (b) common sense. It is quite common to trace areas with remarkably higher income than others, as income inequality exists in most regions worldwide. Our dataset does not seem to deviate from this trend.

Exercise 2.1 (*cont.*)

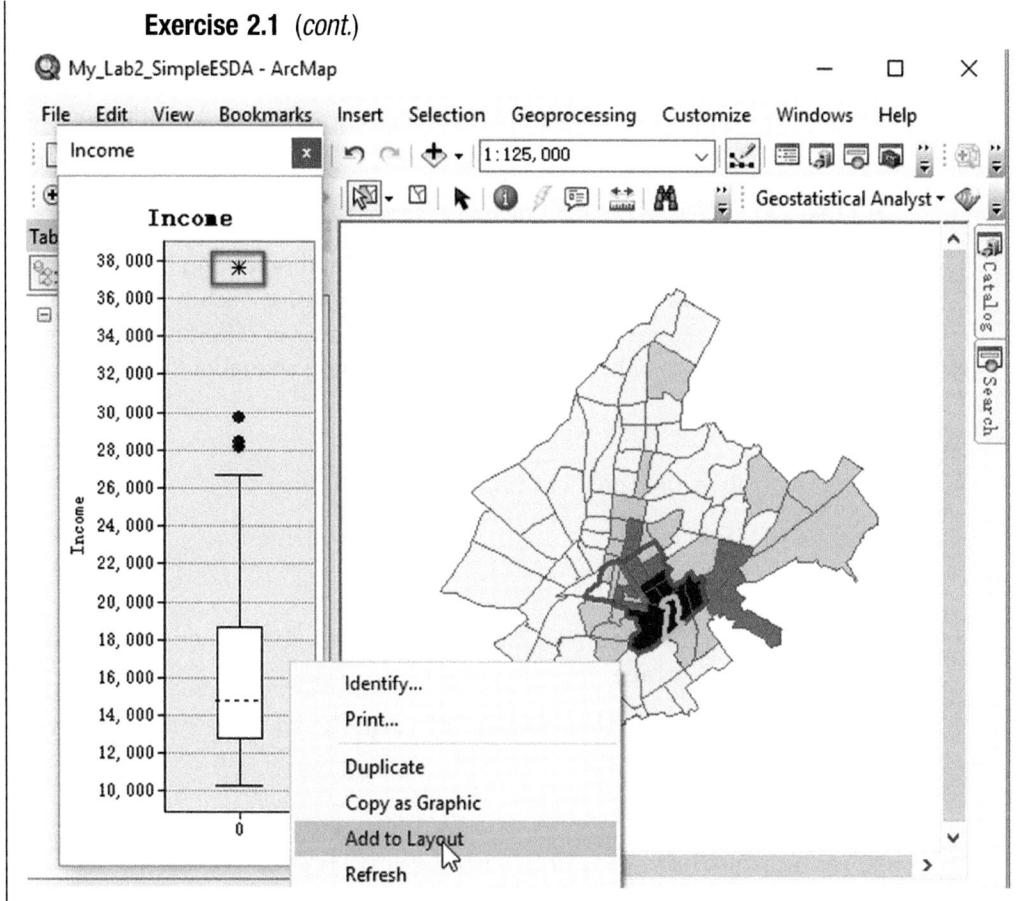

Figure 2.25 Brushing the extreme outlier in the boxplot to locate the postcode in the map. All graphs are saved on the layout view of the .mxd.

Additionally, an Athenian or a visitor would also know that the area that income clusters (including the outlier) is near the central city square (Constitutional Square) where the National Parliament is located. The square is surrounded by many beautiful neoclassical buildings, a city park, the Acropolis archeological site and a residential area, which historically attracted higher-income classes. It seems quite reasonable for these high and extremely high values of income to lie there. For this reason and in the context of the specific project (coffee shop market analysis), we should not remove the observation or transform the dataset, as it provides valuable information. In a different context though, we might have to remove this observation. For example, If we are mostly interested in the social analysis of the average- to low-income regions, then removing the extreme outlier would make the results more realistic. For example, the mean income would be smaller, better reflecting the real economic condition of the majority of the citizens. The standard deviation and other descriptive statistics (e.g., confidence intervals, quartiles) would be different as well. In other words,

Exercise 2.1 (*cont.*)

removing the outlier would portray the socioeconomic profile of the residents better. From the social analysis perspective, the existence of outliers in our case study reveals large income inequality, which is a significant finding.

ACTION: Calculate and render income z-scores. Identify income outliers.

Another way to map income is by calculating and rendering its z-score. By such rendering, we map the deviations of income for each postcode from the mean value. As such, we trace areas with similar or different values from the average income, which also allows for locating spatial outliers.

TOC > RC City > Open Attribute Table

Click at Table Options button > Add Field >

Name = IncZScore

Type = Float

Precision = 5 (stores the total number of digits on both sides of the decimal place

Scale = 3 (stores the number of digits to the right of the decimal place)

OK

RC IncZScore column > Field Calculator > (Click YES if a pop up message appears about calculation outside an edit session) > In the IncZScore field type ([Income]- 16317)/4975.6 > OK > Close table (see Figure 2.26)

See exercise 1.1: Mean = 16317, Standard deviation = 4975.6 ArcToolbox > Spatial Statistics Tools > Rendering > ZScore Rendering (see Figure 2.27)

Input Feature Class = City

Field to Render = IncZScore

Output Layer File = I:\BookLabs\Lab2\Output\IncZScore.lyr

OK

Next identify as outliers those postcodes with IncZScore larger than 2.5.

Main Menu > Selection > Select by Attributes >

Layer = IncZScore (see Figure 2.28)

Method = Create a new selection

DC "IncZScore"

Exercise 2.1 (*cont.*)

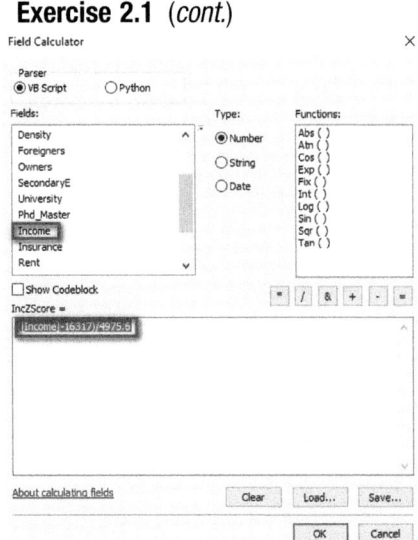

Figure 2.26 Calculating z-score of Income through filed.

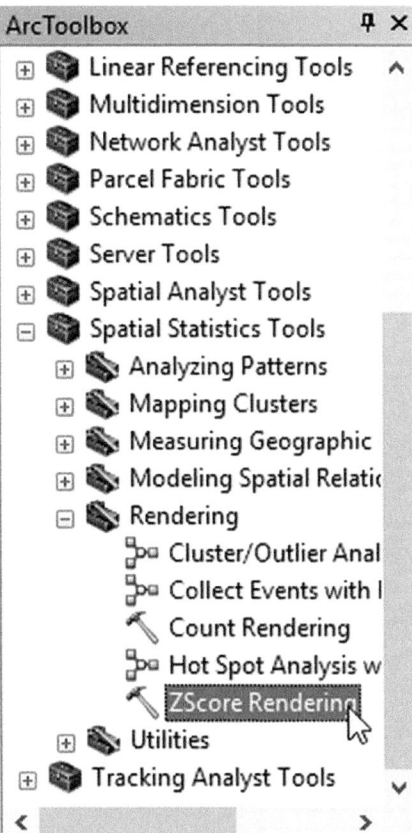

Figure 2.27 Calculating z-score of Income through filed.

Exercise 2.1 (*cont.*)

Figure 2.28 Select by attributes dialog box.

Go to the "SELECT*FROM City WHERE:" window and after "IncZScore" type: >=2.5

OK

Main Menu > Selection > Clear Selected Features

Main Menu > File > Save

Interpreting results: The map in Figure 2.29 depicts the z-scores of income for each postcode. The higher/lower the z-score, the larger the difference between the income value of the postcode with the mean annual income for the entire study region. Red areas have annual incomes larger than two standard deviations from the mean income and cluster in the downtown area. Values larger or less than 2.5 standard deviations reveal potential spatial outliers. Based on this definition, two postcodes with high income can be labeled as spatial outliers (those highlighted with a light blue outline). No outliers for extreme low-incomes are identified. It is obvious that different outliers' definitions lead to slightly differently results (see Figure 2.25). How many and which outliers will finally be retained depends on the analysis.

Exercise 2.1 (*cont.*)

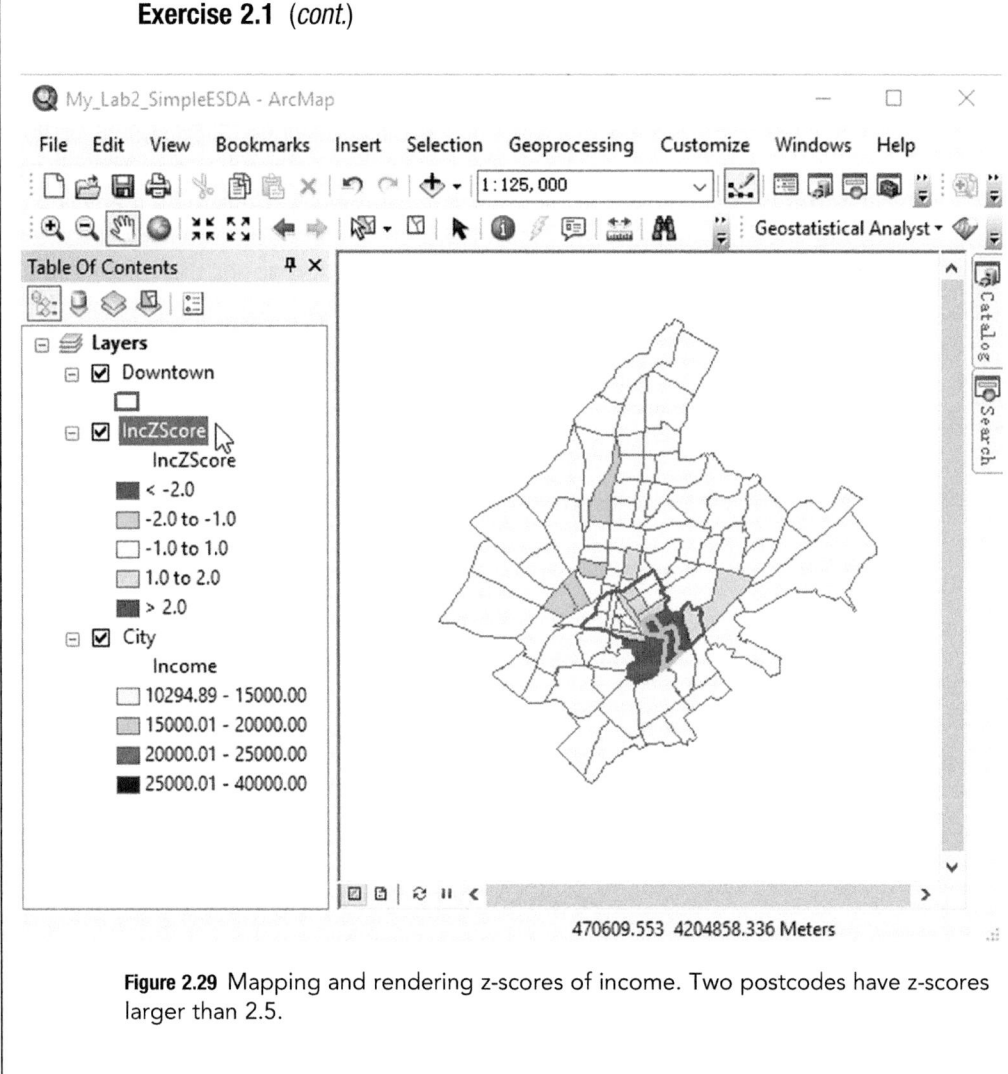

Figure 2.29 Mapping and rendering z-scores of income. Two postcodes have z-scores larger than 2.5.

ArcGIS Tip: If you want to run again this exercise from scratch you should

(a)   Delete City.shp located in I:\BookLabs\Lab2\Output
(b)   Copy City.shp located in I:\BookLabs\Data and paste it in E:\Book-Labs\Lab2\Output (so that the field IncZScore is not contained)

Exercise 2.2 Bivariate Analysis: Analyzing Expenditures by Educational Attainment

In this exercise, we study if educational attainment is related to monthly expenditures (including coffee-related expenses). The following variables are analyzed: Expenses, University and SecondaryE. Expenses is the

Exercise 2.2 *(cont.)*

average monthly expenses per person in euros for everyday costs (e.g., grocery shops, coffee shops). University is the percentage of people living in a postcode who have a bachelor's degree, while SecondaryE is the percentage of people who have completed secondary education. We create a scatter plot and a scatter plot matrix to graphically determine if any linear or nonlinear relations exist among the aforementioned three variables. This type of analysis provides us with initial information about the relative relationships among the variables. It does not allow us to quantify the real effect of one variable on another. To do so, we should apply more advanced methods, such as regression, spatial regression or spatial econometrics (discussed in Chapters 6 and 7). We are still describing and exploring our dataset and not explaining deeper relations, causes and effects.

ArcGIS Tools to be used: Scatter plot, Scatter plot matrix

ACTION: Create scatter plot

Navigate to the location you have stored the book dataset and click My_Lab2_SimpleESDA.mxd

Main Menu > View > Graphs > Create Graph >

Graph type = Scatter plot

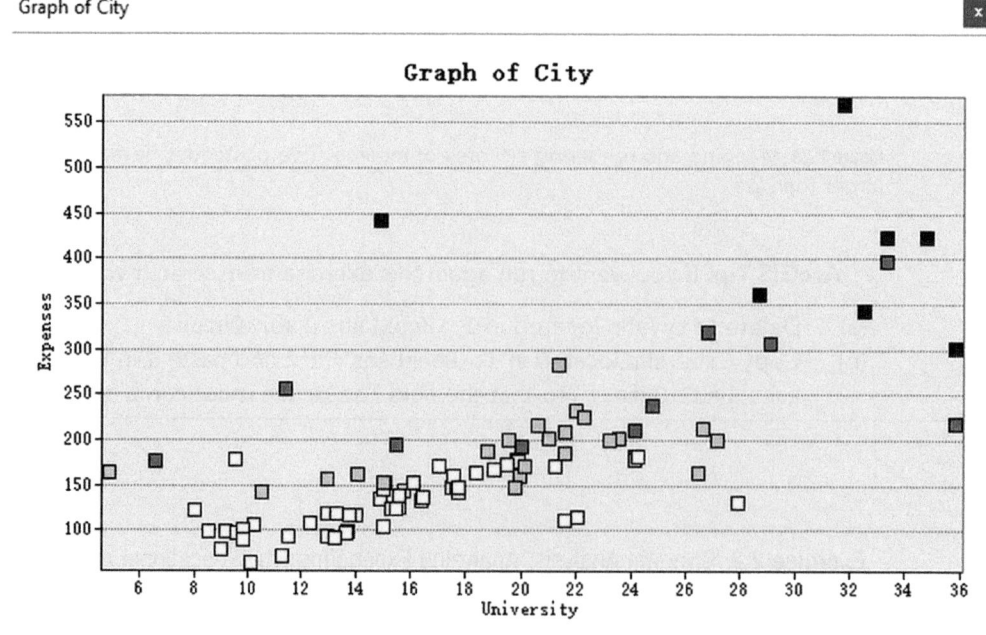

Figure 2.30 Scatter plot of University over Expenses.

Exercise 2.2 *(cont.)*

Layer/Table = City

Y field = Expenses

X field (optional)= University

Uncheck Add to legend > Next > Finish

RC on graph > Add to layout
Get back to the Data View > Brush the dots in the upper-right corner and
observe the selected polygons in the map (see Figure 2.31).

Main Menu > Selection > Clear Selected Features > Close Graph >
Save

Interpreting results: The dots of the scatter plot are colored according to
the income of each postcode (see Figure 2.30). By brushing the points in the

Figure 2.31 Brushing between scatter plot and map.

Exercise 2.2 (*cont.*)

upper-right (high education–high expenditures) of the scatter plot, the related postcodes are highlighted in the map. Selected features reveal two interesting findings: (a) the postcodes are spatially clustered, and (b) most of the postcodes belong to the high-income group (dark red polygons). This reveals colocation among higher education, expenditures and higher income.

   The scatter plot reveals that there is a nearly linear relation between University and Expenses (see Figure 2.30). The higher the percentage of people with university degree in a postcode, the more money they spent on average for monthly expenses. We observe that the postcodes lying on the right part of the graph (high education–high expenditures) are clustered in the city center, where high-income postcodes (dark brown) are also clustered (see Figure 2.31). As a result, we have a first sign of colocation among high-income, high educational attainment and high expenditures.

ACTION: Create scatter plot matrix

Main Menu > View > Graphs > Create Scatter plot Matrix Graph (see
Figure 2.32)>
Layer/Table = City (see Figure 2.33)

1 Field name = Expenses

2 Field name = University

3 Field name = SecondaryE

Check Show Histograms > Next > Finish

RC on graph > Add to layout > Get back to the Data View > Close
Scatter plot Matrix

Main Menu > File > Save

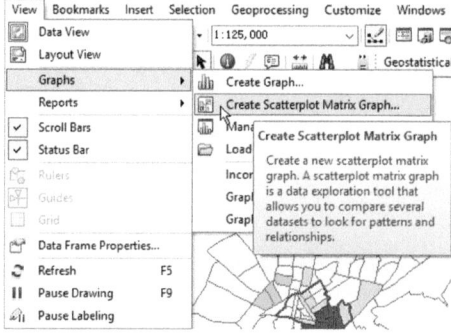

Figure 2.32 Create scatter plot matrix graph.

Exercise 2.2 (*cont.*)

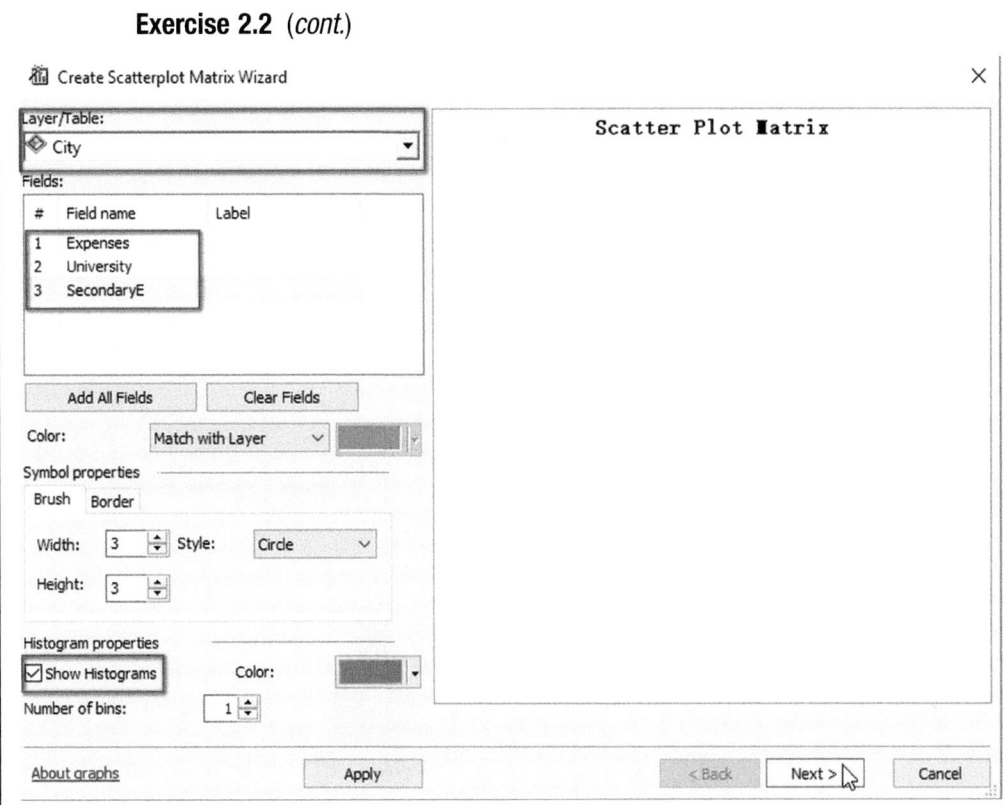

Figure 2.33 Scatter plot matrix wizard. Selection of histograms.

Interpreting results: The scatter plot matrix depicts all combinations among Expenses, University and SecondaryE (see Figure 2.34). The upper-right graph is an enlargement of the selected scatter plot on the matrix (in this case, the lower left). The histograms of each variable are also presented. We observe, that variable Expenses is positively skewed, meaning that people in most of the postcodes spend less than the average. Still, in some postcodes, people spend well above the average. Looking at the scatter plots, there is no apparent relation between secondary education and monthly expenditures (random cloud of points).

On the other hand, there is a positive relationship between having a bachelor's degree and monthly expenditures (see also Figure 2.30). In other words, the level of monthly expenses, including coffee consumption, is not linked to whether people have obtained secondary education but is related to having a university degree. The higher the percentage, the more expenses they are likely to make. As a result, the coffee shop owner would

Exercise 2.2 *(cont.)*

Scatter Plot Matrix

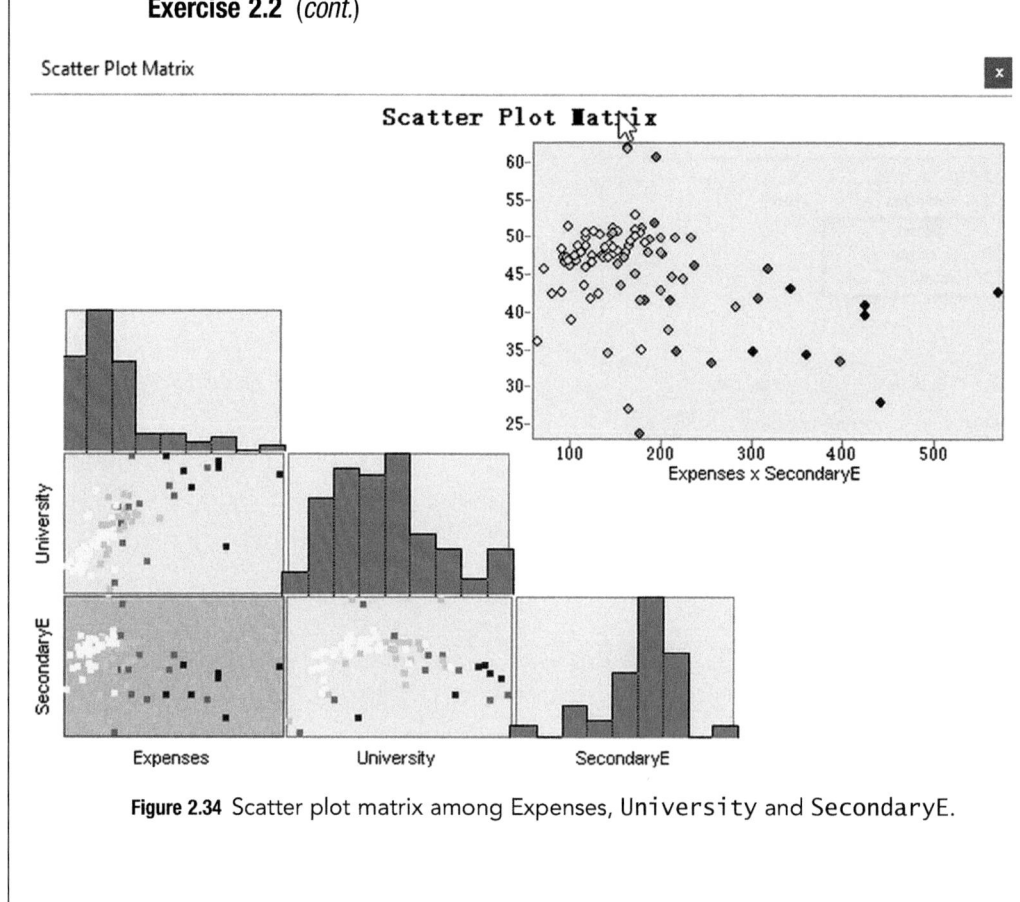

Figure 2.34 Scatter plot matrix among Expenses, University and SecondaryE.

probably prefer to target an area with a high percentage of people possess-
ing a bachelor's degree, as it seems that they are willing to spend more in
this type of market. For additional quantification of the effects of education
and other variables on expenses, see Labs 6 and 7.

Section B   GeoDa

Exercise 2.1 ESDA Tools: Mapping and Analyzing the Distribution of Income

In this exercise, we (a) map Income (yearly average income of the residents
living in each postcode), (b) plot related graphs and (c) apply descriptive
statistics to analyze income's spatial and statistical distribution.

   GeoDa Tools to be used: Choropleth map, Histogram, Boxplots,
Custom breaks

Exercise 2.1 *(cont.)*

Figure 2.35 Choropleth map of income.

ACTION: Create choropleth map of income

Navigate to the location you have stored the book dataset and click the Lab2_SimpleESDA_GeoDa.gda

RC over a polygon on the Map > Change Current Map Type > Custom Breaks > Create New Custom Breaks > Income > OK > Leave New Categories Title as default and press OK (see Figures 1.18, 1.19, 1.20 as well)

Breaks = User Defined

Categories = 4

Write the following values in the break fields: break 1 = 15000 / break 2 = 20000 / break 3 = 25000 > Enter > Close the dialog box > Save

Exercise 2.1 (*cont.*)

Interpreting results: See Section A, Exercise 2.1.

ACTION: Create histogram

Main Menu (see Figure 2.35) > Explore > Histogram > Variable
Settings = Income > OK
RC over the histogram (see Figure 2.36) > Choose Intervals >
Intervals = 10 > OK

RC over the histogram > View > Display Statistics

Select a bin in the graph and examine the highlighted post codes
in the map (see Figure 2.37).

Save

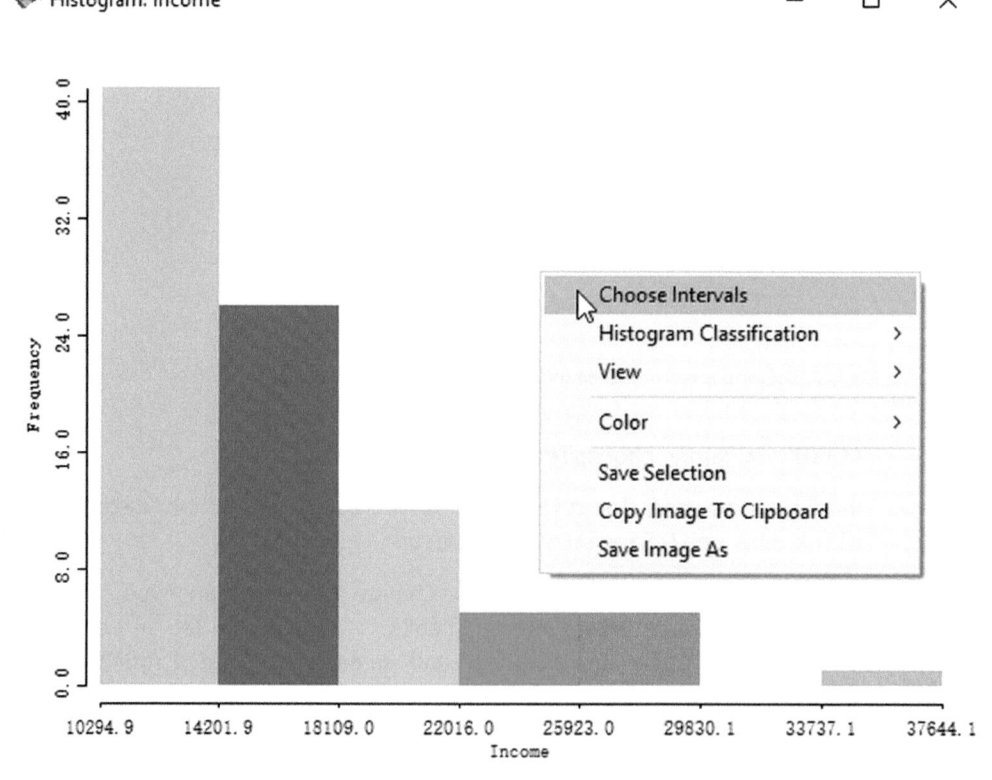

Figure 2.36 Histogram of income.

Exercise 2.1 (*cont.*)

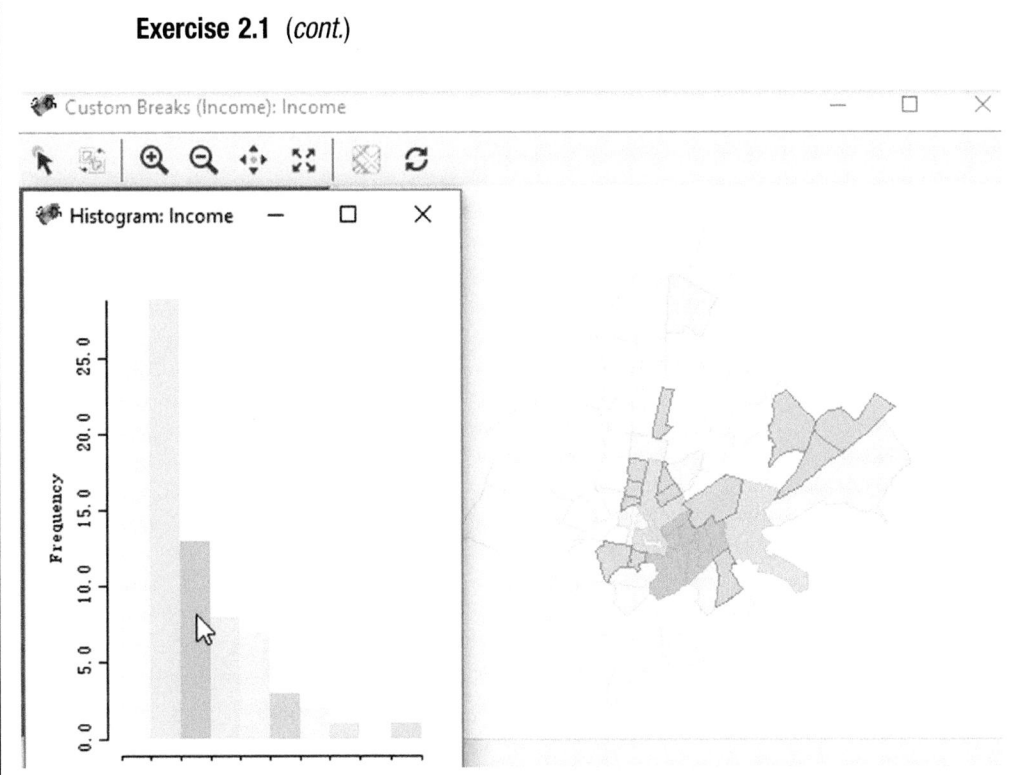

Figure 2.37 Brushing capabilities of ESDA tools.

Interpreting results: See Section A, Exercise 2.1.

ACTION: Create box plot

Main Menu > Explore > Box Plot > First Variable (X) = Income > OK

Brush the extreme outlier in the graph (marked with O) and locate the postcode in the map (see Figure 2.38).

Save

Interpreting results: See Section A, Exercise 2.1.

Exercise 2.1 (*cont.*)

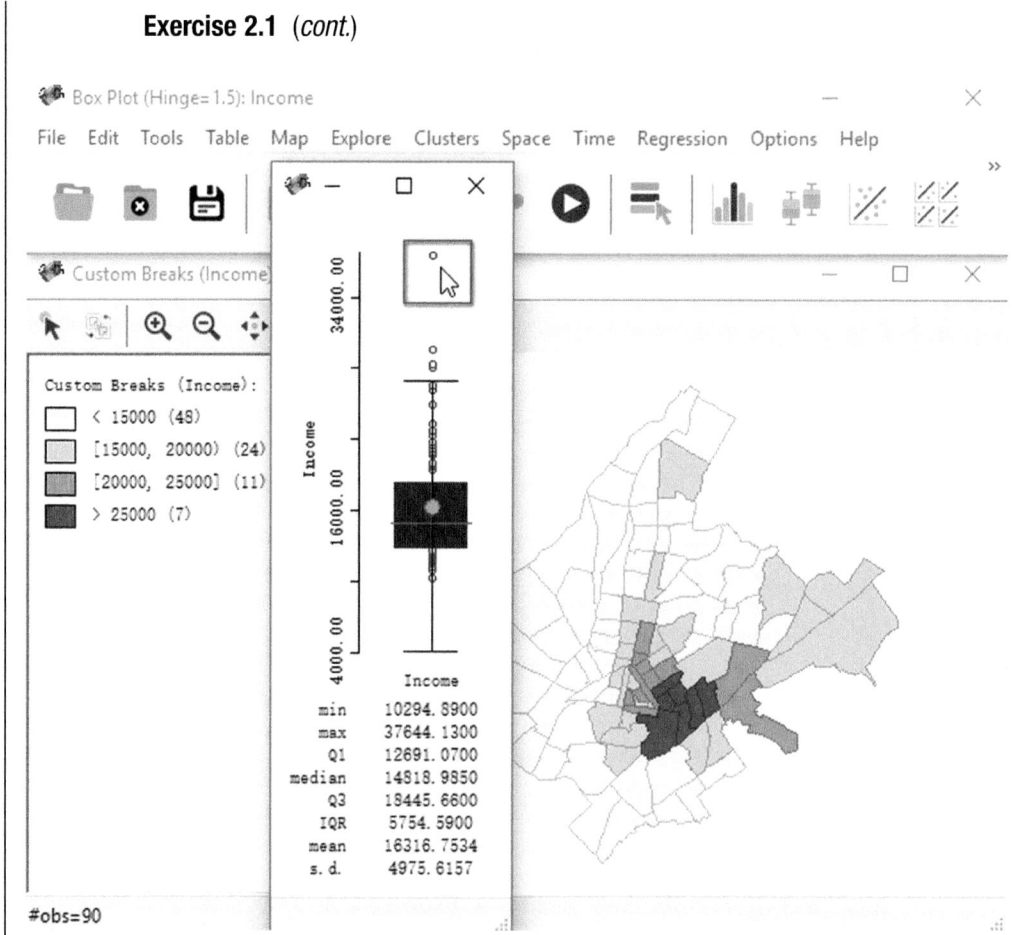

Figure 2.38 Boxplot of income along with statistics. Brushing the extreme outlier in the boxplot highlights the corresponding postcode in the map.

ACTION: Calculate and map income z-scores. Identify income outliers

Main Menu > Table > Calculator >TAB = Univariate

Click Add Variable

Name = IncZScore (see Figure 2.39)

Type = real

Insert before = after last variable

Length = 5

Decimals = 3 (stores the number of digits to the right of the decimal place)

Exercise 2.1 (*cont.*)

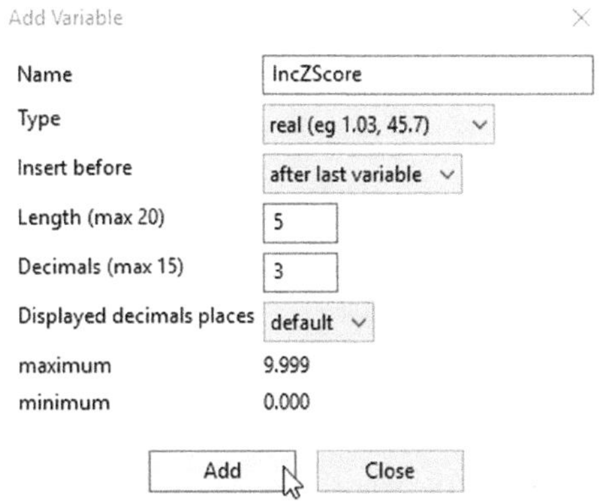

Add Variable   ✕

Name	IncZScore
Type	real (eg 1.03, 45.7) ⌄
Insert before	after last variable ⌄
Length (max 20)	5
Decimals (max 15)	3
Displayed decimals places	default ⌄
maximum	9.999
minimum	0.000

[Add] [Close]

Figure 2.39 Add variable dialog box.

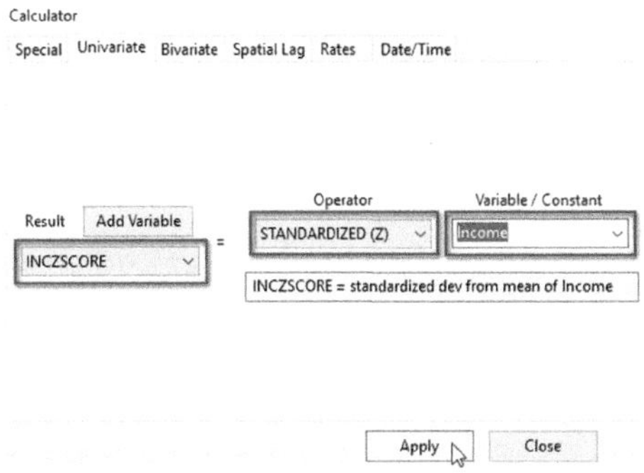

Calculator

Special Univariate Bivariate Spatial Lag Rates Date/Time

Result	Add Variable		Operator	Variable / Constant
INCZSCORE ⌄		=	STANDARDIZED (Z) ⌄	Income ⌄
			INCZSCORE = standardized dev from mean of Income	

[Apply] [Close]

Figure 2.40 Calculating z-score through calculator.

Add
Select INCZSCORE (see Figure 2.40)

Operator = STANDARDIZED (Z)

Variable/Constant = Income

Apply > Close
RC over a polygon on the Map > Change Current Map Type > Custom
Breaks > Create New Custom Breaks > First Variable (X) =
INCZSCORE > OK > New Customer Categories Title: IncZScore > OK

Assoc.Var = INCZSCORE

Breaks = User Defined

Categories = 5

Write the following values in the break fields: break 1 = -2.5 > break 2
= -1 > break 3 = 1 > break 4 = 2.5. Close the dialog box > Save (see
Figure 2.41)

Exercise 2.1 (*cont.*)

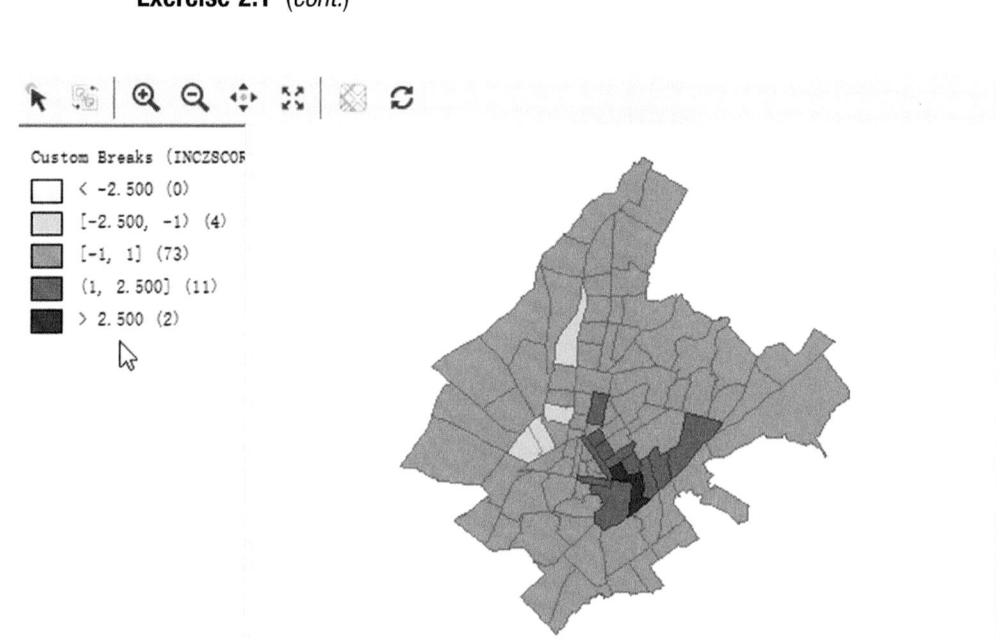

Figure 2.41 Mapping and rendering z-scores of income. Postcodes with Z-scores higher than 2.5 (or smaller than −2.5) are outliers.

Interpreting results: See Section A, Exercise 2.1.

GeoDa TIP: To solve the exercise again remove the filed INCZSCORE from CityGeoDa.shp stored in I:\BookLabs\Lab2\GeoDa. Open the table by selecting the table icon (see Figure 1.16) and then RC INCZSCORE > Delete Variable > INCZSCORE > Delete > Close

---

Exercise 2.2 Bivariate Analysis: Analyzing Expenditures by Educational Attainment

See Section A, Exercise 2.2, for introduction of this exercise.

GeoDa Tools to be used: Scatter plot, Scatter plot matrix

ACTION: Create scatter plot

Navigate to the location you have stored the book dataset and click Lab2_SimpleESDA_GeoDa.gda

Main Menu> Explore > Scatter plot >

Exercise 2.2 (*cont.*)

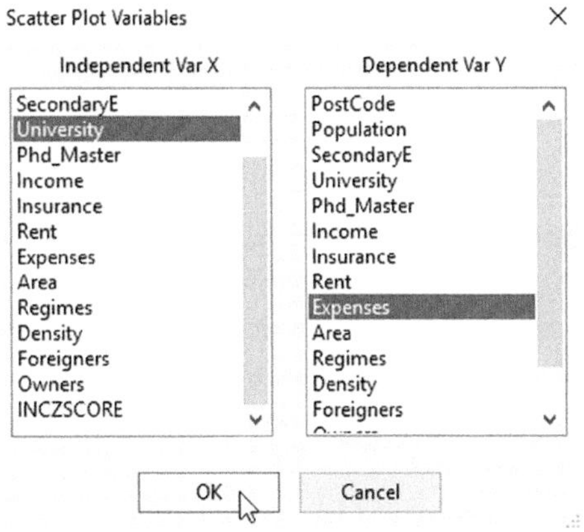

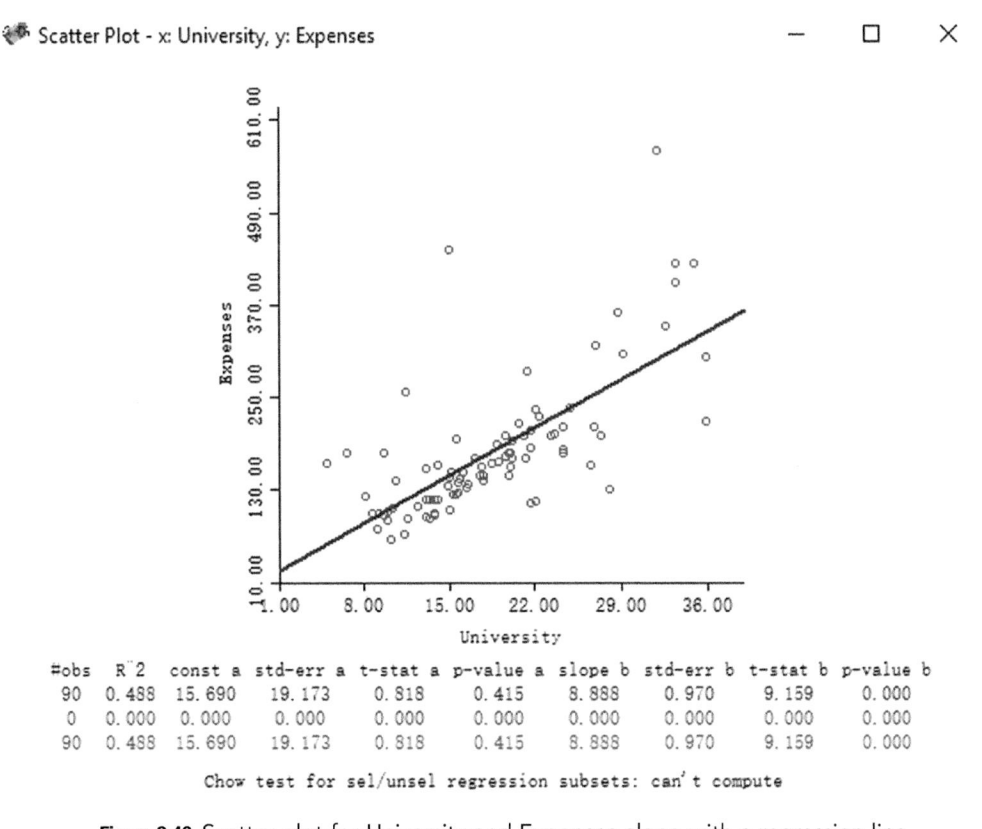

Figure 2.42 Scatter plot dialog box.

Scatter Plot - x: University, y: Expenses   —   □   ✕

#obs	R^2	const a	std-err a	t-stat a	p-value a	slope b	std-err b	t-stat b	p-value b
90	0.488	15.690	19.173	0.818	0.415	8.888	0.970	9.159	0.000
0	0.000	0.000	0.000	0.000	0.000	0.000	0.000	0.000	0.000
90	0.488	15.690	19.173	0.818	0.415	8.888	0.970	9.159	0.000

Chow test for sel/unsel regression subsets: can't compute

Figure 2.43 Scatter plot for University and Expenses along with a regression line superimposed, and related statistics.

Exercise 2.2 (*cont.*)

Independent Var X = University (see Figure 2.42)

Dependent Var Y = Expenses

OK

You should brush the point in the upper right (high education–high expenditures) of the scatter plot to highlight the related postcodes in the map (see Figure 2.43). You can also brush subregions in the map and see if the slope of the regression changes. A systematic change in regression geometry in relation to neighboring areas indicates spatial heterogeneity. For details on regression statistics, see Chapter 6.

Interpreting results: See Section A, Exercise 2.2.

ACTION: Create scatter plot matrix

Main Menu > Explore > Scatter plot Matrix>

Variables = Expenses > Click on the arrow pointing to the right (see Figure 2.44)

Variables = University > Click on the arrow pointing to the right

Variables = SecondaryE > Click on the arrow pointing to the right

Close the dialog box

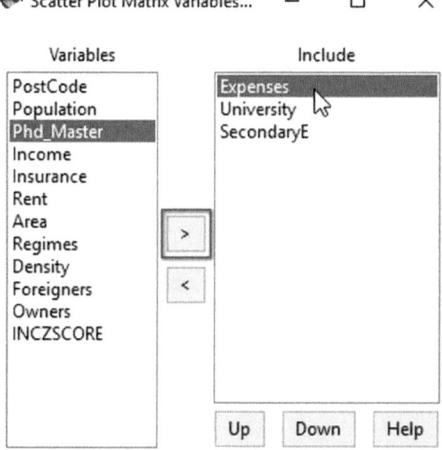

Figure 2.44 Create scatter plot matrix graph.

Exercise 2.2 (*cont.*)

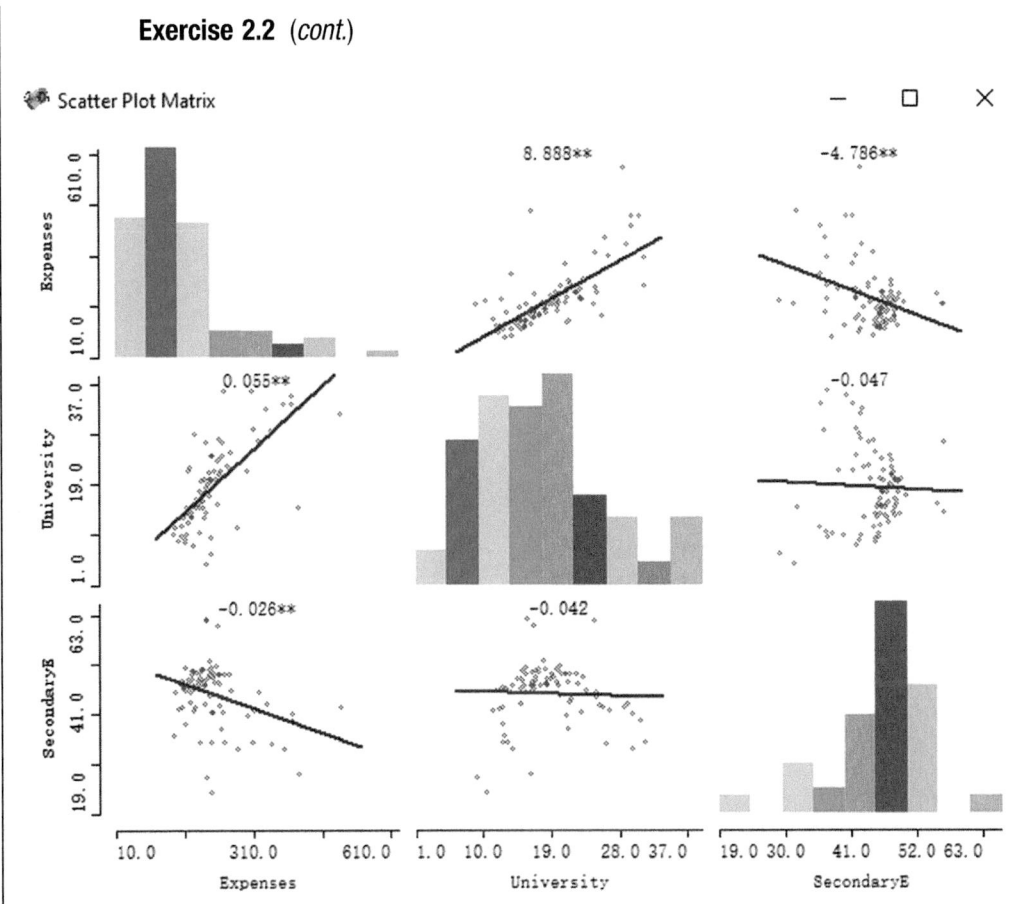

Figure 2.45 Scatter plot matrix among Expenses, University, and SecondaryE. Regression lines and slope values with significance indicated by one asterisk (*, $p < 0.05$) or two asterisks (**, $p < 0.01$) are also presented (see also Figure 2.43, which is a subplot of Figure 2.45).

You can save the graph by right clicking and then select Save Image As.

Interpreting results: See Section A, Exercise 2.2.

3 Analyzing Geographic Distributions and Point Patterns

THEORY

Learning Objectives

This chapter deals with

- Calculating basic statistics for analyzing geographic distributions including mean center, median center, central feature, standard distance and standard deviational ellipse (centrographics)
- Explaining how these metrics can be used to describe spatial arrangements of different sets of point patterns
- Defining locational and spatial outliers
- Introducing the notions of complete spatial randomness, first-order effects and second-order effects
- Analyzing point patterns through average nearest neighbor analysis
- Ripley's K function
- Kernel density estimation
- Randomness and the concept of spatial process in creating point patterns

After a thorough study of the theory and lab sections, you will be able to

- Use spatial statistics to describe the distribution of point patterns
- Identify locational and spatial outliers
- Use statistical tools and tests to identify if a spatial point pattern is random, clustered or dispersed
- Use Ripley's K and L functions to define the appropriate scale of analysis
- Use kernel density functions to produce smooth surfaces of points' intensity over space
- Apply centrographics, conduct point pattern analysis, apply kernel density estimator and trace locational outliers through ArcGIS

3.1   Analyzing Geographic Distributions: Centrography

Centrographic statistics are tools used to analyze geographic distributions by measuring the center, dispersion and directional trend of a spatial arrangement. The centrographic statistics in most common use are the mean center, median center, central feature, standard distance and standard deviational ellipse. Centrographic statistics are calculated based on the location of each feature, which is their major difference with descriptive statistics, which concern only the nonspatial attributes of spatial features.

3.1.1   Mean Center

Definition

Mean center is the geographic center for a set of spatial features. It is a measure of central tendency and is calculated as the average of the x_i and y_i values of the centroids of the spatial features ([3.1]; see Figure 3.1A).

$$\bar{X} = \frac{\sum_{i=1}^{n} x_i}{n}, \bar{Y} = \frac{\sum_{i=1}^{n} y_i}{n} \tag{3.1}$$

   n is the number of spatial objects (e.g., points or polygons)
   x_i, y_i are the coordinates of the i-th spatial object (centroid in case of polygons)
   $\bar{X}, \bar{Y}$ are the coordinates of the mean center

The mean center can be calculated considering weights (3.2). For example, we can calculate the mean center of cities based on their population or income (see Figure 3.1B).

$$\bar{X} = \frac{\sum_{i=1}^{n} w_i x_i}{\sum_{i=1}^{n} w_i}, \bar{Y} = \frac{\sum_{i=1}^{n} w_i y_i}{\sum_{i=1}^{n} w_i} \tag{3.2}$$

where

   w_i is the weight (e.g., income or population) of the i-th spatial object

Why Use

The mean center is used to identify the geographical center of a distribution while the weighted mean center is used to identify the weighted geographical center of a distribution. Mean and weighted centers can be used to compare between the distributions of different types of features (e.g., crime distribution to police station distribution) or the distributions at different time stamps (e.g., crime during the day to crime during the night).

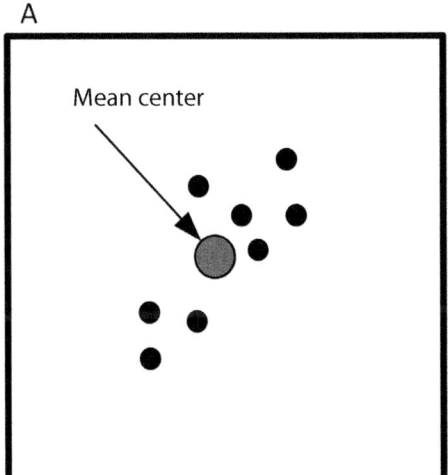
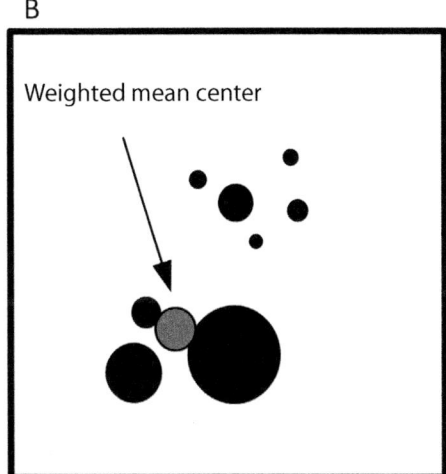

Figure 3.1 (A) Mean center of points' spatial distribution (e.g., cities). The mean center is not an existing point in the dataset. (B) Weighted mean center of a spatial distribution of cities using population as weight. The weighted mean center has been shifted downward relative to the non-weighted mean center (see Figure 3.1A) because the cities in the south are more highly populated.

Interpretation

The mean center offers little information if used on its own for a single geographical distribution. It is used mostly to compare more than one geographic distribution of different phenomena, or in time series data to trace trends in spatial shifts (see Figure 3.2A and B). It should be noted that different geographic distributions might have similar mean centers (see Figure 3.2C). A weighted mean center identifies the location of a weighted attribute. In this respect, it is more descriptive (compared to the mean center), as it indicates where the values of a specific attribute are higher (see Figure 3.1B).

Discussion and Practical Guidelines

The mean center will be greatly distorted where there are locational outliers. Locational outliers should thus be traced before the mean center is calculated. The mean center can be applied to many geographically related problems, such as crime analysis or business geomarketing analysis. Let us consider two examples.

- Crime analysis: Police can identify the central area of high criminality by using a point layer of robberies and calculating the mean center (see Figure 3.1A). This would not be very informative, however, as the mean center would probably lie close to the centroid of the case study area. If

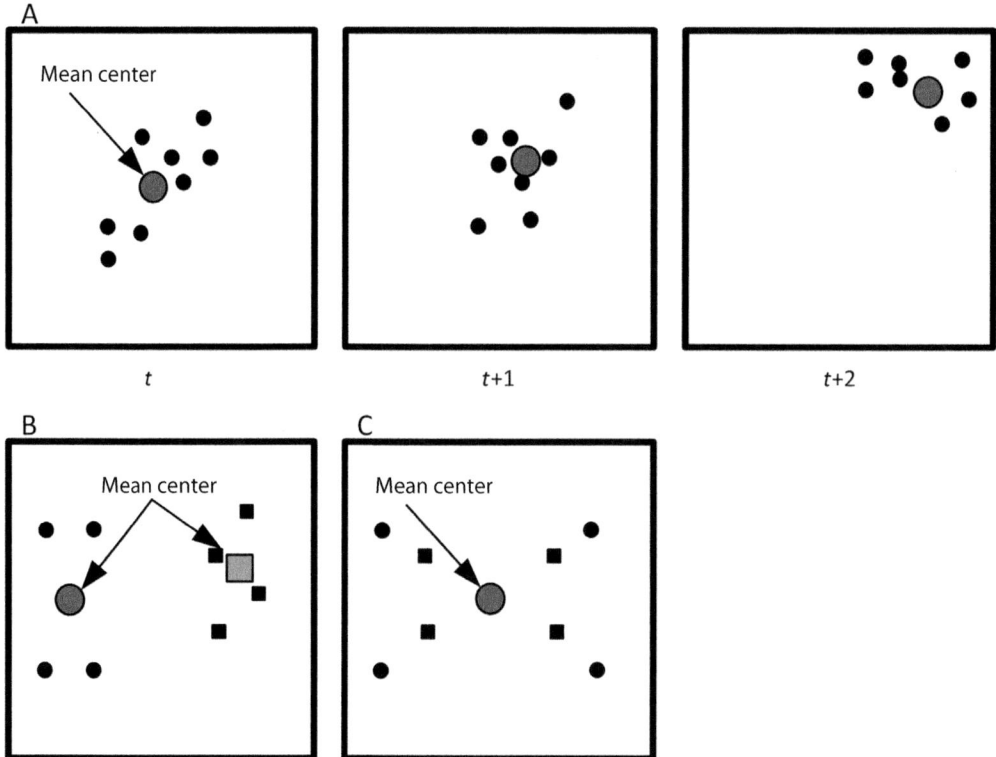

Figure 3.2 (A) Mean center shift for time series data. (B) Different distributions with different mean centers; this is more informative. (C) Different distributions with the same mean center.

data are available for multiple time stamps, the mean center calculation could reveal a shift toward specific directions and thus identify a crime location trend (see Figure 3.2A). Police may further seek to determine why the trend exists and allocate patrols accordingly.

- Geomarketing analysis: Suppose we analyze two different point sets for a city. The first refers to the location of banks (circles) and the second to the location of hotels (squares) (see Figure 3.2B and C). The calculation of the mean center of each point set will probably show different results. We expect that banks will probably be located in the central business district, while hotels closer to the historic center. In this case, the two mean centers will probably lie far apart from each other (see Figure 3.2B). Still, the two mean centers might lie close together if the central business district lies close or engulfs the historic city center (see Figure 3.2C). The mean center is not very informative on its own, but it provides the base upon which more advanced spatial statistics can be built.

3.1.2   Median Center

Definition

Median center is a point that minimizes the travel cost (e.g., distance) from the point itself to all other points (centroids in the case of polygons) in the dataset (see Figure 3.3). It is a measure of central tendency, calculated as shown in (3.3):

$$minimize \sum_{i=1}^{n} d_i \qquad (3.3)$$

where

 n is the number of spatial objects (e.g., points or polygons)
 d_i is the distance of the i-th object to the potential median center

It can also be used with weights (e.g., population, traffic load) calculated as (3.4):

$$minimize \sum_{i=1}^{n} w_i d_i \qquad (3.4)$$

where

 w_i is the weight of the i-th object. Weights can be positive or negative reflecting the pulling or pushing effects of the events on the location of the median center (Lee & Wong 2001 p. 42).

A
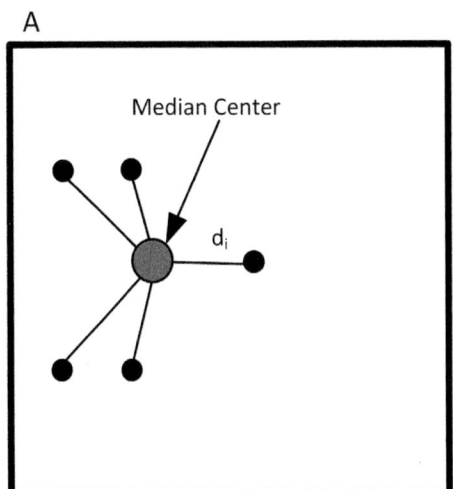

B
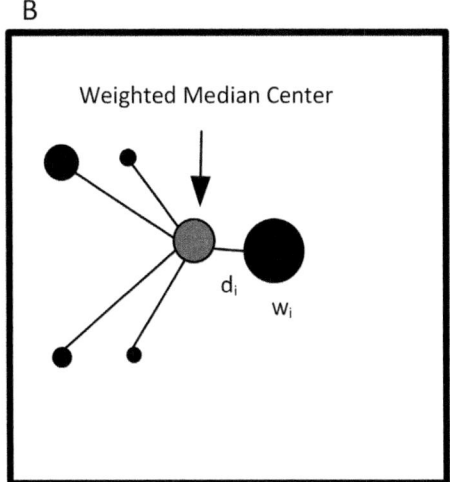

Figure 3.3 (A) Median center. (B) Weighted median center, when using population as weight, shifted to the right.

Why Use
The median center is a measure of the central tendency of spatial data and can be used to find the location that minimizes total travel cost (or weighted cost in the case of a weighted median center).

Interpretation
The median center is a new location and is not necessarily one of the points that exist in the layer. The median center is not as prone to spatial outliers as the mean center is. When there are many spatial objects, the median center is more suitable for spatial analysis. The median center can be also calculated based on weights. For example, to locate a new hospital, we could calculate the median center of postcodes based on their population. The median center will be the location where the total travel cost (total distance) of the population in each postcode to the median center is minimized.

Discussion and Practical Guidelines
There is no direct solution for finding the median center of a spatial dataset. A final solution can be derived only by approximation. The iterative algorithm suggested by Kulin & Kuenne (1962) is a common method used to find the median center (Burt et al. 2009). The algorithm searches for the solution/location that minimizes the total Euclidean distance to all available points. If more than one location minimizes this total cost, the algorithm will calculate only one. Examples related to crime and geomarketing analyses include the following:

- Crime analysis: By calculating the median center of crime, police can locate the point that minimizes police vehicles' total travel cost to the most dangerous (high-crime) areas.
- Geomarketing analysis: Locating a new shop in a way that minimizes the total distance to potential customers in nearby areas.

3.1.3   Central Feature

Definition
Central feature is the object with the minimum total distance to all other features (see Figure 3.4). It is a measure of central tendency. An exhaustive simple algorithm is used to calculate the total distance of a potential central feature to all other features. The one that minimizes the total distance is the central feature (3.5):

$$find\ object\ j\ the\ minimizes\ \sum\nolimits_{i=1}^{n} d_{ji} \qquad (3.5)$$

where

n is the number of spatial objects (e.g., points or polygons)

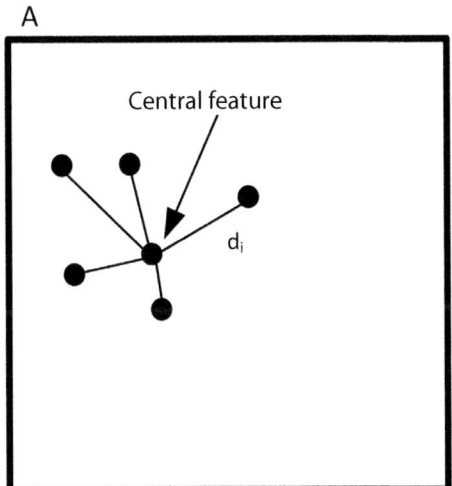
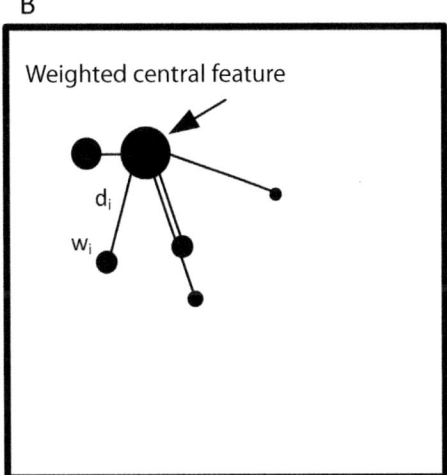

Figure 3.4 (A) The central feature is the existing feature that minimizes total distance. (B) Weighted central feature is the existing feature that minimizes the total distance on some weights (e.g., city population). The points' locations are the same in both graphs.

j is one of the *n* objects selected to be the central feature in an iteration of the algorithm

d_{ij} is the distance of the *i*-th object to the potential central feature *j*. The distance of *j*-th object to itself is zero. Euclidean and Manhattan distance can be used.

It can also be used when weights are selected – e.g., population, traffic load – (3.6):

$$\text{find object } j \text{ the minimizes } \sum_{i=1}^{n} w_i d_{ji} \qquad (3.6)$$

where

w_{ij} is the weight of the *i*-th object to the potential *j*-th central feature

Why Use

To identify which spatial object from those existing in the dataset is the most centrally located.

Interpretation

The central feature is usually located near the center of the distribution, while the weighted central feature can be located far from the center. In this case, the weighted central feature acts as an attraction pole of a large magnitude.

Discussion and Practical Guidelines

The difference with the median center is that a central feature is an existing object of the database, while the median center is a new point. It can be used if we have to select a specific object from the database rather than a new one – for example, to find which postcode is the central feature in a database. Examples related to crime and geomarketing analyses include the following:

- Crime analysis: By calculating the central feature, police can locate which one of the police stations around the city is most centrally located (i.e., a point layer of police stations).
- Geomarketing analysis: To identify the most centrally located postcode when locating a new shop. Using population as the weight, we could locate the central feature that minimizes the distance between the central postcode and the weighted population.

3.1.4   Standard Distance

Definition

Standard distance is a measure of dispersion (spread) that expresses the compactness of a set of spatial objects. It is represented by a circle the radius of which equals the standard distance, centered on the mean center of the distribution (see Figure 3.5). It is calculated as in (3.7):

$$SD = \sqrt{\frac{\sum_{i=1}^{n}(x_i - \bar{X})^2}{n} + \frac{\sum_{i=1}^{n}(y_i - \bar{Y})^2}{n}}. \tag{3.7}$$

n is the total number of spatial objects (e.g., points or polygons)
x_i, y_i are the coordinates of the i-th spatial object
$\bar{X}, \bar{Y}$ are the coordinates of the mean center

The weighted standard distance is calculated (Figure 3.5B; [3.8]):

$$SDw = \sqrt{\frac{\sum_{i=1}^{n}w_i(x_i - \bar{X}_w)^2}{\sum_{i=1}^{n}w_i} + \frac{\sum_{i=1}^{n}w_i(y_i - \bar{Y}_w)^2}{\sum_{i=1}^{n}w_i}} \tag{3.8}$$

where

w_i is the weight (the value e.g., income or population) of the i-th spatial object
$\bar{X}_w$, $\bar{Y}_w$ are the coordinates of the weighted mean center

Why Use

To assess the dispersion of features around the mean center. It is analogous to standard deviation in descriptive statistics (Lee & Wong 2001 p. 44).

A

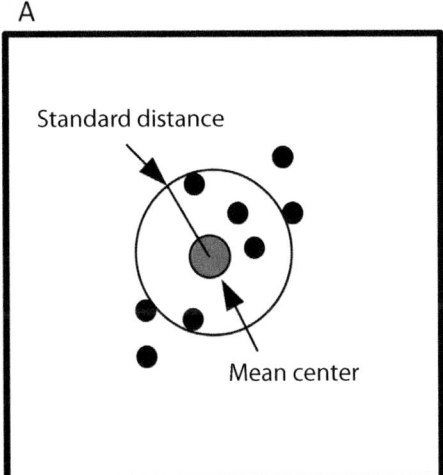

B

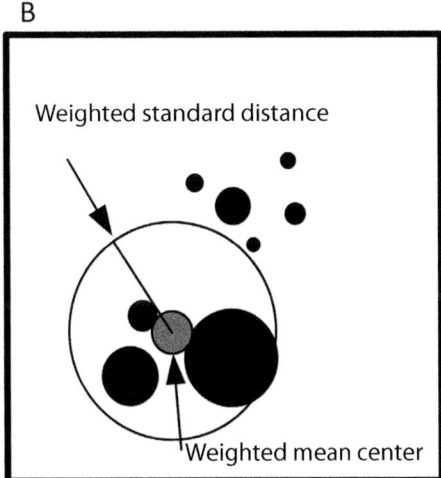

Figure 3.5 (A) Standard distance. (B) Weighted standard distance (weight: city population).

Interpretation

When spatial objects are arranged in such a way that more of them are concentrated in the center with fewer objects scattered towards the periphery (following a Rayleigh distribution) and when we do not use weights, approximately 63% of the spatial objects lie one standard distance from the mean center, and 98% lie within two standard distances from it (Mitchell 2005). The greater the standard distance, the more dispersed the spatial objects are around the mean.

Discussion and Practical Guidelines

Standard distance is a measure of spatial compactness. It is a single measure of spatial objects' distribution around the mean center, similar to statistical standard deviation, which provides a single measure of the values around the statistical mean. Standard distance is more useful when analyzing point layers than when measuring polygons, as polygons are usually human constructed (e.g., administrative boundaries). Examples related to crime and geomarketing analyses include the following:

- Crime analysis: Standard distance can be used to compare crime distributions across several time stamps (e.g., crime during the day relative to crime during the night). If the standard distance during the day is smaller than that at night and is in a different location, this might indicate that police should expand patrols to wider areas during the night (see Figure 3.6A).

- Geomarketing analysis: We could compare the standard distance of different events, such as the spatial distribution of customers of a mall

A

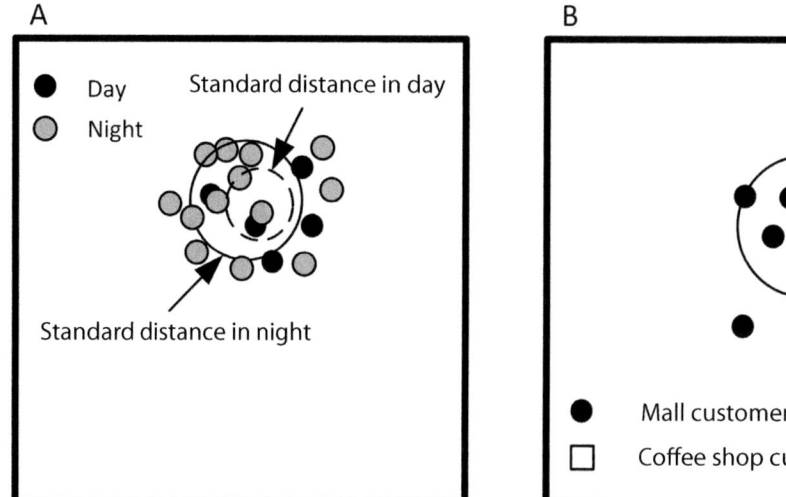

B

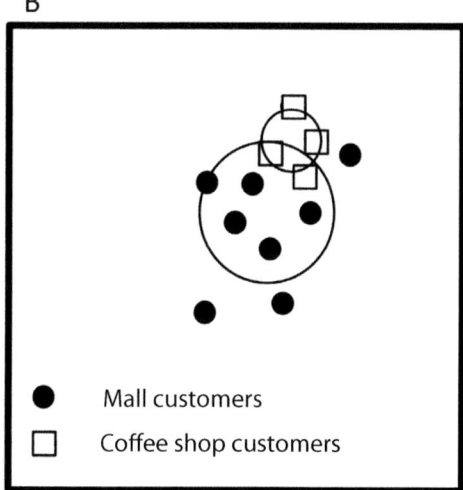

Figure 3.6 (A) Standard distance for the same phenomenon in two different time stamps. (B) Standard distance for two different sets of events. Different standard distance reflects the different dispersion patterns of the two clientele groups.

relative to that of the customers a specific coffee shop. A different standard distance would reveal a more local and less-dispersed pattern for the coffee shop clients compared to those of the mall. An appropriate marketing policy could be initiated based on these findings at the local level (for the coffee shop) and the regional level (for the mall; see Figure 3.6B).

3.1.5   Standard Deviational Ellipse

Definition

Standard deviational ellipse is a measure of dispersion (spread) that calculates standard distance separately in the x and y directions (see Figure 3.7), as in (3.9 and 3.10). The ellipse can also be calculated using the locations influenced by an attribute (weights).

$$SD_x = \sqrt{\frac{\sum_{i=1}^{n}(x_i - \bar{X})^2}{n}} \tag{3.9}$$

$$SD_y = \sqrt{\frac{\sum_{i=1}^{n}(y_i - \bar{Y})^2}{n}} \tag{3.10}$$

n is the total number of spatial objects (e.g., points or polygons)
x_i, y_i are the coordinates of the i-th spatial object
$\bar{X}$, $\bar{Y}$ are the coordinates of the mean center

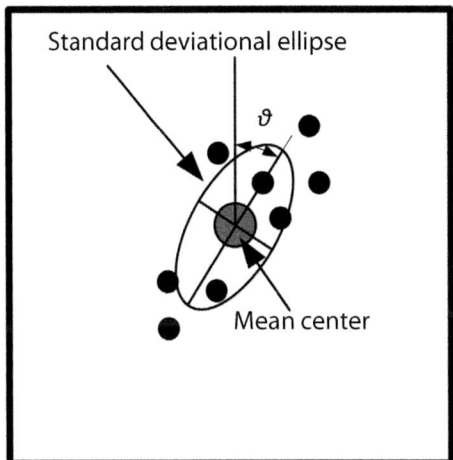

Figure 3.7 Standard deviational ellipse reveals dispersion and directional trend.

These two standard distances are orthogonal to each other and define the standard deviational ellipse. The ellipse is centered on the mean center and is rotated by a particular angle θ form north.

If $x'_i = x_i - \bar{X}$ and $y'_i = y_i - \bar{Y}$ are the deviations of the points form the mean center, the rotation angle is calculated by (3.11):

$$tan\theta = \frac{\left(\sum_i^n x_i'^2 - \sum_i^n y_i'^2\right) + \sqrt{\left(\sum_i^n x_i'^2 - \sum_i^n y_i'^2\right)^2 + 4\left(\sum_i^n x'\sum_i^n y'\right)^2}}{2\sum_i^n x_i'\sum_i^n y_i'} \qquad (3.11)$$

The angle of rotation θ is the angle between north and the major axis. If the sign of the tangent is positive, then the major axis rotates clockwise from the north. If the tangent is negative, then it rotates counterclockwise from north (Lee & Wong 2001 p. 49).

Why Use
The standard deviational ellipse is used to summarize the compactness and the directional trend (bias) of a geographic distribution.

Interpretation
The direction of the ellipse reveals a tendency toward specific directions that can be further analyzed in relation to the problem being studied. Furthermore, we can compare the size and direction of two or more ellipses, reflecting different spatial arrangements. When the underlying spatial arrangement is concentrated in the center with fewer objects lying away from the center, according to a rule of thumb derived from the Rayleigh

distribution for two-dimensional data, a one standard deviational ellipse will cover approximately 63% of the spatial features, two standard deviations will contain around 98% percent of the features, and three standard deviations will cover almost all of the features (99.9%; Mitchell 2005). For a nonspatial one-dimensional variable, these percentages are 68%, 95% and 99%, respectively, for the normal distribution and standard deviation statistics.

Discussion and Practical Guidelines

When we calculate a standard deviational ellipse, we obtain (a) an ellipse centered on the mean center of all objects, (b) two standard distances (the length of the long and short axes) and (c) the orientation of the ellipse. Axis rotation is calculated so that the sum of the squares of the distances between the points and axes is minimized. The standard deviational ellipse is more descriptive and is more widely used than standard distance. Examples include the following:

- Crime analysis: A standard deviational ellipse can reveal a direction. If related to other patterns (e.g., bank or shop locations), it could indicate a potential association (e.g., shop burglaries arranged around a specific route). Police can then better plan patrols around these areas (see Figure 3.8A).
- Geomarketing analysis: Plotting the deviational ellipses calculated on the locations of customers who buy specific products allow us to locate the areas where people prefer one product over another. Marketing policies can then be formulated for these locations (see Figure 3.8B).

3.1.6   Locational Outliers and Spatial Outliers

Definition: Locational Outlier

A **locational outlier** is a spatial object that lies far away from its neighbors (see Figure 3.9). As in descriptive statistics, there is no optimal way to define a location outlier. One simple method is to use the same definition as that in descriptive statistics (see Section 2.2.7). Under this definition, an object whose distance to its nearest neighbor exceeds 2.5 deviations from the mean nearest neighbor average (computed for the entire dataset) is considered the locational outlier, as in (3.12):

Locational Outlier Distance from nearest neighbor $\geq$

(Average Nearesr Neighbor Distance) $+$ *2.5 Standard deviations*   (3.12)

Why Use

Locational outliers should be traced, as they tend to distort spatial statistics outputs. For example, if an outlier exists, the mean center will be

A
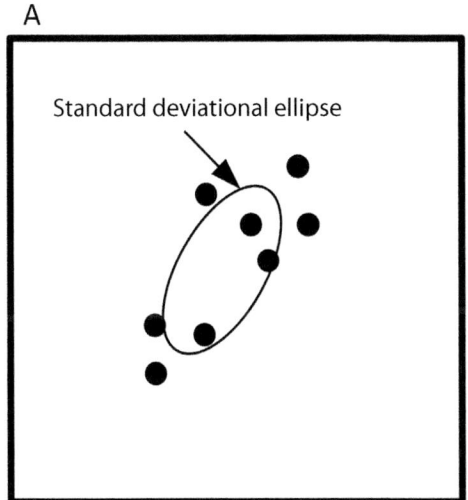

B
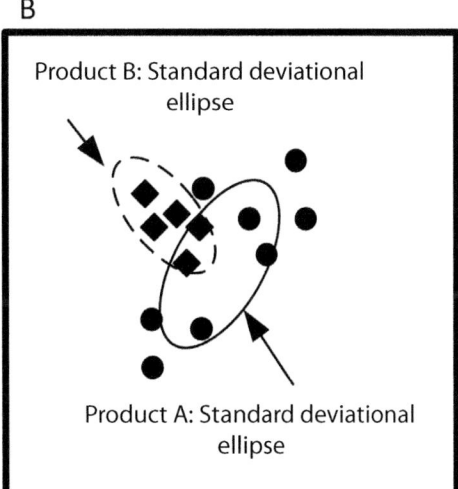

Figure 3.8 (A) Standard deviational ellipse of crimes. A northeast-to-southwest direction is identified. (B) Standard deviational ellipse for two different sets of events. The circles are the locations where customers buy product A, and the squares are the locations where customers buy product B. The directions and dispersion are different, showing that product A penetrates in areas different from those of product B.

significantly different from the mean center when the outlier is not included in the calculations (see Figure 3.9). They may also reveal interesting data patterns (see the Interpretation and Discussion and Practical Guidelines sections that follow).

Interpretation
An outlier typically indicates that we should either remove it from the dataset (at least temporarily) or conduct further research to explain its presence. For example, it may indicate incorrect data entry or a distant location that is abnormal (such as a distant location in which a virus suddenly appears), which should be further studied.

Discussion and Practical Guidelines
Following the definition given earlier, tracing locational outliers requires calculating (a) the nearest neighbor distance of each object (thus creating a distribution of all nearest neighbor distances), (b) the average nearest neighbor distance of all spatial objects, and (c) the standard deviation of the nearest neighbor distances (this is not the standard distance, as discussed in Section 3.1.4).

For polygon features, we can use the area instead of the distance to trace location outliers. We consider locational outliers as those polygons the area of

A

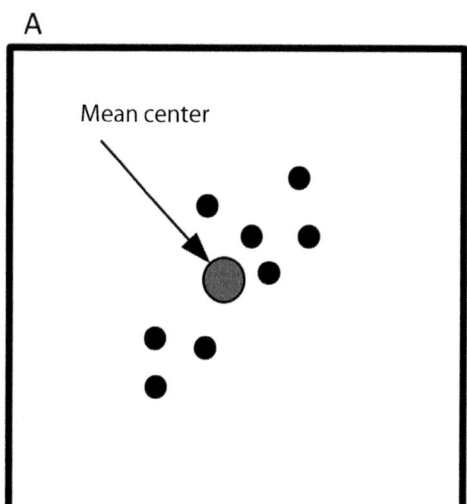

B

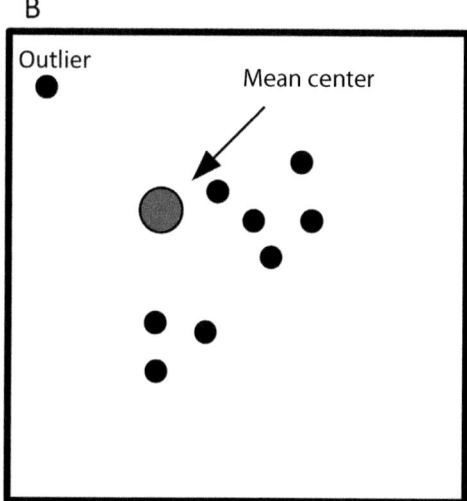

Figure 3.9 (A) Mean center without any locational outlier. (B) Mean center when the outlier is included in the calculation. The mean center is shifted toward the direction of the locational outlier, distorting the real center of the distribution.

which is considerably larger than that of the rest of the objects (again 2.5 or 3 standard deviations away from the mean). In this case, we do not compare nearest neighbors, but we examine the dataset as a whole. Area is related to distance; the larger the area, the more likely it is that the nearest neighbor distance will increase. The calculation of locational outliers using area is not always thorough, as island polygons with small areas may be outliers but will not be classified as such due to their small area value.

Tracing the locational outliers of a polygon shape is essential (especially when calculating spatial autocorrelation; see the next chapter). For example, administrative boundaries, postcodes and school districts are usually human-made. These kinds of spatial data typically have small polygons in the center of a city and larger ones in the outskirts. Thus, the probability of having a locational outlier is high. Spatial statistics tools (in most software packages) estimate locational outliers based on the distances among objects; it is thus better to determine if a polygon is an outlier using distance rather than area. For this reason, polygons should be handled as centroids, and the existence of outliers should be verified based on the distance and distribution of their centroids.

Potential case studies include the following:

- Crime analysis: Identifying if locational outliers in crime incidents exist may reveal abnormal, unexpected behavior that might need additional surveillance. For example, if a credit card that is usually used in specific locations of a city (e.g., shops, restaurants, a house) is also used the same day at a location miles away, that would be a strong indication of fraud.

- Geomarketing analysis: Locational outliers might not be desirable. For example, if a bank wants to locate a new branch, the existence of locational outliers might place the location point in an area that will not be convenient for most of its clientele. The temporary exclusion of the locational outlier will allow for a better branch location. The outlier will then be included in the dataset for other analyses. In other words, when we exclude an outlier, we usually exclude it only for certain spatial statistical procedures, but we include it again because it might be helpful for another type of analysis.

Definition: Spatial Outlier

A **spatial outlier** is a spatial entity whose nonspatial attributes are considerably different from the nonspatial attributes of its neighbors (see Figure 3.10).

Why Use

Spatial outliers reveal anomalies in a spatial dataset that should be further examined.

Interpretation

Spatial outliers can be interpreted differently depending on the problem at hand (see examples and case studies later in this chapter). The basic conclusion we draw when spatial outliers exist is that some locations have attribute values that differ considerably from those of their neighbors. This means that some processes run across space, inferring heterogeneity. Various spatial analysis methods can be used to analyze these processes (see Section 3.2 and Chapter 4).

Discussion and Practical Guidelines

A spatial outlier should not be confused with a locational outlier. To detect locational outliers, we analyze only the distance of a spatial entity to its neighbors. No other attribute analysis takes place. To detect spatial outliers, we study if an attribute value deviates significantly from the attribute values of the neighboring entities. Thus, a spatial outlier does not need to be a locational outlier as well. Additionally, a specific entity may be labeled as a spatial outlier only for a single attribute, while other attribute values might not deviate from the corresponding attribute values of other neighboring entities. Finally, a spatial outlier is not necessarily a global outlier as well, as the spatial outlier is always defined inside a predefined neighborhood.

Let us consider an example. Suppose we study the percentage of flu occurrence within the postcodes of a city (i.e., the percent of people who got the flu in the last month). If we trace a specific postcode with a flu occurrence

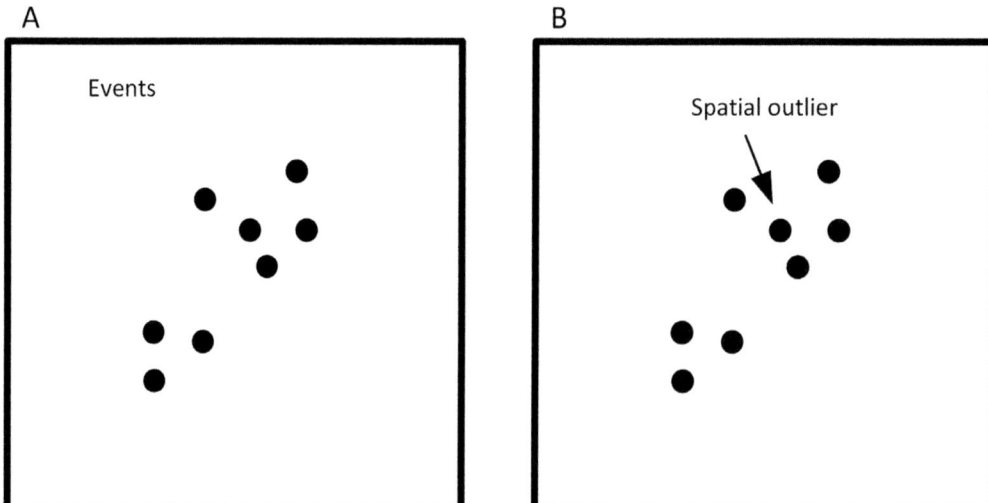

Figure 3.10 (A) Set of events at a set of locations. Each event stands for the location of a house in a neighborhood. (B) If we analyze an attribute (e.g., the size of each house), then a house whose size is considerably different from that of its neighbors is a spatial outlier. For example, the house (event pointed at with an arrow) is a spatial outlier, as its size value is either too large or too small compared to that of its neighbors. As can be observed, a spatial outlier does not have to lie far away from other points.

percentage considerably higher than that of adjacent postcodes, we might label this postcode as a spatial outlier (this might trigger an emergency alert only for this area; Grekousis & Fotis 2012, Grekousis & Liu 2019). This postcode may be centrally located. Then, although it is a spatial outlier, it would not be a locational outlier (neighboring postcodes are in close proximity). If we examine the income attribute of this postcode in relation to the income attributes of the neighboring postcodes, we might find no significant differences in values. This postcode would then be a spatial outlier only for the flu percentage occurrence attribute and not for the income attribute. A spatial outlier does not necessarily mean that the value of the attribute is a global outlier as well. For example, there might be additional postcodes with similarly high flu occurrence percentages but clustered in another area of the city. It is a spatial outlier because it is different (for some attribute) within its neighborhood (so we have to define a neighborhood to trace the spatial outliers). By tracing spatial outliers, we can detect anomalies in many diverse types of data, including environmental, surveillance, health and financial data.

Depicting them in a 3-D graph is a rapid way of detecting spatial outliers. Objects are located on the X-Y plain by their geographic coordinates. The value of the attribute for each spatial object is depicted on the Z axis. Those objects that are significantly higher or lower than their neighbors might be spatial outliers. This method is only graphical. To detect outliers based on numerical results, we can use the Local Moran's *I* index (Section 4.4.1) or the optimized outlier analysis (Section 4.4.2), which are explained in the next chapter.

Spatial outlier detection can be used for the following:

- Crime analysis: One could identify if a postcode has a high crime rate while all the neighboring ones have low rates. This might indicate a ghetto.
- Geomarketing analysis: Identifying a specific postcode where people buy a certain product considerably less often than in adjacent ones might indicate where access to the product is not easy or where a better marketing campaign is required.

3.2 Analyzing Spatial Patterns: Point Pattern Analysis

Definitions
Spatial point pattern *S* is a set of locations $S = \{s_1, s_2, s_3, \ldots s_n\}$ in a predefined region, *R*, where *n* events have been recorded (Gatrell et al. 1996; see Figure 3.11). Put simply, a point pattern consists of a set of events at a set of locations, where each event represents a single instance of the phenomenon of interest (O'Sullivan & Unwin 2010 p. 122).

Event is the occurrence of a phenomenon, a state, or an observation at a particular location (O'Sullivan & Unwin 2010 p. 122).

Spatiotemporal point pattern is a spatial pattern of events that evolves over time. In spatiotemporal point patterns, multiple sets of events occur

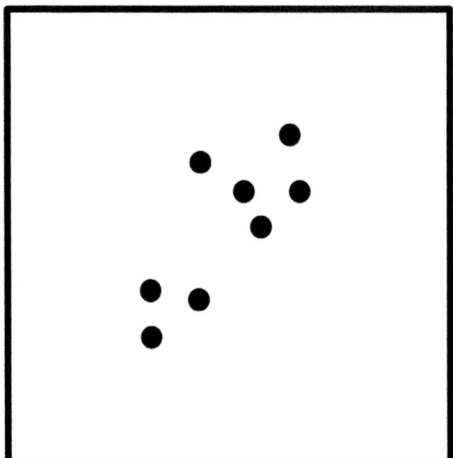

Figure 3.11 Spatial point pattern: Events arranged in a geographic space create a spatial point pattern. A map of events (point pattern) is the realization of a spatial process.

diachronically. These events include several links among places based on physical connections, functional interactions or processes that relate one place to another (Lu & Thill 2003).

Why Use

Point pattern analysis is used mainly to (a) describe the events variation across space and (b) identify relationships among the events by the definition of the spatial process that triggers their arrangement.

Discussion and Practical Guidelines

Sets of point events are widespread in geographical analysis. The locations of shops, crimes, accidents or customers are just a few examples of sets of events that create point patterns in space. Questions can be asked when similar data are available, such as "Where do most of our customers live?" "Where is traffic accident density highest?" and "Are there any crime hot spots in a study area?" Although mapping events provides an initial assessment of their spatial arrangement, more advanced techniques should be used to provide a quantitative evaluation of their geographical distribution. The rationale behind such analysis is deeper. By describing the spatial distribution of events, we attempt to identify the spatial process leading to this formation. The core idea is that there is a spatial process that generates a specific event's arrangement (spatial point pattern) and that this formation is not just the result of a random procedure. In fact, this spatial process is worth further study as the realization of the spatial process in space is the point pattern. Identifying the process allows us to apply the appropriate measures if we want to change the observed pattern (see Figure 3.12).

The term "event" is commonly used in spatial analysis to distinguish the location of an observation from any other arbitrary location within the study area. The term "point" is also often used to describe a point pattern. Events are described by their coordinates $s_i(x_i, y_i)$ and a set of attributes related to the studied phenomenon. The events should be a complete enumeration of the spatial entities being studied and not just a sample (O'Sullivan & Unwin 2010 p. 123). In other words, all available events should be used in point pattern analysis. Although point pattern analysis techniques can be used for samples, the results are very sensitive to missing events. Most point pattern analysis techniques deal only with the location of the events and not with other attributes they might carry (O'Sullivan & Unwin 2010 p. 122). The analysis of attribute value patterns is usually conducted using spatial autocorrelation methods, spatial clustering or spatial regression, as explained in the following chapters (or using geostatistics for field data). The analysis of point patterns may start with centrographic measures, as discussed in Section 3.1, but more sophisticated measures are required, as discussed next.

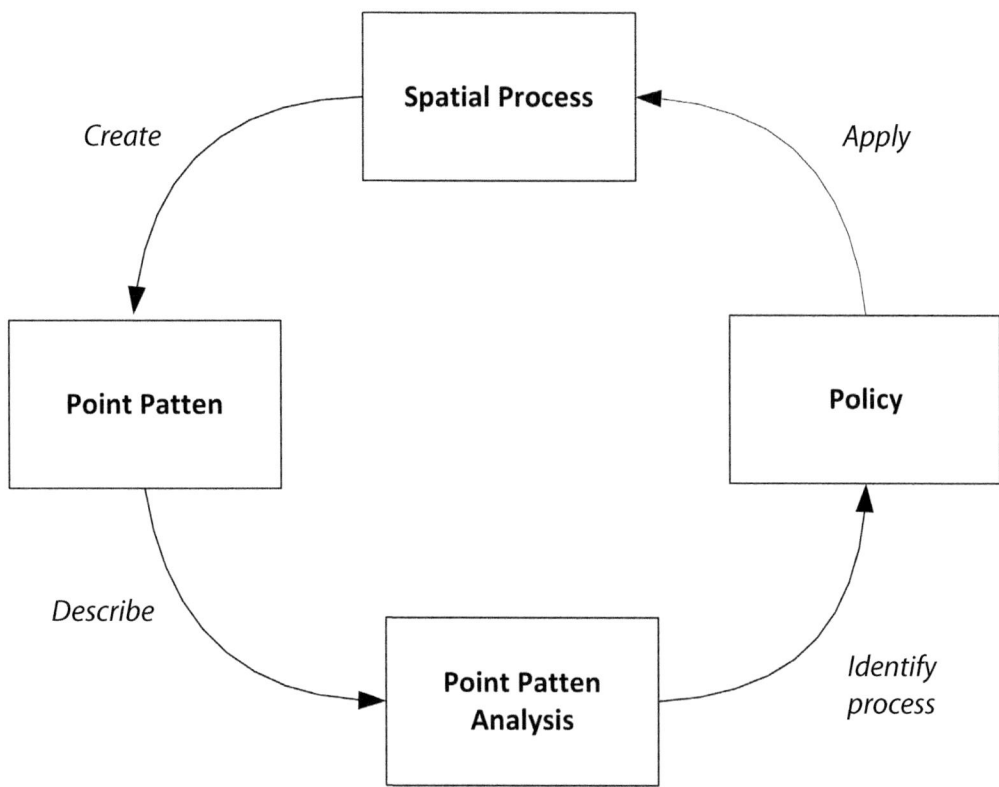

Figure 3.12 A spatial process (not yet known) creates a point pattern that can be described through point pattern analysis. When the spatial process is defined (through the point pattern analysis), measures can be applied to modify the spatial process if needed. For example, the spatial analysis of disease will probably reveal a clustered pattern through an aggregating process. If we locate hot spots, we may apply measures (e.g., vaccination) to specific geographical areas to prevent an expansion of the problem.

3.2.1   Definitions: Spatial Process, Complete Spatial Randomness, First- and Second-Order Effects

Definition

Spatial process is a description of how a spatial pattern can be generated. There are three main types of spatial process (Oyana & Margai 2015 p. 151):

- **Complete spatial randomness process,** also called independent random process (IRP), is a process whereby spatial objects (or their attribute values) are scattered over the geographical space based on two principles (O'Sullivan & Unwin 2010 p. 101):
  - (a) There is an equal probability of event occurrence at any location in the study region (also called first-order stationary).
  - (b) The location of an event is independent of the locations of other events (also called second-order stationary).

- **Competitive process** is a process that leads events to be arranged as far away from each other as possible. The concept is that each event should be located at even, large distance from each neighboring event so that it maximizes its zone of influence/impact. In this respect, events tend to be uniformly distributed.
- **Aggregating process** is a process where events tend to cluster as a result of some pulling action.

There are three main types of spatial arrangement/pattern associated with the above spatial processes (Oyana & Margai 2015 p. 151; see Figure 3.13):

- **Random spatial pattern:** In this type of arrangement, events are randomly scattered all over the study area (see Figure 3.13B). This pattern has a moderated variation and is similar to a Poisson distribution (Oyana & Margai 2015 p. 151). It is the result of an independent random process.
- **Dispersed:** The events are located uniformly around the study area. This is the result of a competitive process, creating a pattern with no or little variation (see Figure 3.13A). Events are located such that they are further away from their neighbors. For instance, the locations of bank branches are more likely to form a dispersed pattern, as there is no reason to have branches of the same bank located near each other.
- **Clustered:** The events create clusters in some parts of the study area, and the pattern has a large variation (see Figure 3.13C). This is the result of an aggregating process. For example, most hotel locations in a city tend to cluster around historical landmarks or major transportation hubs.

First-order spatial variation effect occurs when the values or locations of spatial objects vary from place to place due to a local effect of space (the equal probability assumption of IRP no longer holds; O'Sullivan & Unwin 2003 p. 29). For example, stroke event locations may vary from place to place inside a city

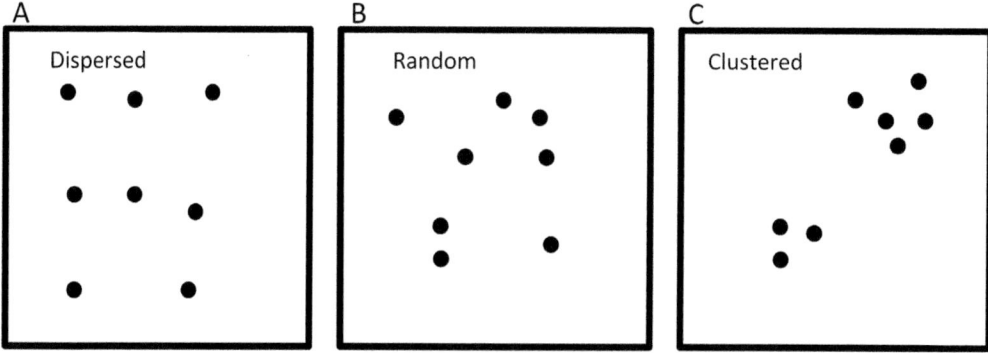

Figure 3.13 (A) In a dispersed pattern, events are scattered in a nearly uniform way. (B) In a random spatial pattern, we cannot identify any clusters or dispersion. (C) In a clustered spatial pattern, clusters will be evident in some parts of the region.

but are expected to be concentrated in places where more of the population is located or in places closer to industrial zones. In other words, events do not occur with the same probability everywhere. This type of spatial effect on the variation of a value or on the location of events is called the first-order effect. First-order effects are mostly associated with density/intensity measures (O'Sullivan & Unwin 2003 p. 112).

Second-order spatial variation effect occurs when there is interaction among nearby locations. Here, the location, or the value of an observation, is highly influenced by a nearby location or the value of a neighboring observation (the independence assumption of IRP no longer holds). For example, immigrants new to a city are more likely to reside in a neighborhood where people of the same ethnicity live, as they would probably feel more comfortable there. In this case, there is a strong local interaction that attracts newcomers; this is a second-order effect of space. Second-order effects are mostly associated with distance measures (O'Sullivan & Unwin 2003 p. 112).

Stationarity is the state in which (a) every location in space has equal probability for an event placement through a spatial process and (b) there is no interaction among events (i.e., independence; O'Sullivan & Unwin 2003 p. 65).

First-order stationary process is a process in which there is no variation in its intensity across space.

Second-order stationary process is a process in which there is no interaction among events.

Intensity of a spatial process is the probability that each small geographical area will receive an event. This probability is the same for every location in a stationary spatial process.

Anisotropic process is a process in which the intensity varies according to specific directions (O'Sullivan & Unwin 2010 p. 108).

Isotropic process, on the other hand, is a spatial process in which directional effects do not exist.

Spatial heterogeneity refers to the non-stationarity of a geographic process (Demšar et al. 2013).

3.2.2   Spatial Process

Each spatial pattern in a particular space and at a specific moment in time is the result of a process occurring within a wider space and time. For a given set of events, it is interesting from the spatial analysis perspective to analyze (a) the observed spatial pattern and (b) the process that created this arrangement (if any). We can summarize the above by posing the following research question: "Can the observed point pattern be the result of a hypothesized spatial process?"

A spatial process can be either (a) deterministic, when inputs and outputs are certain over time, or (b) stochastic, when the outcome is subject to variation and cannot be defined precisely by a mathematical function. Typically, a spatial pattern is the potential realization of some stochastic process. The most common stochastic process is the independent random process (IRP), also known as complete spatial randomness.

First- and second-order effects express the variations in intensity of a spatial process across the study area. A spatial process is first-order stationary if there is no variation in its intensity over space, meaning that there is an equal probability of an event occurring in the study region. A second-order stationary process occurs when there is no interaction among events. Complete spatial randomness is both first- and second-order stationary. Simply put, in complete spatial randomness, there are no first- or second-order effects. When first- and/ or second-order effects exist, the chances of an event occurring at some specific location vary (due to dependence and interactions with the neighboring events); thus, the process is no longer stationary (O'Sullivan & Unwin 2003 p. 65). In zero spatial autocorrelation, no first- or second-order effects exist, and location or value variation is random (no patterns are detected). First- and second-order effects are not always easily distinguished, but tracing their existence is extremely important for the accurate modeling of spatial data (O'Sullivan & Unwin 2003 p. 29)

Complete spatial randomness refers to (a) the specific locations in which objects are arranged and (b) the spatial arrangement of the spatial objects' attribute values. In the first case, the techniques used to detect complete spatial randomness only analyze the location of the spatial objects. Most of these techniques are point pattern analysis methods and are used for point features. In the second case, the focus is on how the values of one or more of the variables for fixed locations are spatially arranged. This approach deals mostly with aerial data (polygon features). Polygons are usually artificially created to express certain boundaries. Therefore, instead of analyzing their distribution as polygons, which would offer no valuable results, we study how their attribute values are arranged in space. Spatial autocorrelation methods are used in this case and will be explained in detail in Chapter 4.

However, what is randomness in a spatial context? In nonspatial inferential statistics, we use representative samples to make inferences for the entire population. To do so, we set a null hypothesis and reject it or fail to reject it using a statistical test. In a spatial context, the null hypothesis used is (typically) complete spatial randomness. Under this hypothesis, the observed spatial pattern is the result of a random process. This means that the probability of finding clusters in our data is minimal, and there is no spatial autocorrelation. To better understand the spatial random process and spatial random pattern, consider the following example. Imagine that you have 100 square paper cards. Each card has a single color: red, green or blue. If we throw (rearrange) all the cards onto the floor, we have our first spatial arrangement (spatial pattern). Is this spatial pattern of colors random, or is it clustered? It is most likely that cards will be scattered, and colors will be mixed. The probability that most of the red cards will be clustered together after our first throw is almost zero. If we repeat the throw 1,000 times, we might identify some clusters of red cards in specific regions occurring, say, in 10 out of the 1,000 throws. This shows that the probability of obtaining a cluster of reds is 1%. In other words, the random process (throwing cards on the floor) generated a random spatial arrangement of objects (spatial pattern) for 99.0% of the trials. Thus, the spatial pattern of the objects is random. The statistical procedure is not as simple as that described here, as it includes additional calculations and assumptions.

For example, to reject or not reject the null hypothesis of complete spatial randomness, we use two metrics: a z-score and p-value (see Section 2.5.5). The z-score is the critical value used to calculate the associated p-value under a standard normal distribution. It reflects the number of standard deviations at which an attribute value lies from its mean. The p-value is a probability that defines the confidence level. When the p-value is very small, the probability that the observed pattern was created randomly is minimal (Mitchell 2005). It indicates the probability that the observed pattern is the result of a random process.

A small p-value in spatial statistical language for the complete spatial randomness hypothesis is translated as follows: It is very unlikely that the observed pattern is the result of a random process, and we can reject the null hypothesis. There is strong evidence that there is an underlying spatial process at play. This is precisely what a geographer is interested in: locating and explaining spatial processes affecting the values of any spatial arrangement or spatial phenomenon.

If the data exhibit spatial randomness, there is no pattern or underlying process. Further geographical analysis is thus unnecessary. If the data do not exhibit spatial randomness, then they do not have the same probability of occurrence in space (first-order effect presence) and/or the location of an event depends on the location of other events (second-order effect presence). In this case, spatial analysis is very important, as space has an effect on event occurrences and their attribute values.

To conclude, when first-order effects exist, the location is a major determinant of event occurrence. When second-order effects exist, the interactions among events are largely influenced by their distance. The absence of a second-order stationary process leads to either uniformity or clustering.

3.3   Point Pattern Analysis Methods

There are two main (interrelated) methods of analyzing point patterns, namely the distance-based methods and the density-based methods.

- **Distance-based** methods employ the distances among events and describe second-order effects. Such methods include the nearest neighbor method

(Clark & Evans 1954; see Section 3.3.1), the *G* and *F* distance functions, the Ripley's *K* distance function (Ripley 1976) and its transformation, the *L* function (see Section 3.3.2).

- **Density-based** methods use the intensity of events occurrence across space. For this reason, they describe first-order effects better. Quadrat count methods and kernel estimation methods are common density-based methods. In quadrat count methods, space is divided into a regular grid (such as a grid of squares or hexagons) of a unitary area. Each unitary region includes a different number of points due to a spatial process. The distribution analysis and its correspondence to a spatial pattern are based on probabilistic and statistical methods. Another, more widely used method is the kernel density estimation (KDE; see Section 3.3.3). This method is better than the quadrat method because it provides a local estimation of the point pattern density at any location of the study area, not only for the locations where events occur (O'Sullivan & Unwin 2003 p. 85).

Another point pattern analysis approach (O'Sullivan & Unwin 2003 p. 127) involves a combination of density and distance, thereby creating proximity polygons. Proximity polygons, such as Delaunay triangulation, and related constructions, such as the Gabriel graph and the minimum spanning tree of a point pattern, display interesting measurable properties. For example, analyzing the distribution of the area among polygons that are close shows how evenly spaced the events in question are. Furthermore, the number of an event's neighbors in Delaunay triangulation and the lengths of the edges are indicators of how the events are distributed in space.

Finally, another way of analyzing events is by using hot spot analysis. This is mainly used to identify if clusters of values of a specific variable are formed in space. It is most commonly used with polygon features and is explained in detail in Section 4.4.2. When considering point features with no associated variables, it is interesting to study the spatial objects' arrangement in terms of intensity levels. This is achieved using optimized hot spot analysis (see Section 4.4.4 for a more in-depth analysis). Strictly speaking, this type of analysis does not assess the type of point pattern or the spatial process that generated it, but it does analyze the location of the events in order to identify any spatial autocorrelation.

3.3.1   Nearest Neighbor Analysis

Definition
Nearest neighbor analysis (also called average nearest neighbor) is a statistical test used to assess the spatial process from which a point pattern has been generated. It is calculated based on the formula (3.13) (Clark & Evans 1954, O'Sullivan & Unwin 2010 p. 142):

$$R = \frac{observed\ mean\ distance}{expected\ mean\ distance} = \frac{\bar{d}_{min}}{E(d)} = 2\bar{d}_{min}\sqrt{n/a} \qquad (3.13)$$

where

$$\bar{d}_{min} = \frac{\sum_{i=1}^{n} d_{min}(s_i)}{n} \qquad (3.14)$$

$$E(d) = 1/\left(2\sqrt{n/a}\right) \qquad (3.15)$$

$\bar{d}_{min}$ is the average nearest neighbors distance of the observed spatial pattern

$d_{min}(s_i)$ the distance of event s_i to its nearest neighbor

n is the total number of events

$E(d)$ is the expected value for the mean nearest neighbor distance under complete spatial randomness

a is the area of the study region, or if not defined the minimum enclosing rectangle around all events

Why Use
To decide if the point pattern is random, dispersed or clustered.

Interpretation
A nearest neighbor ratio R less than 1 indicates a process toward clustering. A value larger than 1 indicates that the pattern is dispersed due to a competitive process. A value close to 1 reflects a random pattern.

Discussion and Practical Guidelines
In practice, this method identifies whether the random, dispersed or clustered pattern better describes a given set of events (also called observed spatial pattern). The statistic tests the null hypothesis that the observed pattern is random and is generated by complete spatial randomness. The method compares the observed spatial distribution to a random theoretical one (i.e., Poisson distribution; Oyana & Margai 2015 p. 153). The test outputs the observed mean distance, the expected mean distance (through a homogeneous Poisson point process), the nearest neighbor ratio R, the p-value and the z-score. The expected mean distance is the mean distance the same number of events would most probably have if they were randomly scattered in the same study area. The p-value is the probability that the observed point pattern is the result of complete spatial randomness. The smaller the p-value, the less likely it is that the observed pattern has been generated by complete spatial randomness. If the p-value is larger

than the significance level, we cannot reject the null hypothesis (as there is insufficient evidence), but we cannot accept it either (for more on how to interpret p-values, see Section 2.5.5). In this case, we have to use other methods (such as those presented later) to determine the observed pattern's type.

A disadvantage of this method is that it summarizes the entire pattern using a single value. It is therefore used mainly as an indication of whether a clustered or a dispersed pattern exists and less often to locate where this process takes place. In addition, this method is highly influenced by the study area's size. A large size might indicate clustering for events for which a smaller size would probably reveal dispersion. Potential case studies include the tracing of clustering in a disease outbreak or identifying if customers of a product are dispersed throughout a region.

3.3.2   Ripley's K Function and the L Function Transformation

Definition
Ripley's K function is a spatial analysis method of analyzing point patterns based on a distance function (Ripley 1976). The outcome of the function is the expected number of events inside a radius of d (Oyana & Margai 2015 p. 163). It is calculated as a series of incremental distances d centered on each of the events in turn (3.16).

$$K(d) = \frac{a}{n^2} \sum_{i=1}^{n} \sum_{j=1, j \neq i}^{n} \frac{I_d(d_{ij})}{w_{ij}} \qquad (3.16)$$

where

> d is an incremental distance value defined by the user (range of distances that the function is calculated).
> d_{ij} is the distance between a target i event and a neighboring j event.
> n is the total number of events.
> a is the area of the region containing all (n) features.
> I_d is an indicator factor. I_d is 1 if the distance between the i and j events is less than d, otherwise it is 0.
> $w_{i,j}$ are weights most often calculated as the proportion of the circumference of a circle with radius d round the target event (Oyana & Margai 2015 p. 163). They are used to correct for edge effects.

To better understand how this function works, imagine that, for each i event (target), we place a circle of d radius and then count how many events lie inside this circle. The total count is the value of the indicator factor I_d. This procedure is used for all events and for a range of distances d (e.g., every 10 m starting from 50 m to 100 m).

In practice, the original Ripley's K function produces large values when the distance increases. Many mathematical transformations of the Ripley's K function exist to account for such a problem. A widely used transformation is the L function (Bailey & Gatrell 1995; O'Sullivan & Unwin 2010 p. 142). This allows for fast computation and also makes the L function linear under a Poisson distribution, which enables easier interpretation, as in (3.17) (Oyana & Margai 2015 p. 163):

$$L(d) = \sqrt{\frac{K(d)}{\pi}} - d \qquad (3.17)$$

Why Use

K and L functions are used, (a) to decide if the observed point pattern is random, dispersed or clustered at a specific distance or a range of distances and (b) to identify the distance at which the clustering or dispersion is more pronounced.

Interpretation

With the L function (Eq. 3.17), the expected value in the case of complete spatial randomness is equal to the input distance d. This means that for a random pattern, the L function equals to zero; for a clustered pattern, it gets a positive value; and for a dispersed pattern, it gets a negative value. To avoid negative and close-to-zero values, we can omit d from Eq. (3.17) and plot L against d to get a graph like the one in Figure 3.14 (ArcGIS applies this approach with a slightly different denominator for K). With this L transformation, the expected value is again equal to distance d, which can now be used as a reference line at a 45° angle. When the observed value of the L function is larger than the expected value for a particular distance or range of distances (see the area above the expected line in Figure 3.14), then the distribution is more clustered than a random distribution.

When the observed value is larger than the upper confidence envelope value, then the spatial clustering is statistically significant (see Figure 3.14; Mitchell 2005). On the other hand, when the observed value of the L function is smaller than the expected (the area below the expected line), then the distribution is more dispersed than a distribution that would be the result of complete spatial randomness. Moreover, when the observed value is smaller than the lower confidence interval envelope, the spatial dispersion is statistically significant for this distance (see Figure 3.14).

The confidence envelope is the area in which the expected values would lie in a random pattern for a specific confidence level (see the Discussion and Practical Guidelines section). To simulate this process, the points are randomly distributed many times (e.g., 99 or 999 times) using the Monte Carlo approach. Each one of these times is called a permutation, and the expected value is

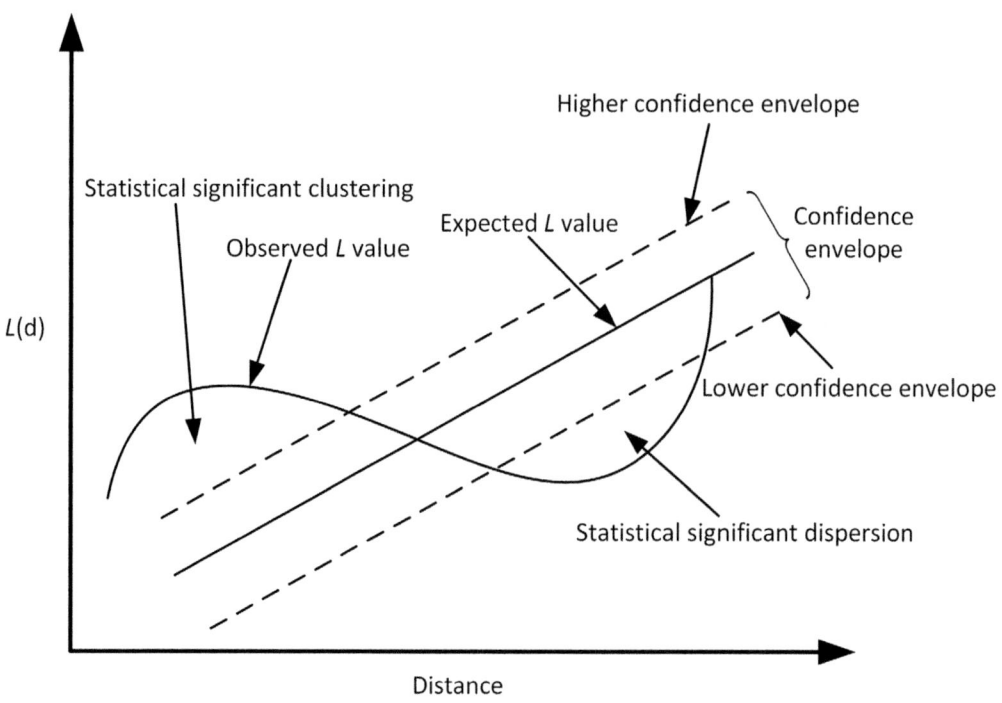

Figure 3.14 L function transformation of Ripley's K function over a range of distances.

calculated for each permutation, creating a sampling distribution. For 999 permutations, the p-value is typically set to 0.001; for 99 permutations, it is set to 0.01. These reflect 99% and 99.9% confidence intervals, respectively. When a value lies outside the envelope area, the null hypothesis for complete spatial randomness is rejected.

Discussion and Practical Guidelines

The process used to calculate the L function occurs in the following steps (Oyana & Margai 2015 p. 164):

1. Calculate the observed value K (using the K function) of the observed set of events. The expected value is obtained by placing a buffer with radius d over a target event and then counting how many events lie inside this zone. Follow the same process for each single event in the dataset. Then calculate the average count of events within a range of distances.

2. Transform the K function estimates to an L function to make it linear.

3. With the L transformation, the expected K value is equal to distance d.

4. Determine the confidence envelope by estimating the minimum and maximum L values using permutations and the related significance levels under the null hypothesis of complete spatial randomness.

5.   Plot L on a graph to reveal if clustering or dispersion is evident at various distances by comparing the expected to the observed values and according to whether the observed values lie inside or outside the confidence envelope.

The K and L functions are sensitive to the definition of the bounding rectangle in which the features are contained. This raises two issues. First, the study region area has a direct effect on the density parameter; second, it gives rise to the edge effect problem (see Section 1.3), in which the events that lie in the edge of the study area tend to have fewer neighbors than those lying in central locations. In the first case, identical arrangements of events are likely to yield different results for various study area sizes. For example, if the size doubles but the points are concentrated in the center, this might result in a clustered pattern. For the exact same event arrangement, however, if the study area is defined by the minimum enclosing rectangle, the pattern might be dispersed or random.

On the other hand, a tight study area can cause the edge effect problem. Events close to the region boundaries tend to have larger nearest neighbor distances, although they might have neighbors just outside the boundaries that lie in closer proximity than those lying inside (O'Sullivan & Unwin 2003 p. 95). Various methods have been developed to account for edge effects (known as edge correction techniques), including guard zones, simulating outer boundaries, shrinking the study area and Ripley's edge correction formula.

The Monte Carlo simulation approach is a more reliable method of accounting for edge effects and the study region area effect. A Monte Carlo procedure generates and allocates n events randomly over the study region hundreds or thousands of times (permutations), creating a sampling distribution. The observed pattern is then compared with the patterns generated under complete spatial randomness through Monte Carlo simulation. The results of complete spatial randomness are also used to construct an envelope inside of which the L value of a random spatial pattern is expected to lie. If there is a statistically significant difference between the observed and the simulated patterns (those lying outside the envelope), then we may reject the null hypothesis that the observed pattern is the result of complete spatial randomness. Since each permutation is subject to the same edge effects, the obtained sampling distribution of the Monte Carlo procedure accounts for both edge and study region area effects simultaneously without the need to apply other corrections (O'Sullivan & Unwin 2010 p. 150).

Spatial patterns change when studied at multiple distances and spatial scales, reflecting the existence of particular spatial processes at specific distance ranges (Mitchell 2005). In other words, a spatial pattern might be random for some range of distances and clustered for others. Ripley's K function illustrates how the spatial clustering or dispersion of features' centroids changes when the neighborhood size changes by summarizing the clustering or dispersion over a range of different distances something helpful for assessing the proper scale of analysis for the problem at hand. This is similar to how the scale of analysis is defined in spatial autocorrelation studies (see Chapter 4). When we analyze a point distribution in different distances, we may trace the distance at which the clustering starts and ends and the distance at which the dispersion begins and ends. Thus, based on the scope of the research, for a single-point dataset, we may address different questions relating to the scale of analysis and the patterns of clustering at various distances (i.e., study how the clustering of crime changes at different distances). In addition, using this type of function does not provide a single description for the pattern in question (as we obtain when we use the average nearest neighbor analysis) but provides deeper insight, as we describe it according to the distance desired.

3.3.3   Kernel Density Function

Definition

Kernel density estimation is a nonparametric method that uses kernel functions to create smooth maps of density values, in which the density at each location indicates the concentration of points within the neighboring area (high concentrations as peaks, low concentrations as valleys; see Figure 3.15). The kernel density at a specific point is estimated using a kernel function with two or more dimensions. The term "kernel" in spatial statistics is used mainly to refer to a window function centered over an area that systematically moves to each location while calculating the respected formula. Kernel functions weigh events according to distance so that those that are closer are weighted more than events further away (O'Sullivan & Unwin 2003 p. 87).

A typical kernel function (Gaussian) looks like a bell, is centered over each event and is calculated, within bandwidth (*h*), based on Eq. (3.18) (Wang 2014 p. 49, Oyana & Margai 2015 p. 167; see Figure 3.15):

$$\hat{f}(x) = \frac{1}{nh^r} \sum_{i=1}^{n} k\left(\frac{d}{h}\right) \tag{3.18}$$

where

   h is the bandwidth (the radius of the search area)
   n is the number of events inside the search radius
   r is the data dimensionality
   k() is the kernel function
   d is the distance between the central event s_i and the *s* event inside the bandwidth

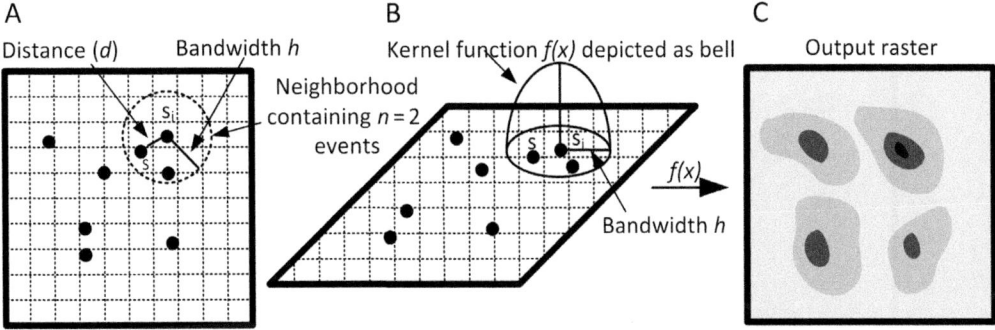

Figure 3.15 (A) Kernel density function is calculated for every single event in turn. (B) We can imagine a kernel as a bell with the same total weight of one unit. The total weight of all bells equals the total number of points (Longley et al. 2011 p. 372). Failing to retain the total population would suggest that there are more or fewer events in the study area than there really are (O'Sullivan & Unwin 2003 p. 87). The shape of the kernel depends on the bandwidth parameter h. A large bandwidth results in a broader and lower height kernel, while a smaller bandwidth leads to a taller kernel with a smaller base. Large bandwidths are more suitable for revealing regional patterns, while smaller bandwidths are more suitable for local analysis (Fotheringham et al. 2000 p. 46). When all events are replaced by their kernels and all kernels are added, then a raster density surface is created, as seen in C. (C) Darker areas indicate higher intensity while lighter areas indicate lower intensity. The output raster allows for a better understanding of how events are arranged in space, as it renders every single cell, creating a smooth surface that also reflects the probability of event occurrence.

The kernel function calculates the probability density of an event at some distance from a reference point (Oyana & Margai 2015 p. 168).

Various kernel types can be plugged into the preceding formula, including normal, uniform, quadratic and Gaussian types. ArcGIS uses the quadratic kernel function described in Silverman (Silverman 1986, Wang 2014 p. 50). The functional form is as shown in (3.19):

$$\hat{f}(x) = \frac{1}{nh^2\pi} \sum\nolimits_{i=1}^{n} \left(1 - \frac{d^2}{h^2}\right)^2 \tag{3.19}$$

Why Use

Kernel density estimation is used to create smooth surfaces that depict, first, the density of events and, second, an estimation of areas of higher or lower event occurrence intensity. Through the raster map visualization of a kernel density estimation, one can quickly locate hot spot and cold spot areas and places of high or lower density (see Figure 3.16). It can also be applied for cluster analysis of weighted or unweighted events in epidemiology, criminology, demography, ethnology and urban analysis.

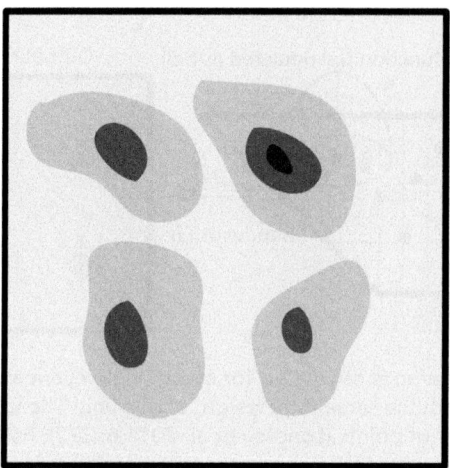

Figure 3.16 Kernel density estimation. A raster output depicts the intensity of events in the study area.

Interpretation

Kernel density estimates the density of the point features, rather than their values (e.g., temperature or height; Silverman 1986, Bailey & Gatrell 1995). The density estimates can be displayed by either surface maps or contour maps that show the intensity at all locations (see Figure 3.15C). Peaks reveal higher concentrations and valleys lower densities.

Discussion and Practical Guidelines

The kernel function provides an interpolation technique for assessing the impact of each event on its neighborhood (as defined by the bandwidth; Oyana & Margai 2015 p. 168). In practice, the kernel function is applied to a limited area around each event defined by the bandwidth h to "spread" its effect across space (see Figure 3.15A). Outside the specified bandwidth, the value of the function is zero. For example, the quadratic kernel function falls off gradually to zero with distance until the radius is reached. Each individual event's impact is depicted as a 3-D surface (e.g., like a bell; see Figure 3.15B). The final density is calculated by adding, for each raster cell, the intersections of the individuals' surfaces.

The main advantage of kernel density estimation is that the resulting density function is continuous at all points along the scale. In addition, kernel density estimation weighs nearby events more heavily than those lying further away using a weighting function, thus applying an attenuating effect in space. This is a major difference with a typical point density calculation, in which density is calculated for each point based on the number of surrounding objects divided by the surrounding area, with no differentiation based on proximity or any other weighting scheme. Kernel density spreads the known quantity to the raster cell inside the specified radius.

A weak point of this technique is the subjectivity in the definition of the bandwidth factor *h*. The selection of an appropriate value for *h* is usually made by trial and error. Large bandwidth *h* values create a less-detailed raster. On the other hand, small values create a raster with a more realistic density (see note on Figure 3.15B; Mitchell 2005). The selection of the appropriate bandwidth depends on the problem at hand and the desired analysis scale. As mentioned in the previous section, both Ripley's *K* and its transformed *L* function provide a graph depicting the distances at which clustering or dispersion is more pronounced. One can first apply the *L* function and then select the distance that better fits the scope of the analysis which also reflects the spatial process at play. Optimized hot spot analysis may also be used to identify the distance at which spatial autocorrelation is more evident and then be used as the bandwidth *h* (see Section 4.4.4). Another approach is to use as the initial bandwidth value the one that results from formula (3.20) (ESRI 2014):

$$h = 0.9 \min \left(SD, \sqrt{\frac{1}{\ln (2)}} D_m \right) n^{-0.2} \tag{3.20}$$

where

    h is the bandwidth (the radius of the search area),
    n is the number of events. If a count field is used then, *n* is the sum of the
        counts
    D_m is the median distance
    SD is the standard distance
    min means that the minimum value between *SD* and $\sqrt{\frac{1}{\ln (2)}} D_m$ will be used in
        (3.20)

Locational outliers should be removed when selecting the appropriate bandwidth value, as they might lead to a very smooth raster layer. Finally, a weight may also be given to each event (e.g., number of accidents at intersections or the number of floors in a building). The weight is the quantity that will spread across the study region through the kernel density estimation and the created surface. This type of analysis might be useful for cross-comparisons with other aerial data – for example, to study respiratory problems of people (weighted or unweighted events) in relation to PM2.5 (particulate matter) by using an air pollutant raster map to identify potential spatial correlations and overlapping.

3.4   Chapter Concluding Remarks

- Point pattern analysis identifies if a point pattern is random, clustered or dispersed. For analyzing the spatial distribution of the attribute values of spatial objects, we apply spatial autocorrelation methods presented in Chapter 4.

- Standard distance is a measure of dispersion (spread). It is a distance expressing the compactness of a set of spatial objects.
- The standard deviational ellipse is more descriptive and more widely used than standard distance.
- Spatial outliers reveal anomalies in a spatial dataset that should be further examined.
- By describing the spatial distribution of events, we attempt to identify the spatial process that leads to this formation.
- Complete spatial randomness process (also called the independent random process, or IRP), is a process wherein spatial objects (or their attribute values) are scattered over the geographical space randomly.
- The first-order spatial variation effect occurs when the values or the location of spatial objects vary from place to place due to local effect of space.
- The second-order spatial variation effect occurs when there is an interaction among nearby locations. The location, or the value of an observation, is highly influenced by a nearby location or the value of a neighboring observation (the independence assumption of IRP no longer holds).
- Stationarity is the state where (a) every location in space has an equal probability of event placement through a spatial process and (b) there is no interaction among events (independence).
- An anisotropic process occurs when intensity varies according to specific directions.
- *L* and *K* functions summarize clustering or dispersion over a range of different distances. This assists in assessing the proper scale of analysis.
- The main advantage of kernel density estimation is that the resulting density function is continuous at all points along the scale.
- Kernel density estimation weights nearby events more heavily than those lying further away using a weighting function, thus applying an attenuating effect in space.
- Kernel density estimates the density of the points rather than their values.
- The kernel function calculates the probability density of an event at some distance based on an observed reference point event.
- A weak point of kernel density estimation is the subjectivity in the definition of the bandwidth factor *h*.

Questions and Answers

The answers given here are brief. For more thorough answers, refer back to the relevant sections of this chapter.

Q1. What are centrographic statistics? Name the most common centrographic spatial statistics tools. What is their main difference from the corresponding descriptive statistics?

A1. Centrographic statistics are tools used to analyze geographic distributions by measuring the center, dispersion and directional trend of a spatial arrangement. The centrographic statistics in most common use are the mean center, median center, central feature, standard distance and standard deviational ellipse. Centrographic statistics are calculated based on the location of each feature, which is their major difference from descriptive statistics, which concern only the nonspatial attributes of spatial features.

Q2. What are the main differences between locational and spatial outliers?

A2. A spatial outlier should not be confused with a locational outlier. To detect locational outliers, we analyze only the distance of a spatial entity to its neighbors. No other attribute analysis takes place. To detect spatial outliers, we study if an attribute value deviates significantly from the attribute values of the neighboring entities.

Q3. What are the main characteristics of a spatial outlier?

A3. A spatial outlier does not need to be a locational outlier. Additionally, a specific entity may be labeled as a spatial outlier only for a single attribute, while other attribute values might not deviate from the corresponding attribute values of other neighboring entities. Finally, a spatial outlier is not necessarily a global outlier as well, as the spatial outlier is always defined inside a predefined neighborhood

Q4. What are the assumptions of complete spatial randomness?

A4. There are two basic assumptions of complete spatial randomness:
   (a) There is an equal probability of event occurrence at any location in the study region (also called first-order stationary).
   (b) The location of an event is independent of the locations of other events (also called second-order stationary).

Q5. What are the main spatial point patterns and the associated spatial processes they are generated from?

A5. Random: In this type of arrangement, events are randomly scattered all over the study area. This is the result of a random process. Dispersed: The events are located uniformly around the study area. This is the result of a competitive process. Clustered: The events create clusters in some parts of the study area, and the pattern has a large variation. This is the result of an aggregating process.

Q6. Why is complete spatial randomness important for spatial statistics?

A6. If the data exhibit spatial randomness, there is no pattern or underlying process. Further geographical analysis is thus unnecessary. If the data do not exhibit spatial randomness, then they do not have the same probability of occurrence in space (first-order effect presence) and/or the location of an event depends on the location of other events (second-order effect presence). In this case, spatial analysis is very important, as space has an effect on event occurrences and their attribute values.

Q7. What are the main methods of point pattern analysis, and what is their calculation based on?

A7. Distance-based methods employ the distances among events and describe second-order effects. Such methods include the nearest neighbor method, the G and F distance functions, the Ripley's K distance function and its transformation, the L function. Density-based methods use the intensity of event occurrence across space. For this reason, they describe first-order effects better (quadrat count methods, kernel estimation). Other methods include proximity polygons by a combination of density and distance (Delaunay triangulation, Gabriel graph, minimum spanning tree) and hot spot analysis to identify if clusters of values of a specific variable are formed in space.

Q8. What is the Ripley's K function, and why it is used? What is its main advantage?

A8. Ripley's K function is a spatial analysis method of analyzing point patterns based on a distance function. It can be used to (a) decide if the observed point pattern is random, dispersed or clustered at a specific distance or a range of distances; and (b) identify the distance at which the clustering or dispersion is more pronounced. Ripley's K function illustrates how the spatial clustering or dispersion of features' centroids changes when the neighborhood size changes by summarizing the clustering or dispersion over a range of different distances. This is very helpful because it assists us in assessing the proper scale of analysis for the problem at hand.

Q9. What is kernel density estimation, and how is it mapped?

A9. Kernel density estimation is a nonparametric method that uses kernel functions to create smooth maps of density values, in which the density at each location indicates the concentration of points within the neighboring area (high concentrations as peaks, low concentrations as valleys). The kernel function provides an interpolation technique for assessing the impact of each event on its neighborhood (as defined by the bandwidth). In practice, the kernel function is applied to a limited area around each event defined by the bandwidth h to "spread" its effect across space. The density estimates can be displayed by either surface maps or contour maps that show the intensity at all locations.

Q10. When should kernel density estimation be used?

A10. Kernel density estimation is used to create smooth surfaces that depict, first, the density of events and, second, an estimation of areas of higher or lower event occurrence intensity. Kernel density estimation can be used in spatial analysis for hot spot and cold spot identification. It can also be applied for clusters analysis of weighted or unweighted events in epidemiology, criminology, demography, ethnology and urban analysis. Through the raster map visualization of a kernel density estimation, one can quickly locate hot spot and cold spot areas and places of high or lower density.

LAB 3
SPATIAL STATISTICS: MEASURING GEOGRAPHIC DISTRIBUTIONS

Overall Progress

Spatial Analysis/Lab Workflow

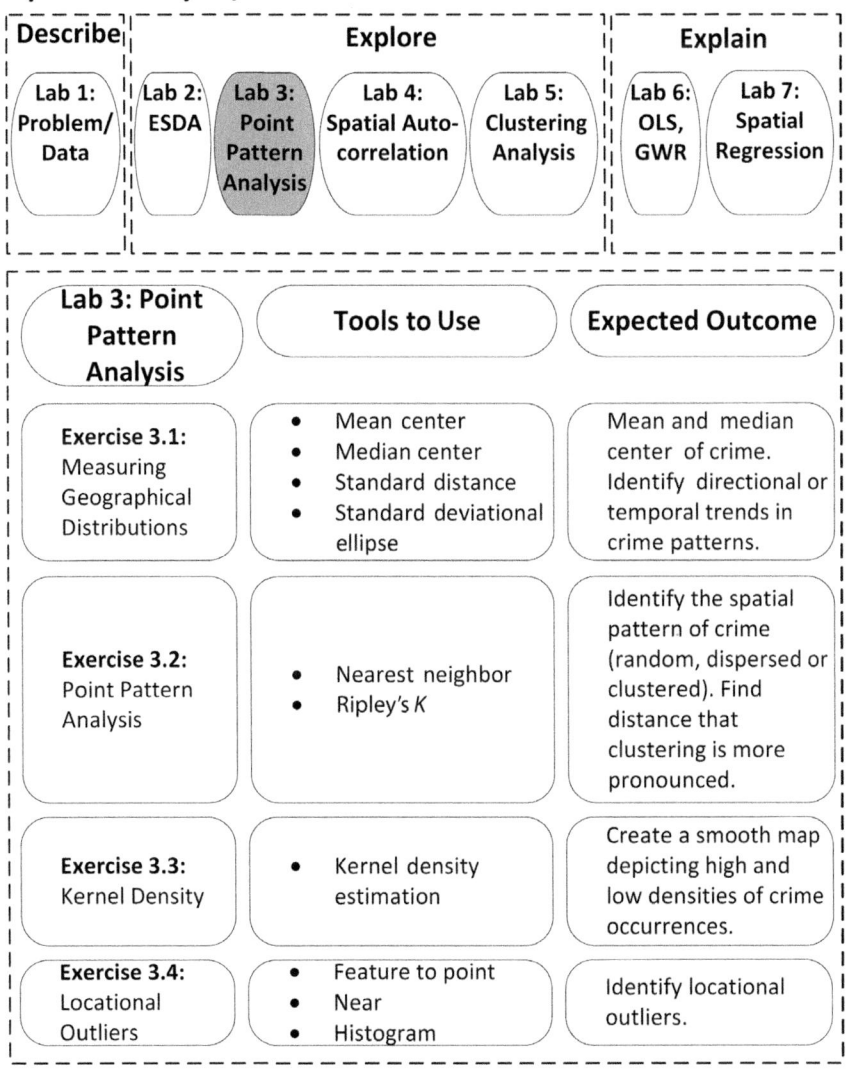

Figure 3.17 Lab 3 workflow and overall progress.

Scope of the Analysis: Crime Analysis

This lab deals with

- **Objective 2:** Locating low-crime areas (see Table 1.2)

In this exercise we apply various spatial statistics (centrographics, nearest neighbor, Ripley's *K*, kernel density estimation) to analyze the spatial patterns of assaults (see Figure 3.17). The crime of assault is considered an act of violence that involves intentional harm. The overall analysis identifies high-crime areas that should be excluded and low-crime areas suitable for locating a coffee shop. Locational outliers are also traced. In Chapter 4, we more thoroughly study other types of crime through spatial autocorrelation and clustering analysis. The following exercises are carried out through ArcGIS only, as GeoDa does not offer point pattern analysis functionalities.

Exercise 3.1 Measuring Geographic Distributions

In this exercise, we calculate the mean and median center of the distribution of assaults and identify directional trends.

ArcGIS Tools to be used: Mean center, Median center, Standard distance, Standard deviational ellipse

ACTION: Calculate the Mean center and Median center

Navigate to the location you have stored the book dataset and click Lab3_SpatialStatistics.mxd

Main Menu > File > Save As > My_Lab3_ SpatialStatistics.mxd

In I:\BookLabs\Lab3\Output

ArcToolBox > Spatial Statistics Tools > Measuring Geographic Distributions > Mean Center (see Figure 3.18)

Input Feature Class = Assaults (see figure 3.19)

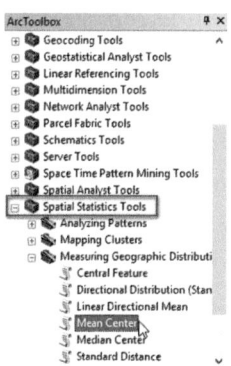

Figure 3.18 Measuring geographic distributions toolbox.

Exercise 3.1 (*cont.*)

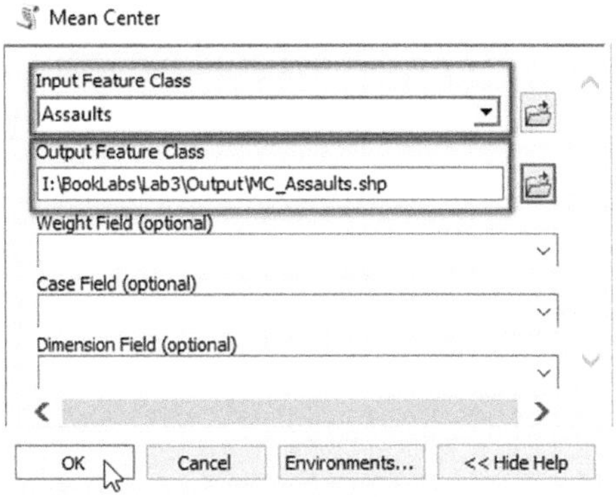

Figure 3.19 Mean center dialog box.

Output Feature Class = MC_Assaults (stored in the output folder of Lab3 as MC_Assaults.shp

Leave all other fields blank/default

OK

Do the same for Median Center

ArcToolBox > Spatial Statistics Tools > Measuring Geographic Distributions > Median Center

Input Feature Class = Assaults

Output Feature Class = MedC_Assaults (stored in the output folder of Lab3 as MedC_Assaults.shp)

Leave all other fields blank/default

OK

Change the symbols of the resulted shapefiles to triangles

TOC > Click on MC_Assaults > Select Triangle 3 > Color = Green > Size = 12 > OK

Click on MedC_Assaults > Select Square 3 > Color = Green > Size = 12 > OK

Interpreting results: The mean center and median center lie close to each other, very close to the western downtown area (see Figure 3.20). A quick graphical inspection shows that assaults occur mainly in the central western, northern and southern postcodes of the city. Central-eastern and eastern postcodes have very few assault events. Areas with a higher

Exercise 3.1 *(cont.)*

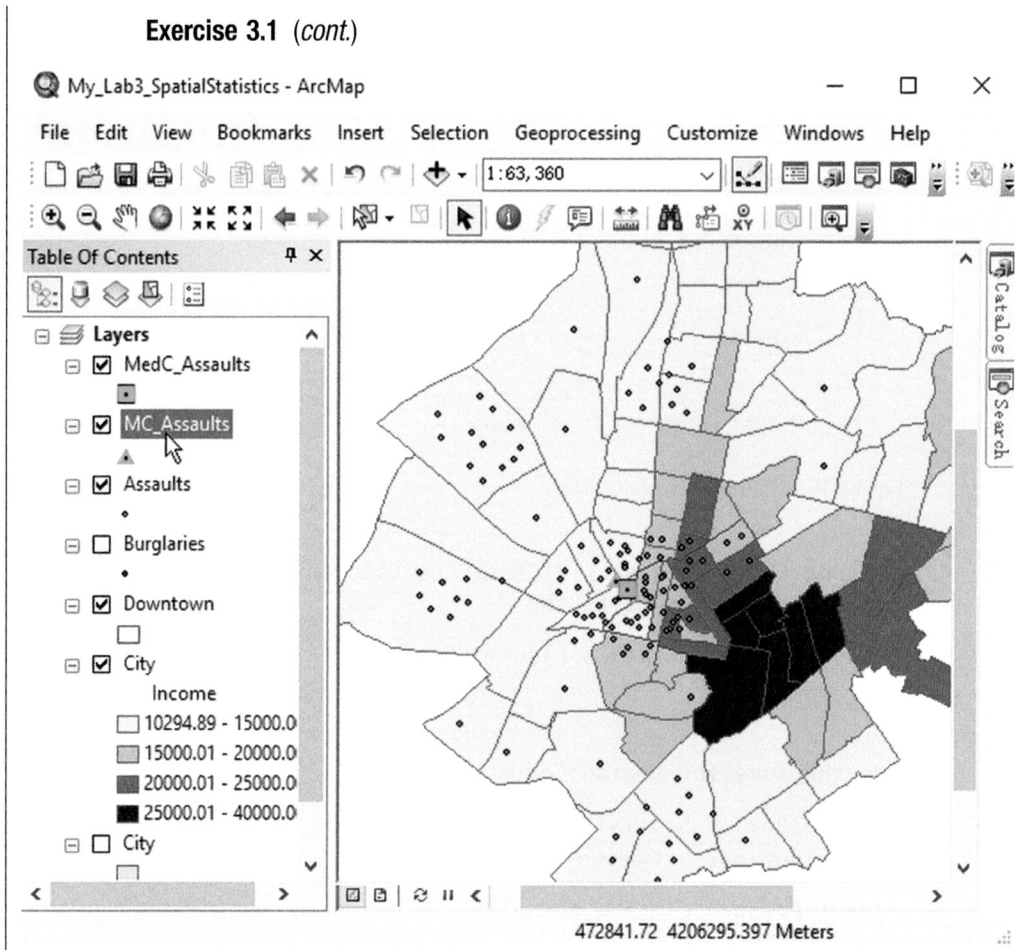

Figure 3.20 Assault events with mean and median center.

concentration of assaults might be removed as location candidates for the new coffee shop. At this point, the mean and median center do not offer much valuable information.

ACTION: Calculate Standard distance and Standard deviational ellipse

ArcToolBox > Spatial Statistics Tools > Measuring Geographic Distributions > Standard Distance

Input Feature Class = Assaults (see Figure 3.21)

Output Feature Class = SD_Assaults (stored in the output folder of Lab3 as I:\BookLabs\Lab3\Output\SD_Assaults.shp)

Leave all other fields blank/default

OK

Exercise 3.1 (*cont.*)

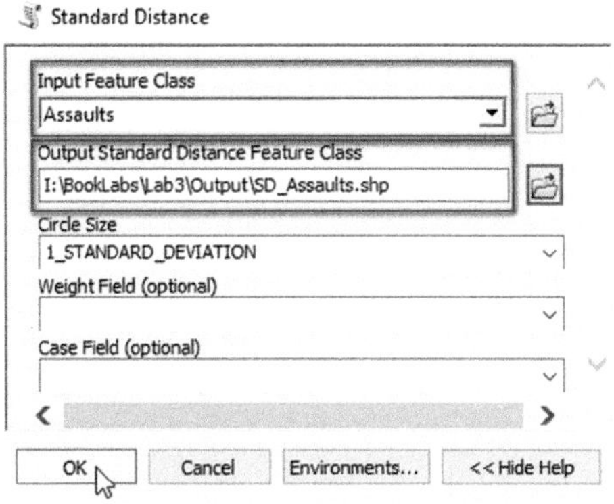

Figure 3.21 Standard distance dialog box.

Change the color of the standard distance shapefile

TOC > Click on SD_Assaults > Outline Color Color = Green > OK

Calculate how many events lie inside one Standard deviation distance

Main Menu > Selection > Select by location >

Target Layer = Assaults (see figure 3.22)

Source layer = SD_Assaults

Spatial selection method = are within the source layer file

OK

TOC > RC Assaults (see figure 3.23) > Open Attribute Table

Check in the lower right corner of the table: 77 out of 129 selected (which is 59.68%)

Close the table

Main Menu > Selection > Clear Selected Features

Calculate Standard Deviational Ellipse

ArcToolBox > Spatial Statistics Tools > Measuring Geographic Distributions > Directional Distribution

Input Feature Class = Assaults (see Figure 3.24)

Output Feature Class = SDE_Assaults (stored in the output folder of Lab3 as I:\BookLabs\Lab3\Output\SDE_Assaults.shp)

Exercise 3.1 *(cont.)*

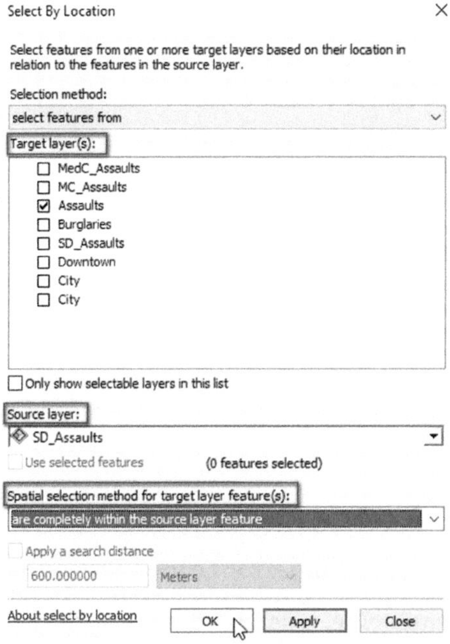

Select By Location   ✕

Select features from one or more target layers based on their location in
relation to the features in the source layer.

Selection method:

| select features from   ⌄ |

Target layer(s):

☐ MedC_Assaults
☐ MC_Assaults
☑ Assaults
☐ Burglaries
☐ SD_Assaults
☐ Downtown
☐ City
☐ City

☐ Only show selectable layers in this list

Source layer:

◈ SD_Assaults   ▾

☐ Use selected features   (0 features selected)

Spatial selection method for target layer feature(s):

are completely within the source layer feature   ⌄

☐ Apply a search distance

600.000000   Meters   ⌄

About select by location   [OK ⌐]   [Apply]   [Close]

Figure 3.22 Select be location tool.

```
Leave all other fields blank/default

OK
```

Calculate how many events lie inside one Standard deviational
ellipse distance

```
Main Menu > Selection > Select by location >

Target Layer = Assaults

Select layer = SDE_Assaults

Spatial selection method = are within the source layer file

OK

TOC > RC Assault > Open Attribute Table
```

Check in the lower right corner of the table: 75 out of
129 selected (58.13%)

```
Main Menu > Selection > Clear Selected Features

Close attribute table

Main Menu > File > Save
```

Exercise 3.1 (*cont.*)

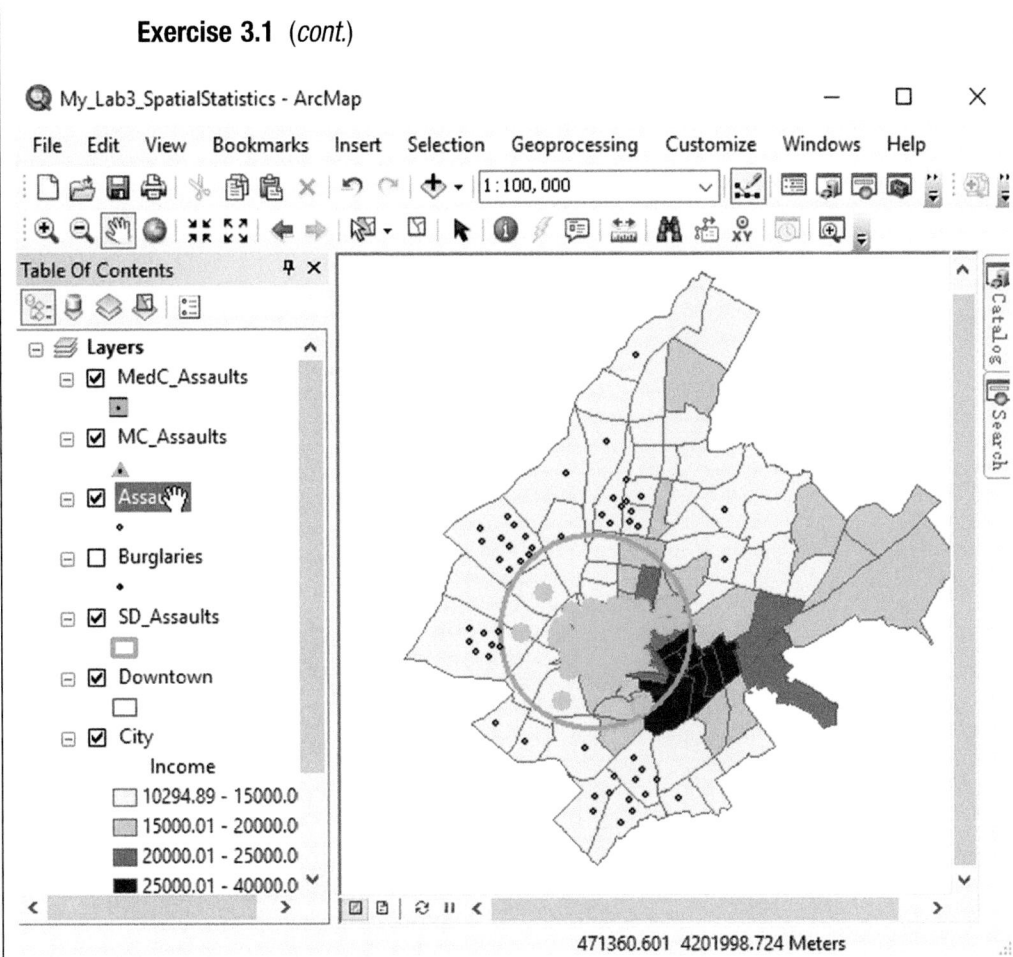

Figure 3.23 Standard distance (SD = 1,537.10 m) of assaults centered over the mean center; 77 out of 129 (highlighted) lie within one standard distance from the mean center.

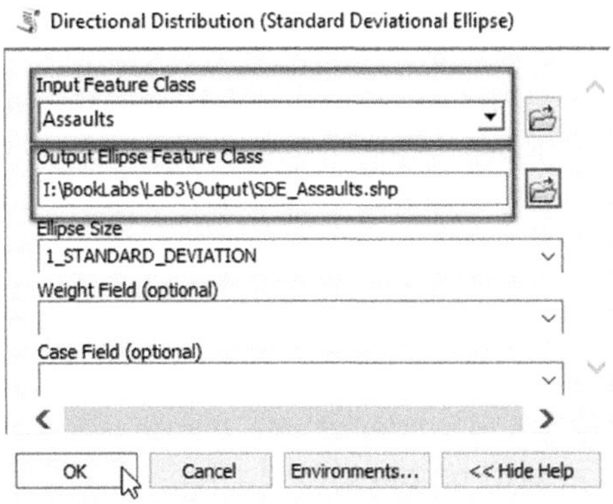

Figure 3.24 Standard deviational ellipse dialog box.

Exercise 3.1 (*cont.*)

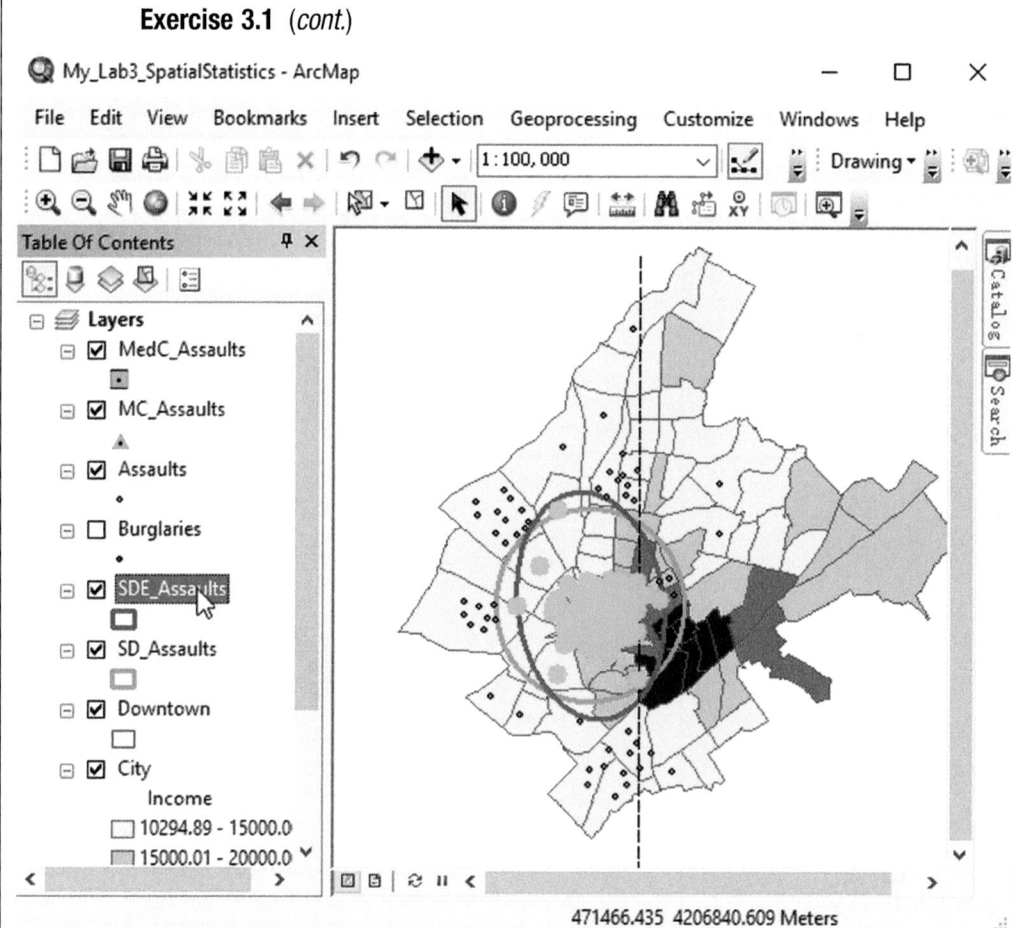

Figure 3.25 Assault events plotted along with standard distance and standard deviational ellipse. The ellipse reveals a south-to-north directional trend. The dashed line splits the study area in two. The majority of assaults lie in the left-hand side, while the right-hand side has only a few assault occurrences. The overall pattern shows substantial heterogeneity.

Interpreting results: The standard distance is 1,537.10 m (attribute table of SD_Assaults, last column; see Figure 3.23); 77 out of 129 events (59.68%) lie less than one standard distance from the mean center, a relatively small area compared to the city, revealing the assaults concentration. Areas further away have a lower risk of assault and are probably safer and suitable for locating the coffee house.

We also observe that there is a directional trend in the event distribution, which is depicted by using the standard deviational ellipse (see Figure 3.25). The results reveal a south-to-north tendency. Additionally, a one standard deviational ellipse covers 58.13% (75 out of 129) of the assaults. Although

Exercise 3.1 (*cont.*)

this is similar to the percentage produced by the standard distance, it is more informative because it also provides the direction, allowing for a better tracing of areas that might be excluded due to higher crime rates (see Figure 3.25). Another interesting finding is that the study area seems to be split in half. If we consider the north-to-south axis passing through the center of the downtown area (denoted by the red outline), most of the assaults lie on the left-hand side of the axis, while only a few lie on the right-hand side. This reveals heterogeneity in the west-to-east direction (see Figure 3.25). Furthermore, a graphical inspection (see Figure 3.25) shows clusters of events on the left-hand side. The centrographics used provide a description regarding central tendencies and directional trends but fail to spot the areas in which crime clusters away from the center. To better describe the assault point pattern, we apply more advanced tools in the following exercises.

Exercise 3.2 Point Pattern Analysis

In this exercise, we go beyond just describing the geographic distribution of assaults (in Exercise 3.1). We analyze the spatial pattern to identify whether it is clustered, random or dispersed. We also investigate if any specific spatial process generates the observed pattern and if this pattern is of any particular importance to our project. Finally, we determine the scale of our analysis.

ArcGIS Tools to be used: Average nearest neighbor, Ripley's K
ACTION: Average Nearest Neighborhood

Navigate to the location you have stored the book dataset and click

My_Lab3_ SpatialStatistics.mxd

ArcToolBox > Spatial Statistics Tools > Analyzing Patterns > Average Nearest Neighbor

Input Feature Class = Assaults (see Figure 3.26)

Distance Method = Euclidean Distance

Check Generate Report

Leave all other fields blank/default

OK

Exercise 3.2 *(cont.)*

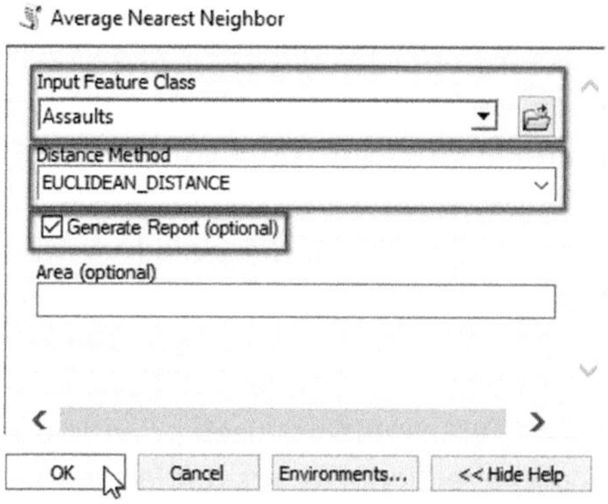

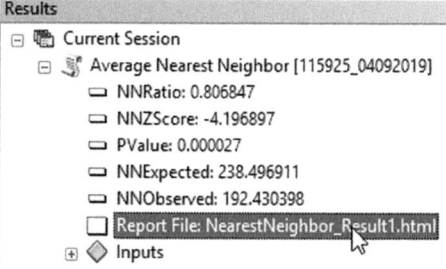

Figure 3.26 Average nearest neighbor dialog box.

To open the report, go to Results window. If Results window is not activated:

Main Menu > Geoprocessing > Results

Current Session > Click at the plus sign at the left of Average Nearest Neighbor > Double Click Report File: NearestNeighbor_-Result.html (see Figure 3.27)

Figure 3.27 Average nearest neighbor dialog box. Results are: Nearest Neighbor Ratio = 0.806846, z-score = −4.196897, p-value = 0.000027.

Report opens

The z-score (−4.19) is less than −2.58, indicating a p-value smaller than 0.01 (far left of the graph; see Figure 3.28). There is a less than 1% likelihood that this clustered pattern could be the result of complete spatial randomness.

Close report and the Results window

Save

Exercise 3.2 *(cont.)*

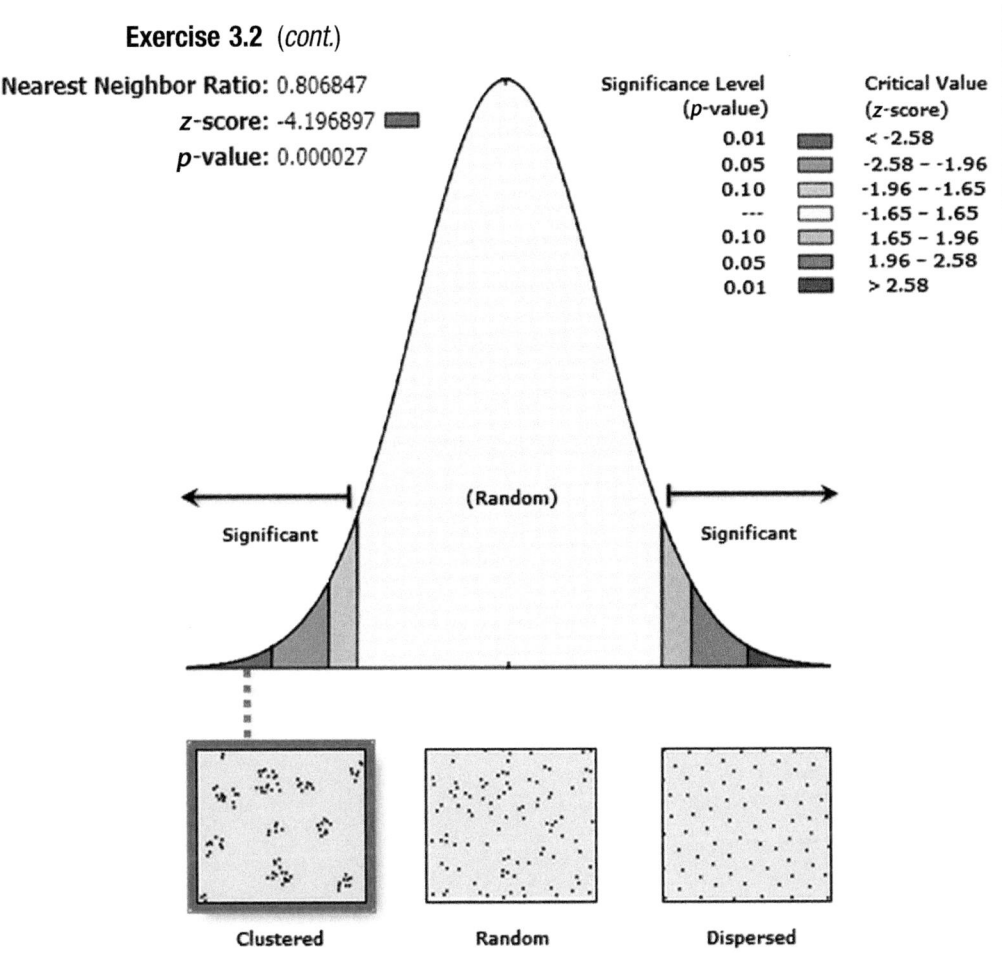

Figure 3.28 Graphical explanation of the average nearest neighbor statistic output.

Interpreting results: The nearest neighbor ratio is 0.806 (less than 1) and is statistically significant at the 0.01 *p*-value level, indicating a process toward clustering (see Figure 3.28). Given the z-score of −4.1968971298 and the *p*-value of 0.000027, there is a less than 1% likelihood (significance level α) that this clustered pattern could be the result of complete spatial randomness. In other words, we have a 99% probability (confidence level) that the distribution of assaults forms a clustered pattern (see Figure 3.28). Thus, crime is not distributed randomly across space, and some places have a higher probability of assaults occurrence than others. Nearest neighbor analysis does not trace where these clusters are. These areas should be located. To this end, we apply kernel density estimation (see Exercise 3.3) by defining an appropriate bandwidth *h*. One method of doing so is by using Ripley's *K* function and its *L* transformation, as shown next.

Exercise 3.2 (*cont.*)

ACTION: Ripley's K Function

Before we use Ripley's Function we create the boundary of the study area (needed for the tool).

Main Menu > Geoprocessing > Dissolve

Input Feature Class = City (see Figure 3.29)

Output Table = I:\BookLabs\Lab3\Output\Boundaries.shp

Leave all other fields blank/default

OK

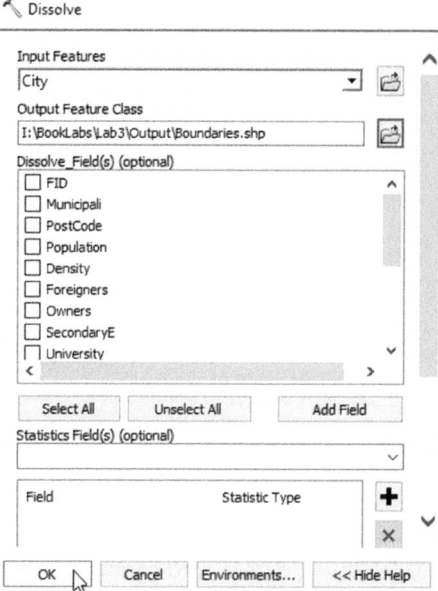

Figure 3.29 Dissolve GIS operation dialog box, needed to create the case study boundaries.

TOC > Click the polygon symbol of Boundaries > Select Hollow > Outline Width = 1 > OK

Now we proceed with K-Ripley's function.

ArcToolBox > Spatial Statistics Tools > Analyzing Patterns > Multi-Distance Spatial Cluster Analysis

Input Feature Class = Assaults (see Figure 3.30)

Output Table = I:\BookLabs\Lab3\Output\RipleyTableAssaults

Exercise 3.2 (*cont.*)

Number of Distance Bands = 10

Compute Confidence Envelope = 99_PERMUTATIONS

Check Display Results Graphically

Boundary Correction Methods = SIMULATE_OUTER_BOUNDARY_VALUES (This method is used to simulate points outside the study area so that the number of neighbors close to the boundaries is not underestimated.)

Study Area Method = USER_PROVIDED_STUDY_AREA_FEATURE_CLASS

Study Area Feature Class = Boundaries

Leave all other fields blank/default

OK

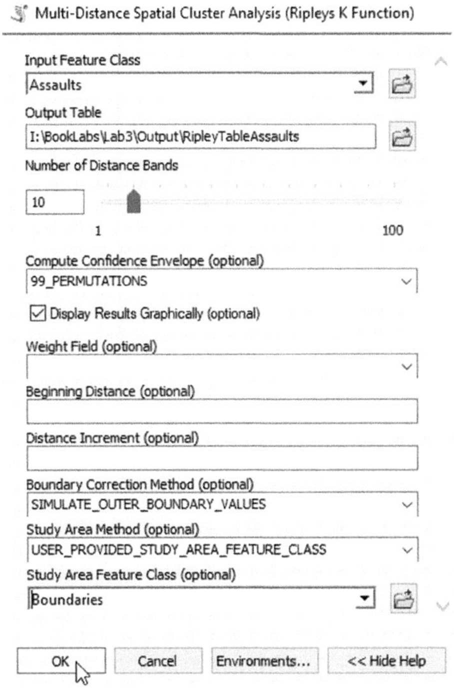

Figure 3.30 Ripley's *K* function dialog box.

Results are

k-Function Summary

Distance*	L(d)	Diff	Min L(d)	Max L(d)
201.88	501.77	299.89	181.75	277.62

Exercise 3.2 (*cont.*)

403.75	1009.37	605.62	384.11	496.62
605.63	1414.56	808.92	600.95	705.46
807.51	1717.79	910.28	813.69	914.76
1009.39	1944.48	935.09	1005.37	1132.73
1211.26	2091.78	880.52	1212.24	1344.59
1413.14	2225.09	811.95	1402.31	1542.64
1615.02	2367.96	752.95	1606.73	1751.01
1816.89	2543.14	726.24	1807.94	1943.34
2018.77	2738.94	720.17	1981.50	2144.12

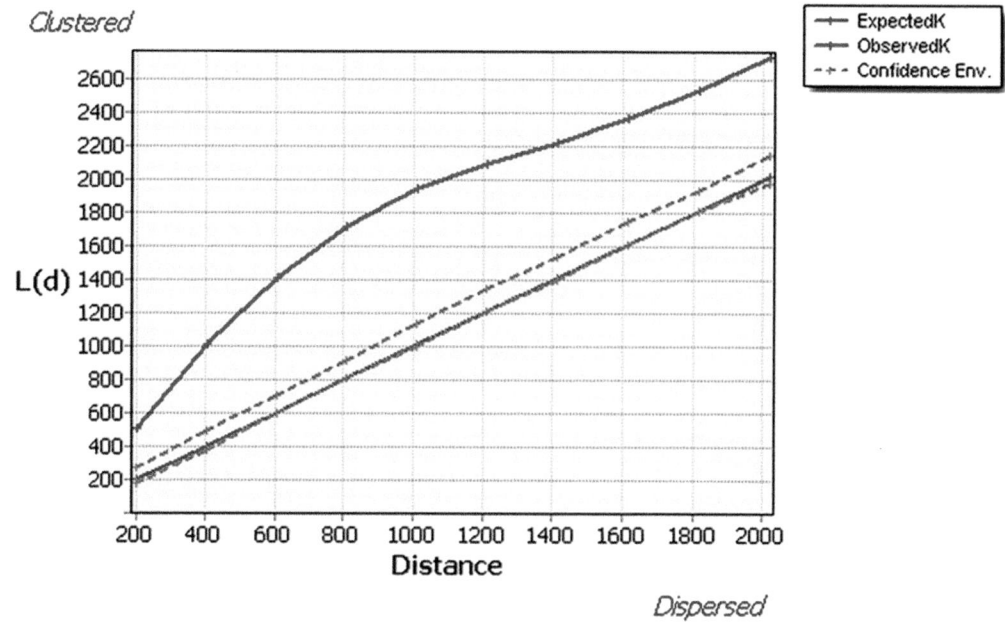

Figure 3.31 Ripley's *K* function graph. Distances range from 200 m to 2,000 m, with 10 intervals of around 202 m each. The observed *K* values are indicated by the red line.

RC on graph > Add to Layout > Return to Data View > Close graph

Results are saved on the RipleyTableAssaults table.

TOC > RC RipleyTableAssaults > Open

Main Menu > File > Save

Exercise 3.2 (*cont.*)

Interpreting results: The observed *K* value is larger than the expected value for all distance values within the range (200 m–2,000 m), indicating that the distribution of the events tends to be a clustered pattern rather than a random one (see Figure 3.31). The observed *K* value is larger than the upper confidence envelope (MaxL(d)) value for all distances, showing that the spatial clustering for each distance is statistically significant. The Ripley's *K* function indicates that there is a statistically significant clustering of assaults at various distances. The larger the difference between the observed and the expected value, the more intense the clustering at that distance (Diff). As such, the spatial clustering of assaults is more evident at 1,009.39 m (where the difference is maximized; see results listed earlier). From the analysis perspective, this means that, if we would like to analyze this point pattern further, a good scale of analysis (distance at which spatial weights are calculated for spatial statistics) would be 1 km. We could also use a smaller distance, such as 807.51 m, which is the second largest difference. It is not easy to define which distance bandwidth is the most appropriate; with the Ripley's *K* function, however, we obtain a measure of the distances at which clustering seems to be more pronounced. In the next exercise, the 1 km bandwidth is applied, but you can experiment with other bandwidths, using the preceding table as a guide and then analyzing the results in the context of the specific analysis. Therefore, there is no correct bandwidth value; the choice largely depends on the problem, the scope and the regional or local type of the analysis.

Exercise 3.3 Kernel Density Estimation

In this exercise, we use kernel density estimation to create a smooth map of density values, in which the density at each location indicates the concentration of assaults (high concentrations of assaults as peaks, low concentrations of assaults as valleys; see Figure 3.36). Kernel density tools reflect the probability of event occurrence, which is very useful for identifying the areas that have a high or low risk of assault.

   ArcGIS Tools to be used: Kernel density estimation, Reclassify, Raster to Polygon

ACTION: Kernel density estimation

Navigate to the location you have stored the book dataset and click

Exercise 3.3 *(cont.)*

My_Lab3_ SpatialStatistics.mxd

ArcToolBox > Spatial Analyst Tools > Density > Kernel Density

Input point = Assaults (see Figure 3.32)

Population field = NONE

Output Raster = E:\BookLabs\Lab3\Output\KDEAssault1km

Output cell size = Let default

Search radius = 1000

Leave the other fields as default

OK

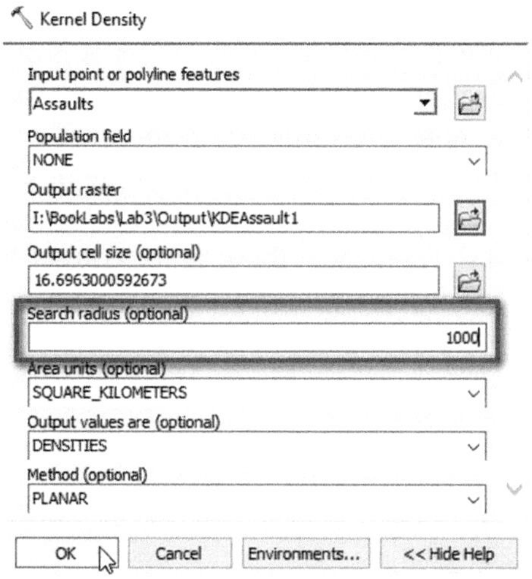

Figure 3.32 Kernel density dialog box.

TOC > Drag the layer above the City layer (You should click first the List By Drawing Order button) (see Figure 3.33)

Main Menu > File > Save

Interpreting results: The kernel density raster output at a 1 km bandwidth is depicted in Figure 3.33 (bandwidth estimated by the Ripley's *K* function in Exercise 3.2). This map highlights areas at higher and lower risk of assault occurrence. Several assault hot spots are clearly identified. Light grey areas (valleys) indicate low-intensity crime areas. The darker the grey color (peaks),

Exercise 3.3 *(cont.)*

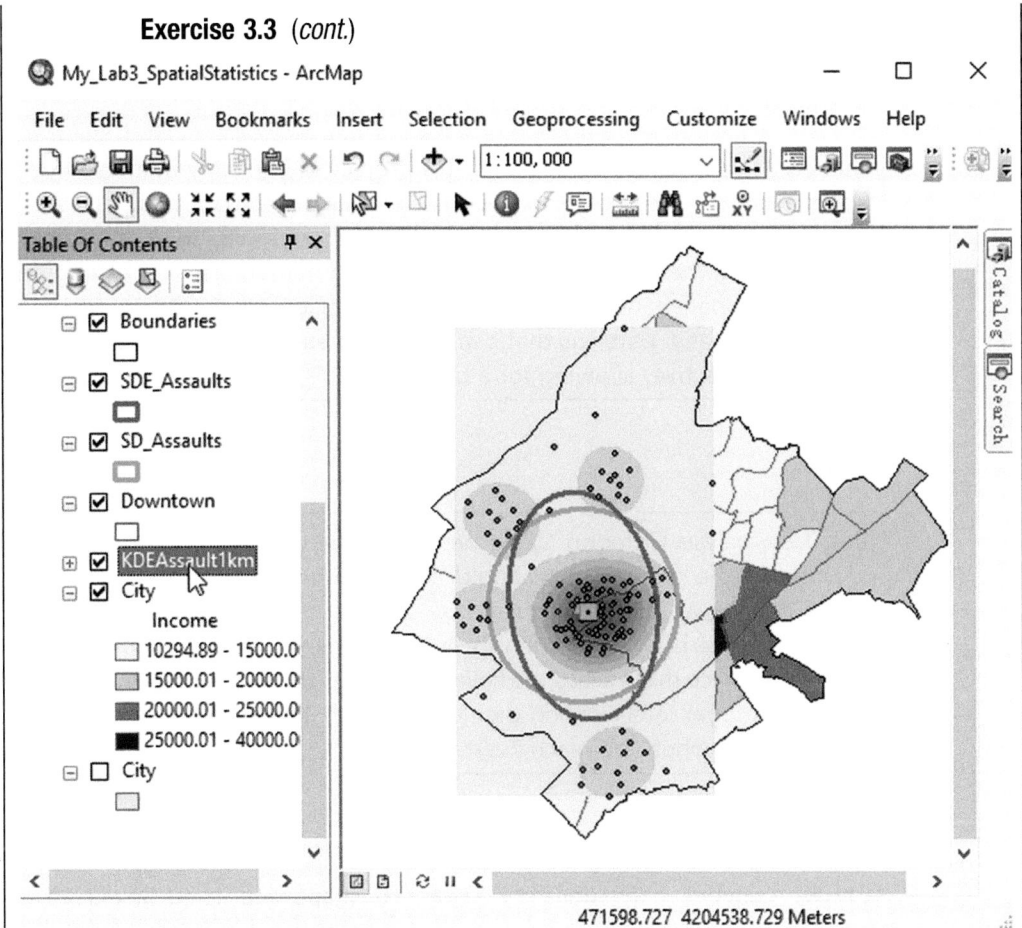

Figure 3.33 Kernel density estimation map of crimes. Crime events, mean center, median center, standard distance and standard deviational ellipse are also overlaid.

the higher the intensity of assaults and the probability of crime occurrence. The crime pattern resembles a polycentric model. There is one large central crime hot spot covering almost half of the downtown area (its western part) and four smaller hot spots allocated around the main hot spot at its western, northern and southern sides. The raster file does not cover several parts of the map. In fact, the intensity is not estimated for the eastern postcodes of the city because no assault events were recorded there. However, the outcome of the kernel density applies only to its coverage area and cannot be generalized for the uncovered area. In the context of the one-year time-span the crime data refer to, we can state that no assaults were reported at specific places, and density maps cannot be produced. Nevertheless, the current crime data clearly indicate areas of crime clusters and areas where crime is likely to emerge in the future. Areas of high crime intensity should be avoided when locating the coffee shop (see Boxes 3.1 and 3.2).

Exercise 3.3 (cont.)

> **Box 3.1** It is beyond the scope of this project to explain why these hot spots exist in these specific areas. Further analysis could investigate the demographic and socioeconomic profiles of these areas, the urban patterns formed, the living conditions and many other factors. This example shows that spatial analysis assists in identifying, locating and quantifying interesting spatial patterns that can be further studied from a multidisciplinary perspective, allowing for a better analysis of various phenomena.

> **Box 3.2** Analysis Criterion C2 to be used in synthesis Exercise 5.4: The location of the coffee shop should not lie within areas of high assault densities. [C2_AssaultsHighDensity.shp].
>
> We regard as high assault density areas all those with a KDE value larger than 20. To trace these areas, we have to reclassify the raster image to group values in integer intervals and then convert it to shapefile. You may complete this task when you reach Exercise 5.4.

ACTION: Convert high-density areas to shapefile (see Box 3.2)

ArcToolBox > Spatial Analyst Tools > Reclass > Reclassify

Input Raster = KDEAssault1km (see Figure 3.34)

Reclass field = VALUE

Classify > Classes = 2 > Break Values = 20 (only change first row) > OK

Output raster > In the window that opens create a new Geodatabase (by selecting the relevant icon) within Lab3\Output and name it Raster.gdb > DC Raster.gdb and type RecKDE (This saves the file to I:\BookLabs\Lab3\Output\Raster.gdb\RecKDE)

OK

Exercise 3.3 *(cont.)*

✎ Reclassify

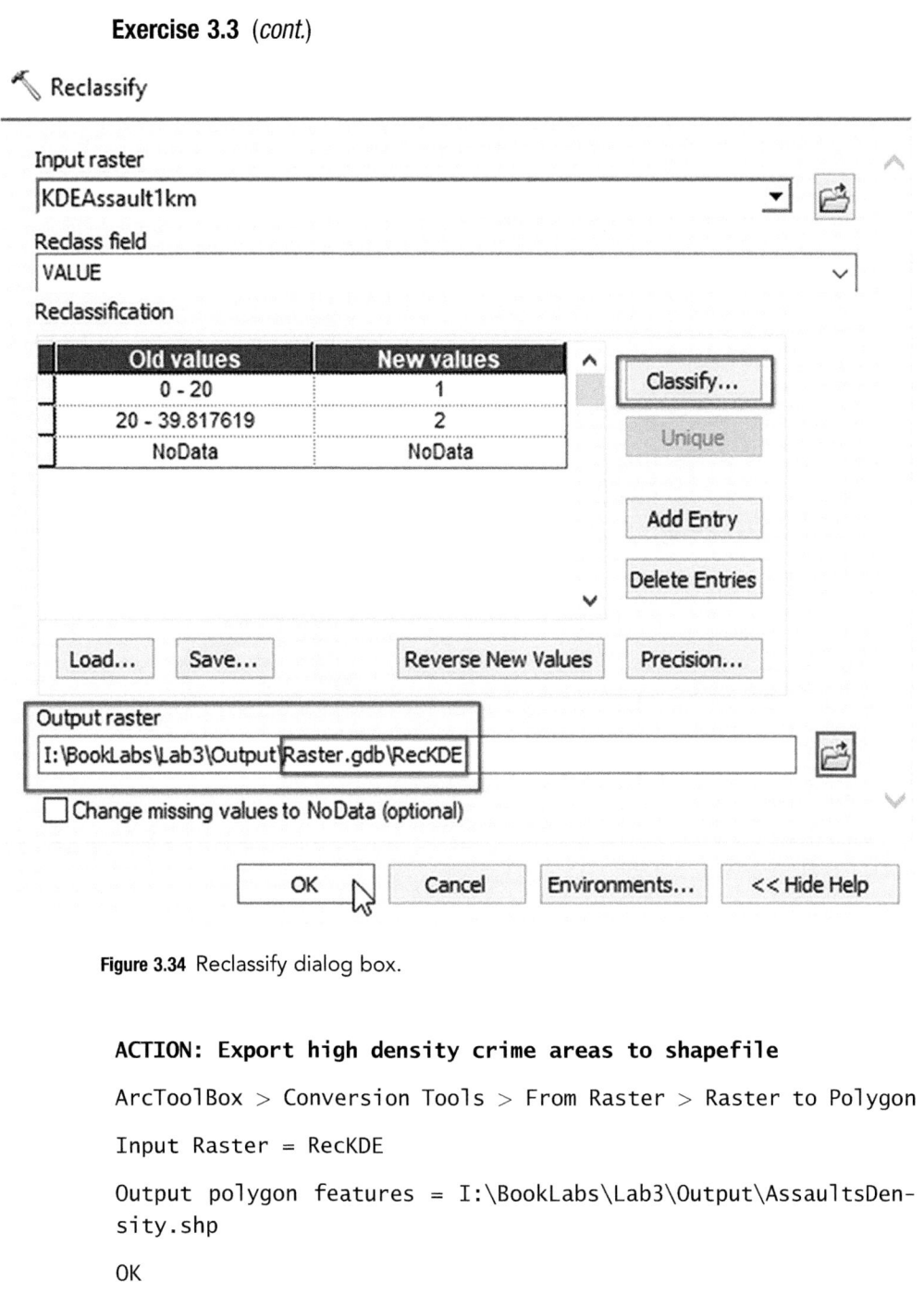

Input raster

KDEAssault1km

Reclass field

VALUE

Reclassification

Old values	New values
0 - 20	1
20 - 39.817619	2
NoData	NoData

Classify...

Unique

Add Entry

Delete Entries

Load... Save... Reverse New Values Precision...

Output raster

I:\BookLabs\Lab3\Output Raster.gdb\RecKDE

☐ Change missing values to NoData (optional)

OK   Cancel   Environments...   << Hide Help

Figure 3.34 Reclassify dialog box.

ACTION: Export high density crime areas to shapefile

ArcToolBox > Conversion Tools > From Raster > Raster to Polygon

Input Raster = RecKDE

Output polygon features = I:\BookLabs\Lab3\Output\AssaultsDen-sity.shp

OK

Exercise 3.3 (*cont.*)

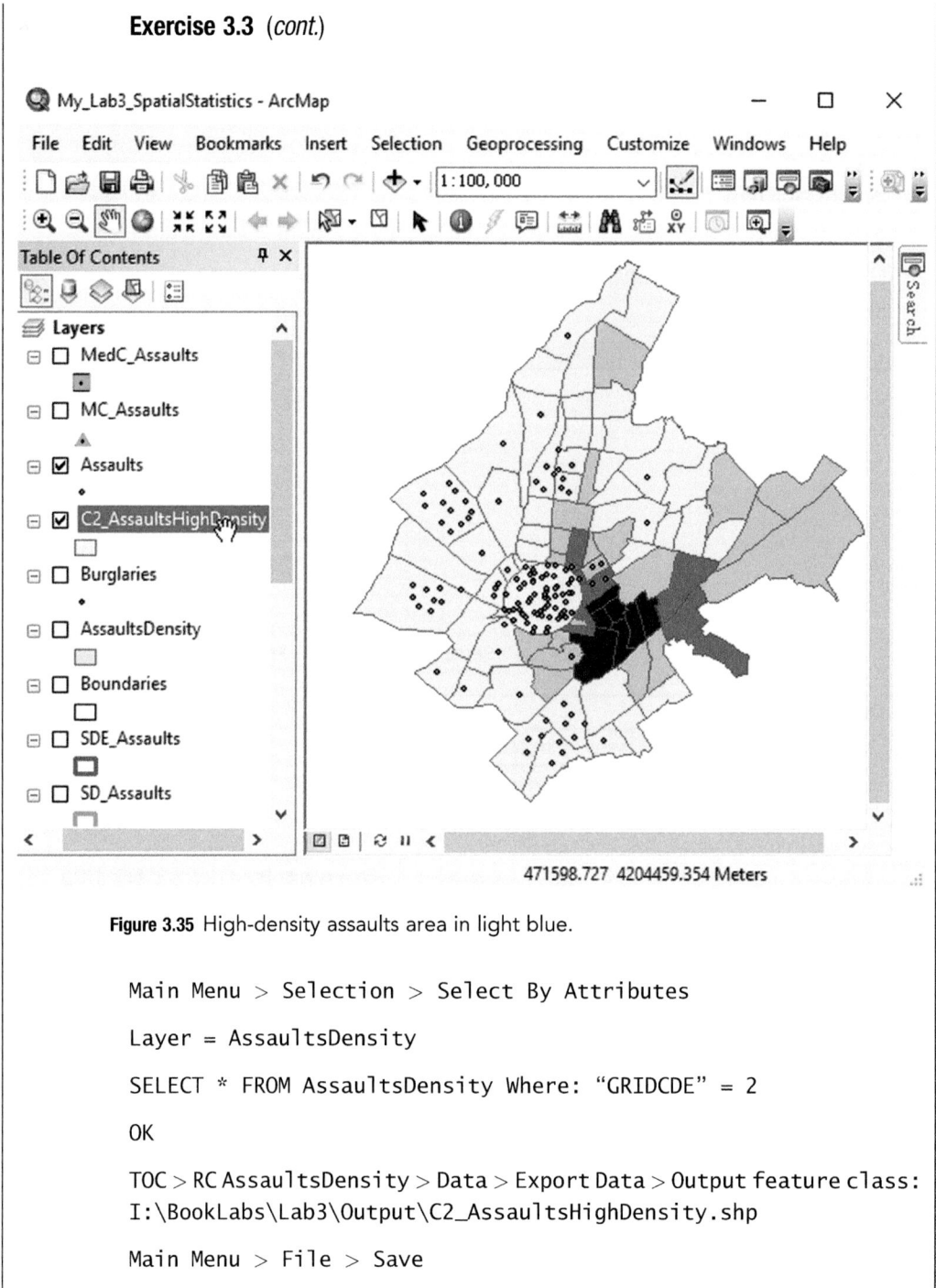

Figure 3.35 High-density assaults area in light blue.

Main Menu > Selection > Select By Attributes

Layer = AssaultsDensity

SELECT * FROM AssaultsDensity Where: "GRIDCDE" = 2

OK

TOC > RC AssaultsDensity > Data > Export Data > Output feature class:
I:\BookLabs\Lab3\Output\C2_AssaultsHighDensity.shp

Main Menu > File > Save

Exercise 3.4 Locational Outliers

In this exercise, we determine if any postcodes can be characterized as locational outliers. The outcome of this exercise will be used in Lab 4 for spatial autocorrelation estimation. First, we calculate the centroids of the polygons, and then we calculate the distance of every centroid (point) to its nearest one. ArcGIS considers outliers to be those objects that lie more than three standard distances away from the closest one.

ArcGIS Tools to be used: Feature To Point, Near, Histogram.

ACTION: Calculate centroids

Navigate to the location you have stored the book dataset and click

My_Lab3_ SpatialStatistics.mxd

ArcToolBox > Data Management Tools > Features > Feature To Point

Input Features = City (see Figure 3.36)

Output Feature Class = I:\BookLabs\Lab3\Output\CityCentroids.shp

OK

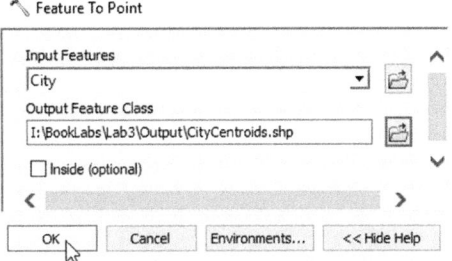

Figure 3.36 Feature to point dialog box.

ACTION: Calculate nearest neighbor distances

ArcToolBox > Analysis Tools > Proximity > Near

Input Features = CityCentroids

Exercise 3.4 (*cont.*)

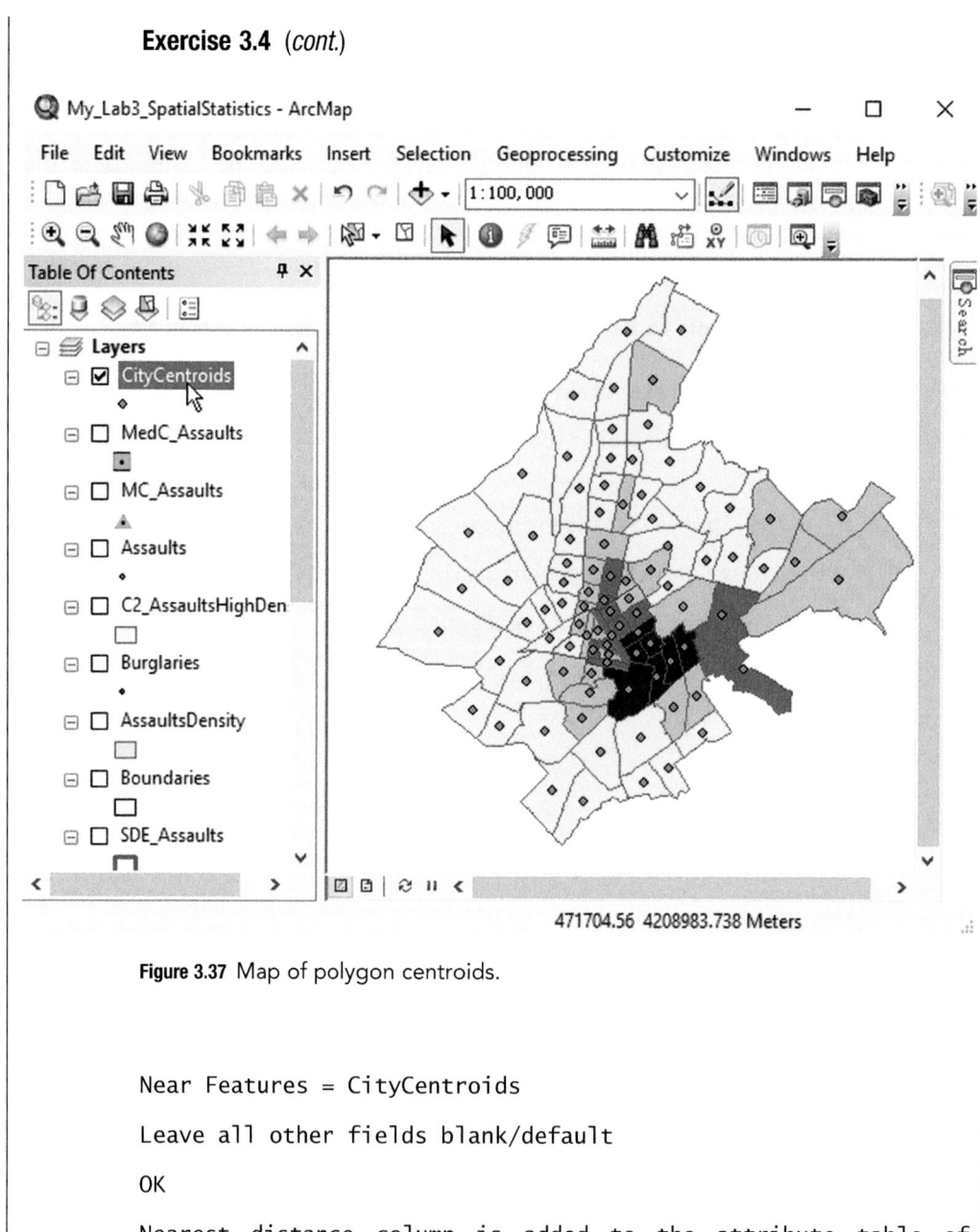

Figure 3.37 Map of polygon centroids.

Near Features = CityCentroids

Leave all other fields blank/default

OK

Nearest distance column is added to the attribute table of CityCentroids (see Figure 3.38)

Exercise 3.4 *(cont.)*

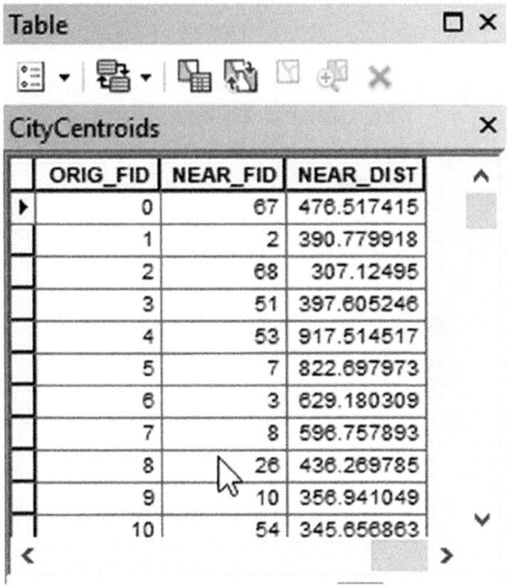

ORIG_FID	NEAR_FID	NEAR_DIST
0	67	476.517415
1	2	390.779918
2	68	307.12495
3	51	397.605246
4	53	917.514517
5	7	822.697973
6	3	629.180309
7	8	596.757893
8	26	436.269785
9	10	356.941049
10	54	345.656863

Figure 3.38 Nearest neighbor ID and nearest distance for each point are added to the attribute table of the `CityCentroids` layer.

ACTION: Calculate nearest neighbor distance average and standard deviation

TOC > RC CityCentroids > Open Attribute Table > RC on NEAR_DIST column > Statistics

Close Statistics table

Close Table

Main Menu > File > Save

Interpreting results: The distribution of the variable "Nearest neighbor distance" (the distance of each point to its nearest neighbor) is slightly skewed, with an average nearest neighbor distance of 468.20 m and a standard deviation of 206.03 m (see Figure 3.39). We regard as locational outliers those lying more than three standard deviations from the mean, which is 468.20 +3*206.03 = 1,086.29 m. In this dataset, no object (post

Exercise 3.4 (cont.)

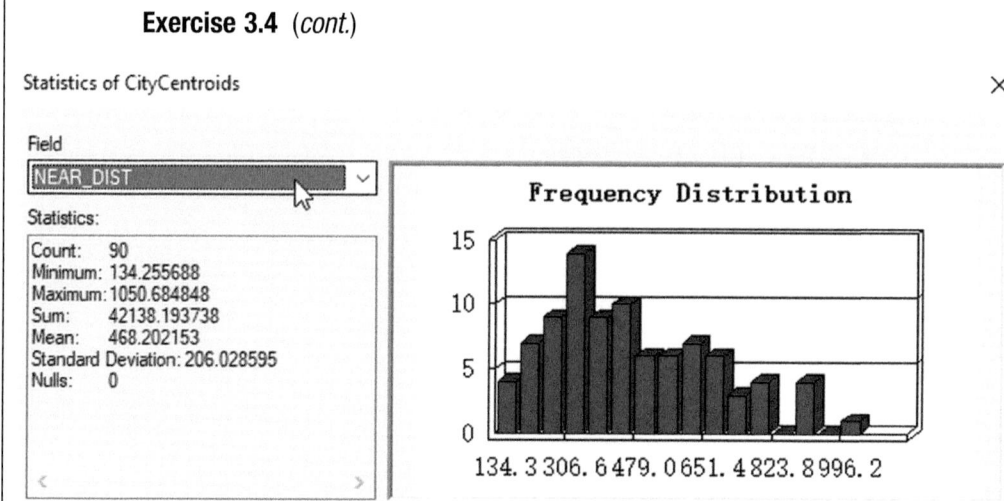

Figure 3.39 Frequency distribution of the variable "Nearest neighbor distance."

code) lies more than 1,086.69 from its nearest neighbor, so no locational outlier is identified. The existence of locational outliers can be easily detected by using the Optimized Hot Spot Analysis Tool (see the next chapter). However, this tool does not indicate which objects are outliers; it indicates only how many outliers exist. Sometimes, this is sufficient. However, if we need to examine the location of these outliers, the preceding analysis is preferred.

4   Spatial Autocorrelation

THEORY

Learning Objectives

This chapter deals with

- Spatial autocorrelation and its importance to geographical problems
- Global and local spatial autocorrelation techniques like Moran's I, Getis-Ord G and Geary C
- Tracing spatial clusters of high values (hot spots) or low values (cold spots)
- Tracing spatial outliers
- Optimized hot spot analysis
- Interpreting the statistical significance of results
- Incremental spatial autocorrelation used to define the appropriate scale of analysis
- The multiple comparison problem and spatial dependence
- Introducing Bonferroni correction and the false discovery rate
- Spatiotemporal autocorrelation analysis using bivariate and differential Local Moran's I index
- Presenting step-by-step examples using ArcGIS and GeoDa

After a thorough study of the theory and lab sections, you will be able to

- Distinguish between global and local spatial autocorrelation
- Understand why spatial autocorrelation analysis is relevant to geographical analysis
- Apply local and global indices of spatial autocorrelation like local Moran's, Getis-Ord G_i and G_i^*
- Use Moran's I scatter plot to identify patterns
- Identify hot spots or cold spots
- Identify and locate spatial outliers
- Use bivariate and differential Local Moran's I to identify if spatiotemporal autocorrelation exists and if changes cluster over time

- Apply these tools using ArcGIS
- Interpret the results from both the statistical significance and spatial analysis standpoints

4.1   Spatial Autocorrelation

Definition

Spatial autocorrelation is the degree of spatial dependency, association or correlation between the value of an observation of a spatial entity and the values of neighboring observations of the same variable. The terms "spatial association" and "spatial dependence" are often used to reflect spatial autocorrelation as well.

Why Use

Spatial autocorrelation is examined to determine if relationships exist among the attribute values of nearby locations and if these values form patterns in space.

Interpretation

According to the first law of geography, objects in a neighborhood tend to have more similarities and interactions than those lying further away. This is what we call "spatial dependency." To measure spatial dependency, we use spatial autocorrelation metrics. Put simply, spatial autocorrelation measures how much the value of a variable in a specific location is related to the values of the same variable at neighboring locations.

The spatial autocorrelation concept is similar to that of the statistical correlation used for nonspatial variables. Still, there is a major difference. While statistical correlation refers to two distinct variables with no reference to location, spatial autocorrelation refers to the value of a single variable at a specific location in relation to the values of the same variable at neighboring locations.

In statistical correlation, if two variables tend to change in similar ways (e.g., higher income correlated to higher educational attainment), we have positive correlation. Likewise, if similar values of a variable (either high or low) in a spatial distribution tend to collocate, we also have positive spatial autocorrelation. Positive spatial autocorrelation is the state where "data from locations near one another in space are more likely to be similar than data from locations remote from one another"(O'Sullivan & Unwin 2010 p. 34). In other words, autocorrelation (or self-correlation) exists when an attribute variable of a spatial dataset, correlates with itself at specific distances, called lags. This means that location affects the values of the variable in such a way that promotes values clustering in specific areas. A typical example of positive spatial autocorrelation is the income distribution within a city. Households with higher incomes generally tend to cluster in specific regions of the city, while households with lower incomes tend to cluster in other regions.

With negative spatial autocorrelation, on the other hand, neighboring spatial entities tend to have different values. This is similar to negative correlation, where high values of one variable indicate low values of the other. When there is no spatial autocorrelation, there is a random distribution of the values in relation to their locations, with no apparent association among them.

Discussion and Practical Guidelines

Spatial autocorrelation analysis is extremely important in geographical studies. If spatial autocorrelation did not exist, geographical analysis would be of little interest (O'Sullivan & Unwin 2010 p. 34). Think about it. We perform geographical analysis because we assume that location matters. If it did not, geography would be irrelevant. In most cases, phenomena do not vary randomly across space. For example, population concentrates on cities, income concentrates on cities, temperature displays small fluctuations inside a small area, and rain is uniform for a relatively small area. A student searching for a seat in an auditorium is most likely to sit next to a friend (who is already sitting). If you visit a restaurant, you will sit at an empty table. All these facts reveal nonrandom patterns. This is why geography is worth studying. If spatial arrangements were random, the global population could be located in every single location of the world with the same probability. If this were the case, more people would be living in the Antarctic or at heights above 5,000 m. If the temperature were random, you might be able to experience 30°C weather while standing outside the front door of your house and 1°C weather by jumping into the backyard. If rain were random, you might get wet while sunbathing at a beach in Santorini (Greece) during summer while the fellow right next to you lies on the sunbed sweating underneath the sun. In an auditorium, you will hardly see a student sit in a position that is already occupied. In a restaurant, it is rare (though not unusual) to sit at a table with people entirely unknown to you.

The aforementioned examples show that location matters and that a certain state influences what follows in a nonrandom way. They also remind us of the first- and second-order effects (see Section 3.2.1). By studying location, we reveal trends and patterns regarding the phenomenon at hand – for example, the spatial distribution of household income, the pattern of a disease outbreak, the relationship between residential location and mental well-being (Liu et al. 2019), or the linkage between built environment and physical health (Wang et al. 2019). Spatial autocorrelation is quite common in geographical analysis. This does not necessarily mean that it will occur across the entire study area (known as global autocorrelation). Spatial autocorrelation is sometimes evident only in subregions of the study area (known as local autocorrelation). In any case, spatial autocorrelation reveals the nonrandom distribution of the phenomenon being studied.

The nonrandom geographical distribution of the values of the variables under study has significant effects on the accuracy of classical statistics. In conventional statistics, the observed samples are assumed to be independent. In the presence of spatial autocorrelation, this assumption is violated. Observations are now spatially clustered or dispersed. This typically means that classical statistical tools are no longer valid. For example, linear regression would lead to biased estimates or exaggerated precision (Gangodagamage et al. 2008 p. 34). Bias leads to overestimate or underestimate of a population parameter, and exaggerated precision leads to a higher likelihood of getting statistically significant results when, in reality, we should have gotten less (de Smith 2018 p. 122). In addition, spatial autocorrelation infers redundancy in the dataset. Each newly selected sample is expected to provide less new information, affecting the calculation of confidence intervals (O'Sullivan & Unwin 2010). That is why we should use spatial statistics when analyzing spatial data and perform a spatial autocorrelation analysis before conducting any conventional statistical analysis.

Several diagnostic measures can be used to identify spatial autocorrelation. Those that estimate spatial autocorrelation by a single value for the entire study area are named **global spatial autocorrelation measures**. The most commonly used are

- Moran's I index
- General G-Statistic
- Geary's C index

As mentioned, it is unlikely that any spatial process will be homogeneous in the entire area due to the nonuniformity and noncontinuity of space. The magnitude of spatial autocorrelation may vary from space to space due to spatial heterogeneity. To estimate spatial autocorrelation at the local level, we use **local measures** of spatial autocorrelation, like

- Local Moran's I index
- Getis-Ord G_i and G_i^* statistics

With such metrics, we describe spatial heterogeneity (in the distribution of the values of a variable) as they identify hot or cold spots, clusters and outliers.

Spatial autocorrelation (either positive or negative) is a key concept in geographic analysis. A test of global or local spatial autocorrelation should be conducted prior to any other advanced statistical analysis when dealing with spatial data. Note that correlation does not necessarily imply causation; it implies only association. Relationships of cause and effect should always be established only after thorough analysis in order to avoid erroneous linkages (see Section 2.3.4 for more discussion on this).

4.2   Global Spatial Autocorrelation

4.2.1   Moran's *I* Index and Scatter Plot

Definition
Moran's *I* index computes global spatial autocorrelation by taking into account feature locations and attributes values (of a single attribute) simultaneously (Moran 1950). It is calculated by the following formula (4.1):

$$I = \frac{n}{\sum_i^n \sum_j^n w_{ij}} \frac{\sum_i^n \sum_j^n w_{ij}(x_i - \bar{x})(x_j - \bar{x})}{\sum_i^n (x_i - \bar{x})^2} \qquad (4.1)$$

where

> n is the number of the spatial features
> x_i is the attribute value of feature i, (remember that a variable is also called attribute in the spatial analysis context)
> x_j is the attribute value of feature j
> $\bar{x}$ is the mean of this attribute
> $w_{i,j}$ is the spatial weight between feature i and j
> $\sum_i^n \sum_j^n w_{ij}$ is the aggregation of all spatial weights

The tool calculates the mean $\bar{x}$, the deviation from the mean $(x_i - \bar{x})$ and the data variance $\frac{\sum_{i=1}^n (x_i - \bar{x})^2}{n}$ (denominator). Deviations from all neighboring features are multiplied to create cross-products (the covariance term). Then, the covariance term is multiplied by the spatial weight. All other parameters are used to normalize the value of the index. For example, the aggregation of spatial weights is used to normalize for the number of adjacencies. By the same means, the variance is used to ensure that the value index will not be large just because of a large variability in x values (O'Sullivan & Unwin 2010 p. 206).

Why Use
Global Moran's *I* is used as a metric for global spatial autocorrelation. It is mostly used for aerial data, along with ratio or interval data.

Interpretation and Moran's *I* Scatter Plot
Moran's *I* index is an inferential statistic. It is interpreted based on the expected value calculated (Eq. 4.2) under the null hypothesis of no spatial autocorrelation (complete spatial randomness) and is statistically evaluated using a *p*-value and a *z*-score (just as any common inferential statistic – see Section 2.5). The expected value for a random pattern is (4.2):

$$E(I) = \frac{-1}{n-1} \qquad (4.2)$$

where n denotes the number of spatial entities.

The expected value is the value that would have resulted if the specific dataset were the result of complete spatial randomness. The more spatial objects there are, the more the expected value tends to zero. The observed index value is the Moran's I index value calculated for the specific dataset through Equation (4.1). Positive Moran's I index values (observed) significantly larger than the expected value $E(I)$ indicate clustering and positive spatial autocorrelation (i.e., nearby locations have similar values). Negative Moran's I index values (observed) significantly smaller than the expected value $E(I)$ indicate negative spatial autocorrelation, meaning that the neighboring locations have dissimilar values. Values close to the expected value indicate no autocorrelation.

The difference between the observed and expected values has to be evaluated based on a z-score and a p-value. Through these metrics, we assess if this difference is statistically significant.

- If the p-value is large (usually $p > 0.05$), the results are not statistically significant, and we cannot reject the null hypothesis. The interpretation in statistical jargon is that *we cannot reject the null hypothesis that the spatial distribution of the values is the result of complete spatial randomness due to a lack of sufficient evidence.*
- A small p-value (usually $p < 0.05$) indicates that we can reject the null hypothesis of complete spatial randomness and accept that spatial autocorrelation exists:
  - ➤ In such a case, when the z-value is positive, there is positive spatial autocorrelation and a clustering of high or low values. Nearby locations will have similar values on the same side of the mean (O'Sullivan & Unwin 2010 p. 206).
  - ➤ If the z-value is negative, there is negative spatial autocorrelation and a dispersed pattern of values. Nearby locations will have dissimilar attribute values on the opposite sides of the mean (i.e., a feature with a high value repels other features with low values).

Let us consider an example. Imagine a spatial arrangement of 25 spatial objects (e.g., postcodes) with attribute values of either 1 (white) or 0 (black; see Figure 4.1).

- In a **perfectly dispersed** pattern, squares are located so that each one has neighbors of the opposite value. Spatial autocorrelation exists, as the squares have a competitive spatial relationship. If one square is black, the neighbor is white. Moran's I gets a negative value, smaller than the expected value, and there is negative spatial autocorrelation (see far left in Figure 4.1).

- If the squares are **grouped** (as in the far right in Figure 4.1), then clustering occurs. There is again spatial autocorrelation, but it is positive. The spatial relationship is that similar values tend to cluster. Moran's *I* gets a positive value, significantly larger than the expected one.
- When the values are **randomly scattered**, then Moran's *I* value is close to the expected values, and there is zero spatial autocorrelation (see central section in Figure 4.1).
- In intermediate states, there is no perfectly uniform status. However, an indication of clustering, dispersion or randomness can be assessed depending on the index value. Positive higher index values show a tendency toward clustering. Lower negative index values show a tendency toward dispersion. Note that we might obtain a low positive Moran's *I* value while also observing local clusters (see the following discussion). In all cases, we should evaluate the difference between the observed and expected values through *p*-values and *z*-scores, as mentioned before.

Moran's *I* scatter plot is used to visualize the spatial autocorrelation statistic (see Figure 4.2). It allows for a visual inspection of the spatial associations in the neighborhood of each observation (data point). In other words, it provides a representation used to assess how similar an attribute value at a location is to its neighboring ones. Data points are points that have as coordinates the values of the variable *X* for the *x*-axis and the spatial lag of the variable *X* (*Lag-X*) for the *y*-axis. *Lag-X* is the weighted average values of *X* in a specified neighborhood. Both *X* and *Lag-X* variables are used in a standardized form. As such, the weighted average is plotted at the center of the graph on the coordinates (0, 0). Distances in the plot are expressed as number of standard deviations from

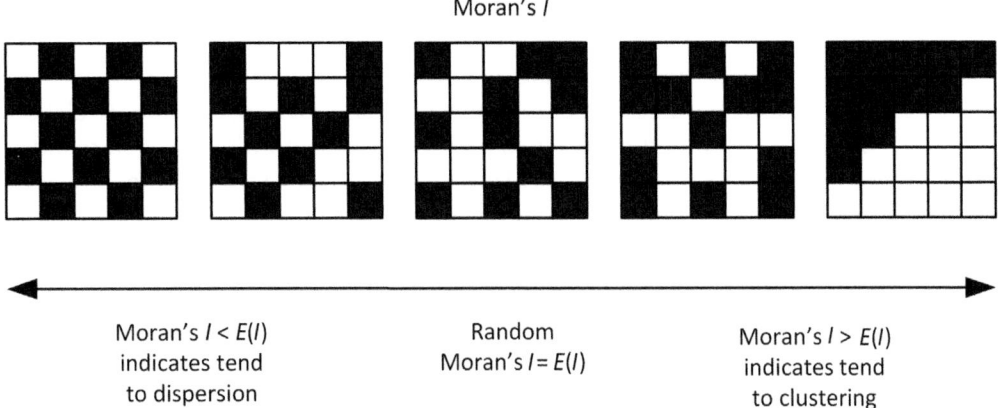

Figure 4.1 Global Moran's *I* and spatial autocorrelation.

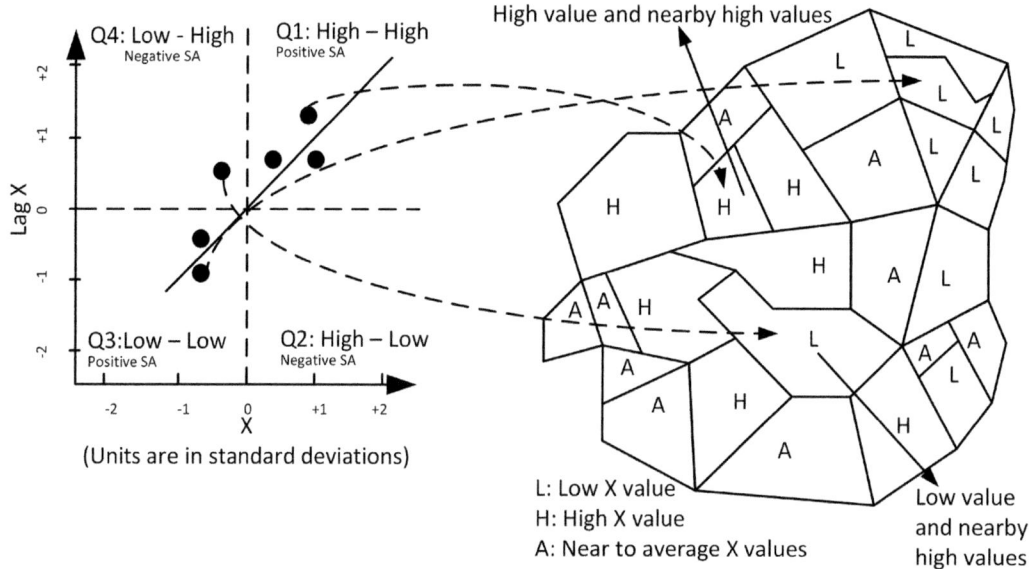

Figure 4.2 Moran's *I* scatter plot. The four quadrants divide the space into four types of spatial autocorrelation. Each dot in the scatter plot stands for one polygon in the map (in this graph, fewer dots are depicted for clarity's sake). Polygons with high values surrounded by polygons with high values are placed in the upper-right quadrant (Q1). Not all surrounding polygons need to have high values, but the more polygons with similar values there are, the stronger their associations.

the origin (0, 0). The produced scatter plot identifies which type of spatial autocorrelation exists according to the place where a dot lies (the dot standing for a spatial entity, such as a polygon). In the case of a polygon layer, a dot in the upper-right corner (Q1) indicates a polygon that has high *X* and high *Lag-X* (also called "High-High"). In other words, this polygon has a high value of *X* and is surrounded by other polygons that also have high values of *X*. That is why the *Lag-X* (average values of these neighboring polygons) is also high. In this case, there is positive spatial autocorrelation. If a dot lies in the lower-left corner (Q3), the polygon has a low value of *X* and is surrounded by polygons with low values of *X* (i.e., Low-Low). We thus again have positive spatial autocorrelation. A dot in the upper-left corner (Q4) indicates a polygon with low *X* surrounded by polygons with high *X* (i.e., Low-High). This is negative spatial autocorrelation and a strong indication of outlier presence. Finally, a dot in the lower-right corner (Q2) indicates a polygon with high *X* surrounded by polygons with low *X* (i.e., High-Low). There is negative spatial autocorrelation and an indication of outlier presence again.

Dots can also be compared to a superimposed regression line. The slope of the regression line over the data points equals the Moran's *I* index value when calculated using binary and row standardized weights. The closer a dot is to the line, the closer the polygon is to the general spatial autocorrelation trend. The further away a dot lies from the line, the greater the deviation of this spatial unit from the general trend. Points that deviate greatly from the regression line can be regarded as outliers. Potential outliers with respect to the regression line may function as leverage points that distort the Moran's *I* value. Such observations should be examined further, as in any case of an outlier's presence.

Moran's *I* index values can be bounded to the range −1.0 to +1.0 when the weights are row standardized (see Section 1.9). For most real-world problems, it is hard to find perfectly dispersed (−1) or clustered (+1) patterns. An index score higher than 0.3 is an indication of relatively strong positive autocorrelation, while a score lower than −0.3 is an indication of relatively strong negative autocorrelation (O'Sullivan & Unwin 2010 p. 206).

If we do not row standardize the weights, Moran's *I* Index might have values beyond the [−1,1] boundaries. This typically indicates problems with the tool's parameter settings. Examples of such problems are as follows:

- Values of the attribute in question are skewed. Check the histogram to see if this is the case.
- Some features do not have neighbors or have relatively few. The conceptualization of the spatial relationships or distance band should be checked to fix this problem. In skewed distributions, each object must have at least eight neighbors. This is not always sufficient (in skewed distributions), however, and has to be determined by the user.
- Selecting inverse distance often performs well but may produce very small values, something that should be avoided.
- Row standardization is not applied but should be (i.e,. row standardization is suggested most of the times when data refer to polygons).

Discussion and Practical Guidelines

Moran's *I* index is a global statistic and assesses the overall pattern of a spatial dataset. By contrast, local spatial autocorrelation metrics focus on each spatial object separately within a predefined neighborhood. Statistically significant results for the local Moran's *I* (e.g., detecting clustering) do not imply statistically significant results for the Global Moran's *I*. Although clusters may exist and be evident at the local level, these clusters may remain unnoticed when we examine the pattern at the global level. Global statistics are more effective when there is a consistent trend across the study area. If global statistics fail to reveal a pattern in the spatial distribution, this does not mean that local statistics will perform in similar ways. On the contrary, we should use them to find localized trends and patterns hidden at the global level.

It should also be mentioned that we use neighborhoods for Global Moran's *I* calculation, but this does not make the statistic local. The term "global" implies that a single value for the index is produced for the entire pattern. The term "local" means that a value is produced for each spatial object separately. We can thus map local Moran's *I*, as each spatial object has a local Moran's *I* value, but we cannot map Global Moran's *I*. To map Global Moran's *I*, we use the Moran's *I* scatter plot.

The neighborhood is also defined in Global Moran's *I*, for two main reasons:

(a) To reduce computational cost: The more objects there are on the data, the more time is required to compute the results.

(b) As we use a distance function (e.g., distance decay), the further away the objects lie, the less impact they have on each other. As a result, there is no need to calculate the index when weights are practically close to zero. Selecting the right cutoff distance is not trivial. It is suggested to start from a distance so that each object has at least one neighbor. In the case of skewed data, each object has to have at least eight neighbors. Alternatively incremental spatial autocorrelation is a useful method of determining the appropriate cut of distance, as explained in Section 4.3.

The following are some practical guidelines:

- Results are reliable if we have at least 30 spatial objects.
- Row standardization should be applied if necessary. Row standardization is common when we have polygons.
- When the *p*-value is low (usually $p < 0.05$), we can reject the null hypothesis of zero spatial autocorrelation:
  (a) If the *z*-score is positive, there is positive spatial autocorrelation (clustering tendency).
  (b) If the *z*-score is negative, there is negative spatial autocorrelation (dispersion tendency).

Finally, potential case studies for which Moran's *I* index could be used include

- Examining if income per capita is clustered (socioeconomic analysis)
- Analyzing consumption behavior (geomarketing analysis)
- Analyzing house values (economic analysis)

4.2.2   Geary's C Index

Definition

Geary's C index is a statistical index used to compute global spatial autocorrelation (Geary 1954) and is calculated by the following formula (4.3):

$$C = \frac{(n-1)}{2\sum_i^n \sum_j^n w_{ij}} \frac{\sum_i^n \sum_j^n w_{ij}(x_i - x_j)^2}{\sum_i^n (x_i - \bar{x})^2} \tag{4.3}$$

where

    n is the total number of spatial objects
    x_i is the attribute value of feature *i*, x_j is the attribute value of feature *j*
    $\bar{x}$ is the mean of this attribute

w_{ij} is the spatial weight between feature i and j

$\sum_i^n \sum_j^n w_{ij}$ is the aggregation of all spatial weights

Why Use
To trace the presence of global spatial autocorrelation in a spatial dataset.

Interpretation
Geary's C index varies between 0 and 2. A value of 1 typically indicates no spatial autocorrelation. Values significantly smaller than 1 indicate positive spatial auto-correlation, while values significantly larger than 1 indicate negative spatial autocorrelation (O'Sullivan & Unwin 2010 p. 211).

Discussion and Practical Guidelines
Moran's *I* offers a global indication of spatial autocorrelation, while Geary's C is more sensitive to differences in small neighborhoods (Zhou et al. 2008 p. 69). As such, when we search for global spatial autocorrelation, Moran's *I* is usually preferred over Geary's C.

4.2.3   General G-Statistic

Definition
General G-Statistic is a statistical index used to compute global spatial auto-correlation. The General G-Statistic detects clusters of low values (cold spots) or high values (hot spots) and is an index of spatial association (Getis & Ord 1992, O'Sullivan & Unwin 2010 p. 223).

$$G(d) = \frac{\sum_i^n \sum_j^n w_{ij}(d)x_i x_j}{\sum_i^n \sum_j^n x_i x_j}, \forall j \neq i \tag{4.4}$$

where

n is the total number of observations (spatial objects)
x_i is the attribute value of feature i
x_j is the attribute value of feature j
d is the distance that all pairs (x_i, x_j) lie within
w_{ij} is the spatial weight between feature i and j
$\forall j \neq i$ indicates that features i,j cannot be the same

Why Use
This index is used to distinguish if the positive spatial autocorrelation detected is due to clustering of high values or due to clustering of low values. When clusters of low values coexist with clusters of high values in the same study area, they tend to counterbalance each other. Moran's *I* is more suitable to trace this association on the global scale.

Interpretation

The General G-Statistic is inferential, and its results are interpreted based on a rejection (or not) of the null hypothesis that there is complete spatial randomness and, thus, clusters do not exist. A z-score and a p-value are calculated along with the expected index value. The expected value is the value that would result if the spatial distribution of the values (of the variable being studied) were the outcome of complete spatial randomness. The difference between the observed and expected values is evaluated based on a z-score and a p-value that test if this difference is statistically significant.

- When the p-value is small (usually $p < 0.05$), the null hypothesis is rejected, and there is statistically significant evidence for clustering:
    - (a) If the z-value is positive, the observed General G-Statistic value is larger than expected, indicating a concentration of high values (hot spots).
    - (b) If the z-value is negative, the observed General G-Statistic value is smaller than expected, indicating that low values (cold spots) are clustered in parts of the study area. In both cases, this is an indication of positive autocorrelation.

In cases where the weights are binary or less than 1, the index value is bounded between 0 and 1. This happens because the denominator includes all (x_i, x_j) pairs, regardless of their vicinity. The numerator will be always less than or equal to the denominator. The final outcome regarding spatial association should be concluded only after an examination of the p-value and the z-score.

Discussion and Practical Guidelines

The General G-Statistic measures the overall (hence "general") clustering of all pairs (x_i, x_j) within a distance d of each other (Getis & Ord 1992). Moran's I index cannot distinguish between high- and low-value clustering, but the General G-Statistic can. On the other hand, the General G-Statistic is appropriate only when there is positive spatial autocorrelation, as it detects the presence of clusters. If the General G-Statistic does not produce statistically significant results, we cannot reject the null hypothesis of complete spatial randomness. Negative spatial autocorrelation might exist through a competitive process (i.e., where high values and low values for the same variable are nearby). As a result, the General G-Statistic is more an index of spatial association or an index of positive spatial autocorrelation, rather than a pure spatial autocorrelation index.

Here are some practical guidelines to follow:

- General G-Statistic works only with positive values.
- A binary weights matrix is more appropriate for this statistic. It is thus recommended to use fixed distance band, polygon contiguity, k-nearest neighbors or Delaunay triangulation that produce binary weighting schemes. For example, if we set a fixed distance of 2 km, each object

inside this distance will be a neighbor and have a weight of 1. All other objects lying further away than 2 km will not be neighbors and have zero weight. In this case, row standardization is not necessary, as the weights already lie in a 0-to-1 range.

- When we use binary weighting and fixed distance, the size of the polygons might matter. For example, if large polygons tend to have lower values for the attribute being studied (e.g., population density) than the small polygons, we might obtain higher observed values of the General G-Statistic because more smaller polygons are creating pairs at the same distance set. We would thus obtain higher z-values and stronger clustering results than what the real situation would justify.

- It is more common to use the local version of the General G-Statistic index, as it provides the exact locations of the clusters. There are two versions of the Local G-Statistic: the G_i and the G_i^*.

Finally, potential case studies include

- Analyzing PM2.5 in an urban environment (environmental analysis)
- Analyzing educational patterns (demographic analysis)
- Analyzing house rents (economic analysis)

4.3   Incremental Spatial Autocorrelation

Definition
Incremental spatial autocorrelation is a method based on Global Moran's *I* index to test for the presence of spatial autocorrelation at a range of band distances (ESRI 2015).

Why Use
It is used to approximate the appropriate scale of analysis (appropriate analysis distance). Instead of arbitrary selecting distance bands, this method identifies an appropriate fixed distance band for which spatial autocorrelation is more pronounced. In other words, it allows us to identify the farthest distance at which an object still has a significant impact on another one. After the appropriate scale of analysis is established, local spatial autocorrelation indices and other spatial statistics can be calculated more accurately.

Interpretation
Incremental spatial autocorrelation is used to calculate Global Moran's *I* for a series of incremental distances (see Figure 4.3). For each distance increment, the method produces Global Moran's *I*, Expected *I*, variance, a z-score and a p-value. With this method, we can plot a graph of z-scores over an increasing distance.

z-score peaks reflect distances at which a clustering process seems to be occurring. The higher the z-score, the stronger the clustering process at that distance. By locating the peak in the graph, the z-score and the corresponding distance, we can better define the distance band, which can be used in many spatial statistics such as hot spot analysis (see Section 4.4.2). The distance in the first peak of the z-score graph is often selected as the appropriate scale for further analysis (but this is not always the case; see the following discussion).

Discussion and Practical Guidelines

Selecting the appropriate scale of analysis for the problem at hand is one of the most challenging tasks in spatial analysis. The scale of analysis defines the size and shape of the neighborhoods for which spatial statistics are calculated and is closely related to the problem in question (see Section 1.3). Researchers and analysts lacking an in-depth understanding of spatial statistics often tend to apply spatial statistics tools based on the predefined default values (i.e., distance band) of the software being used. The uncritical selection of spatial parameters (e.g., the distance of analysis) leads to incorrect estimates and conclusions. The difficulty is not running a software tool but setting it up properly and interpreting the results based on statistical theory. The graph produced by incremental spatial autocorrelation allows for a statistically sound estimation of the scale of the analysis, which is superior to an arbitrary or intuitive selection.

It is quite common that more than one peak may occur. Different distance bands (peaks) might reveal underlying processes on different scales of analysis. Hypothetically, unemployment clustering statistically significant at 100 m and 1,000 m peaks reflect patterns of clustering at both the census block level and the postcode level. If we are interested only in the census block level, we could apply the 100 m distance in our analysis. Greater distances reflect broader, regional trends (e.g., east to west), while smaller distances reflect local trends

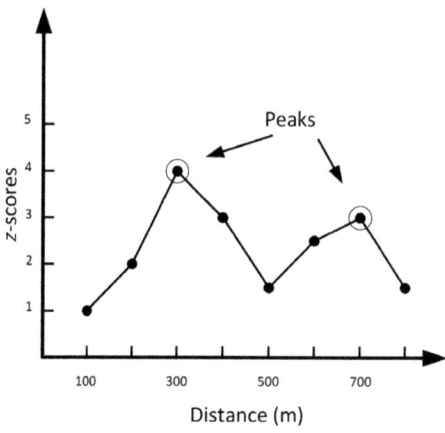

Figure 4.3 z-scores over incremental distance.

(e.g., between neighborhoods). For example, the scale of analysis is usually small when the analysis is of children going to school (usually close to their homes), while the scale is larger when analyzing commuting patterns.

There are some practical guidelines regarding this method. An initial distance (the value from which distances will start incrementing) should be set. The initial default distance value should ensure that each object has at least one neighbor. In this case, if locational outliers (objects lying far away from others) exist, the initial distance calculated may be large enough. A large initial distance value results in a graph with no peaks, simply because a peak lies before the beginning distance. In addition, an increment distance should also be set. We can use either a distance increment that fits the needs of the study (i.e., 100 m for a local study) or try the average distance to each feature's nearest neighbor (usually the software's default value). Locational outliers may distort this value, leading to very large increments that might not be representative. For this reason, before running incremental spatial autocorrelation, we should check if locational outliers exist.

To define the appropriate scale of analysis, the following procedure can be applied:

Step 1: Check for locational outliers (see Section 3.1.6 and Exercise 3.4). If no locational outliers exist, perform incremental spatial autocorrelation as described earlier. If locational outliers exist, go to step 2.

Step 2: Select all features except outliers and perform *Incremental Spatial Autocorrelation* (only for the selected features).

Step 3: Locate a peak and keep the relevant distance

Step 4: Create a *Spatial Weights Matrix* for the entire dataset (including the locational outliers) with the distance defined in step 3 (see Section 1.9). Set the *Number of Neighbors* to a value so that each object has at least this number of neighbors.

Step 5: Run spatial statistic (e.g., Local Moran's *I* index) using the *Spatial Weights Matrix* created in step 4.

When locational outliers are removed, the z-scores graph might change significantly, yielding a completely different scale of analysis. Through this procedure, *Spatial Weights* are calculated based on the distance threshold that results when outliers are removed. For objects that have no neighbors at this distance, the *Number of Neighbors* parameter will be used instead. This practically means that outliers will be treated differently but will be included in the study so that they do not negatively impact the rest of the objects. As such, local spatial autocorrelation indices will be calculated based on the final weights matrix for all objects in the dataset.

Other practical guidelines include the following (ESRI 2015):

- There might be more than one peak. Each one reflects spatial autocorrelation at different distances. We will typically select the one with the highest z-score, which is usually the first. We may also select a peak (distance) that better reflects the regional or local perspective for the problem at hand.

- Select as the beginning distance the one that will ensure that each object has at least one neighbor.
- Look for locational outliers before setting the initial distance. A locational outlier inflates distance metrics, with negative impacts on the graph. Remove outliers, and run the incremental spatial autocorrelation tool.
- If there are no peaks, we can use smaller or larger distance increments. If we still cannot locate any peaks, we should avoid the incremental spatial autocorrelation method. We should rely on other criteria or our common sense based on previous knowledge to define the appropriate distance band. We can also use optimized hot spot analysis, which enables distance band definition even if no peaks exist (see Section 4.4.3).
- The final distance selected should provide an adequate number of neighbors for each feature and an appropriate scale of analysis.

4.4 Local Spatial Autocorrelation

Global indices of spatial autocorrelation identify whether there is clustering in a variable's values, but they do not indicate where clusters are located. To determine the location and magnitude of spatial autocorrelation, we have to use local indices instead. Local Moran's I and local Getis-Ord G_i^* are the most widely used local indices of spatial autocorrelation.

4.4.1 Local Moran's I (Cluster and Outlier Analysis)

Definition
Local Moran's I is an inferential spatial statistic used to calculate local spatial autocorrelation. For n spatial objects in a neighborhood (Anselin 1995), the local Moran's I of the i object is given as (4.5):

$$I_i = \frac{x_i - \bar{X}}{m_2} \sum_j w_{ij} \left(x_j - \bar{X} \right) \tag{4.5}$$

$$m_2 = \frac{\sum_i \left(x_j - \bar{X} \right)^2}{n} \tag{4.6}$$

where

n is the total number of observations (spatial objects).
x_i is the attribute value of feature i.
x_j is the attribute value of feature j.
$\bar{X}$ is the mean of this attribute.

$w_{i,\,j}$ is the spatial weight between feature i and j.

m_2 is a constant for all locations. It is a consistent but not unbiased estimate of the variance. (Anselin 1995 p. 99).

Tip: Keep in mind that m_2 is actually a scalar that does not affect the significance of the metric as it is the same for all locations (Anselin 2018). In some cases, the formula of this scalar may slightly change. For example, ArcGIS and GeoDa uses $n - 1$ in the denominator instead of n (de Smith et al. 2018).

Why Use

Local Moran's I can be used for an attribute to (a) identify if a clustering of high or low values exists and (b) to trace spatial outliers (Grekousis & Gialis 2018).

Interpretation

Local Moran's I index is interpreted based on the expected value, a pseudo p-value, and a z-score under the null hypothesis of no spatial autocorrelation (complete spatial randomness). The expected value for a random pattern is (Anselin 1995 p. 99):

$$E(I_i) = \frac{-1\sum_j w_{ij}}{n-1} \tag{4.7}$$

where

n denotes the number of spatial entities

w_{ij} is the spatial weight between feature i and j

The expected value is the value that would have resulted if the specific attribute's values geographical distribution were the result of complete spatial randomness. The observed index value is the local Moran's I index value given by Equation (4.5).

Positive local Moran's I index values (observed) significantly larger than the expected value indicate potential clustering and positive spatial autocorrelation. Negative local Moran's I index values (observed) significantly smaller than the expected value indicate the potential presence of spatial outliers and negative spatial autocorrelation. Values close to the expected value indicate no autocorrelation. To finalize a conclusion regarding spatial autocorrelation's presence, we should evaluate the previously mentioned difference (expected vs. observed) based on the z-score and the p-value. Using these two metrics, we assess if this difference is statistically significant.

- If the p-value is large (usually $p > 0.05$), the results are not statistically significant (even if the difference is large), and we cannot reject the null hypothesis (see Table 4.1). The interpretation is that, *due to a lack of sufficient evidence, we cannot reject the null hypothesis that the spatial distribution of the values is the result of complete spatial randomness.*

Table 4.1 Interpretation of Moran's I p-values and z-scores. z-score values are indicative and can be differentiated based on data.

p-value	z-score	Interpret
$>\alpha$ (e.g., 0.05)		Can not reject the null hypothesis of complete spatial randomness.
$<\alpha$ (e.g., 0.05)	$z < 0$	Negative spatial autocorrelation. Low negative values (e.g., $z < -2.60$) is an indication of spatial outlier presence.
	$z > 0$	Positive spatial autocorrelation: clustered pattern. Large positive values (e.g., $z > 2.60$) indicate intense clustering of either low (cold spots) or high (hot spots) values.

- A small p-value (usually $p < 0.05$) indicates that we can reject the null hypothesis of complete spatial randomness and accept that spatial autocorrelation exists. In this case:

  (a)   If the z-value is positive, we have positive spatial autocorrelation and clustering.

  b)   il the z-value is negative, we have negative spatial autocorrelation, and spatial outliers may exist especially for low z-score values.

A high positive z-score (e.g., greater than 2.60) for a spatial entity means that the neighboring spatial entities have similar values. If the values are high, then High-High clusters are formed, meaning that spatial entities with high values (for a specific variable) are surrounded by spatial entities of high values (of the same variable). If the values are low, then "Low-Low clusters" are formed, meaning that spatial entities with low values are surrounded by spatial entities of low values. Note that the spatial clusters formed in High-High or Low-Low arrangements depict only the core of a real cluster. This happens because the statistical value for each location is calculated based on the neighboring values. Thus, the locations (e.g., polygons) at the periphery of a cluster might not be assigned to a High-High or Low-Low cluster.

A low negative z-score, (e.g., less than -2.60) for a spatial entity indicates dissimilar nearby attribute values and potential spatial outliers. If the spatial entity has a low attribute value, then it is surrounded by features with high values, creating a Low-High arrangement. If a spatial entity has a high attribute value, then it is surrounded by features with low values, creating a High-Low arrangement.

Discussion and Practical Guidelines

Even with complete spatial randomness, clustering or outliers might exist due to randomness. To overcome this problem, we use a Monte Carlo random permutation procedure. Permutations are used to estimate the likelihood of generating, via complete spatial randomness, a spatial arrangement of values similar to the observed one. Using Monte Carlo, we generate multiple random patterns and then compare the results to those of the local Moran's I of the

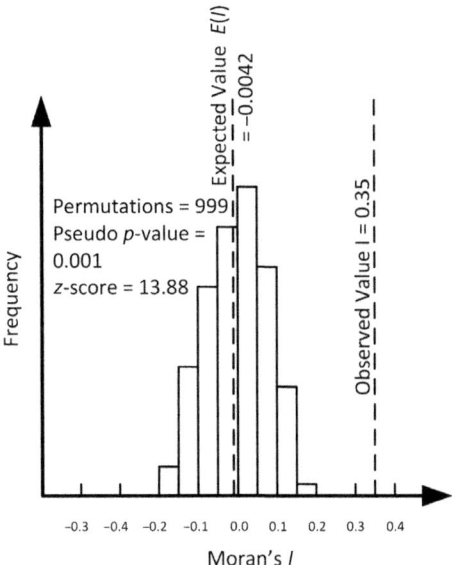

Figure 4.4 Permutation reference distribution for 999 random permutations under complete spatial randomness. The results of this example suggest that the observed value *I* = 0.35 is highly significant and not a result of spatial randomness, as it lies far away from the rest of the values (with a z-score of 13.88) and the expected (theoretical) Moran's *I* value *E(I)* = −0.0042. The pseudo *p*-value is 0.001, indicating that none of the 999 random patterns' local Moran's *I* values surpassed the observed value.

original dataset. By inspecting the reference distribution, we assess how unusual the observed value would be in relation to this randomized benchmark (see Figure 4.4).

For each permutation (i.e., of a total of 999), the values (of the attribute variable) are randomly rearranged around each feature, and the local Moran's *I* index is calculated. A reference distribution of the local Moran's *I* index values is then created (see Figure 4.4). The reference distribution should be centered at around zero, as it is supposed to be the result of complete spatial randomness with no spatial autocorrelation. The range of local Moran's *I* index values, which vary due to randomness, is depicted in the x-axis. If the local Moran's *I* observed value lies far away from the reference distribution and in relation to the z-score obtained (which quantifies the distance from the mean), we can reject the null hypothesis of complete spatial randomness and accept that the spatial autocorrelation observed is statistically significant.

A pseudo *p*-value ($p = \frac{R+1}{M+1}$) is calculated as the proportion of how many times (*R*) the computed local Moran's *I* values generated by the permutations are equal to or larger than the observed local Moran's *I* to the number of permutations (*M*; Anselin 2018). Typically, for 999 permutations, the pseudo *p*-value is set to 0.001; for 99 permutations, it is set to 0.01. A pseudo *p*-value cannot be interpreted as a typical p-value, as it is the summary of the reference distribution (Anselin 2018). A pseudo p-value of 0.01 (i.e., $p = \frac{0+1}{99+1} = 0.01$), for example, means that none (zero) of the 99 random patterns yielded a local Moran's I value equal to or more extreme than the observed data. In other words, no pattern exhibited clustering (or dispersion) equal to or larger than the observed one.

Practical guidelines include the following (ESRI 2018a):

- Results are reliable for at least 30 spatial objects.
- We cannot perform this test for points events (e.g., points as crime incidents, without any attribute fields attached). Nevertheless, we can aggregate data into polygons and then continue with the analysis in the usual fashion (see discussion in Section 4.4.3).
- Each feature has to have at least one neighbor.
- No feature has to have all features as neighbors.
- When values are skewed, each feature should have around eight neighbors or more.
- The conceptualization of spatial relationships, distance bands and distance functions used should be done carefully.
- The false discovery rate (see Section 4.6) can be used to account for multiple comparison problems and spatial dependence.

Potential Case Studies Include
- Analyzing unemployment distribution
- Analyzing income inequalities
- Analyzing house values

4.4.2   Optimized Outlier Analysis

Definition
Optimized outlier analysis is a procedure used to optimally select the parameters of the local Moran's I index (ESRI 2018a). Similar to local Moran's I, it locates clusters of either high or low values and traces spatial outliers.

Why Use
Optimized outlier analysis is used to overcome the difficulties of setting the parameters of the local Moran's I index. Optimized outlier analysis performs an automated preliminary analysis of the data to ensure optimal results. The method is used to

(a)   Identify how many locational outliers exist (if any).
(b)   Estimate the distance band at which the spatial autocorrelation is more pronounced (scale of analysis) through incremental spatial autocorrelation.
(c)   Adjust for spatial dependence and multiple testing through the false discovery rate correction method (see Section 4.6).

(d)   Handle point events with no variables attached. These events are automatically aggregated into weighted features within some regions (i.e.,
grid; see discussion in Section 4.4.3). The weighted variable is then
analyzed in the usual fashion.

Interpretation

Optimized outlier analysis applies the local Moran's I index, and the results can
be interpreted accordingly (see Section 4.4.1).

Discussion and Practical Guidelines

Optimized outlier analysis applies incremental spatial autocorrelation to define
the scale of analysis (see Section 4.3). The distance band selected is the one in
which a peak occurs in the related graph. If multiple peaks are found, the
distance related to the first peak is usually selected. If no peaks occur, the
optimized outlier analysis applies a different procedure. The spatial distribution
of the features is analyzed by calculating the average distance so that each
feature has K neighbors. K is defined as 5% of the total number of (n) features
($K = 0.05 \times n$). K is adjusted so that it ranges between 3 and 30. If the average
distance that ensures K neighbors for each feature is larger than one standard
distance, the distance band is set to one standard distance (ESRI 2018a). If it is
not, the K neighbor average distance reflects the appropriate scale of analysis.
Finally, optimized outlier analysis is effective even for data samples. It is also
effective in case of oversampling, as the associated tools have more data with
which to compute accurate results.

4.4.3   Getis-Ord G_i and G_i^* (Hot Spot Analysis)

Definition

Getis-Ord G_i index and G_i^* index (pronounced G-i-star) comprise a family of
statistics that identify statistically significant clusters of high values (hot spots)
and clusters of low values (cold spots) and are used as measures of spatial
association (Getis & Ord 1992, Ord & Getic 1995, O'Sullivan & Unwin 2010
p. 219). The process is also named hot spot analysis. The G_i index is given as
(4.8):

$$G_i(d) = \frac{\sum_j w_{ij}(d) x_j}{\sum_{j=1}^{n} x_j}, j \neq i \tag{4.8}$$

where

   d is the estimated range of observed spatial autocorrelation
   $\sum_j w_{ij}(d)$ is the sum of weights for $j \neq i$ within distance d
   n is the total number of observations
   x_j is the attribute value of feature j

Getis-Ord G_i^* index is given as (4.9):

$$G_i^*(d) = \frac{\sum_j w_{ij}(d)x_j}{\sum_{j=1}^{n} x_j} \tag{4.9}$$

Note that in contrast to G_i, in G_i^* the restriction $\forall j \neq i$ is lifted. In other words, the index takes into account the attribute value x_i in location i.

Why Use
Hot spot analysis identifies if low or high values of a variable are spatially clustered and create cold spots or hot spots respectively.

Interpretation
For each polygon, a z-score value is calculated along with a p-value to assess the statistical significance of the results. The null hypothesis is that *There is complete spatial randomness of the values associated with the features*. Having a high positive z-score and a small p-value is an indication of spatial clustering of high values (i.e., a hot spot), whereas having a low negative z-score with a small p-value reveals the presence of cold spots (spatial clustering of low values). In both cases, there is positive spatial autocorrelation. The higher the z-score (either positive or negative), the more intense the clustering at hand. z-scores values near to zero typically indicate no spatial clustering. When p-values are larger than 0.05 (or larger than another established significance level), the null hypothesis cannot be rejected, and the results are not statistically significant. Nonsignificant results mean that there is no indication of clustering, as the process at hand might be random. The results can be rendered in a map with three confidence level classes (99%, 95% or 90%) for hot spot polygons, three classes (99%, 95% or 90%) for cold-spot polygons and another class for rendering polygons with nonsignificant results.

Discussion and Practical Guidelines
G_i^* is used more widely than G_i. The use of G_i^* is also linked to hot spot analysis. Hot spot analysis is mainly used to identify if clusters of values of a specific variable are formed in space (Grekousis 2018). It is most commonly used with polygon features. For point features, though, it is more helpful to study the intensity of the objects rather than a specific attribute. In this respect, we use hot spot analysis to identify if hot spots or cold spots of events' intensity exist. Such analysis should begin by aggregating points into some regions (e.g., postcodes, census tracts). This can be easily done by overlaying the relevant (administrative) polygon layers and applying typical GIS techniques (such as spatial join to count how many points lie within each polygon). Alternatively, we can create a grid (in the absence of a polygon layer, or to avoid at some extent the modifiable areal unit problem – see Chapter 1) by using a fishnet tool and then perform spatial join. The grid size should be set in such a way that

most of the grids have more than one incident. The new attribute field created, containing the number of points per polygon/grid, can be used as the attribute field to be analyzed, similarly to any other polygon layer.

Some practical guidelines for the G_i^* (ESRI 2018a):

- Results are reliable if we have at least 30 objects.
- Fixed distance is recommended for the G_i^* index. The appropriate distance should be determined by using incremental spatial autocorrelation or optimized hot spot analysis (see Section 4.4.4). In case of locational outliers, a fixed distance band can be combined with a minimum number of neighbors per spatial feature. In this case, when the fixed distance band leaves some polygons with no neighbors, the minimum number of neighbors ensures that all polygons will have at least a specific number of neighbors.
- The false discovery rate (see Section 4.6) can also be used here, as in the case of Local Moran's I.
- This index cannot be applied for point objects (e.g., crime incidents), without any attribute fields attached. However, we can aggregate data by using spatial join to polygons and then continue with the analysis.
- The spatial relationships, distance bands and distance functions used should be conceptualized carefully.

Potential case studies include

- Human geography/demographics: Are there any areas where the unemployment rate form spatial clusters?
- Economic geography: How is income spatially distributed? Are there cold or hot spots?
- Health analysis: Are there any unusual patterns to heart attacks?
- Voting pattern analysis: Do people in favor of a specific party cluster together?

4.4.4   Optimized Hot Spot Analysis

Definition
Optimized hot spot analysis is a procedure used to optimally select the parameters of the Getis-Ord G_i^* index (ESRI 2018a). Similar to Getis-Ord G_i^*, it locates spatial clusters of low values (cold spots) and spatial clusters of high values (hot spots).

Why Use
Optimized hot spot analysis is used to overcome the difficulties of setting the parameters of the Getis-Ord G_i^* index. Optimized hot spot analysis performs hot spot analysis using the Getis-Ord G_i^* index in an automated way to ensure optimal results. The method is used to

(a)   Identify how many locational outliers exist (if any).

(b)   Estimate the distance band at which the spatial autocorrelation is more pronounced (scale of analysis).

(c)   Adjust for spatial dependence and multiple testing through the false discovery rate correction method (see Section 4.6).

(d)   Handle point events with no variables attached. These events are automatically aggregated into weighted features within some regions (i.e., grid; see discussion in Section 4.4.3). The weighted variable is then analyzed in the usual fashion.

Interpretation

Optimized hot spot analysis applies the Getis-Ord G_i^* index and results can be interpreted as the Getis-Ord G_i^* results are interpreted (see Section 4.4.3).

Discussion and Practical Guidelines

Optimized hot spot analysis applies incremental spatial autocorrelation to define the scale of analysis (see Section 4.3). This is done in the same way as that described for optimized outlier analysis (see Section 4.4.2).

As mentioned, for point features with no other attributes attached, optimized hot spot analysis aggregates the points into zones and identifies potential event concentration (clustering) or dispersion across space. It can thus be regarded as an alternative approach to point pattern analysis that indicates if spatial autocorrelation among point events is evident (see Section 3.2.1). Finally, the scale of analysis resulting from the optimized hot spot analysis can be applied in kernel density estimation as an alternative way to select the bandwidth h (see Section 3.2.4).

4.5   Space–Time Correlation Analysis

4.5.1   Bivariate Moran's *I* for Space–Time Correlation

Definition

Bivariate Moran's *I* measures the degree to which a variable in a specific location is correlated with the spatial lag (average value at nearby locations) of a different variable (Anselin 2018). A special case of Bivariate Moran's *I* (and a more useful one) occurs when a single variable (instead of two different variables) is used for two different time stamps. It measures the degree of the spatiotemporal correlation of a single variable.

Why Use

Bivariate Moran's *I* index is used to assess how the linear association (positive or negative) of two distinct variables varies in space. The extension of Bivariate

Moran's *I* for space–time correlation is used to trace if spatiotemporal auto-correlation exists for the same variable.

Interpretation
In the case of row standardized weights, Bivariate Moran's *I* for space–time correlation would lie between -1 and 1, with values close to zero indicating no correlation, values close to 1 strong positive correlation and values close to -1 strongly negative correlation. The significance of the statistic is determined through a permutations approach.

Discussion and Practical Guidelines
Before discussing the bivariate extension to time (single variable), let us briefly describe how the standard Bivariate Moran's *I* index works through the Bivariate Moran scatter plot. As explained before, Bivariate Local Moran's *I* analyzes two distinct variables. For instance, suppose we study the potential association between income at a specific location and land prices in surrounding areas. A Bivariate Moran's *I* scatter plot would relate the income values for each location (Income, horizontal axis) to the average land value at nearby locations (W_land, vertical axis; see Figure 4.5A). However, bivariate spatial correlation does not take into account the inherent correlation between the two variables (i.e., income and land price at the same location; Anselin 2018). Leaving unaccounted this correlation makes Bivariate Moran's *I* hard to interpret; this often leads to incorrect conclusions, as the statistic may overestimate the spatial effect of the correlation, which might be merely the result of the same location correlation (Anselin 2018). Thus, Bivariate Moran's *I* is more useful when time is included.

More analytically, a particular case of Bivariate Moran's *I* spatial correlation occurs when the correlation is calculated for variable *X* in a location with *Lag-X* within a time interval. *Lag-X* is the average value of *X* in a nearby location but in a previous time stamp (see Figure 4.5B). This is the Bivariate Moran's *I* for space–time correlation. Conceptually, this approach explains how neighboring values in a previous period affect the present value (Anselin 2018). To put it slightly differently, this approach would explain how the value at a location in a subsequent time is affected by the average values at nearby locations in a previous time. It can be regarded as the inward diffusion from the neighbors at a specific point in time to the core in the future. Switching the selection of variable settings in the scatter plot axes produces a scatter plot with $X(t-1)$ in the x-axis and the $lagX(t)$ in the y-axis. This approach studies how a location in a previous time affects the values of nearby locations in the future. It can be seen as an outward diffusion originating from the core at a specific time to the neighbors in the future (Anselin 2005). Although this approach is formally correct, the results might be misleading (Anselin 2018), mainly because the notion of spatial autocorrelation refers to how neighbors affect the value of a central location and not the contrary (Anselin 2018). These approaches are

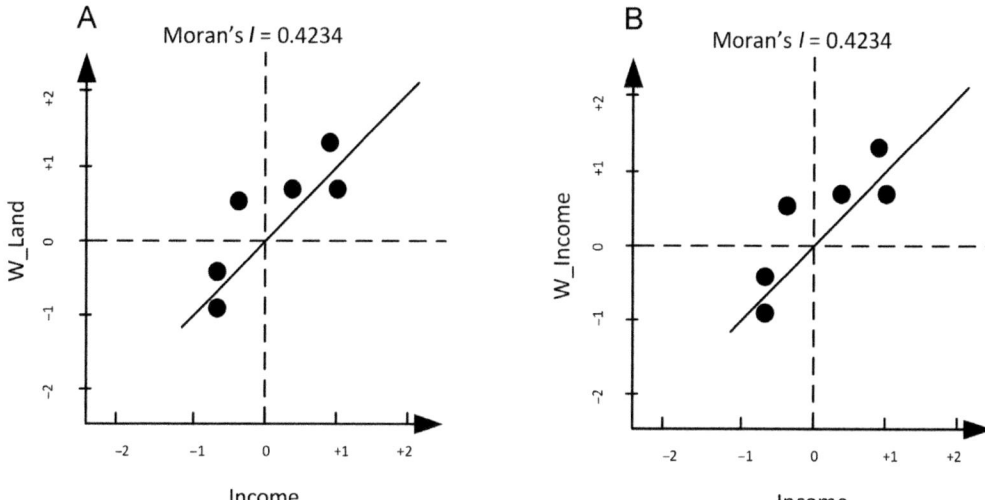

Figure 4.5 (A) Bivariate Moran's *I* for income and land price. The graph indicates that income (*x*-variable) is correlated with the weighted average value of land (*y*-W_Land) within its neighborhood. All variables are expressed in standardized forms, with zero mean and a variance of one. Spatial weights are also row standardized (Anselin 2018). (B) Spatiotemporal Bivariate Moran's *I*. Moran's *I* calculates the correlation of variable *X* in a location with *Lag-X* with the previous time stamp. The proper interpretation is that a value at a location for income (*x*-variable) is correlated with the weighted income value of its neighbors in a previous time (Anselin 2018).

slightly different, and the best one to use depends on the problem at hand and the underlying process being studied.

4.5.2   Differential Moran's *I*

Definition
For two time stamps, **Differential Moran's *I*** tests whether a variable's change at a specific location is related to the change of the same variable in neighboring locations.

Why Use
Differential Global or Local Moran's *I* is used to identify if changes over time are spatially clustered.

Interpretation
As with the interpretation of Moran's *I*, if a high change in a variable's value between two time stamps for a specific location is accompanied by a high change of the same variable in the surrounding area, there is positive spatial autocorrelation of the High-High type (i.e., hot spots) (see Figure 4.6). In other words, the change of a variable's value in a specific location follows a

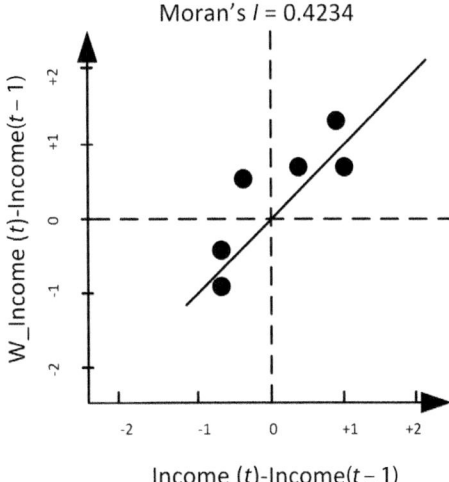

Moran's I = 0.4234

Figure 4.6 Differential Moran's I. The x-variable is the difference in the variable between two time stamps. The y-variable is the spatial lag of this difference (weighted difference calculated as the average difference for the nearby locations). A high value of the statistic indicates that changes in the variable cluster over time (and space).

trend similar to the change of the same variable in the neighboring area. If low changes in a variable's value between two time stamps are surrounded by low changes, then a Low-Low type of cluster is formed (i.e., cold spots). If a low change in a variable's value between two time stamps for a specific location is accompanied by a high change of the same variable in the surrounding area, there is negative spatial autocorrelation of the Low-High type. A High-Low type of negative spatial autocorrelation emerges when high values are surrounded by low values. In other words, the change in a location does not follow a trend similar to that followed by the neighboring area. Spatiotemporal outliers can also be traced in the case of High-Low or Low-High formations.

Discussion and Practical Guidelines

Differential Moran'sI is more descriptive in a spatiotemporal context than is mapping the local Moran's I index of a variable for each time stamp.

4.5.3   Emerging Hot Spot Analysis

Definition

Emerging hot spot analysis identifies spatiotemporal clusters in a spatial dataset (point counts or attribute values) using the Getis-Ord G_i^* statistic (hot spot analysis; ESRI 2017).

Why Use
A cluster may exist throughout the entire period, diminish after a specific time stamp, emerge after some other time stamp or disappear at some other point in time. Emerging hot spot analysis is used to trace such types of different hot spots or cold spots through time.

Interpretation
Emerging hot spot analysis groups locations according to the definitions as presented in Table 4.2 (ESRI 2017).

Discussion and Practical Guidelines
Emerging hot spot analysis is very useful for locating fluctuations in the density, distribution and total count of events in temporal point data. For example, crime events might concentrate on specific locations during the day and on other locations during the night. Through appropriate analysis, measures and related policies may be implemented to better handle the problems being studied.

4.6 Multiple Comparisons Problem and Spatial Dependence

Multiple Comparisons Problem
The multiple comparisons problem, also known as the multiple testing problem, is the problem of getting false significant results in multiple hypothesis testing. Local spatial statistics rely on tests conducted for every single spatial feature in the dataset. As multiple inferences (tests) are drawn for the same set of spatial features, there is a probability that some results will be declared statistically significant by chance, something that should be controlled (Mitchell 2005, Caldas de Castro & Singer 2006, ESRI 2018c). The multiple comparisons problem is a Type I error (see Section 2.5.5). In this type of error, we reject the null hypothesis when, in fact, it is true. In the context of geography, the more spatial objects, the more likely it is that some will be misclassified as statistically significant when a hypothesis is tested.

For example, if we run a test to detect spatial outliers and a spatial feature gets a p-value of 0.04, this spatial feature would be a spatial outlier based on statistically significant results at the 95% confidence level. However, there would be a 5% chance that this feature is not an outlier. When we run multiple statistical tests for a few spatial features, the multiple comparisons problem is not very severe. For many objects, however, the multiple comparisons problem is significant. Detecting spatial outliers in 10,000 spatial features infers 10,000 hypothesis tests and 10,000 p-values (one test and one p-value per object). For a 95% confidence level, the likelihood for each object to pass the test is 5%. In other words, 500 objects might be found to be significant by chance, largely altering the conclusions to be drawn.

Table 4.2 Emerging hot spot analysis groups.

Pattern	Description when statistically significant
No pattern	There is no indication for any hot or cold spots through the entire study period
New hot spot	This location is a new hot spot at the last time stamp. Before that, there was no pattern in this location.
Consecutive hot spot	This location is a hot spot for the final time steps.
Intensifying hot spot	This location is a hot spot for 90% of the time intervals including the last step. In addition, in each time step, the intensity of clustering is increasing.
Persistent hot spot	This location is a hot spot for at least 90% of the time intervals but with no notable fluctuations in the intensity of clustering.
Diminishing hot spot	A location that is a hot spot for at least 90% of the time intervals including the last one. The intensity of clustering is decreasing overall.
Sporadic hot spot	A location that is a hot spot for less than 90% of the time intervals. For none of the time intervals has this location been a statistically significant cold spot.
Oscillating hot spot	A hot spot for the final time step that has also been cold spot in some other time intervals.
Historical hot spot	This location is not a hot spot for the most recent time intervals. Still, it has been traced as statistically significant hot spot for at least 90% of the past time intervals.
New cold spot	This location is a new cold spot at the last time stamp. Before that, there was no pattern in this location.
Consecutive cold spot	This location is a cold spot at the final time steps.
Intensifying cold spot	This location is a cold spot for 90% of the time intervals including the last step. In addition, in each time step, the intensity of clustering is increasing.
Persistent cold spot	This location is a cold spot for at least 90% of the time intervals but with no notable fluctuations in the intensity of clustering.
Diminishing cold spot	A location that is a cold spot for at least 90% of the time intervals including the last one. The intensity of clustering is decreasing overall.
Sporadic cold spot	A location that is a cold spot for less than 90% of the time intervals. For none of the time intervals has this location been a statistically significant hot spot.
Oscillating hot spot	A cold spot for the final time step that has also been a hot spot in some other time intervals
Historical hot spot	This location is not a cold spot for the most recent time intervals. Still, it has been traced as statistically significant cold spot for at least 90% of the past time intervals.

Spatial Dependency

According to the first law of geography, spatial entities that are closer tend to be more similar than those lying further away. This is what we call spatial dependency (see Section 1.3). In local spatial statistics, spatial dependency is highly likely to seem more evident than it really is. The reason is that local spatial statistics are calculated using the neighboring values of each spatial feature (using a spatial weights matrix). However, features that are near each other are likely to share common neighbors as well, leading to an overestimation of spatial dependence and an artificial inflation of statistical significance.

Dealing with Multiple Comparisons Problem and Spatial Dependence

Two approaches can be used to handle the multiple comparisons problem and spatial dependence:

Bonferroni correction (Bonferroni 1936): This correction divides the alpha significance level by the number of tests (in spatial analysis this equals the number of features). For example, for ten tests and alpha = 0.05, tests with p-values smaller than 0.05/10 = 0.005 are statistically significant. In other words, the p-value at which a result is declared statistically significant is stricter.

False discovery rate (FDR) correction: FDR correction has been particularly influential in statistics and has been applied to other research areas as well – e.g., genetics, biochemistry (Benjamini & Hochberg 1995, Benjamini 2010). False discovery rate correction is used to account for both spatial dependency and the multiple comparisons problem. It lowers the p-value at which a statistic is regarded as significant. FDR correction estimates the number of objects misclassified (false positive error, rejects the null hypothesis) for a given confidence level and then adjusts the critical p-value. Statistically significant p-values (less than alpha) are ranked from smallest (strongest) to largest (weakest). FDR calculates the expected error in rejecting the null hypothesis (false positive) and, based on this estimate, the weakest objects are eliminated. Within the spatial statistics context, applying FDR correction reduces the number of features with statistically significant p-values.

Many statisticians recommend ignoring both the multiple comparisons problem and the spatial dependence problem. For a small number of spatial objects (say, fewer than 100), few objects are likely to be misclassified, so correction may not be necessary. As the number of objects increases, correction should be considered. As software tools for applying corrections are readily available, it is more rational to utilize them and then compare the results with the non-corrected outputs. For example, applying FDR correction in hot spot analysis will probably reduce the features assigned to clusters relative to hot spot analysis without correction. It is advised to test which features are not included, along with their attributes and their neighbors. Finally, we should keep in mind that, even with corrections, we might still experience false results. The question of whether the FDR or Bonferroni correction should be applied in the calculation of spatial statistics depends on the problem, the knowledge of the study area, the intuition of the researcher and the results produced with and without corrections.

4.7   Chapter Concluding Remarks

- Spatial autocorrelation is the degree of spatial dependency, association or correlation between the value of an observation of a spatial entity and the values of neighboring observations of the same variable.
- A major difference with statistical correlation is that, while statistical correlation refers to two distinct variables with no reference to location (or for the same location), spatial autocorrelation refers to a value of a single variable at a specific location in relation to the values of the same variable to its neighboring locations.
- *Lag-X* is the weighted average values of *X* in a specified neighborhood.
- There are four types of arrangement in a Moran's *I* scatter plot: High-High and Low-Low, expressing positive spatial autocorrelation, and High-Low and Low-High, indicating negative spatial autocorrelation.
- Obtaining statistically significant results for the Local Moran's *I* (e.g., detecting clustering) does not mean that we will obtain statistically significant results for the Global Moran's *I* as well.
- When calculating spatial autocorrelation, "global" implies that a single value for the index is produced for the entire pattern, while "local" means that a value is produced for each spatial object separately.
- While Moran's *I* index cannot distinguish between high- or low-value clustering, the General G-Statistic can. On the other hand, the General G-Statistic is appropriate only when there is positive spatial autocorrelation, as it detects the presence of hot or cold spots.
- General G-Statistic is more an index of spatial association, or an index of positive spatial autocorrelation, than a pure spatial autocorrelation index.
- In incremental spatial autocorrelation, by locating the peak in the graph, the z-score and the corresponding distance, we can better define the distance band to be used in many spatial statistics such as hot spot analysis.
- More than one peak may occur. This is not wrong. Different distance bands (peaks) might reveal underlying processes at different scales of analysis.
- Smaller distances are often more suitable for geographical analysis at the local scale.
- Before running incremental spatial autocorrelation, we should check if locational outliers exist and remove them if necessary.
- Local Moran's *I* is used to identify if clusters or outliers exist in the spatial dataset. That is why this method is also called "cluster and outlier analysis."
- When calculating local spatial autocorrelation, permutations are used to estimate the likelihood of generating, through complete spatial randomness, a spatial arrangement of values similar to the observed one.

- Hot spot analysis cannot be used to locate outliers.
- Hot spot analysis cannot be directly applied to point objects (e.g., crime incidents) without any attribute fields attached. However, we can aggregate data by using spatial join to polygons and then continue the analysis in the usual fashion.
- Optimized hot spot analysis is a procedure used to optimally select the parameters of the Getis-Ord G_i^* index.
- It is easier to use optimized hot spot analysis instead of the Getis-Ord G_i^* index, as long as we comprehend the outputs.

Questions and Answers

The answers given here are brief. For more thorough answers, refer back to the relevant sections of this chapter.

Q1. What is spatial autocorrelation? What types of spatial autocorrelation exist? Which are the most commonly used metrics?

A1. Spatial autocorrelation is the degree of spatial dependency, association or correlation between the value of an observation of a spatial entity and the values of neighboring observations of the same variable. There are two types of spatial autocorrelation, namely global and local. Global spatial autocorrelation measures autocorrelation by a single value for the entire study area. To estimate spatial autocorrelation at the local level, we use local measures of spatial autocorrelation. The most common global spatial autocorrelation measures are the Moran's I index and the General G-Statistic. Local measures of spatial autocorrelation are the Local Moran's I index, the Getis-Ord G_i and the Getis-Ord G_i^* statistic.

Q2. Why is spatial autocorrelation important to geographical analysis and spatial statistics?

A2. Spatial autocorrelation analysis is extremely important in geographical studies. If spatial autocorrelation did not exist, geographical analysis would be of little interest. In conventional statistics, the observed samples are assumed to be independent. In the presence of spatial autocorrelation, this assumption is violated. Observations are now spatially clustered or dispersed. This typically means that classical statistical tools are no longer valid. For example, linear regression would lead to biased estimates or exaggerated precision. As such, spatial statistics should be used instead.

Q3. What is incremental spatial autocorrelation, and why is it used?

A3. Incremental spatial is a method based on Global Moran's I index to test for the presence of spatial autocorrelation at a range of band distances. It is used to approximate the appropriate scale of analysis. Instead of arbitrarily selecting distance bands, this method identifies an appropriate fixed distance band, for which spatial autocorrelation is more pronounced. In other words, it allows us to identify the farthest distance at which an object still has a significant impact on another one. After the appropriate scale of analysis is established, local spatial autocorrelation indices and other spatial statistics can be calculated more accurately.

Q4. Why is a Moran's *I* scatter plot used?

A4. It is used to visualize the spatial autocorrelation statistic. It allows for a visual inspection of the spatial associations in the neighborhood of each observation (data point). In other words, it provides a representation used to assess how similar an attribute value at a location is to its neighboring ones. The slope of the regression line over the data points equals the Moran's *I* index value when calculated using binary and row standardized weights. The produced scatter plot identifies which type of spatial autocorrelation exists according to the place where a dot lies (the dot standing for a spatial entity, such as a polygon).

Q5. How can we identify a spatial outlier with a Moran's scatter plot? What does a Low-High arrangement mean?

A5. We can identify a spatial outlier by inspecting a Moran's *I* scatter plot in locations where High-Low or Low-High concentrations exist. A Low-High arrangement means that a spatial object depicted as a dot in the scatter plot has a low value (for the variable studied) and is surrounded by spatial objects with high values. It is probably a spatial outlier, but further analysis should be carried out to confirm it.

Q6. Why is it necessary to set up the right scale of analysis?

A6. The scale of analysis defines the size and shape of the neighborhoods for which spatial statistics are calculated. The scale of analysis is closely related to the problem in question. Hypothetically, unemployment clustering statistically significant at 100 m and 1,000 m peaks reflects patterns of clustering at both the census-block level and the postcode level. If we are interested only in the census-block level, we could apply the 100 m distance in our analysis. Greater distances reflect broader, regional trends (e.g., east to west), while smaller distances reflect local trends (e.g., between neighborhoods). If we use a large scale of analysis, when we are looking at a local level, we might generalize and lose hidden spatial heterogeneity.

Q7. What is cluster and outlier analysis? How can we interpret the results of the index used?

A7. It is an analysis applying local Moran's *I* to (a) identify if a clustering of high or low values exists and (b) to trace spatial outliers. If the p-value is large (usually $p > 0.05$), the results are not statistically significant. A small p-value (usually <0.05) indicates that we can reject the null hypothesis of complete spatial randomness and accept that spatial autocorrelation exists. In this case, when z-value is positive, we have positive spatial autocorrelation and clustering. If z-value is negative, we have negative spatial autocorrelation and a dispersed pattern of values. A high negative value is an indication of a spatial outlier.

Q8. What is a High-High or Low-Low cluster in a Local Moran's I?

A8. A high positive z-score (e.g., greater than 2.60) for a spatial entity means that the neighboring spatial entities have similar values. If the values are high, then High-High clusters are formed, meaning that spatial entities with high values (for a specific variable) are surrounded by spatial entities of high values (of the same variable). If the values are low, then Low-Low clusters are formed, meaning that spatial entities with low values are surrounded by spatial entities of low values.

Q9. What is hot spot analysis, and what are a cold spot and a hot spot? Can this analysis be used with point data?

A10. Hot spot analysis identifies if low or high values of a variable are spatially clustered and create cold spots or hot spots respectively. In case of point features, it might be more interesting to study the intensity of the objects rather than a specific attribute. In this respect, we use hot spot analysis to identify if hot spots or cold spots of events' intensity exist. Such analysis should begin by aggregating points into some regions (e.g., postcodes, census tracts).

Q10. What are the main benefits of using optimized hot spot analysis?

A10. Optimized hot spot analysis performs hot spot analysis using the Getis-Ord G_i^* index in an automated way to ensure optimal results. The method is used to

(a) Identify how many locational outliers exist (if any).

(b) Estimate the distance band at which the spatial autocorrelation is more pronounced (scale of analysis).

(c) Adjust for spatial dependence and multiple testing through the false discovery rate correction method.

(d) Handle point events with no variables attached. These events are automatically aggregated into weighted features within some regions (i.e., grid). The weighted variable is then analyzed in the usual fashion.

LAB 4
SPATIAL AUTOCORRELATION

Overall Progress

Spatial Analysis/Lab Workflow

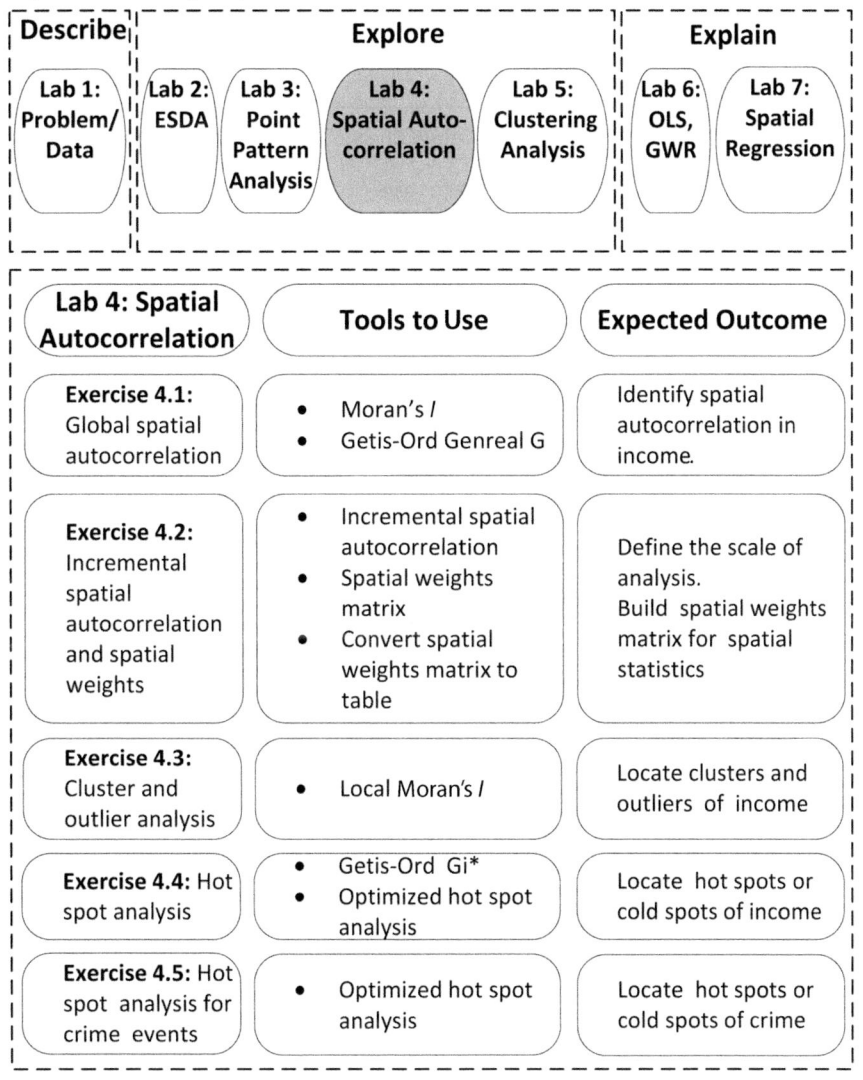

Figure 4.7 Lab 4 workflow and overall progress.

Scope of the Analysis

This lab deals with

- **Objective 1:** Locating high income areas (see Table 1.2)
- **Objective 2:** Locating low crime areas

We further analyze the income distribution in the city to identify whether spatial clustering and spatial autocorrelation exist and also to locate income hot spots through spatial statistics (see Figure 4.7). Moreover, we will study the spatial autocorrelation patterns of crime by locating cold and hot spots.

Section A ArcGIS

Exercise 4.1 Global Spatial Autocorrelation

In this exercise, we calculate the global spatial autocorrelation of income using the Moran's *I* index and the Getis-Ord General G-Statistic.

ArcGIS Tools to be used: Spatial Autocorrelation (Moran's I),

High/Low Clustering (Getis-Ord General G))

ACTION: Calculate Global Moran's I

Navigate to the location you have stored the book dataset and click on Lab4_SpatialAutocorrelation.mxd

Main Menu > File > Save As > My_Lab4_SpatialAutocorrelation.mxd

In I:\BookLabs\Lab4\Output

ArcToolBox > Spatial Statistics Tools > Analyzing Patterns > Spatial Autocorrelation (Moran's I)

Input Feature Class = City (see Figure 4.8)

Input Field = Income

Generate Report = Check the box

Conceptualization of Spatial Relationships = INVERSE_DISTANCE (See Chapter 1 for theory.)

Distance: EUCLIDEAN_DISTANCE

Standardization = ROW (See Chapter 1 for theory. We should use ROW when we have polygons and data aggregated at this level.)

Distance Band or Threshold Distance = Leave blank. This is a cutoff distance for Inverse Distance and Fixed Distance conceptualization methods. Features outside the specified cutoff value for a target feature`are ignored. The tool uses by default the distance ensuring that each single feature has at least one

Exercise 4.1 (*cont.*)

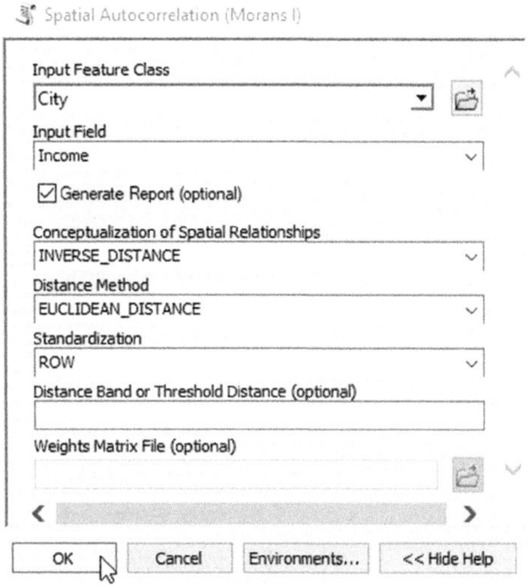

Figure 4.8 Global Moran's *I* tool.

neighbor. This distance is not necessarily the appropriate one. We can use this value to begin with and progressively increase it to test how values of the index will vary. See Chapter 1 for theory.

OK

Main Menu > Geoprocessing > Results > Current Session > Spatial Autocorrelation > DC on MoransI_Result.html (see Figure 4.9)

Interpreting results: The Moran's *I* Index is 0.61 (see Figure 4.9). Given the z-score of 10.86 and the *p*-value of 0.000000, there is a less than 1% likelihood (significance level α) that this clustered pattern is the result of random chance. In other words, we have a 99% probability (confidence level) that the distribution of income forms a clustered pattern. A slightly different way to interpret the results is as follows: The spatial arrangement of the income values has a tendency to cluster, and there is a likelihood of less than 1% that this pattern is the result of random chance. The distance threshold defined by the tool is set to 1050.79 m, so every postcode has at least one neighbor. Global Moran's provides a first indication of income clustering. However, we cannot locate where the clustering occurs just by using this index. Moreover, although we calculated spatial autocorrelation, we have not yet defined the appropriate scale of analysis.

Exercise 4.1 *(cont.)*

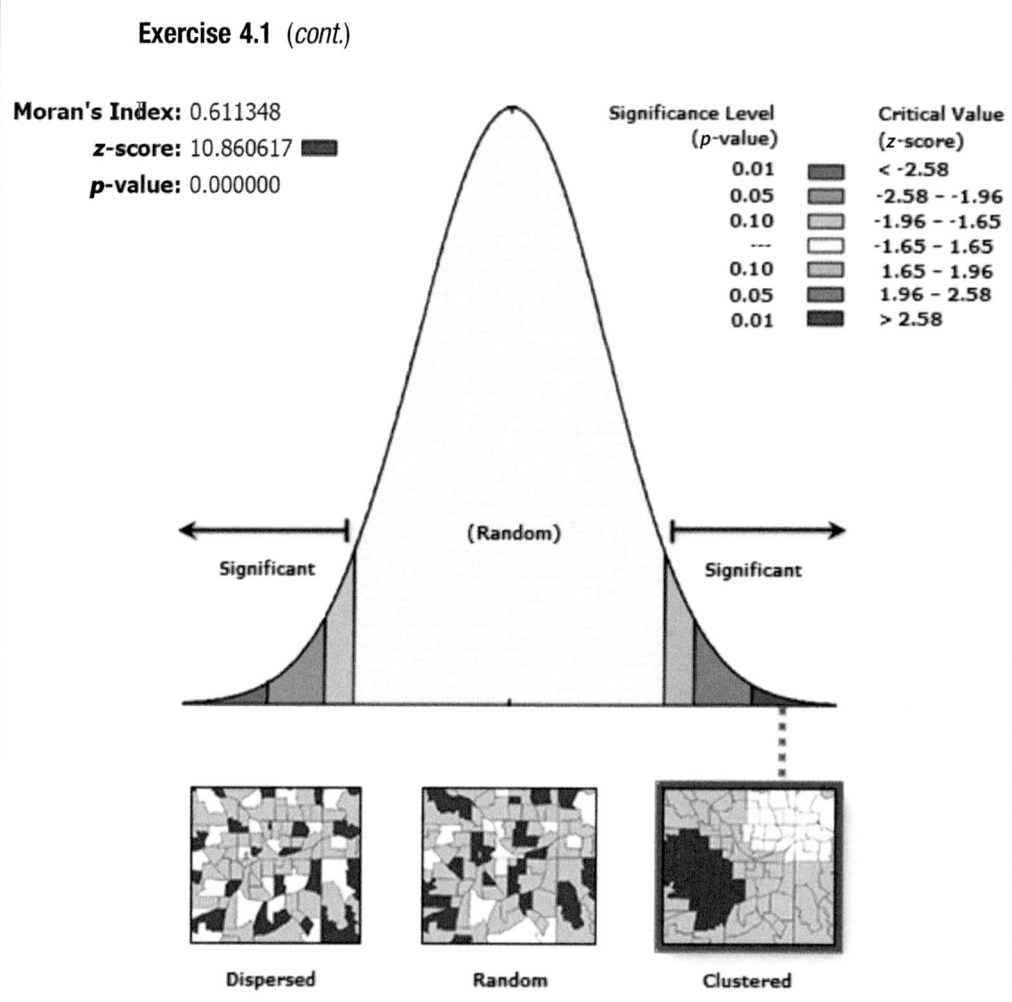

Moran's Index: 0.611348

   z-score: 10.860617

   p-value: 0.000000

Significance Level (*p*-value)		Critical Value (*z*-score)
0.01		< -2.58
0.05		-2.58 – -1.96
0.10		-1.96 – -1.65
---		-1.65 – 1.65
0.10		1.65 – 1.96
0.05		1.96 – 2.58
0.01		> 2.58

Figure 4.9 Global Moran's *I* report.

ACTION: Calculate High/Low Clustering (Getis–Ord General G)

ArcToolBox > Spatial Statistics Tools > Analyzing Patterns > High/Low Clustering (Getis–Ord General G)

Input Feature Class = City (see Figure 4.10)

Input Field = Income

Generate Report = Check the box

Conceptualization of Spatial Relationships = INVERSE_DISTANCE (See Chapter 1 for theory)

Exercise 4.1 (*cont.*)

Distance: EUCLIDEAN_DISTANCE

Standardization = ROW

Distance Band or Threshold Distance = Leave blank

OK

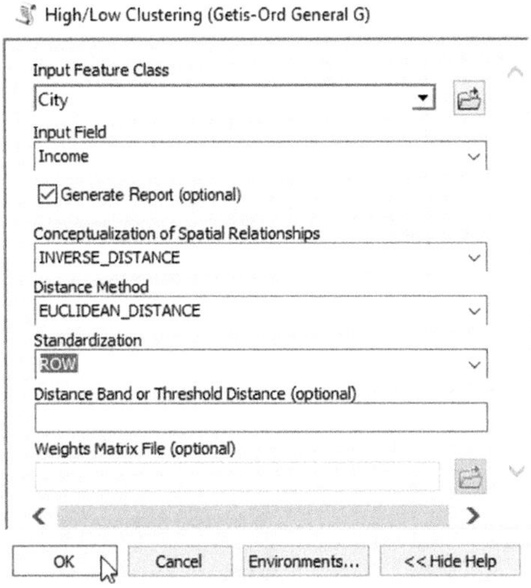

Figure 4.10 Getis-Ord General G tool.

Main Menu > Geoprocessing > Results > Current Session > High/ Low Clustering (Getis-Ord General G)> DC on GeneralG_Results .html (see Figure 4.11)

Main Menu > File > Save

Interpreting results: The Getis-Ord General G Index is 0.01 (see Figure 4.11). Given the z-score of 7.052 (positive value) and the p-value of 0.000020, there is a less than 1% likelihood (significance level α) that this clustered pattern of high values is the result of random chance. In other words, we have a 99% probability (confidence level) that the distribution of income forms a clustered pattern of high values. As does the Global Moran's I, the Getis-Ord General G provides a first indication of income clustering.

Exercise 4.1 (*cont.*)

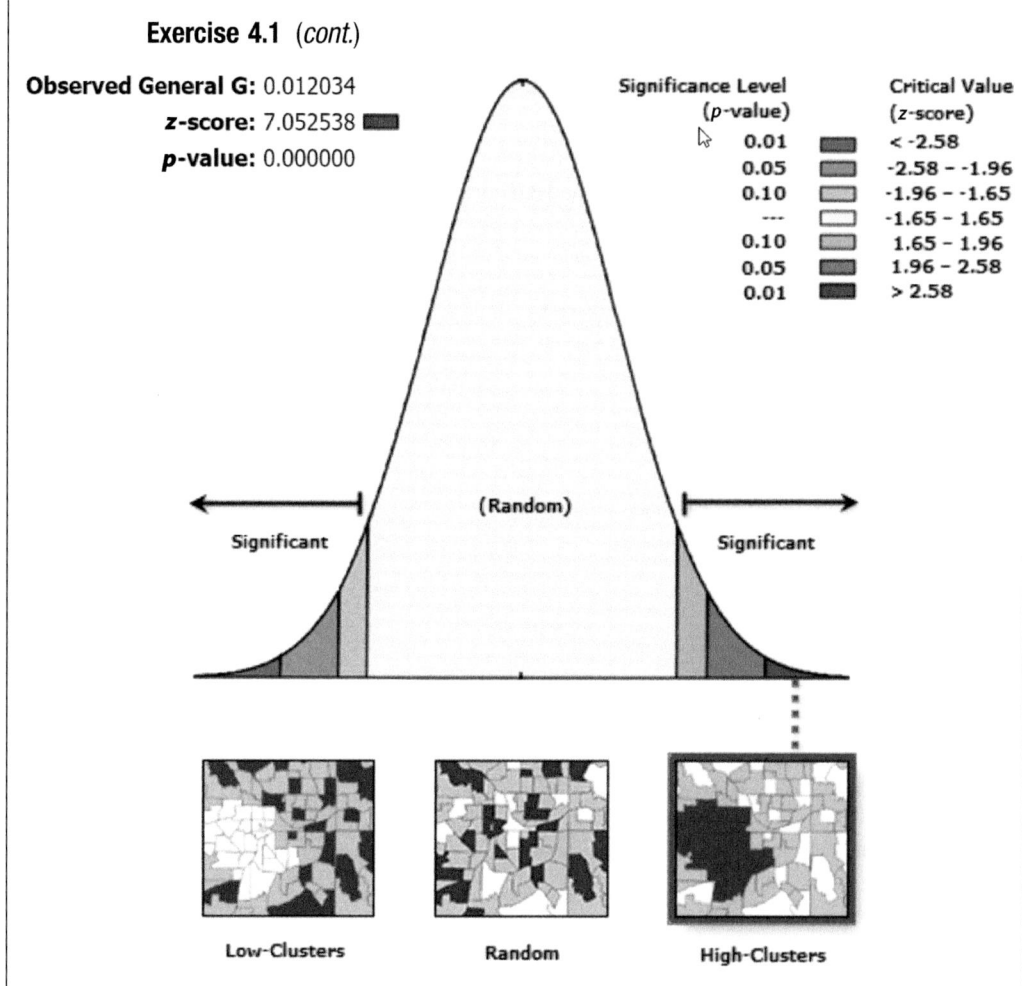

Figure 4.11 Getis-Ord General G report.

However, we cannot locate where the clustering occurs just by using this index. Moreover, although we calculated spatial autocorrelation, we have not yet defined the appropriate scale of analysis.

Exercise 4.2 Incremental Spatial Autocorrelation and Spatial Weights Matrix

In this exercise, we calculate the incremental spatial autocorrelation of income to define an appropriate scale of analysis. Based on this scale, the spatial weights matrix (see Section 1.9) is calculated, which is needed for local spatial statistics.

Exercise 4.2 (*cont.*)

ArcGIS Tools to be used: Incremental Spatial Autocorrelation, Generate Spatial Weights Matrix, Convert Spatial Weights Matrix to Table

ACTION: Incremental Spatial Autocorrelation

Navigate to the location you have stored the book dataset and click

My_Lab4_SpatialAutocorrelation.mxd

ArcToolBox > Spatial Statistics Tools > Analyzing Patterns > Incremental Spatial Autocorrelation

Input Features = City (see Figure 4.12)

Input Field = Income

Number of distance bands = 10

Beginning Distance = Leave blank

Distance Increment = Leave blank

Distance = EUCLIDEAN

Row Standardization = Check the box

Figure 4.12 Incremental spatial autocorrelation dialog box.

Exercise 4.2 (*cont.*)

Output Table = I:\BookLabs\Lab4\Output\Increment

Output Report File = I:\BookLabs\Lab4\Output\Increment.pdf

OK

Main Menu > Geoprocessing > Results > Current Session > Incremental Spatial Autocorrelation > DC on Output Report File: Increment.pdf

Global Moran's I Summary by Distance

Global Moran's I Summary by Distance

Distance	Moran's Index	Expected Index	Variance	z-score	p-value
1051.00	0.554941	-0.011236	0.003096	10.174893	0.000000
1157.38	0.515557	-0.011236	0.002437	10.670662	0.000000
1263.76	0.471874	-0.011236	0.002063	10.635359	0.000000
1370.14	0.453057	-0.011236	0.001729	11.166550	0.000000
1476.52	0.415797	-0.011236	0.001389	11.457432	0.000000
1582.90	0.369182	-0.011236	0.001170	11.120921	0.000000
1689.27	0.330625	-0.011236	0.001012	10.745499	0.000000
1795.65	0.296338	-0.011236	0.000852	10.537571	0.000000
1902.03	0.272382	-0.011236	0.000747	10.378124	0.000000
2008.41	0.241786	-0.011236	0.000641	9.991983	0.000000

First Peak (Distance, Value): 1157.38, 10.670662

Max Peak (Distance, Value): 1476.52, 11.457432

Distance measured in Meters

Interpreting results: Prior to incremental spatial autocorrelation, we should trace if locational outliers exist. We conducted this analysis in Exercise 3.4 and concluded that no locational outliers existed (the theoretical discussion in Section 4.3 explains how to handle locational outliers).

We observe that there are two peaks (see Figure 4.13): one at 1,157 m and one at 1,476 m. The Moran's *I* values for these distances are 0.51 and 0.41, respectively (with high z-scores), indicating intense clustering. Both distances reveal a form of clustering. As mentioned in the theoretical section, there is not a single correct distance at which to perform our analysis, as the

Exercise 4.2 (*cont.*)

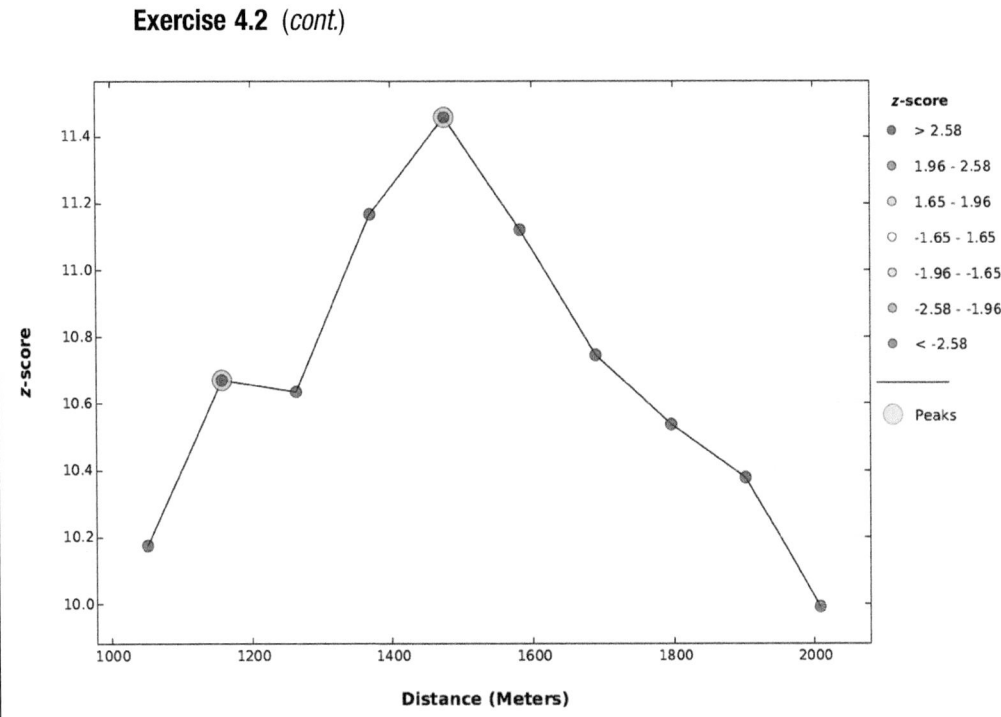

Figure 4.13 Incremental spatial autocorrelation graph. z-scores are plotted over incremental distances. Peaks are highlighted with a larger circle.

scale largely depends on our problem. It is quite common to select the first peak. As such, the scale of analysis of income can be set to 1,150 m (rounded 1157.38). The overall conclusion is that there is spatial autocorrelation of income and an underlying clustering process.

ACTION: Generate Spatial Weights Matrix

ArcToolBox > Spatial Statistics Tools > Modeling Spatial Relationships > Generate Spatial Weights Matrix

Input Feature Class = City (Navigate to I:\BookLabs\Data\City. shp) (see Figure 4.14)

Unique ID Field = PostCode

Output Spatial Weights Matrix File =

I:\BookLabs\Lab4\Output\CityWeights.swm

Conceptualization of Spatial Relationships = FIXED_ DISTANCE

Distance Method = EUCLIDEAN

Exercise 4.2 (*cont.*)

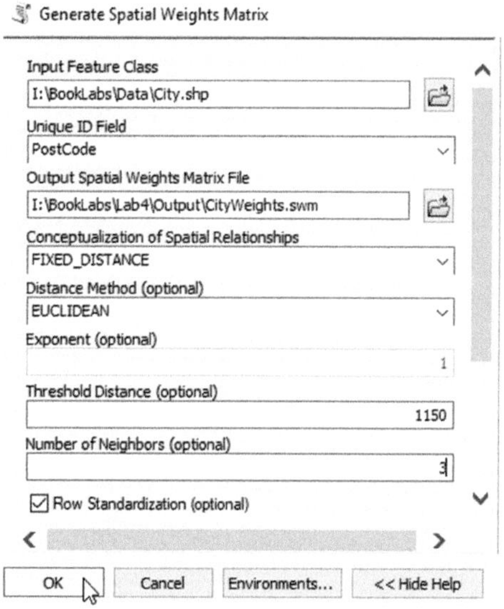

Figure 4.14 Generate spatial weights matrix dialog box.

Exponent = 1

Threshold Distance = 1150 (as produced from incremental spatial autocorrelation in exercise 4.1)

Number of Neighbors = 3

Row Standardization = Check

OK

ArcGIS tip: The Number of neighbors parameter (see Figure 4.14) is available only from the Generate Spatial Weights Matrix tool. The *k*-nearest neighbors option is also used in exploratory regression (analyzed in Chapter 6) to assess regression residuals. It takes a default value of 8.

Interpreting results: We set FIXED_DISTANCE as the function with which to conceptualize space, as this method is more appropriate for hot spot analysis (see Figure 4.14). The Threshold Distance is set to 1150 as the appropriate scale of analysis and is the result of the first part of this exercise (scale of analysis through incremental spatial autocorrelation). Objects lying further away than this distance will not be included in the calculation of the weights function. As we set a cutoff value for the FIXED_DISTANCE, some features may have no neighbors at this distance. To calculate the weights

Exercise 4.2 (*cont.*)

matrix, however, we should have at least some minimum number of neighbors for all features. To ensure that each object has at least a minimum number of neighbors, we use the parameter Number of Neighbors. By this combination (threshold and number of neighbors), features with no neighbors (or fewer than three) inside the threshold value will finally be attached to their neighbors. In other words, the threshold value is temporarily extended to ensure that each feature will have at least a minimum number of neighbors defined. In general, a spatial weights matrix is automatically generated when we apply spatial statistics by defining a conceptualization method. Nevertheless, to have more control over the weights, it is recommended to create a user-defined spatial weights matrix that can be applied thereafter. If uniformity or the isotropic environment is violated in a part of our case study, we might need to change the weights. For instance, two objects might have large weights, indicating high interaction and small distance. Due to a natural barrier (e.g., river, lake, island polygons), these objects might be close, but their interaction might be low. In such a case, we could edit the weight matrix accordingly.

ACTION: Convert Spatial Weights Matrix to Table

A typical spatial weights matrix in ArcGIS has three columns (see Figure 4.15): ID (unique ID of the spatial object), NID (the ID of the neighboring object with which there is a relationship) and Weight (the value of the weight that quantifies the spatial relationship). Nonexistent spatial relationships (weight = 0) are not included in the matrix to keep the table short. The output file is in a unreadable format. To read and edit the weights for each set of spatial objects, we must convert the .swm file into a table using the Convert Spatial Weights to Table tool.

ArcToolBox > Spatial Statistics Tools > Utilities > Convert Spatial Weights Matrix to Table

Input Spatial Weights Matrix File =

I:\BookLabs\Lab4\Output\CityWeights.swm

Output Table = I:\BookLabs\Lab4\Output\CityWeights

OK

TOC > List By Source > RC CityWeitghs > Open

Close table

Main Menu > File > Save

Exercise 4.2 (*cont.*)

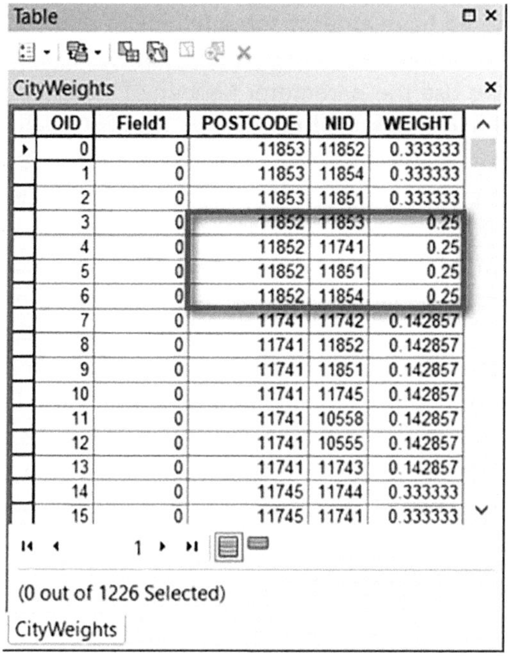

OID	Field1	POSTCODE	NID	WEIGHT
0	0	11853	11852	0.333333
1	0	11853	11854	0.333333
2	0	11853	11851	0.333333
3	0	11852	11853	0.25
4	0	11852	11741	0.25
5	0	11852	11851	0.25
6	0	11852	11854	0.25
7	0	11741	11742	0.142857
8	0	11741	11852	0.142857
9	0	11741	11851	0.142857
10	0	11741	11745	0.142857
11	0	11741	10558	0.142857
12	0	11741	10555	0.142857
13	0	11741	11743	0.142857
14	0	11745	11744	0.333333
15	0	11745	11741	0.333333

(0 out of 1226 Selected)

CityWeights

Figure 4.15 Spatial weights table.

Interpreting results: NID is the ID of the neighboring object, and WEIGHT is the calculated weight (see Figure 4.15). For example, postcode 11852 has four neighbors (11853,11741,11851,11854) within a fixed distance of 1,150 m (the distance between the polygon centroids); that is why the weight is 0.25 on each.

Exercise 4.3 Cluster and Outlier Analysis (Anselin Local Moran's *I*)

In this exercise, we calculate the local spatial autocorrelation of income using Local Moran's *I* to identify if clusters and outliers exist.

ArcGIS Tools to be used: Cluster and Outlier Analysis

ACTION: Cluster and Outlier Analysis

Navigate to the location you have stored the book dataset and click

Exercise 4.3 (*cont.*)

My_Lab4_ SpatialAutocorrelation.mxd

ArcToolBox > Spatial Statistics Tools > Mapping Clusters > Cluster and Outlier Analysis

Input Feature Class = City (see Figure 4.16)

Input Field = Income

Output Feature Class = I:\BookLabs\Lab4\Output\LocalMoranI.shp

Conceptualization of Spatial Relationships =

GET_SPATIAL_WEIGHTS_FROM_FILE

Weights Matrix File = I:\BookLabs\Lab4\Output\CityWeights.swm

Apply False Discovery Rate (FDR) Correction = Check

OK
TOC > RC LocalMoranI > Open Attribute Table

Close Table

Main Menu > File > Save

Figure 4.16 Local Moran's *I* dialog box.

Exercise 4.3 *(cont.)*

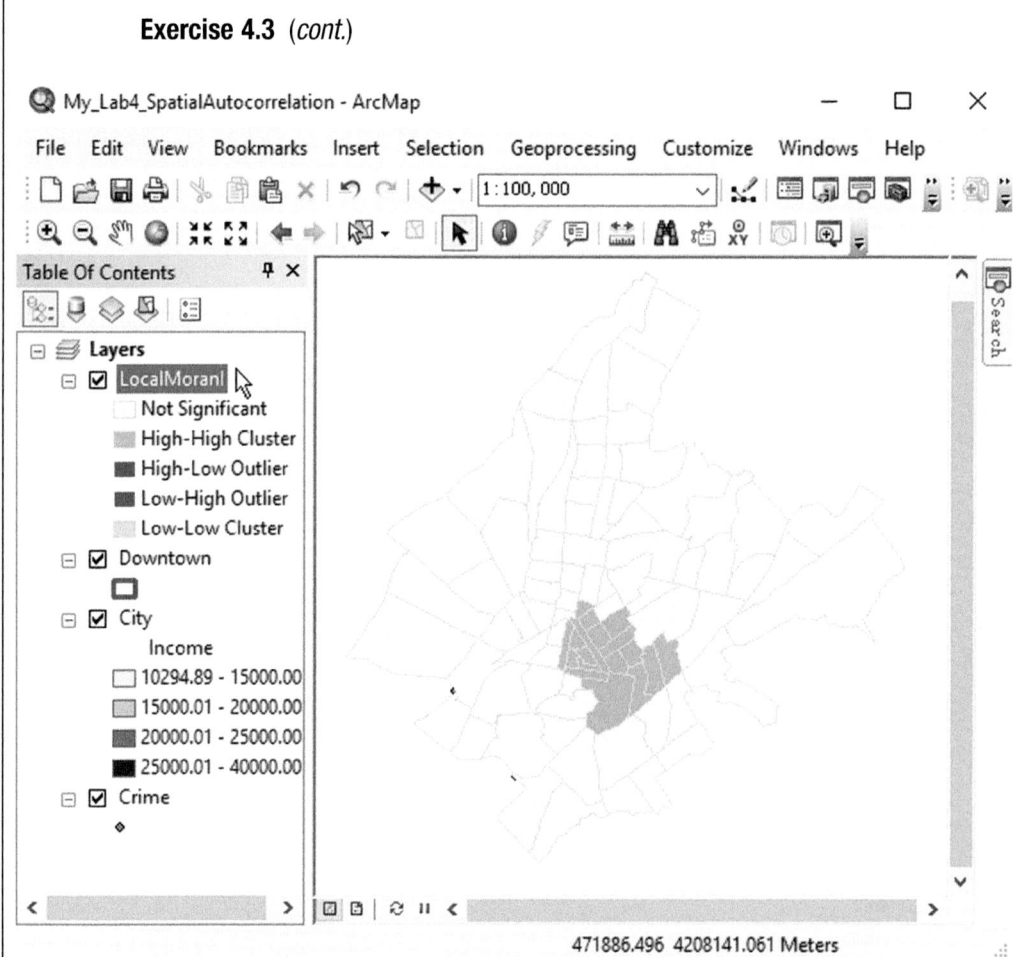

Figure 4.17 Local Moran's *I* output map.

Table

LocalMoranI

FID	Shape *	PostCode	Income	LMiIndex	LMiZScore	LMiPValue	COType
66	Polygon	10432	17303.37940	0.051067	0.382064	0.702414	
67	Polygon	10433	16837.43258	0.014275	0.135738	0.892028	
68	Polygon	10437	11588.62801	-0.118458	-0.622711	0.533474	
69	Polygon	10673	28401.35222	2.716906	15.401938	0	HH
70	Polygon	10674	37644.13135	5.188319	23.367887	0	HH
71	Polygon	10675	28192.54567	3.373068	14.634014	0	HH
72	Polygon	10676	25006.13072	2.373099	9.052616	0	HH
73	Polygon	10677	21019.82373	0.410108	2.790492	0.005263	HH
74	Polygon	10678	21579.56497	0.454739	3.163032	0.001561	HH
75	Polygon	10679	22395.23767	1.080097	7.227728	0	HH

1 (0 out of 90 Selected)

LocalMoranI

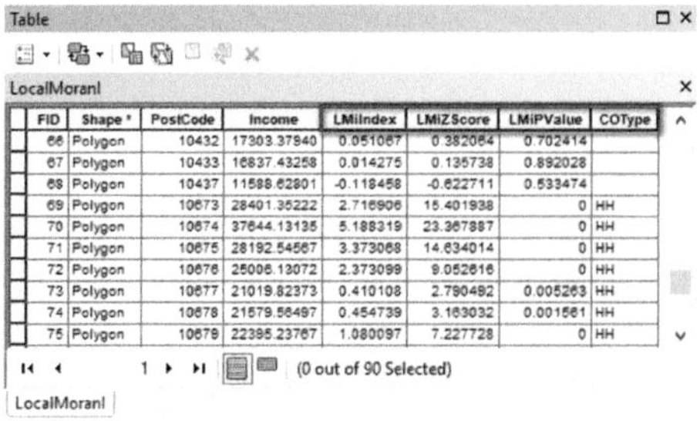

Figure 4.18 Local Moran's *I* table with local Moran's *I* index value, *z*-score, *p*-value and type of cluster assigned to each object.

Exercise 4.3 (*cont.*)

Interpreting results: A new layer is added to the table of contents (see Figure 4.17). The COType field in the Output Feature Class indicates if a postcode is an outlier or if it belongs to a cluster (see Figure 4.18). If the postcode has high income and is surrounded by postcodes with low incomes, it is marked as HL. If the postcode has low income and is surrounded by postcodes with high income, it is marked as LH. If postcodes are clustered, the COType field is HH for a statistically significant cluster of high-income values and LL for a statistically significant cluster of low-income values. The attribute table also shows the local Moran's *I* value, the z-score and the *p*-value. In this example and with FDR applied, income is positively spatially autocorrelated, and a statistically significant clustering of High-High values is observed in the center of the city at the 99% confidence level. No outliers or clusters of low values are detected elsewhere. In other words, people with high incomes tend to live in the red areas located in and around the downtown area.

Exercise 4.4 Hot Spot Analysis (Getis-Ord Gi*I) and Optimized Hot Spot Analysis

In this exercise, we calculate local spatial autocorrelation to identify income hot spots and cold spots using the local Getis-Ord G_i^* index.

    ArcGIS Tools to be used: Hot Spot Analysis, Optimized Hot Spot Analysis

ACTION: Hot Spot Analysis

Navigate to the location you have stored the book dataset and click

My_Lab4_ SpatialAutocorrelation.mxd

ArcToolBox > Spatial Statistics Tools > Mapping Clusters > Hot Spot Analysis

Input Feature Class = City (see Figure 4.19)

Input Field = Income

Output Feature Class = I:\BookLabs\Lab4\Output\HotSpotIncome.shp

Conceptualization of Spatial Relationships =

Exercise 4.4 *(cont.)*

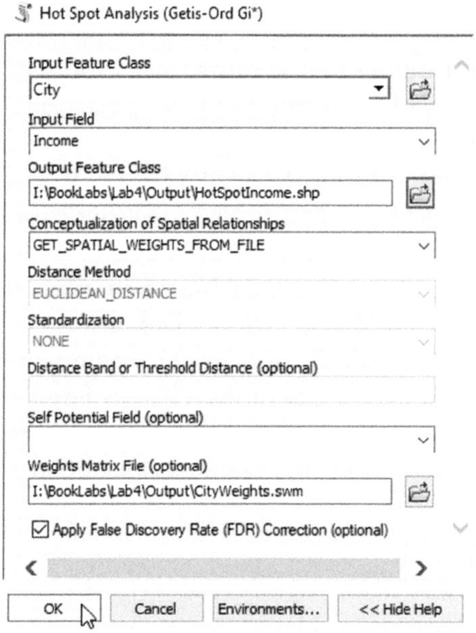

Figure 4.19 Hot spot analysis dialog box.

```
GET_SPATIAL_WEIGHTS_FROM_FILE

Self Potential Filed = Leave blanc

Weights Matrix File = I:\BookLabs\Lab4\Output\CityWeights.swm
```

(We can also directly use the FIXED_DISTANCE_BAND in the conceptualization method and add the distance threshold. Still, this option does not allow for specifying a minimum number of nearest neighbors. It is advised to use a spatial weights matrix from file – see Exercise 4.2).

Apply False Discovery Rate (FDR) Correction = Do not check (Check what happens when FDR is checked and refer back to theory)

OK

Interpreting results: A new layer is added to the table of contents (see Figure 4.20). The Gi_Bin field in the Output Feature Class indicates whether there is a hot spot or a cold spot and the related confidence level. In our example, we locate a statistically significant hot spot in and around the city center and a statistically significant cold spot in the western part of the city.

Exercise 4.4 (*cont.*)

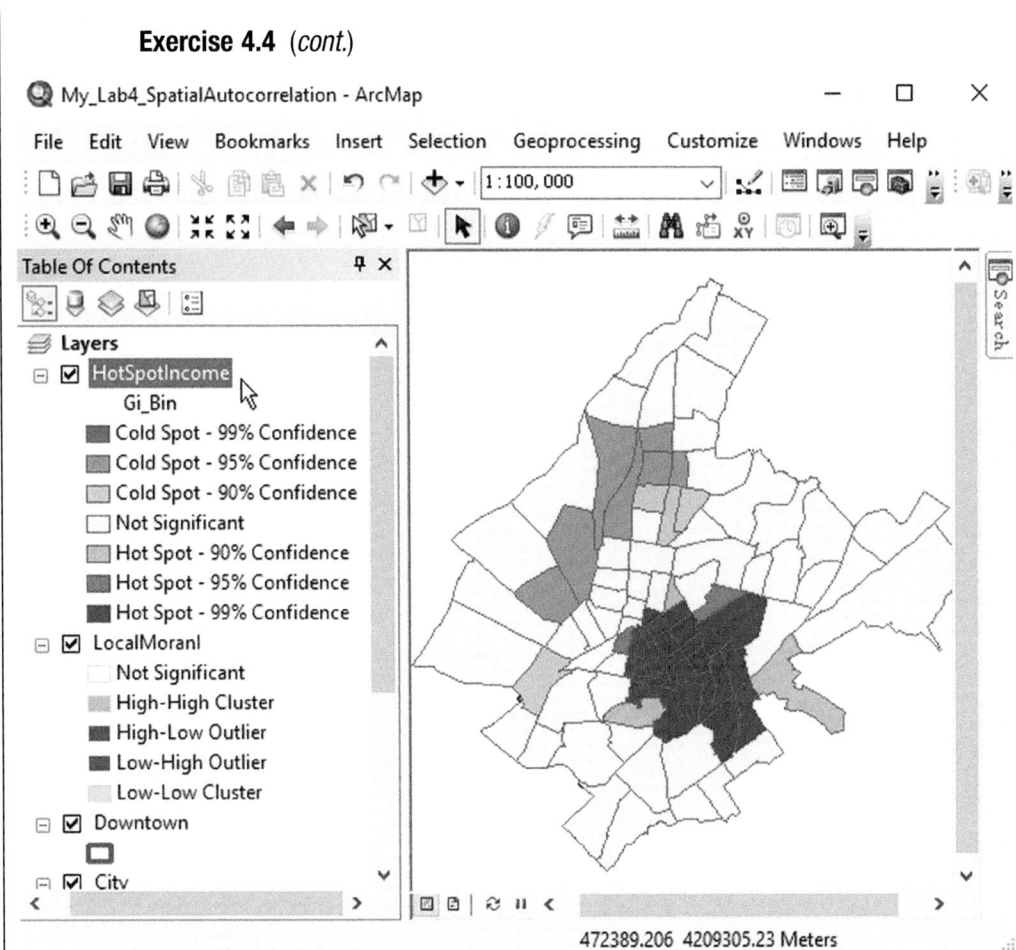

Figure 4.20 Hot spot analysis output map indicating cold (blue) and hot (red) spots.

Nonsignificant results mean that there is no indication of income clustering in these postcodes. An income cold spot means that polygons with low income values are surrounded by polygons with low income values. An income hot spot means that polygons with high income values are surrounded by polygons with high values. We notice that the hot spot analysis identifies a cluster of low values in addition to those identified in the cluster and outlier analysis in Exercise 4.3. This means that it is better to run both statistics and evaluate the results comparatively. As we are looking for high-income areas, the red areas might be more appropriate as locations for the coffee shop.

ACTION: Optimized Hot Spot Analysis

ArcToolBox > Spatial Statistics Tools > Mapping Clusters > Optimized Hot Spot Analysis

Exercise 4.4 (*cont.*)

```
Input Features = City (see Figure 4.21)

Output Features =

I:\BookLabs\Lab4\Output\C3_IncomeOpitmizedHotSot.shp

Analysis Field = Income

OK
```

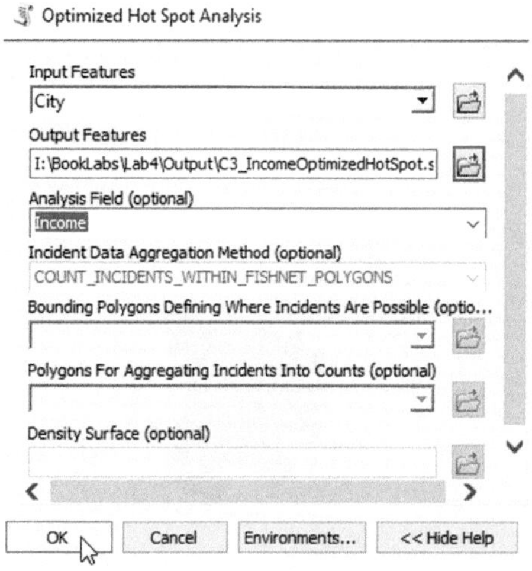

Optimized Hot Spot Analysis

Input Features
City
Output Features
I:\BookLabs\Lab4\Output\C3_IncomeOptimizedHotSpot.s
Analysis Field (optional)
Income
Incident Data Aggregation Method (optional)
COUNT_INCIDENTS_WITHIN_FISHNET_POLYGONS
Bounding Polygons Defining Where Incidents Are Possible (optio...
Polygons For Aggregating Incidents Into Counts (optional)
Density Surface (optional)

OK   Cancel   Environments...   << Hide Help

Figure 4.21 Optimized hot spot analysis dialog box.

```
*********************Initial Data Assessment*********************

Making sure there are enough weighted features for analysis....

     - There are 90 valid input features.

Evaluating the Analysis Field values....

     - INCOME Properties:

          Min:       10294.8949

          Max:       37644.1314

          Mean:      16316.7536

          Std. Dev.:  4947.8961

Looking for locational outliers....

     - There were no outlier locations found.
```

Exercise 4.4 (*cont.*)

************************Scale of Analysis************************
Looking for an optimal scale of analysis by assessing the intensity of clustering at increasing distances....

- The optimal fixed distance band is based on peak clustering found at 1157.3791 Meters

***********************Hot Spot Analysis************************
Finding statistically significant clusters of high and low INCOME values....

- There are 38 output features statistically significant based on an FDR correction for multiple testing and spatial dependence.

***************************Output******************************
Creating output feature class:

I:\BookLabs\Lab4\Output\C3_IncomeOptimizedHotSpot.shp

- Red output features represent hot spots where high INCOME values cluster.

- Blue output features represent cold spots where low INCOME values cluster.
The above results can be also found through:

Main Menu > Geoprocessing > Results > Current Session > Optimized Hot Spot Analysis > Messages

Close Results

Main Menu > File > Save

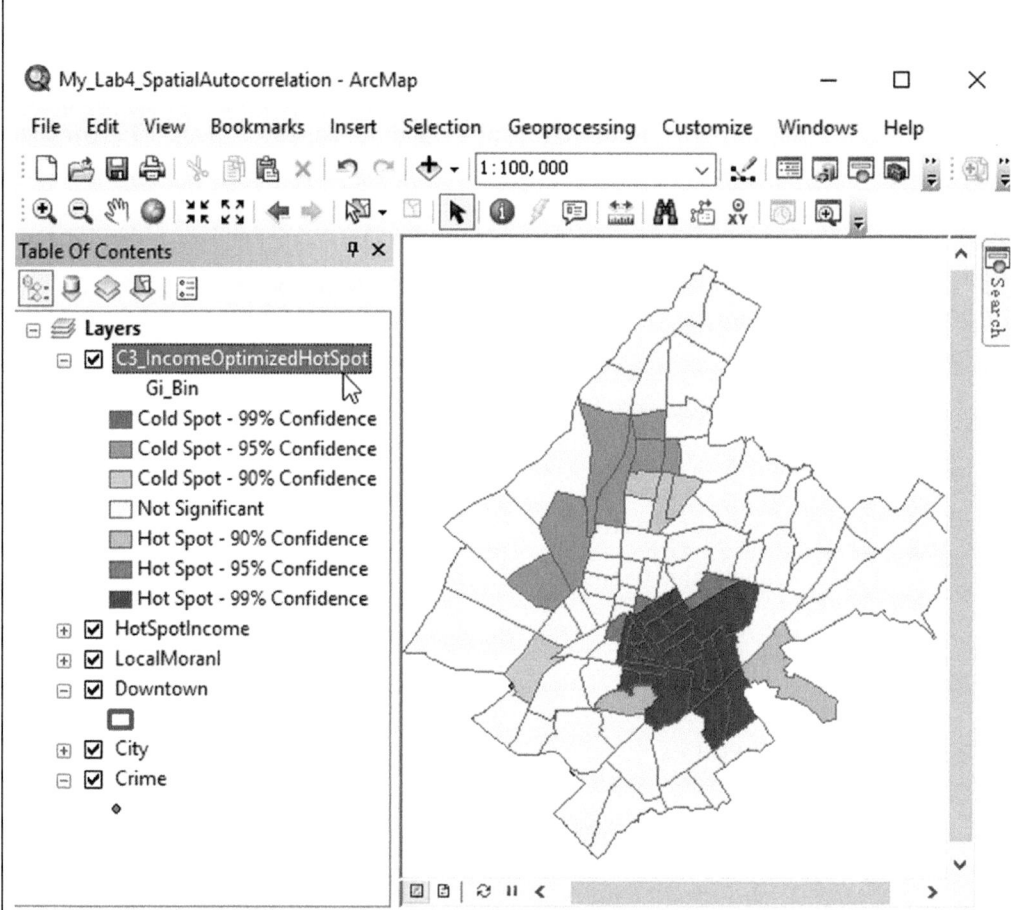

Exercise 4.4 (*cont.*)

Figure 4.22 Optimized hot spot analysis output map.

Interpreting results: Optimized hot spot analysis runs the entire procedure in an automated way, so it saves significant analysis time. The results are similar to those of the hot spot analysis (because the hot spot analysis used a similar procedure but in a non-automated way; see Figure 4.22). The tool reports the scale of analysis (1,157 m) as well as the presence (or not) of locational outliers. Hot spots of income are potential areas for the location of the new coffee shop (see Box 4.1).

Exercise 4.4 (*cont.*)

Box 4.1 Analysis Criterion C3 to Be Used in Synthesis Lab 5.4: The location of the coffee shop should lie within income hot spots areas. [C3_IncomeHotSpot.lyr]

TOC > RC C3_IncomeOptimizedHotSpot > Save As Layer File > C3_IncomeHotSpot.lyr > Save

---

Exercise 4.5 Optimized Hot Spot Analysis for Crime Events

In this exercise, we perform optimized hot spot analysis of 539 crime events for a period of two years to identify hot spots and cold spots of crime incidents (see Figure 4.24).

   ArcGIS Tools to be used: Optimized Hot Spot Analysis, Hot spot analysis

ACTION: Hot Spot Analysis

Navigate to the location you have stored the book dataset and click

My_Lab4_ SpatialAutocorrelation.mxd

TOC > List By Drawing Order > Drag Crime.shp on the top of all layers.

ArcToolBox > Spatial Statistics Tools > Mapping Clusters > Optimized Hot Spot Analysis

Input Feature Class = Crime (see Figure 4.23)

Output Feature Class =

I:\BookLabs\Lab4\Output\C4_CrimeOptimizedHotSpot.shp

Leave other blank / default

OK

Exercise 4.5 *(cont.)*

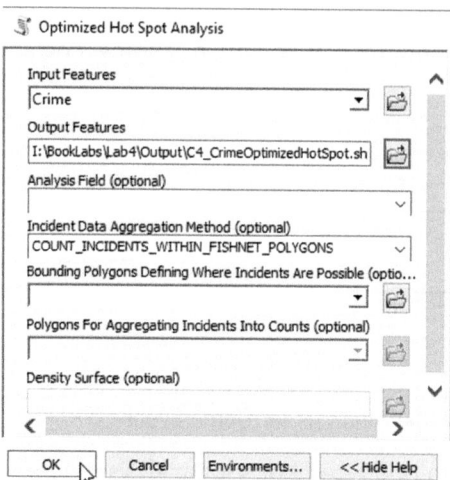

Figure 4.23 Optimized hot spot analysis dialog box.

*******************Initial Data Assessment*******************

Making sure there are enough incidents for analysis....

- There are 768 valid input features.

Looking for locational outliers....

- There were 5 outlier locations; these will not be used to compute the polygon cell size.

*******************Incident Aggregation*******************

Creating fishnet polygon mesh to use for aggregating incidents....

- Using a polygon cell size of 233.0000 Meters

Counting the number of incidents in each polygon cell....

- Analysis is performed on all polygon cells containing at least one incident.

Exercise 4.5 (*cont.*)

Evaluating incident counts and number of polygons....

- The aggregation process resulted in 457 weighted polygons.

- Incident Count Properties:

    Min: 1.0000

    Max: 5.0000

    Mean: 1.6805

    Std. Dev.: 0.8868

*************************Scale of Analysis***********************

Looking for an optimal scale of analysis by assessing the intensity of clustering at increasing distances....

- The optimal fixed distance band is based on peak clustering found at 933.0000 Meters

*************************Hot Spot Analysis**********************

Finding statistically significant clusters of high and low incident counts....

- There are 82 output features statistically significant based on an FDR correction for multiple testing and spatial dependence.

******************************Output******************************

Creating output feature class:

I:\BookLabs\Lab4\Output\C4_CrimeOptimizedHotSpot.shp

- Red output features represent hot spots where high incident counts cluster.

- Blue output features represent cold spots where low incident counts cluster.

Exercise 4.5 (cont.)

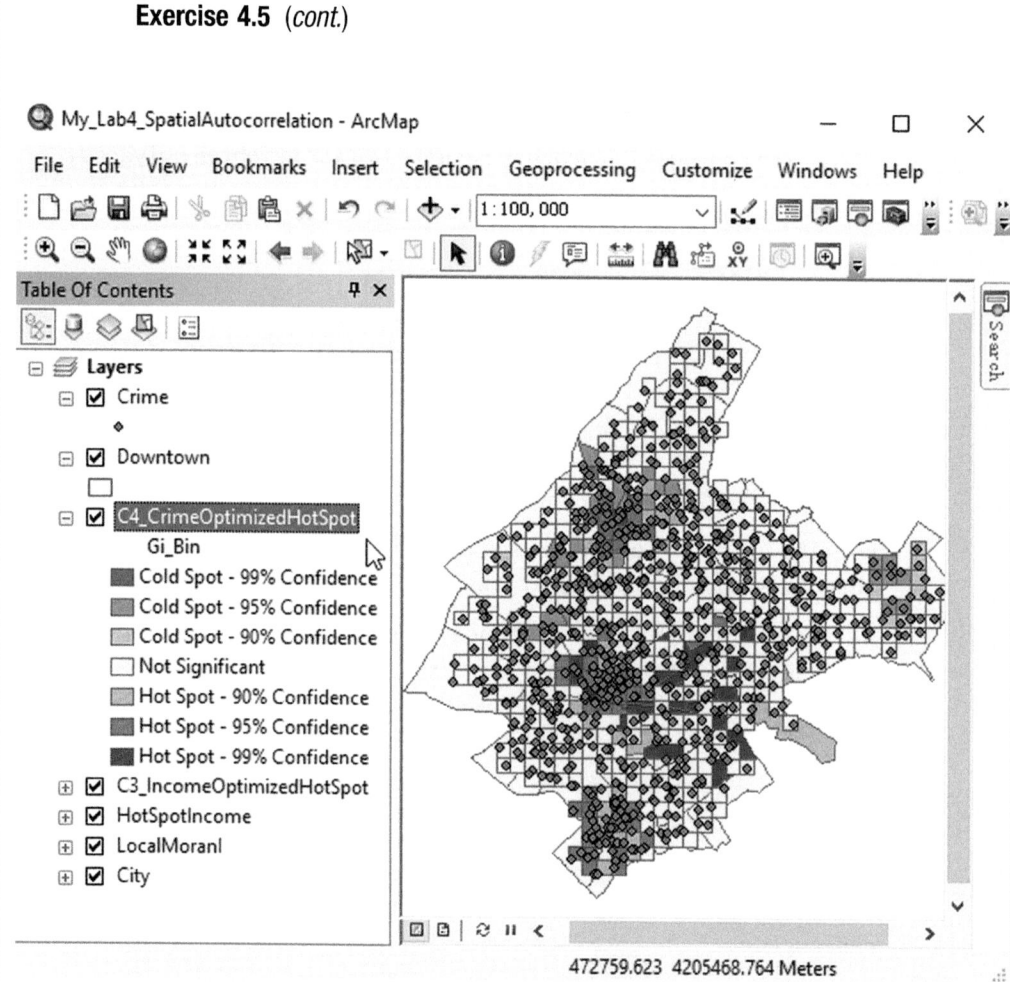

Figure 4.24 Optimized hot spot analysis dialog box.

The above results can be found through:

Main Menu > Geoprocessing > Results > Current Session >
Optimized Hot Spot Analysis > Messages

Close Results
Main Menu > File > Save

Exercise 4.5 (*cont.*)

Interpreting results: Optimized hot spot analysis for point entities aggregates events into polygon cells (see Figure 4.24). To calculate the polygon cell size, five locational outliers are removed. Finally, crime events are aggregated on polygon cells with a size of 233 m. The optimal fixed distance is identified at 700 m, and FDR correction is applied. Crime events are scattered all over the study area, but one cold spot and three hot spots of crime are identified, as shown on the map (see Figure 4.24). The distance (700 m) at which autocorrelation is more pronounced reveals that the hot spots are quite large (relative to the case study area's size) and that crime is a significant problem in three regions of the city. These hot spots also reflect the center of the real clusters of crime, and, in this sense, crime might be evident in polygons adjacent to the hot spot polygons as well. Crime hot spots should be excluded as candidates for the new coffee shop's location (see Box 4.2).

Box 4.2 Analysis Criterion C4 to Be Used in Synthesis Lab 5.4: The location of the coffee shop should not lie within crime hot spots areas. [C4_CrimeHotSpot.lyr]

TOC > RC C4_CrimeOptimizedHotSpot > Save As Layer File > C4_CrimeHotSpot.lyr > Save

---

Section B GeoDa

Exercises 4.1 and 4.2 Global Spatial Autocorrelation and Spatial Weights Matrix

In this exercise, we calculate the global spatial autocorrelation of income using the Moran's *I* index and the Getis-Ord General G-Statistic. Before doing so, we should create the spatial weights matrix. Unlike the exercises in Section A, Exercises 4.1 and 4.2 are presented in reverse order because GeoDa requires that the spatial weights be created by the user, while ArcGIS allows for automatic calculation when the spatial autocorrelation tools are executed.

Exercises 4.1 and 4.2 (*cont.*)

   GeoDa Tools to be used: Weights Manager , Univariate Moran's I

ACTION: Calculate Global Moran's I

Navigate to the location you have stored the book dataset and click the Lab4_SpatialAutocorrelation_GeoDa.gda

Main Menu > Tools > Weights Manager > Create >

Select ID Variable = PostCode (see Figure 4.25)

TAB = Distance Weight

TAB = Distance band > Specify bandwidth > Leave default value:1050.6848

Check the "Use inverse distance". Set Power to 1.

Create

File name = CityGeoDa (see Figure 4.26)

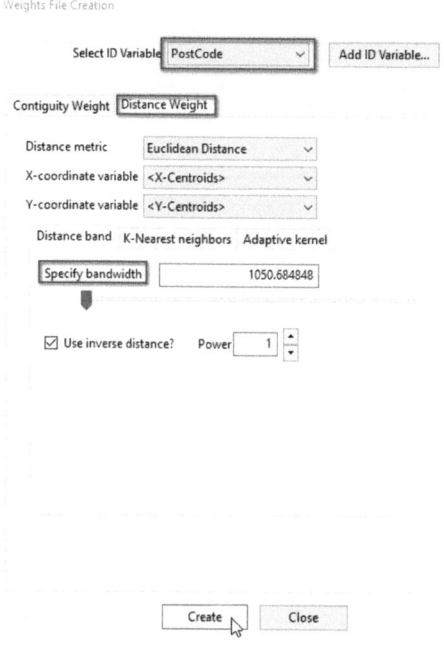

Figure 4.25 Calculating spatial weights dialog box.

Exercises 4.1 and 4.2 (*cont.*)

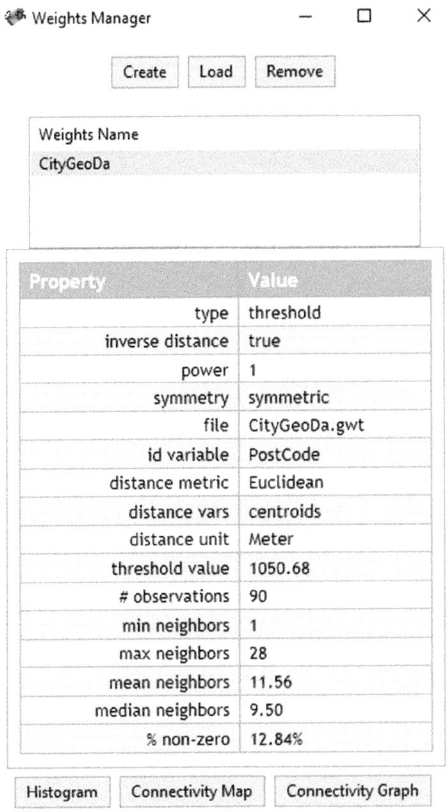

Property	Value
type	threshold
inverse distance	true
power	1
symmetry	symmetric
file	CityGeoDa.gwt
id variable	PostCode
distance metric	Euclidean
distance vars	centroids
distance unit	Meter
threshold value	1050.68
# observations	90
min neighbors	1
max neighbors	28
mean neighbors	11.56
median neighbors	9.50
% non-zero	12.84%

Figure 4.26 Weights manager showing the CityGeoDa spatial weights file.

Save as type = gwt (inside folder Lab4/GeoDa)

The weights manager dialog box is updated

Save > Close > Close Weights Manager window

ACTION: Calculate Global Moran's I

Main Menu > Space > Univariate Moran's I >

First Variable (X) = Income (see Figure 4.27)

Weights = GityGeoDa

OK

Exercises 4.1 and 4.2 (*cont.*)

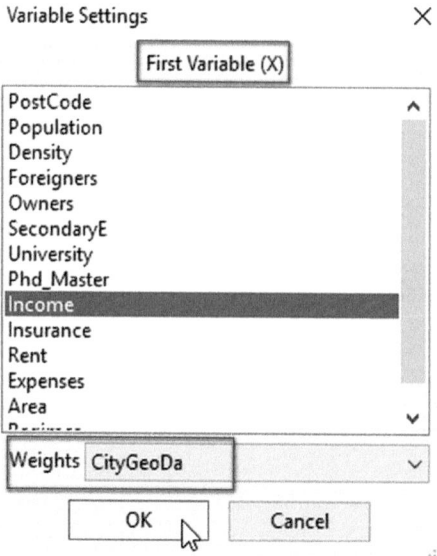

Figure 4.27 Setting the variable and weights file for Moran's *I*.

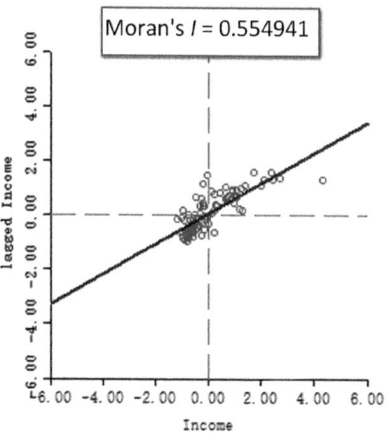

Figure 4.28 Moran's *I* scatter plot with a Moran's *I* index of 0.555. Graph reveals high positive autocorrelation.

Exercises 4.1 and 4.2 (*cont.*)

You can save the graph as image file if you wish.

Permutations are used to estimate how likely it is that a spatial arrangement of values similar to that we observe would be produced through complete spatial randomness. We use Monte Carlo and the following procedure.

RC on the scatter plot > Randomization > 999 permutations (see Figure 4.29)

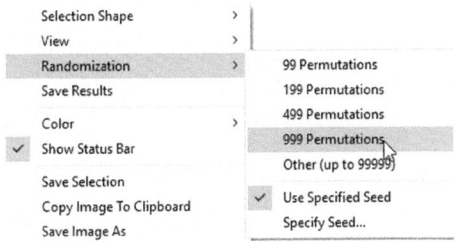

Figure 4.29 Mont Carlo simulation with 999 permutations.

Interpreting results: The results of this example suggest that the observed value I = 0.5549 is highly significant and not a result of spatial randomness, as it lies far away from the rest of the values (see Figure 4.30) (z-score = 9.8410; the green vertical line at the right depicts the Moran's I statistic value). The rest of the values depict the distribution that results from complete spatial randomness. The number of permutations, the setting of the pseudo p-value significance level, the expected (theoretical) Moran's I value $E(I)$, the observed Moran's I (I), the standard deviation, the z-value and the mean of the reference distribution are also presented. With 999 permutations, the pseudo p-value is set to 0.001, indicating that none of the 999 random patterns' local Moran's I values surpassed the observed value. As such, the spatial arrangement of the income values has a tendency to cluster. The distance threshold defined by the tool is set to 1,050.79 m so that every postcode has at least one neighbor. This is a first indication of income clustering. However, we cannot locate where the clustering occurs just by using this index. Moreover, although we calculated spatial autocorrelation, we have not yet defined the appropriate scale of analysis. GeoDa does not offer an automated tool for incremental spatial autocorrelation as ArcGIS does; this analysis is therefore not carried out here. As the scale of analysis, we will use the outcome of the incremental spatial autocorrelation as presented in Exercise 4.2 in Section A, which is 1150 m.

Exercises 4.1 and 4.2 *(cont.)*

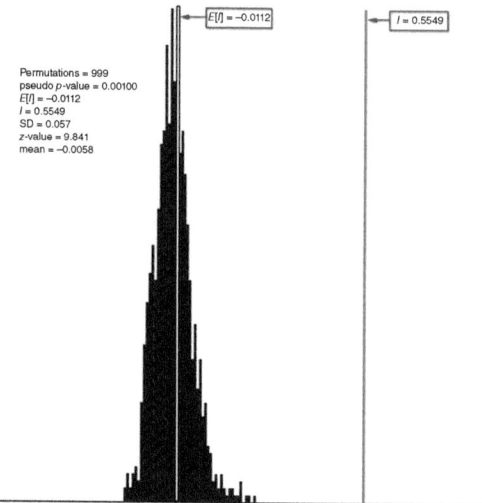

Figure 4.30 Mont Carlo reference distribution for 999 permutations.

Tip: Using the random seed for permutations might cause the results to differ slightly among different computers or sometimes even when using the same machine.

Exercise 4.3 Cluster and Outlier Analysis (Anselin Local Moran's *I*)

In this exercise, we calculate the local spatial autocorrelation of income using Local Moran's *I* to identify clusters and outliers. As mentioned in the previous exercise, the scale of analysis is 1,150m. Before we calculate the Local Moran's *I*, we should recalculate the spatial weights to reflect the adopted scale of analysis.

   GeoDa Tools to be used: Weights Manager, Univariate Moran's I, Moran's scatter plot

Exercise 4.3 *(cont.)*

ACTION: Weights Manager

Navigate to the location you have stored the book dataset and click Lab4_SpatialAutocorrelation_GeoDa.gda

Main Menu > Tools > Weights Manager > Create

Select ID Variable = PostCode

TAB = Distance Weight

TAB = Distance band > Specify bandwidth = 1150

Check "Use inverse distance". Set Power to 1.

Create

File name = CityGeoDa1150

Save as type = gwt (inside folder Lab4/GeoDa)

Save > OK > Close > Close Weights Manager window

ACTION: Cluster and Outlier Analysis

Main Menu > Space > Univariate Local Moran's I > Income > Weights = CityGeoDa1150 > OK

Check: Significance Map

Check: Cluster Map

Check: Moran Scatter Plot

OK

Save Project (see Figure 4.31)

Interpreting results: If a postcode has high income and is surrounded by postcodes with low income, it is marked as High-Low. If the postcode has low income and is surrounded by postcodes with high income, it is marked as Low-High. Where postcodes are clustered, they are labeled as High-High for a statistically significant cluster of high-income values and Low-Low for a statistically significant cluster of low-income values.

In this example and without FDR correction, income is positively spatially autocorrelated, and a statistically significant clustering of high values is observed in the center of the city at the 99% confidence level. In other words, people with high incomes tend to live in the red areas located in and around

Exercise 4.3 (*cont.*)

Figure 4.31 Clusters and significance map for three levels of significance (0.05, 0.01, 0.001). Unlike with the ArcGIS output in Figure 4.17, we have not applied FDR correction, and more postcodes are thus statistically significant. FDR can be applied manually in GeoDa.

the downtown area. A cluster of low values is detected in the western parts of the city (this cluster is not identified with ArcGIS due to FDR corrections). One outlier of Low-High values and five outliers of High-Low values are also located in the study area.

Exercise 4.4 Hot Spot Analysis (Getis-Ord Gi*I)

In this exercise, we calculate the local spatial autocorrelation of income using the local Getis-Ord G_i^* index to identify hot spots and cold spots (optimized hot spot analysis is not carried out as GeoDa does not offer such a tool).

GeoDa Tools to be used: Local G*

ACTION: Local G*

Navigate to the location you have stored the book dataset and click Lab4_SpatialAutocorrelation_GeoDa.gda

Main Menu > Space > Local G* > Income > Weights = CityGeoDa1150 > OK

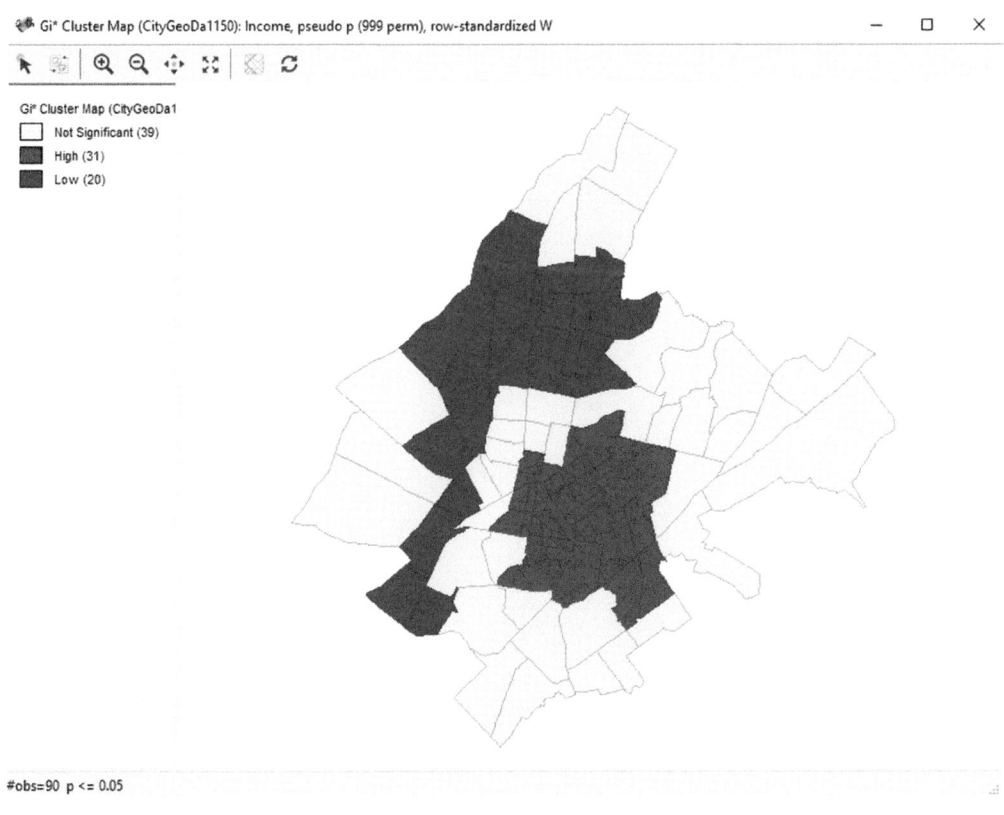

Figure 4.32 Hot spot analysis output map indicating cold (blue) and hot (red) spots significant at the $p <= 0.05$ level. There are minor differences from the results shown in Figure 4.20 using ArcGIS due to the slightly different weights matrix used. However, the main conclusions regarding the presence of hot and cold spots remain unchanged.

Exercise 4.4 *(cont.)*

Check: Cluster Map

Check: using row-standardized weights

Save project

Interpreting results: We locate a statistically significant hot spot in and around the city center and a statistically significant cold spot in the western part of the city (see Figure 4.32). Nonsignificant results mean that there is no indication of income clustering for these postcodes. A cold spot of income means that polygons with low values of income are surrounded by polygons with low values of income. A hot spot of income means that polygons with high values of income are surrounded by polygons with high values. As we are looking for areas with high income, the red areas might be more appropriate as locations for the coffee shop.

   Remark: Optimized hot spot analysis is not offered in GeoDa, and Exercise 4.5 is presented only in Section A.

5 Multivariate Data in Geography
Data Reduction and Clustering

Learning Objectives

This chapter deals with multivariate statistical methods for data reduction and clustering, commonly used in geographical analysis, such as

- Principal component analysis
- Factor analysis
- Multidimensional scaling
- Hierarchical clustering
- *k*-means clustering
- Regionalization (SKATER, REDCAP)
- Density-based clustering (DBSACN, HDBSCAN, OPTICS)
- Similarity analysis (Cosine similarity)

After a thorough study of the theory and lab sections, you will be able to

- Understand why multivariate data and statistics are essential in geographical analysis as for example, in geodemographics
- Understand that observations in multivariate datasets are points in a multidimensional data space
- Understand what principal components are and how they can be mapped in a GIS environment
- Map multidimensional datasets to a 2-d or 3-d representation by multidimensional scaling
- Understand why hierarchical clustering is important to identify the structure of clusters
- Use the *k*-means algorithm in a geographical problem
- Evaluate the importance of taking into account spatial constraints in clustering (regionalization)
- Use density-based clustering to analyze large datasets of point entities
- Apply similarity analysis to identify common characteristics (profiles) on your spatial entities
- Perform Principal Component Analysis, Multidimensional Scaling and Hierarchical clustering in Matlab

- Conduct *k*-means clustering, similarity analysis, and spatial clustering in ArcGIS
- Conduct *k*-means clustering and spatial clustering in GeoDa

5.1   Multivariate Data Analysis

Definitions

Multivariate data are data with more than two values recorded for each observation (O'Sullivan & Unwin 2003 p. 316). A typical representation of a multivariate dataset *A*, with *n* observations and *p* variables is through a $n \times p$ matrix (5.1):

$$A = \begin{bmatrix} a_{1,1} & \cdots & a_{1,p} \\ \vdots & \ddots & \vdots \\ a_{n,1} & \cdots & a_{n,p} \end{bmatrix} \tag{5.1}$$

where columns represent the *p* variables and lines represent the *n* observations.

Multivariate data exist in a multidimensional space where the number of dimensions equals the number of variables. For example, a case study area with 64 spatial units (e.g., postcodes) and 50 variables is a $p = 50$-dimensional dataset consisting of $n = 64$ observations (see Table 5.1).

Multivariate statistical analysis is a collection of statistical methods to analyze multivariate data.

Multivariate statistical analysis methods use various statistical distance metrics (e.g., Euclidean, Minkowski, Manhattan, see Chapter 1) to express dissimilarity (or similarity) among observations and project them into a new multidimensional space. The main advantage of Euclidean distance in a multidimensional space is that it is interpreted more easily compared to other distance metrics. Whichever statistical distance is used, small distances reflect similarity among observations, while large distances reveal dissimilarity. Values are stored at a matrix called dissimilarity matrix.

Table 5.1 Multivariate dataset with spatial reference (postcodes polygons).

P 1		2	3	4		50
n	Postcode ID	Population	Income	Unemployment	...	Medical Expenses
1	13231	12,568 ($a_{1,2}$)	25,000 ($a_{1,3}$)	12% ($a_{1,4}$)		3,200 ($a_{1,50}$)
2	12137	2,4585	17,250	6%		550
3	12461					
...	...	...	...	...	...	...
64	12242	$a_{64,2}$				$a_{64,50}$

Dissimilarity matrix is a square and symmetric matrix that stores the pairwise dissimilarities among n observations (data points) calculated through any statistical distance metric d (5.2). The diagonal cells are defined as zero (distance with itself) while the off-diagonal cells store the pairwise dissimilarities.

$$Dissimilarity\ matrix = \begin{bmatrix} observations & 1 & 2 & 3 & \dots & n \\ 1 & 0 & d(1,2) & d(1,3) & \dots & d(1,n) \\ 2 & d(2,1) & 0 & d(2,3) & \dots & d(2,n) \\ 3 & d(3,1) & d(3,2) & 0 & \dots & d(3,n) \\ \dots & \dots & \dots & \dots & 0 & \dots \\ n & d(n,1) & d(n,2) & d(n,3) & \dots & 0 \end{bmatrix} \quad (5.2)$$

For example, for any given multivariate dataset A, we can calculate the statistical distance of any two observations c, d using the Euclidean distance norm extended to a multidimensional space (5.3):

$$dist(c,d) = \sqrt{(a_{c1} - a_{d1})^2 + (a_{c2} - a_{d2})^2 + \dots (a_{cp} - a_{dp})^2} \quad (5.3)$$

where c,d are observations, and 1,2, ... p are variables.

In practice, for any two observations (lines), we subtract the values among the same variables (columns) and then square and add them to finally get the square root, just as if they were coordinates of a point. For the example given in Table 5.1, the statistical distance between postcodes 13231 and 12137 is (in practice, most of the times we standardize data before calculating distance – see Section 2.4):

$$dist(13231, 12137)$$
$$= \sqrt{(12568 - 24585)^2 + (25000 - 17250)^2 + \dots (3200 - 550)^2} = 500$$

and an indicative dissimilarity matrix would be

$$Dissimilarity\ matrix = \begin{bmatrix} & 13231 & 12137 & 12461 & \dots & 12242 \\ 13231 & 0 & 500 & 350 & \dots & 100 \\ 12137 & 500 & 0 & 150 & \dots & 160 \\ 12461 & 350 & 150 & 0 & \dots & 180 \\ \dots & \dots & \dots & \dots & 0 & \dots \\ 12242 & 100 & 160 & 180 & \dots & 0 \end{bmatrix}$$

Although most of the methods to analyze multivariate data are purely statistical, spatial extensions have been proposed to reflect the underlying geography. For this reason, this chapter presents basic multivariate statistical methods, from a spatial perspective, for data reduction and data clustering as

- Principal Component Analysis (PCA),
- Factor Analysis (FA)
- Multidimensional Scaling (MDS)
- Cluster Analysis (for classifying observations)

- Regionalization when clustering is made with spatial constraints
- Similarity analysis

Why Use

Vast amounts of data are collected daily from various sources such as satellite and environmental sensors, web geo-location services and social media (Grekousis et al. 2013b). Integrating these data to existing datasets, derived from national censuses or other depositories, offers a wealth of information, if analyzed wisely (Grekousis et al. 2019a). Multivariate techniques delve into this endless pool of data to discover patterns and unexpected trends or behaviors, and extract hidden knowledge valuable for spatial analysis and spatial planning.

Multivariate statistical analysis methods presented in this chapter are used to

(a) Eliminate collinearity
(b) Reduce the dimensions of multivariate data (group variables)
(c) Uncover latent variables
(d) Map observations to lower dimensions
(e) Cluster objects to homogenous groups (group observations)

Discussion and Practical Guidelines

Selecting the most appropriate variables for any geographical analysis is not a trivial problem. Including all available variables would probably lead to serious multicollinearity issues – i.e., many variables would provide relatively less new added information. Multicollinearity exists among two or more variables in a dataset when they are highly correlated (see Chapter 6). In addition, a large number of variables may lead to overrepresentation in some categories. For example, we might have eight lifestyle variables related to education and how people spend their free time on outdoor activities. If only two variables refer to education, this may lead to the creation of a distance matrix that will emphasize free time differences rather than educational differences. For this reason, a careful inspection of the variables to be selected is necessary so that variables are balanced and multicollinearity is kept to low levels. On the other hand, selecting a few variables, based on our previous experience or intuition, may lead to information loss, as variables excluded might reveal important hidden (latent) information.

In this respect, data reduction methods as PCA, FA and MDS are essential in multivariate data analysis, as they reduce the number of columns in a dataset. PCA and FA's main advantage is that they reveal latent variables uncovering hidden interactions. Moreover, by removing multicollinearity, components and factors are uncorrelated to each other. As a result, they can be used as independent explanatory variables in a subsequent regression analysis (Wang 2014). MDS's main characteristic is that it maps observations to two or three dimensions, providing a graphical representation of similarity for objects clustered together and dissimilarity for objects lying apart.

Clustering analysis, on the other hand, has the advantage of grouping observations while retaining the number of variables, which makes a significant difference from the previous data reduction methods. Clustering analysis is important in a geographical context, as it reveals various hidden underlying spatial processes at play. When clustering takes into account spatial constraints, regionalization methods are used. Regionalization methods' main advantage is that they produce homogenous clusters of spatial features that are also adjacent. These methods are fundamental in decision making and spatial planning (Grekousis et al. 2013c). Finally, similarity analysis is beneficial when we need to rank spatial features according to how similar or dissimilar they are.

A typical problem in multivariate datasets is that variables are not always on the same measurement scale and of the same units. Some variables might refer to extremely small values, while others refer to large ones. For example, in a household survey, the variable "Number of children" gets values usually less than 10 while the "Annual household income" variable might be some thousands of dollars. Large values tend to dominate the results, and for this reason, values should be standardized to z-scores using the mean value of each variable and its standard deviation (see Section 2.4; O'Sullivan & Unwin 2003 p. 325). By standardizing values, variables are no longer dependent on the measurement scale and are comparable to each other (Wang 2014 p. 144). Normalization is another method used to rescale data in the same values range and is widely used before any statistical distance is calculated. See Section 2.4 for more details on differences between normalization and standardization. Standardization is occasionally preferred over normalization as it better retains the importance of each variable due to the non-bounding limitation. For example, in case of outliers, normalized data are squeezed at a small range, and as such, when dissimilarities (through statistical distances) are calculated, they contribute less to the final values.

Multivariate methods have been mainly originated from classical statistical analysis, but their usage in geographical analysis is extensive. The reason is that geographical studies heavily rely on census, socioeconomic or other large multivariate datasets and consequently deal with either variable reduction or data clustering. In this respect, these methods are of crucial importance in order to better analyze data and are necessary in spatial analysis.

5.2 Principal Component Analysis (PCA)

Definition
Principal component analysis (PCA) is a technique used to summarize multivariate data in fewer interpretable variables called principal components.

A **principal component** is a linear combination of the original (or standardized) values of the variables and is calculated by extracting the eigenvectors and eigenvalues of the variance–covariance matrix or the correlation matrix (5.4) of matrix A (5.1), (Penn State University 2018).

$$PC_i = \vec{X_j} \times \vec{e_i} = [X_1 \quad X_2 \quad X_3 \quad \ldots \quad X_p] \times \begin{bmatrix} e_{i1} \\ e_{i2} \\ e_{i3} \\ \ldots \\ e_{ip} \end{bmatrix} = e_{i1}X_1 + \cdots e_{ip}X_p \quad (5.4)$$

for i = 1 to p (number of components-variables),
for j = 1 to n (number of observations),

where $\vec{X_j}$ is the vector of the j-th observation of the matrix A : $\vec{X_j} = [X_1 \quad X_2 \quad X_3 \quad \ldots \quad X_p]$, it is equivalent to $\vec{X_j} = [a_{j1} \quad a_{j2} \quad a_{j3} \quad \ldots \quad a_{jp}]$ (see Eq. 5.1).

$\vec{e_i}$ is the eigenvector of the $i - th$ component, $\begin{bmatrix} e_{i1} \\ e_{i2} \\ e_{i3} \\ \ldots \\ e_{ip} \end{bmatrix}$.

The values e_{i1}, e_{i2} ... of each eigenvector are called principal component coefficients or loadings (Wang 2014 p. 144).

Why Use

PCA reduces the dimensions (variables) of a multivariate dataset to a set of independent uncorrelated variables, called the principal components, to make analysis more comprehensive and interpretation easier, while at the same time preserving most of the information (variation) existing in the dataset (Cangelosi & Goriely 2007). The principal components can substitute the original variables in any subsequent analysis.

Interpretation

Eigenvectors are extracted from the variance–covariance matrix of A (see Section 2.3.3). We do not get into details on how eigenvectors and eigenvalues are computed through the variance–covariance matrix, but we focus on their geometric meaning. PCA projects data from their original dimensions to new ones so that the variation of the data is better explained. The eigenvectors are used to construct the new axes called principal components, which correspond to the direction (in the original space) with the largest variance in the data

(Hamilton 2014). Each eigenvector has a corresponding eigenvalue (also named latent roots) that expresses the variability of each corresponding principle component (O'Sullivan & Unwin 2003 p. 345). A principal component with a low eigenvalue does not explain a lot of data variation. The eigenvalue can be also used to draw the standard deviational ellipse as shown in Figure 5.1D (O'Sullivan & Unwin 2003 p. 345). The original values can now be projected to the new dimensions based on the scale factor (eigenvalue) and the new space defined by the components (new axes).

The number of principal components is equal to the number of the original variables in the dataset. Each component explains a certain amount of the original variation of the variables. Components are ordered according to their eigenvalues, so the first one explains the largest variability (largest eigenvalue) of the dataset, the second component explains the second largest variability and so on (O'Sullivan & Unwin 2003 p. 345). "Explaining the variability (or variation) in the data" stands for the percentage revealing the amount of information retained of the original dataset after the transformation has been applied. A 60% variability explained means that by data reduction through PCA transformation, we kept 60% of the initial information, or we lost 40% of the original information. The components are structured so that they are uncorrelated with each other, and this is achieved by the orthogonal transformation applied by PCA in the multidimensional space (for this reason, principal components are often treated as dependent variables for regression analysis and are also used in cluster analysis, as they do not exhibit multicollinearity). In other words, the second component is orthogonal to the first one, the third one is orthogonal to the second one and so on. As the orthogonal arrangment of components is not easily comprehensible in a multidimensional space, let us describe a more straightforward example in the two-dimensional space.

Suppose we analyze the average annual "Income" and the average "House size" of 10 spatial units (e.g., postcodes). Using a scatter plot in the two-dimensional space (see Figure 5.1A), we observe a diagonal trend between these two variables (see Figure 5.1B). We can calculate the variance σ^2 in the x-direction and the variance σ^2 in the y-direction as a measure of the values spread. Still, the horizontal and vertical variance does not accurately explain the clear diagonal trend. Instead of calculating the variance for the x- and y-axes, it is better to rotate them so that the x-axis captures the maximum of the variance of the data points cloud (observations; see Figure 5.1C). The y-axis will remain orthogonal to x, capturing another proportion of the variance. The new axes are the first and the second principal components. The eigenvalues are also used to create the standard deviational ellipse (see Chapter 2), which contains the majority of the data points (see Figure 5.1D). The center of the ellipse is the mean center of the data points. The major axis lies on the first component, and the minor axis – which is orthogonal to the major one – lies on

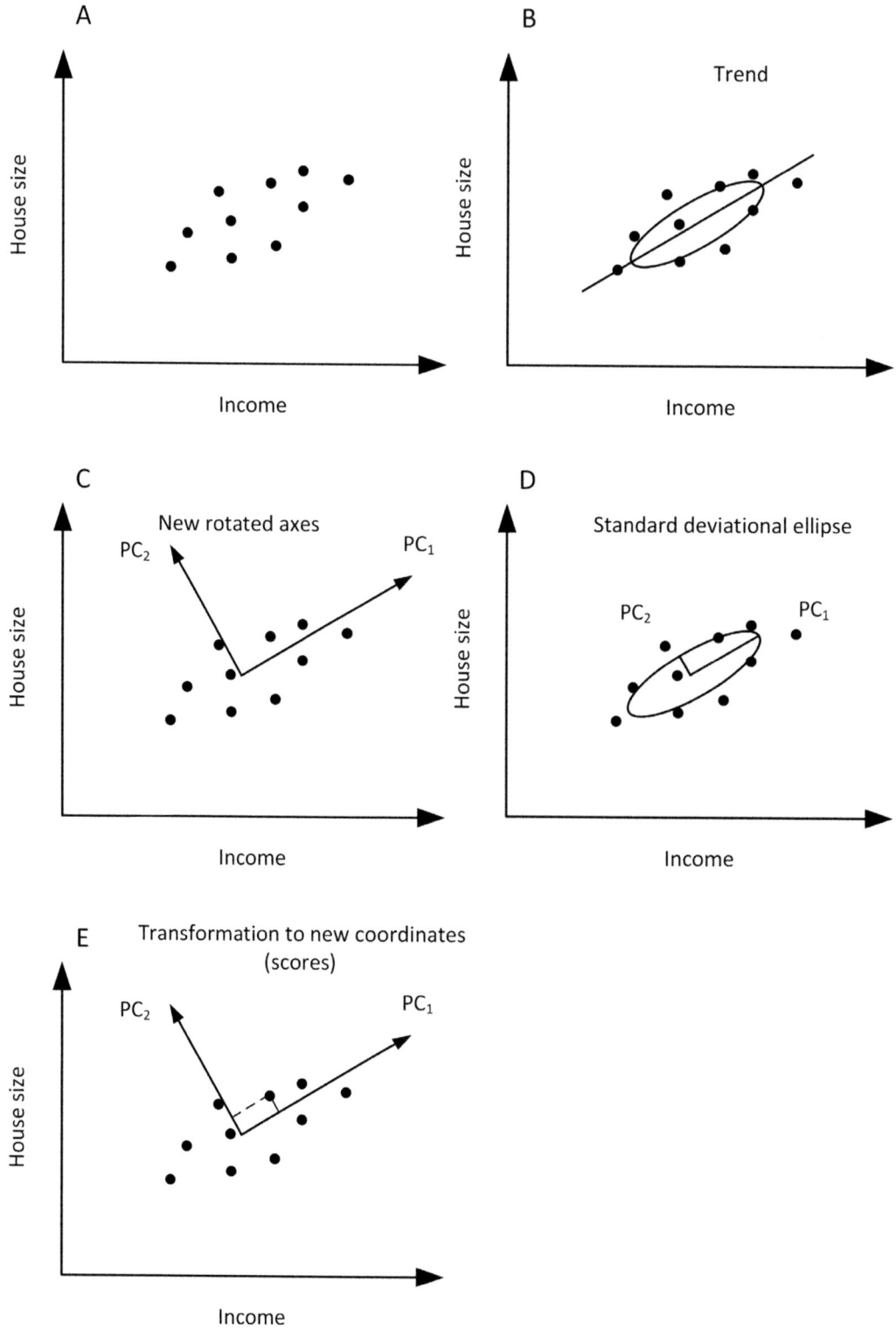

Figure 5.1 PCA graphical example.

the second component. The length of each axis is the square root of the corresponding eigenvalue. Original data can now be projected to the new space, and their scores (new coordinates) can by calculated applying formula (5.4). In this example, the scores of the first two principal components are the coordinates of the original data to the new axis (see Figure 5.1E).

Each subsequent component is orthogonal to the previous one, thus explaining less variability (Troy 2008). Therefore, only a few of the first components should be kept: those that explain most of the variability. To calculate the percentage of the explained variability of each component, we use the corresponding eigenvalue. The eigenvalue of each component equals the variance of the data (how much the data vary) along this component (O'Sullivan & Unwin 2003 p. 346, (Wang 2014). The proportion of the overall data variance accounted by the i-th component is given in (5.5):

$$ExplVar_i = \frac{\lambda_i}{\lambda_1 + \cdots + \lambda_p} \qquad (5.5)$$

where λ_i is the eigenvalue of the i-th component and p is the total number of components.

The larger the eigenvalue, the more significant a component is, as it captures a lot of the original data variability. There are many different ways to determine the number of meaningful components to be retained, including the cumulative percentage of the total variance, the SCREE plot, the Kaiser–Guttman test, the broken stick model, Cattell's cross-validation, bootstrapping techniques and Bartlett's test for equality of eigenvalues (Cangelosi & Goriely 2007). For additional methods, one might refer to the works of Jolliffe (2002) and Jackson (1991).

We describe here the first three methods:

- **Cumulative percentage of the total variance**: In this method, the total variation explained by the final components should be larger than a threshold value (e.g. 70%). Still, for a high-dimensional dataset achieving a high cumulative value might result in retaining too many components that are hard to interpret.
- **SCREE plot**: A SCREE plot is created to depict the fraction of the variance explained over each additional component. The location at the graph where additional components do not significantly change the explained variance is called "elbow point," as there is a sharp change in the slope. Components after the elbow point can be discarded (see also the elbow method in Section 5.5.2 and Figure 5.2).
- **Kaiser–Guttman test**: If the principal components are derived from a covariance matrix, then those eigenvalues larger than the average of all eigenvalues are retained. In the case where the principal components are derived from a correlation matrix, the average of eigenvalues is 1; therefore, eigenvalues larger than 1 are retained.

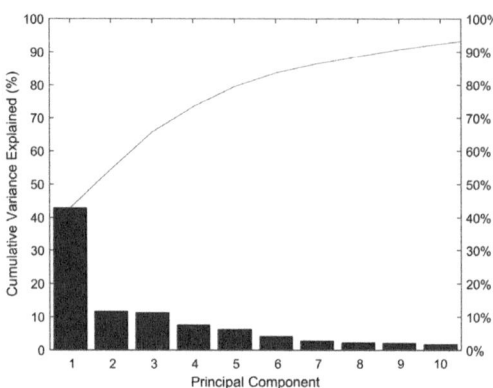

Figure 5.2 Scree plot of explained variance over the number of components.

The selection of the appropriate principal components to be retained depends on the problem studied and the research questions. Too many components make interpretation hard, while too few components may be insufficient if the explained variance is small. A low explained variance might be the result of many highly skewed variables. Moreover, mixing too many conceptually different variables may lead to low pairwise correlations. In this case, principal components can hardly capture a large share of variance. When pairwise correlations among variables do not exceed 0.3 (or are not less than -0.3), PCA may not reach a sufficient level of explained variance. The reason is that PCA transforms correlated variables into a new component that summarizes them. In this sense, a good starting point before applying PCA is to check pairwise correlations and potentially remove those variables that have moderate to low correlation (for example, remove those with $r < 0.50$). However, setting aside some variables may substantially alter PCA results, and as such, it depends on the scope of the analysis to finally decide which variables to retain.

If variables are not measured in the same units or when variables' values exhibit large differences in their range, data (*A* matrix) should be standardized. Standardization is commonplace in social analysis, as social data are characterized by large differences on their scales.

Discussion and Practical Guidelines (Workflow and Spatial Data)

Within a geographical context, PCA is deployed in five steps (see Box 5.1):

Step 1. *Standardize dataset (if needed).* Apply descriptive statistics to check if the variables are of different scales (or if different units are used). If so, standardizing data by calculating z-scores is needed (see Section 2.4). To standardize data, we can also use the pairwise correlation matrix which is scale independent. The covariance matrix is scale dependent and should be used if data are of similar scales.

Step 2. *Compute: (a) the eigenvectors (loadings of principal components), (b) the principal component scores and (c) the eigenvalues-variance (latent roots).*

Step 3. *Select the number of principal components to retain using an appropriate method (e.g., by constructing a SCREE plot).*

Step 4. *Interpret principal components.* PCA is a descriptive technique and is not based on hypothesis testing. As such, PCA interpretation largely depends on how we describe the corresponding components. We label each component by disregarding those loadings below a certain threshold. A rule of thumb is to keep those loadings larger than 0.3 or smaller than −0.3. We have to underline that principal components are likely to produce unmeaningful results. Although, produced components are nicely organized variables sometimes this does not equate to something in reality – a major drawback of PCA. Scatter plots may also be used for the first two principal components.

Step 5. *Map scores in geographical space.* This step is not included when PCA is applied for nonspatial data. Mapping does not refer to a typical scatter plot of the first two principal components scores. It refers to the assignment of scores to spatial units as an extra variable that can be subsequently analyzed by spatial analysis techniques (e.g., spatial autocorrelation). Thus, by PCA, we keep the best components that explain most of the data variation and then map them to further discover any spatial associations.

The primary criticism of PCA is that it transforms data on orthogonal space. Still, the real space cannot be accurately transformed in this way, as heterogeneity and autocorrelation exist. In addition, the results produce a global summary of the data and are not presented spatially by the method itself.

In case of vector data, PCA runs on the entire dataset, and geographical effects are not taken into account in the calculations (Demšar et al. 2013). The mapping of scores of vector data in a GIS environment does not make PCA a spatial analysis method, as it does not take into account spatial concepts (distance, location, neighborhood) or spatial heterogeneity and autocorrelation in its construction.

In case of raster data, components are calculated based on measurements referring to each cell. This approach is common in remote sensing when different raster datasets with the same spatial reference and extent are combined to produce composite indices. For example, we may combine a land use/land cover image, a raster for soil variables, a raster of temperature, a raster of CO_2 emissions and a raster of socioeconomic and census data. By applying PCA, we detect the principal components, and we map each one of them as a new raster file. In this way, we depict the spatial distribution of each component's scores directly. PCA dealing with raster data is named raster PCA and is typically applied in a GIS environment. Raster PCA handles better raster spatial data but does not account for any spatial effects. Raster data PCA is suitable for combining social data (usually in vector format) with environmental data (typically in raster format) to produce composite indices usually derived from the first few principal components (Demšar et al. 2013).

To account for spatial heterogeneity and spatial autocorrelation when dealing with spatial data, several approaches have been proposed as the geographically weighted PCA.

Geographically weighted PCA (GWPCA) (Fotheringham et al. 2002, Charlton et al. 2010, Harris et al. 2011) calculates a local PCA model for every single location based on geographically weighted data of a user-defined neighborhood. Principal components, eigenvalues and eigenvectors are calculated for every single spatial unit and thus can be mapped and further spatially analyzed (Demšar et al. 2013). GWPCA can be used to produce local composite indices because each local principal component describes the relationships of the original variables at the specific location. Furthermore, GWPCA can be used as an interpolation technique to obtain eigenvalues and eigenvectors at unobserved locations. By using GWPCA, we may additionally estimate scores at locations where data do not exist by generating spatial surfaces of eigenvalues and eigenvectors (Harris et al. 2011). Finally, GWPCA can be used prior to geographically weighted regression (see Section 6.5) to produce uncorrelated compound variables.

Numerical Example

Let's see a numerical example that illustrates the five PCA steps as presented earlier. Only a snapshot of the final results is desribed per step, as the focus lies on the procedure and the output interpretation. Suppose we have a dataset (matrix A) describing $n = 50$ neighborhoods (observations) with $p = 5$ variables (columns): Income, Housing Conditions, Crime, Health, Pollution. To perform PCA, we follow the next steps:

Step 1. *Standardize this dataset.* The first two rows of the standardized matrix are presented in Table 5.2.

Step 2. *Compute: (a) the eigenvectors (loadings of principal components), (b) the principal component scores and (c) the eigenvalues-variance (latents).*

Results are

(a)   Table 5.3 presents the loadings of the first two principal components.

(b)   Principal component scores calculated based on Equation (5.4):

$$PC_i = \vec{X_J} \times \vec{e_i} = [X_1 \quad X_2 \quad X_3 \quad \ldots \quad X_p] \times \begin{bmatrix} e_{i1} \\ e_{i2} \\ e_{i3} \\ \ldots \\ e_{ip} \end{bmatrix} = e_{i1}X_1 + \cdots e_{ip}X_p$$

Table 5.2 Standardized matrix.

	Variables				
	X_1	X_2	X_3	X_4	X_5
Observations (neighborhoods)	Income	Housing Conditions	Crime	Health	Pollution
1	−0.146	−0.899	−0.945	−0.106	−0.123
2	0.300	−0.087	0.468	−0.210	0.463
...	...	...	...	...	...
50	...	...	...	...	...

Table 5.3 Multivariate dataset. Loadings with bold indicate moderate to strong correlations that assist in interpreting principal components.

	Eigenvectors (e_i) (Loadings)	
Variables	First principal component	Section principal component
Income	**0.78**	**−0.42**
Housing Conditions	**0.62**	**−0.35**
Crime	0.12	**0.59**
Health	0.32	0.18
Pollution	0.05	**0.45**

The general formula for the first principal component score is

$$PC1 = 0.78 \times \text{Income} + 0.62 \times \text{Housing Conditions} + 0.12 \times \text{Crime} + 0.32 \times \text{Health} + 0.05 \times \text{Pollution}$$

The values of the variables are derived from the standardized matrix (Table 5.2). The score of the first principal component for the first observation is

$$PC_i = \begin{bmatrix} -0.146 & -0.899 & -0.945 & -0.106 & -0.123 \end{bmatrix} \times \begin{bmatrix} 0.78 \\ 0.62 \\ 0.12 \\ 0.32 \\ 0.005 \end{bmatrix}$$

$$PC1 = 0.78 \times (-0.146) + 0.62 \times (-0.899) + 0.12 \times (-0.945) + 0.32 \times (-0.106) + 0.05 \times (-0.123) = -0.825$$

The score of the second principal component for the first observation is:

$$PC2 = -0.42 \times (-0.146) + (-0.35) \times (-0.899) + 0.59 \times (-0.945) + 0.18 \times (-0.106) + 0.45 \times (-0.123) = -0.256$$

Table 5.4 Scores table.

Observations	PC1	PC2	PC3	PC4	PC5
1	−0.825	−0.256	. . .	. . .	. . .
2	. . .	. . .	. . .	. . .	. . .
. . .	. . .	. . .	. . .	. . .	. . .
50	. . .	. . .	. . .	. . .	. . .

Table 5.5 Eigen values and explained variance.

Component: (i)	Eigenvalue (λ_i) (latent)	Explained variance: $\lambda_i/Sum = \lambda_i/5$	Cumulative explained variance
1	3.23	0.646	0.646
2	1.36	0.272	0.918
3	0.23	0.046	0.964
4	0.13	0.026	0.990
5	0.05	0.010	1.000
Sum	5	1.000	

Likewise, we calculate the scores of all components (no matter if they are finally included) for all observations. Scores are stored in a matrix with as many rows as the observations and as many columns as the variables (in this example, 50×5; see Table 5.4). The scores matrix is the representation of standardized A in the principal component space.

(c)   Eigenvalues-variance (latents).

The eigenvalues, the explained variance, and the cumulative explained variance are presented in Table 5.5. Based on Equation (5.5), the first principal component explains 64.6% of the dataset's variance, while both first and second principal components explain 91.8% (cumulative explained variance of each row is the addition of explained variance of each row and above; see Table 5.5). Note that since we standardized data, the variance of each variable should equal 1. As a result, the sum of all eigenvalues (total variation) should be $p = 5$ (as many as the variables).

Step 3. *Select the number of principal components.*
Based on the explained variance (see Table 5.5), we create the SCREE plot to select the number of principal components to finally keep (see Figure 5.2). From the plot graphical inspection, there is a sharp change in slope (elbow criterion) in component 3. Still, we do not keep this component, as it only accounts for an additional 4.6% variance explained (see Table 5.5). It is more rational to retain only the first two principal components, as they capture more than 90% of the total variance. It

would not be wrong, though, if we included the third component. It depends on the loadings and on the scopes of analysis.

Step 4. *Interpret principal components.*

To label components, we define two thresholds. Variables with loadings larger than 0.3 are considered positively correlated, while variables with loadings smaller than -0.3 are considered negatively correlated (see Table 5.3). The first principal component is a measure of wealth as it is highly correlated with both income and the housing conditions (the rest variables have lower loadings and thus are not included in the description of the component). In other words, neighborhoods with high income are more likely to exhibit good housing conditions. The positive relation (positive sign) reveals that an increase in one variable will lead to an increase of the other.

The second principal component is a measure of deprivation. There is a negative correlation between income and housing conditions and a positive correlation between crime and pollution. Neighborhoods with low income and inadequate housing conditions tend to have high crime rates and more pollution.

The scores of the first and second principal components of each observation (see Table 5.4) can be plotted in a scatter plot (for example, for the first observation, the coordinates at the scatter plot are $(-0.825, -0.256)$). Plotting the coordinates assists on further describing the first two principal components.

Step 5. *Map scores in geographical space.*

Create a choropleth map depicting the scores of the first (or second) principal component in a GIS environment.

Box 5.1 Matlab. As ArcGIS and GeoDa do not offer tools for PCA, we present a small example through Matlab. Suppose we have a set of socio-economic variables (Data) for the postcodes of the City and we want to narrow data to a more meaningful dataset. We can easily perform PCA in Matlab. Go to the Matlab folder of Lab 5 to find data and code to complete a PCA through the steps explained before (standardization, computing: eigenvectors, eigenvalues and principal component scores, calculating explained variance and scree plot and plotting a scatter plot of the first and second principal component). Run PCA.m.

5.3   Factor Analysis (FA)

Definition

Factor analysis is a data dimension reduction techinique that describes a set of observed variables using fewer unobserved (latent) variables called factors.

Why Use

Factor analysis is used when we need to reduce the existing variables by using fewer factors that better represent the original ones, thus simplifying the data structure and the overall analysis (Wang 2014 p. 143).

Interpretation

The factors attempt to explain the variation of the original data, and they can be viewed as broad concepts that describe a set of observations. Although FA is closely related to PCA, it follows a different approach. PCA transforms the original observed variables, and thus, it can be considered as a mathematical transformation using linear combinations of the original variables (Demšar et al. 2013, Wang 2014). On the other hand, factor analysis attempts to capture the variations of the observed variables on the assumption of latent variables' existence, the factors, associated with error terms and for this reason can be regarded a statistical process. As FA needs much subjective judgment, it is highly controversial in statistical circles. Factor analysis is not that common in geographical analysis lately.

5.4   Multidimensional Scaling (MDS)

Definition

Multidimensional scaling (MDS) is a technique that reduces the dimensionality of an N-dimensional dataset ($N > 2$) into two or three dimensions while preserving at some extent the relationships (similarities or dissimilarities) among the observations (O'Sullivan & Unwin 2003 p. 340).

Why Use

MDS is useful for visualizing the similarities/dissimilarities of a complex dataset by mapping them into two or three dimensions.

Interpretation

An MDS algorithm maps objects from an N-dimensional space to a two- or three-dimensional space configuration so that between N-dimensional objects, distances are retained as much as possible. Given pairwise dissimilarities, the algorithm reconstructs a map that preserves distances. The closer two objects they lie in the two or three-dimensional space, the closer they lie in the N-dimensional space as well.

Discussion and Practical Guidelines

Given pairwise dissimilarities, the algorithm reconstructs a map in the two- or three-dimensional space, called ordination, which preserves as much as possible the original distances (O' Sullivan et al. 2003 p. 340). The closer two objects lie in the ordination, the closer they lie to the N-dimensional space as

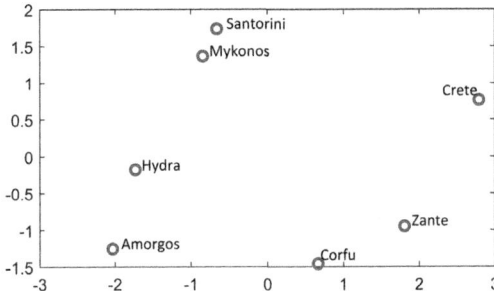

Figure 5.3 MDS ordination. Reducing from five to two dimensions.

well. The ordination is used to effectively identify interesting relationships among objects that are not obvious when using other statistical metrics. Locating objects with common characteristics as sets of points forming clusters in the ordination is a matter of an eye blink (see Figure 5.3). Dissimilarities can also be located by identifying which objects lie far away from the others. MDS calculates the new positions of the objects based on a distance or dissimilarity metric (i.e., Euclidean, Manhattan) through a matrix of pairwise distances (or dissimilarities) derived from the original dataset. MDS technique creates then a new point configuration whose inter-point distances approximate to the original dissimilarities.

As distortions from the 3-D plane to the 2-D plan are inevitable (i.e., projecting points in earth's surface to a map), MDS also infers distortions to the original N-dimensional data. This distortion is called stress. The method proceeds iteratively by mapping original data to two or three dimensions so that the stress function is minimized. Although stress might be large in some cases, MDS gains much acceptance especially for concept mapping where ideas or other interesting concepts are spatialized to reflect similarities and differences.

Let's see a brief example about seven Greek touristic islands, described by five variables (dimensions): the number of tourists visited the island the previous summer, the average money spent per tourist, the average nights per stay, the percentage of tourists coming from Italy and the percentage of tourists coming from the UK (see Table 5.6, Box 5.2). To identify any similarities and dissimilarities among the islands pertaining to the tourism industry we conduct MDS (see Figure 5.3).

The stress value is 0.063, reflecting a relatively low distortion. Results show that Santorini and Mykonos are more similar in comparison to Hydra and Amorgos. Also, Corfu and Zante have more similarities. Crete is entirely different from Amorgos (inspect Table 5.6 in comparison to the ordination in Figure 5.3). By MDS and ordination, we get a quick view of similarities and dissimilarities among the observations that would not be apparent by just inspecting Table 5.6.

Table 5.6 Five-dimensional dataset (five variables) related to the tourist industry of the Greek islands.

ID	Island name	Tourists (in hundred thousand)	Money spent (in thousand Euros)	Average nights per stay	Percentage of tourists from Italy	Percentage of tourists from the UK
1	Mykonos	2	1.6	7	15	25
2	Crete	5	0.7	14	10	70
3	Santorini	2.5	1.7	5	14	30
4	Corfu	3.5	0.9	10	30	50
5	Zante	4.3	1	14	25	65
6	Hydra	2.2	0.9	2	19	10
7	Amorgos	1.5	0.6	3	25	5

Box 5.2 Matlab. You can easily reproduce the preceding representation and analysis in Matlab. Go to the Matlab folder of Lab 5 to find data and code. Run MDS.m.

5.5   Cluster Analysis

Cluster analysis is a process where objects (observations) of a dataset are grouped into a number of clusters. Clusters are formed on the basis that objects within a cluster are as similar as possible (in respect to their attributes/characteristics), while objects belonging in different clusters are as dissimilar as possible.

The formation of the clusters is based on a distance matrix between all observations in the dataset (inter-observation distance – see Eq. 5.2), the dissimilarity matrix (see Eq. 5.2). Most of the time, before calculating the dissimilarity matrix, the data should be rescaled so that variables do not depend on the measurement scale and are comparable with each other. Failing to rescale data leads to assigning disproportionally more importance in variables with significantly larger values with respect to the other ones. Standardization is preferred to normalization, as it better retains the importance of each variable due to the non-bounding limitation. For example, in case of outliers, normalized data are squeezed at a small range, and as such, when dissimilarities (through statistical distances) are calculated, they contribute less to the final values. Still, rescaling is not always desirable. In case we have data of similar scales, proportions (e.g., percentages) or we want to assign weights to the variables with larger values, we might not consider normalizing, adjusting or standardizing. The decision on which rescaling type to apply depends on the type of the clusters we wish to shape and the dataset available. (see Section 2.4).

Cluster analysis removes observations from a multivariate matrix A (5.1) in contrast to the previous methods (PCA, FA, MDS) that remove variables. In this section, two nonspatial major clustering techniques are presented, namely the hierarchical clustering and the partitioning clustering (k-means clustering). Section 5.6 presents spatial clustering methods.

5.5.1   Hierarchical Clustering

Definition
Hierarchical clustering is an unsupervised method of grouping data by build-ing a nested hierarchy of clusters (O'Sullivan & Unwin 2003 p. 328). It is based on the creation of a tree-based representation of the data, which is called a dendrogram. There are two approaches to perform hierarchical clustering, namely agglomerative (bottom-up) and divisive (top-down).

Methods
Agglomerative hierarchical clustering starts from the bottom, assigning each object (also called leaf nodes) to a single separate cluster (see Figure 5.4). At this stage, each cluster contains only one object (member). In the second cycle, pairs of objects that are more similar (see similarity/dissimilarity measures later in this section) are grouped, creating clusters with two objects each. In the third cycle, each cluster (formed in the previous cycle) is joined with the one that is more similar, and a new set of clusters is created using a linkage method

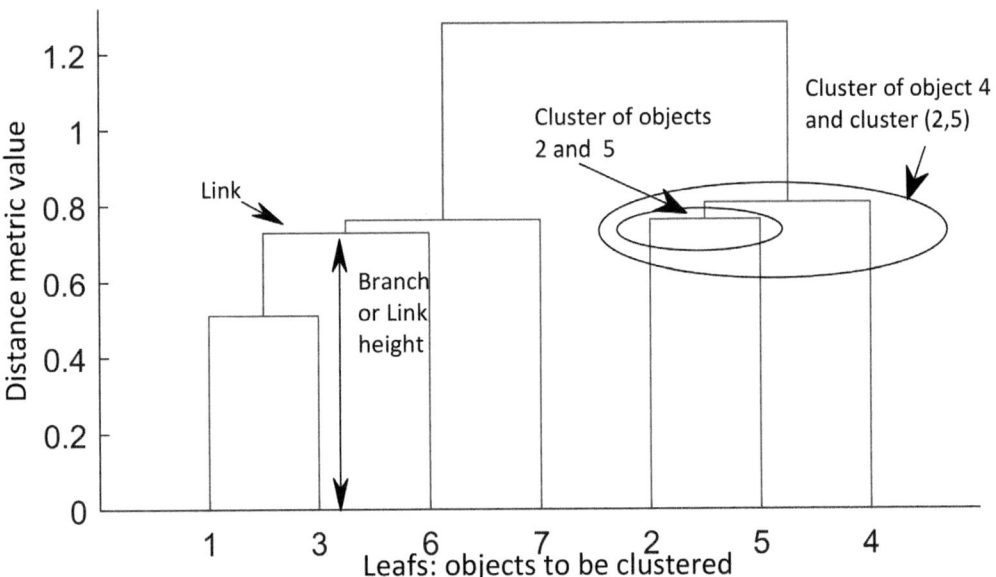

Figure 5.4 Dendrogram components for agglomerative hierarchical clustering algorithm.

(cluster of clusters). The procedure goes on until all clusters are clustered in one at the top of the tree.

Divisive hierarchical clustering operates in the opposite direction than the agglomerative method does. It works in a top-bottom manner, considering in its first cycle all objects belonging at a single cluster. At each following cycle, the most heterogeneous cluster is split into two clusters, until each object belongs to its own cluster.

Dendrogram

Whichever approach used, results are presented in a treelike diagram called dendrogram, depicting the hierarchical relationships between clusters (see Figure 5.4). The leaves of the tree, represent the observations (first-level: each object belongs to its own cluster). Going from bottom up, each two clusters create a new cluster in a higher level. Clusters that are merged create branches. The height of each branch (between the two original clusters and the merged new one) represents the distance between the two original clusters. The height is also called cophenetic distance. The taller the branch, the larger the distance of the merged clusters. The horizontal line that connects the two branches is called link. We only inspect the height of a dendrogram and not the horizontal distance which is only used for arranging the clusters in the horizontal direction.

Distance Metrics

The agglomerative hierarchical algorithm is based on the creation of (a) a dissimilarity matrix created through a measure of distance for calculating pairwise distances and (b) a linkage method for calculating inter-cluster distances.

Dissimilarity Matrix (Pairwise Distance)

Dissimilarity matrix is a distance square symmetric matrix with its elements corresponding to the pairwise distances (also called inter-observations distance) between the observations. Any distance metric can be used such as (see Chapter 1)

- Euclidean distance
- Manhattan distance
- Minkowski distance
- Pearson correlation distance
- Kendal correlation distance (for ranked based correlation analysis)

Euclidean and Pearson correlation distances are quite common methods in socio-economic analysis. When the Euclidean distance is applied, clustering is based on whether objects have similar values are not. On the other hand, Pearson correlation distance considers two observations similar if they are highly correlated. A high correlation between objects does not necessarily mean that their Euclidean is distance is small as well. There are cases where observations are highly correlated but lie far apart regarding their Euclidean distance. Pearson correlation distance is particularly helpful when we are interested in not clustering magnitude but relationships. For instance, in a consumer segmentation analysis, we may want to target subgroups that have the same attitude/preferences (e.g., buy similar things) but have different socioeconomic profiles. In this case, we are interested in finding highly correlated variables, and for this reason, correlation-based distances are more suitable. Keep in mind that in a complex analysis, we should test our dataset with more than one distance metrics to get a better insight.

Linkage Methods (Inter-Cluster Distance)

Apart from calculating the pairwise distances, we calculate the distances among clusters (inter-cluster distances). The inter-cluster distance is used to merge similar clusters in each subsequent step of the algorithm. In this case, clusters have more than one member and comparison is not straightforward as in the case of calculating pairwise distances among single observations. Linkage methods to calculate the inter-cluster distance between two clusters include:

- Single linkage: The distance between two clusters equals the minimum distance of an observation in one cluster with another observation in the other cluster (nearest neighbors between clusters). It produces unbalanced clusters and is not that widely used (Wang 2014 p. 149). Single linkage is appropriate when the two clusters are well separated.

- Average linkage: The distance between two clusters is the average of all pairwise distances (all pairs of observations) in the two clusters. The average linkage is based more on a measure of central location.

- Centroid linkage: The distance between two clusters is the Euclidean distance between the centroids (means) of the two clusters. Like average linkage, it provides results based on central location. This method is appropriate when outliers exist (Wang 2014 p. 151).

- Complete linkage: The distance between two clusters equals the most distant observations between these clusters (furthest neighbor). This method forms clusters with similar diameters ensuring that all observations inside a cluster lie within the maximum distance. It is appropriate for producing compact clusters.

- Ward's linkage: The distance between two clusters is the sum of squared deviations from each point to the centroid of each cluster. This method attempts to minimize the within-cluster variance and to produce nearly spherically shape clusters with a similar number of observations.

Standardization of data is also necessary in cases where the scale or the units that measurements are made for each variable are different and should be applied before any distance matrix calculations.

Choosing the Number of Clusters

For many problems, the number of clusters to partition a dataset is predefined. For instance, in a geomarketing analysis, we may want to segment

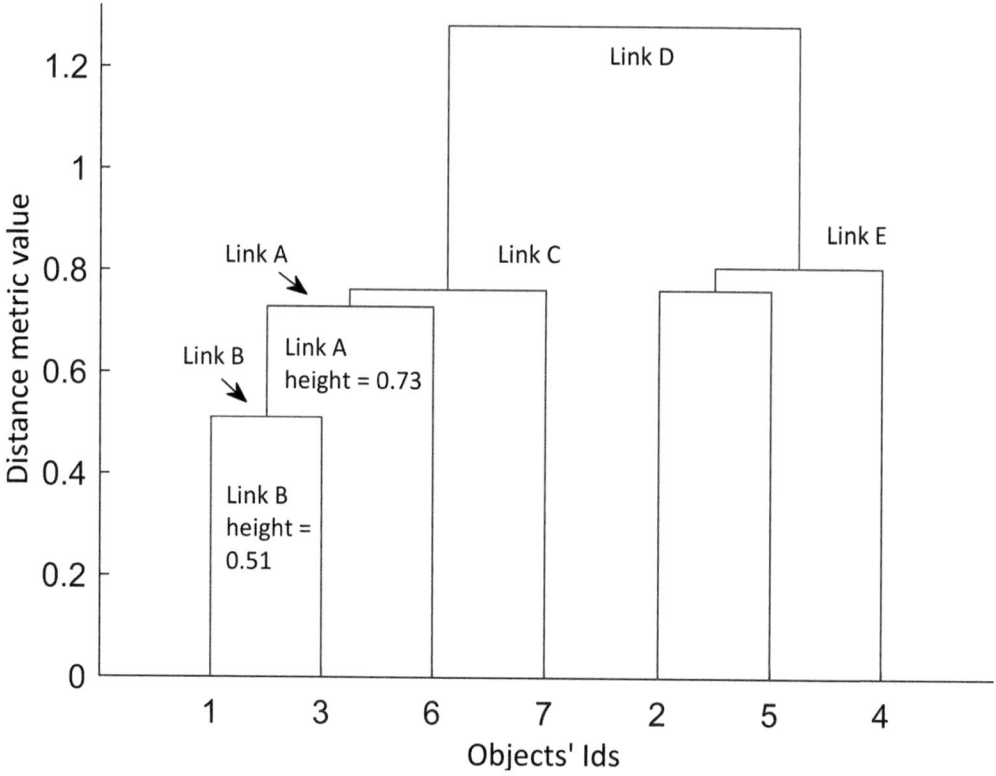

Figure 5.5 Inconsistency coefficient and links.

georeferenced data (i.e., socioeconomic variables referred to postcodes) into two clusters namely, *high spenders* and *low spenders*. However, in most real case studies, it is hard to determine in advance the optimal number of clusters that best partitions a dataset. In a wider aspect, each clustering method should finally produce clusters that are (Griffith et al 1997)

- Similar at their size (containing relatively similar number of objects)
- Not overlapping in shape (distinct boundaries among clusters)
- Internally homogeneous (observations values are close together inside each cluster)

More partitions than necessary lead to clusters overlapping with small between clusters differences (meaning that some clusters might be quite similar). Fewer clusters than necessary lead to nonhomogenous clusters having large within cluster dissimilarities.

For the hierarchical clustering, we may select the optimal number of clusters by inspecting the dendrogram's overall appearance (shape) and verifying its consistency. By verifying consistency, we compare the height of each link with the height of the links in lower levels. When heights are similar, clusters merged are not that distinct and create a uniform new cluster, as the distances among points before and after merging are similar. In other words, these clusters are consistent. If heights between links are large, then clusters merged are not consistent, as objects of the merging clusters lie far apart from each other. In such case, it is better not to merge these two clusters, as the new one results in objects with high dissimilarity. This is the cutoff point to consider that the optimal number of clusters lies and keep the clusters right before it (by trimming the dendrogram). A method to locate this point is by using the inconsistency coefficient (Jain & Dubes 1988). Inconsistency coefficient compares the height of each link with the adjacent links that lie below at a certain depth. Depth is usually one or two levels below in the hierarchy tree (see Figure 5.5). A high inconsistency coefficient reveals that the merging clusters are not homogenous, while a low inconsistency coefficient reveals that clusters can be merged.

By inspecting the inconsistency coefficient, we can select a cutoff value that splits the dendrogram into two parts and keep those clusters that lie below the cutoff level (Figure 5.6).

Verify Cluster Dissimilarity

Apart from defining the optimal number of clusters, we should also verify the quality of partitions regarding their similarity/dissimilarity. A method to verify cluster similarity is the cophenetic correlation coefficient, which computes the linear correlation coefficient between the cophenetic distances (height of link that two objects are first merged) and the original distances among objects (Sokal & Rohlf 1962). A strong cophenetic correlation reveals that the dendrogram represents the dissimilarities among observations accurately, while a low cophenetic correlation reveals an invalid clustering. The cophenetic correlation coefficient is very useful when we compare clustering results using different distance measures or different linkage methods. For each different distance (or linkage method), we can calculate the cophenetic correlation coefficient and keep the distance metric (or linkage method) that yields the higher one. We have to emphasize that clustering results should be meaningful according to our research scopes. For this reason, sometimes we might apply a distance metric that produces slightly less cophenetic correlation coefficient than another distance metric, still leading to an easier (or more rational) clusters interpretation (see Box 5.3).

Box 5.3 Matlab. You can easily reproduce the preceding representation and analysis in Matlab by using the first two numeric columns of Table 5.6. Go to the Matlab folder of Lab 5 to find data and code. Run HC.m.

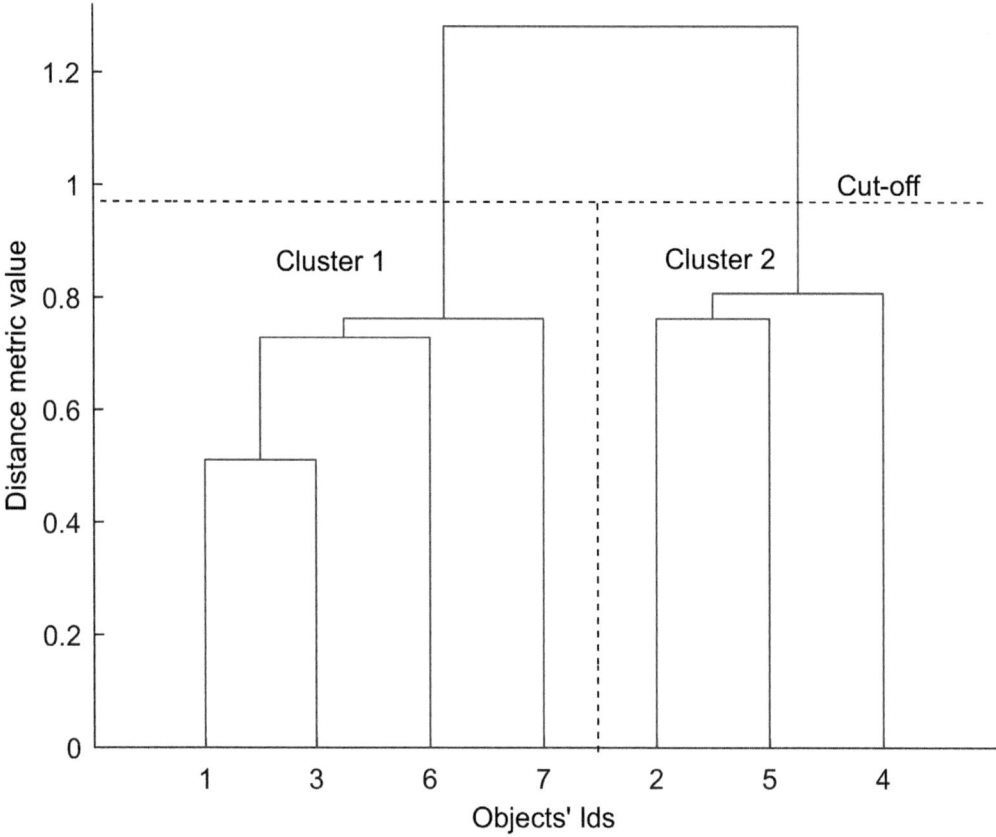

Figure 5.6 Cutoff inconsistency value of 0.8. By setting this cutoff value, we keep clusters that do not exceed this inconsistency. Two clusters are finally created: Cluster 1 consisting of objects [1,3,6,7] and Cluster 2 consisting of objects [2,5,4]. If these objects refer to spatial entities (e.g., postcodes) we can easily map the two clusters using two distinctive colors by assigning for example 1 (in a new column labeled "CLUSTER" in the attribute table) for objects belonging to cluster A, and 0 for those belonging to cluster B.

5.5.2   *k*-Means Algorithm (Partitional Clustering)

Definition

k-means is a clustering algorithm that partitions a dataset of n observations $(x_1, x_2, \ldots, x_n)$ into k clusters $C = \{c_1, c_2, \ldots, c_k\}$ by minimizing the within-cluster sum of squares (MacQueen 1967, Pena et al. 1999; see Figure 5.7). The goal is to identify the cluster centers μ_i, $i = 1 \ldots k$ that minimize the function (5.6):

$$argmin \sum_{i=1}^{k} \sum_{x \in c_i} \|x - \mu_i\|^2 \tag{5.6}$$

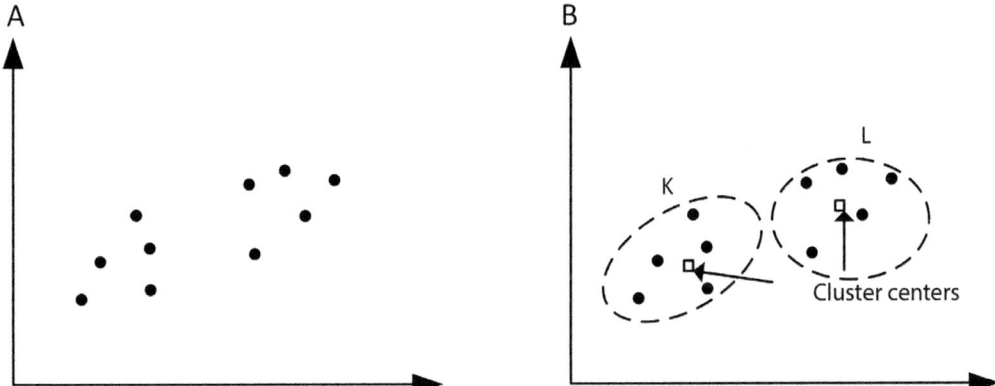

Figure 5.7 (A) Observations before clustering. (B) Observations clustered into two groups, namely K and L. The *k*-means algorithm attempts to form clusters in which (a) objects within the cluster are as close as possible (close to the cluster center) and (b) objects in the cluster are as far as possible from objects of other clusters.

where

c_i is the set of observations that belong to cluster *i*
μ_i is the mean of observations in c_i, and
$k \leq n$

The *k*-means is an unsupervised machine-learning algorithm, as it does not require training (set of pre-classified features that offer some initial knowledge of the clustering schema). Computing a *k*-means algorithm involves the following steps (Arthur & Vassilvitskii 2007):

1) **The initialization step**. The selection/calculation of the initial cluster centers μ_i (also called seeds) should be *k* in total.

$$\mu_i = some\ value,\ i = 1\ldots k \tag{5.7}$$

2) **The assignment step**. Each object is assigned to the closest (in the data space) cluster center.
3) **The calculation of the new cluster center**. For each cluster created, the mean center is calculated.

$$\mu_i = \frac{1}{|c|} \sum_{j \in c_i} x_j \tag{5.8}$$

where |*c*| is the total number of objects in cluster *c*.
The new cluster centers may not be existing observations.

4) **The reassignment step**. The algorithm returns to step 2, and each object is reassigned to the closest center. Steps 2 and 3 are repeated until reassignments at step 4 are stabilized (when the assignments do not

change from one iteration to the other) or a maximum number of iterations is met.

In practice, to find the optimal clusters, we should try all possible combinations. As this is infeasible, especially when the number of variables and the observations is large, we turn to heuristic algorithms. The k-means heuristic algorithm does not guarantee reaching a global minimum, but it yields a near-global minimum. The global minimum is the real minimum value of the dataset. A near-global minimum means that the algorithm reaches a solution that we hope it is close enough to the global minimum. We anticipate that a robust algorithm would approximate the global optimal solution.

Why Use
k-Means algorithm is used to group a set of observations to clusters with similar characteristics (homogeneous).

Interpretation
Objects of the same cluster are more similar compared to objects in other clusters. When referring to spatial data, clusters can be visualized by using typical GIS maps. Polygons belonging to the same cluster are rendered with the same color.

Discussion and Practical Guidelines
Three main topics should be defined prior to the k-means algorithm being run: (a) the initialization method, (b) the number of clusters k that the dataset will be partitioned into and (c) the variables to include.

- **Select the initialization method:** There are various initialization methods. The most commonly used are the Random Partition and the Forgy methods (Pena et al. 1999). The Random Partition method assigns a cluster randomly to each object and then calculates the cluster centers. The Forgy method chooses k objects from the dataset as the initial cluster centers (also called seeds) and then assigns each of the rest objects to the closest seed. Another method (algorithm) called k-means ++ selects an object randomly as the first cluster center (Arthur & Vassilvitskii 2007). Each subsequent cluster center is selected using a probability that is proportional to the squared distance of each observation to the nearest cluster center. Those seeds lying further away in the data space from the previously selected seeds are favored.

- **Select the number of clusters k:** To run the k-means algorithm, we should first set the number of clusters k. Most of the time, the number of clusters that a dataset should be partitioned is not known in advance, and a circularity emerges. We need to know k, to partition the dataset into k-clusters, but we also need to identify which is the appropriate number of clusters that partitions the dataset in an optimal way. The choice of the appropriate number of clusters is not trivial and strongly depends on the problem at hand and the data available. Various methods exist to define the number of clusters that a dataset should be partitioned to produce well-separated and homogenous clusters. We can rely on previous knowledge or specific requirements for the number of clusters to be used. We may also conduct several segmentations with incremental numbers of clusters and keep as the more appropriate k, the one that yields the most meaningful results in respect to our analysis.

Another approach is using the elbow method. The elbow method plots the percentage of variance explained (ratio of the between-cluster variance to the total variance) as a function of increasing number of clusters (this is similar to PCA and the use of a scree plot; see Figure 5.2). The point at the graph that adding an extra cluster does not explain much more of the data can be considered as the appropriate number of k (see Figure 5.8). Put simply, the first cluster explains a lot of the variance, and the second cluster adds extra information explaining additional variance, but at some point, an additional cluster will offer less new added variance explained, and the graph will significantly change its slope. This point is called the "elbow" (due to the sharp slope change). From this point onward, additional clusters do not significantly improve the variance explained. We select as the optimal k the one that corresponds to the elbow point.

The Caliński–Harabasz pseudo F-statistic is another method to select the appropriate number of clusters. It is a measure of separation between the clusters, reflecting within-cluster similarity and between-cluster

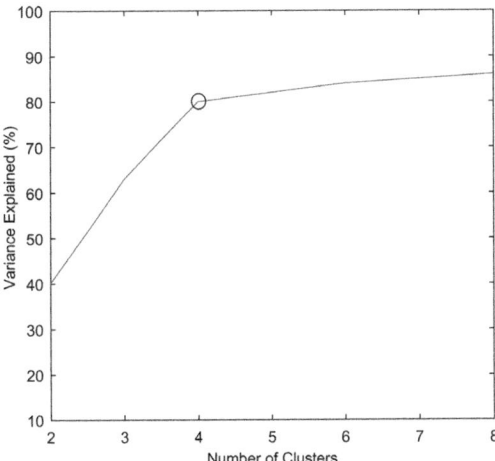

Figure 5.8 Elbow criterion. The location where the graph has a sharp change in its slope is called the elbow (here marked with a circle), and this point indicates the appropriate number of clusters k ($k = 4$).

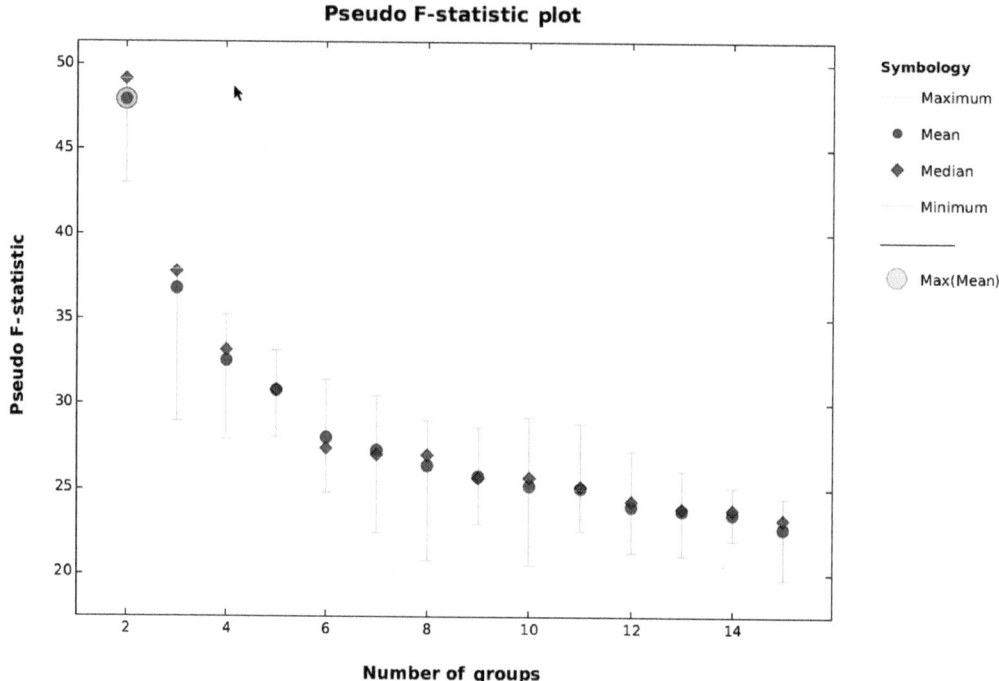

Pseudo F-statistic plot

Symbology

Maximum

● Mean

◆ Median

Minimum

⊙ Max(Mean)

Figure 5.9 Pseudo F-statistic criterion to select the number of clusters. The highest pseudo F-statistic value signifies large separation among clusters and indicates an appropriate number of clusters. In this example, the pseudo F-statistic is highest for $k = 2$.

difference (Caliński & Harabasz 1974). The most approprate k is the one with the highest pseudo F-statistic value (see Figure 5.9; ESRI 2019). Akaike Information Criterion (AIC; Akaike 1974) and Bayesian information criterion (BIC) (Schwarz 1978) can also be used to determine the optimal k. In both cases, the lowest value reflects the optimal clustering (see Box 5.4).

Box 5.4 The preceding criteria belong to a wider category of measures named validation measures/criteria (for more details, see Halkidi et al. 2001). The purpose of calculating validation measures is twofold. First, validation measures evaluate the results of a clustering algorithm (shape/correctness of clusters) and, second, assist in assessing the optimal number of clusters (Grekousis & Hatzichristos 2013). This process is also called cluster validity. Validation measures can be also used to compare results across different clustering algorithms. Two measurement criteria used to determine the optimal number of groups and the clustering scheme in general are the compactness and the separation of the clusters. Objects in

---

Box 5.4 (*cont.*)

each cluster should be compact (as close to each other as possible). Variance is a standard measure of compactness. Moreover, clusters should lie away as far as possible and be well separated. In this respect, a reliable validation measure should consider both the compactness and the separation of a cluster.

- **Select variables:** Out of a large number of variables that a database contains only a small fraction is necessary to define the cluster structure (Brusco & Cradit 2001). The choice of which variables to retain and which "masking" variables to eliminate is not easy. A masking variable is a variable that does not define a true cluster structure and may obscure the clustering analysis (Brusco & Cradit 2001). There are two broad approaches to trace masking variables, namely variable weighting and variable selection (Gnanadesikan et al. 1995). Variable weighting method assigns a weight to each cluster by minimizing a measure of stress. Variable selection method assigns a weight of 1 to those selected and zero to the masking variables. The simplest way to select a variable is to rely on previous knowledge or on conceptual relevance with the scopes of the study (Grekousis & Thomas 2012). For example, in geodemographical clustering, education would be more rational in comparison to person's height. A simple approach is to start with a relatively small number of variables, conceptually related to the problem, and to add new ones successively. If the validation measures and the clustering scheme improves, then we can keep the newly added variables; otherwise, we can drop them. Variables that exhibit multicollinearity may also be removed.

  An additional method is to use the *R*-squared statistic (R^2), which is calculated for each variable (5.9):

  $$R^2 = \frac{TSS - ESS}{TSS} \tag{5.9}$$

  where

  > *TSS* (total sum of squares) is the total sum of squared deviations from the global mean value of a variable.
  > *ESS* (explained sum of squares) is the sum of squared deviations from the mean value of the group it belongs.

  R^2 reflects how much of the original variation of the variable (for the entire dataset) has been retained after clustering (ESRI 2019). The higher the R^2, the better a variable divides the original dataset into meaningful clusters. On the other hand, a variable with low R^2 would mean low variance retained in the groups.

Finally, variables should be rescaled before calculating the dissimilarity matrix so they do not depend on the measurement scale and are comparable with each other. Failing to rescale data leads to assigning disproportionally more importance in variables with significantly larger values with respect to the other ones. Standardization is preferred to normalization, as it better retains the importance of each variable due to the nonbounding limitation (see Section 2.4).

Another aspect we should consider prior to any clustering is the outlier detection. Outliers' existence strongly affects clustering results. To trace outliers, the methods presented in Chapters 2 and 3 can be applied. From the clustering point of view, outliers can be treated as clusters of a small number of objects that lie far away from the rest of the clusters (Jiang et al. 2001). The existence of outliers occasionally unveils interesting patterns that should be further investigated.

Finally, the k-means algorithm does not imply any spatial constraints, and the algorithm yields clusters not necessarily contiguous in space. For spatially contiguous clusters, we apply regionalization methods explained in the next section.

5.6   Regionalization

Definition
Regionalization consists of a set of methods that cluster multivariate spatial data based on spatial constraints. It is a procedure of grouping a large number of spatial objects into a desirable smaller number of homogeneous clusters that also occupy contiguous regions in space (Assunção et al. 2006). The spatial clusters are also named regions. The entire procedure can also be named as zone design (Openshaw 1977).

Why Use
Regionalization methods are applied to cluster spatial features into groups with similar characteristics that are also spatially contiguous. From the spatial planning and policy perspective, regionalization is an important process where neighborhoods, census tracts, postcodes, districts or counties are grouped to form wider homogenous regions that policies related to social, educational, health, environmental or financial issues are applied.

Interpretation
Regionalization methods produce clusters in which (a) features within clusters are as similar as possible, (b) between-cluster difference is as large as possible and (c) clusters are composed of contiguous spatial entities.

Discussion and Practical Guidelines

A wide variety of geographical applications exist, especially in planning and policy analysis, where we need to create clusters of contiguous spatial entities that form wider homogenous, compact and cohesive regions (Alvanides & Openshaw 1999, Stillwell et al. 1999, Photis & Grekousis 2012). Examples include redesigning the provinces of a state or creating zones that national funds would be allocated to equally promote sustainable development. We might also redesign the neighborhoods of a city based on their demographical characteristics for designing new school districts. In marketing analysis, we may wish to create homogenous regions of economic activities to promote targeted policies or conduct market segmentation and identify regions that specific products have better penetration.

Furthermore, the predefined administrative boundaries (that data are aggregated) may not reflect well the scopes of a study. If the original data are available at the smallest aggregation level, then we can design new zones reflecting on the specific needs of the analysis. For instance, if a supermarket chain retains a large customer database including address, demographical variables and products preference, then new zones can be designed for tailored analysis, thus avoiding the ecological fallacy and the modifiable area unit problem (see Section 1.3) that may arise if data are aggregated to a higher-level predefined zone (i.e., postcode). In particular, zone designing in various geographical scales also allows for testing the modifiable areal unit problem.

There is another large class of problems for which regionalization methods are useful. When we study rare events (e.g., crime events, such as homicides, or health issues, such as AIDS), we might encounter the problem of relatively few events existing in our case study area. This is also called the small population (numbers) problem (Wang 2014 p. 193). In this type of problem, aggregating events to predefined zones (e.g., postcodes or census tracts) would most likely lead to a large majority of the zones with no events contained and with a small number of events (maybe one or two on average) on the remaining ones. Applying typical spatial statistics would yield unreliable estimates. For example, spatial autocorrelation, or hot spot analysis, cannot be performed when most zones contain zero events. With regionalization, though, as similar areas are merged to new regions, spatial autocorrelation is less of a concern in the newly defined zones (Wang 2014).

Additionally, ordinary least squares (OLS) regression is not suitable for this type of problem, as two basic assumptions of OLS are violated, namely the homogeneity of error variance (because errors of prediction are larger in polygons with fewer events) and the normal error distribution (see also Chapter 6; Wang 2014 p. 194). Although there are many different approaches to deal with the small population problem (e.g., use counts instead of rates, Poisson regression instead of OLS, floating catchment area, kernel density estimation, locally weighted average and adaptive spatial filtering), regionalization methods are considered very efficient (Wang 2014 p. 194).

There are many approaches to deal with regionalization (Assunção et al. 2006, Duque et al. 2007). The SKATER method and the REDCAP method are presented below.

5.6.1   SKATER Method

Definition
SKATER (Spatial "K"luster Analysis by Tree Edge Removal) is an algorithm using a connectivity graph, built by a minimum spanning tree, to create spatially constrained clusters based on a set of variables for a predefined number of clusters (Assunção et al. 2006).

Why Use
The SKATER algorithm is used to perform spatially constrained multivariate clustering.

Interpretation
The spatial features belonging to the same cluster are both contiguous and homogenous.

Discussion and Practical Guidelines
There are various spatial constraints used – like contiguity edges, contiguity edges-corners, k-nearest neighbors, Delaunay triangulation and predefined spatial weights (see Sections 1.6 and 1.8). Once a spatial constraint is set, a proximity matrix is created. The minimum spanning tree algorithm is then applied (using the proximity matrix) to create a connectivity graph and a minimum spanning tree that represents (a) the relationships among neighboring features and (b) the features similarity. Each feature is represented as a node in the tree, and it is connected to other features through branches called edges. Each edge gets a weight that it is proportional to the similarity between the features it connects (ESRI 2019).

The SKATER algorithm prunes the graph to get contiguous clusters (ESRI 2019). The algorithm starts by cutting the tree into two parts that form two well-separated clusters (minimizes dissimilarity in the new clusters). Then, it divides each part separately creating new clusters up to the point that the total number of clusters set initially is reached. Each division is made so that the separation between clusters and the similarity within clusters is maximized.

The size of clusters can be set by either a count (i.e., what is the minimum or maximum number of features that a cluster should have) or the sum of a variable (i.e., the total human population). For example, in case of districts, we can create larger administrative zones where each one should have at least

10 districts and a population ranging from 500,000 to 2,000,000 people. There is always the chance that such constraints cannot be met for all clusters because of the way that the minimum spanning tree has been constructed or because the maximum and minimum constraints are very close to each other. In this case, the clusters that do not fulfill the criteria should be mentioned.

As SKATER algorithm is a heuristic algorithm (algorithms used to find solutions when traditional methods are too slow or when finding the optimal solution is infeasible), it cannot guarantee an optimal solution. This practically means that for different algorithm runs, different solutions are likely to occur, and subsequently, a spatial feature might belong to different regions in each run (a "run" involves the entire process of the algorithm with all iterations completed and should not be confused with an iteration of the algorithm). To account for this problem, we calculate the probability of cluster membership for each feature using permutations of random spanning trees. By defining a number of permutations, we define the number of random spanning trees to be created. SKATER runs for each different spanning tree and the frequency that each spatial object is assigned to each cluster is recorded (e.g., 99 times out of 100, an object is clustered in cluster A). A high membership probability would typically mean that a feature is highly likely to belong to the specific cluster finally assigned by the SKATER algorithm (or else a feature is assigned to the same group in most of the permutations). A low probability would typically indicate that the specific feature is switching into different groups in each permutation and the final assignment is not reliable. The number of objects not well assigned should be kept to a minimum.

In general, three basic settings have to be defined to run SKATER, namely the spatial constraints method, the optimal number of clusters and the variables to include (ESRI 2019). Here are some guidelines:

- The polygon contiguity options (contiguity edges, contiguity edges corners) are not appropriate in the case of island polygons (noncontiguous polygons). In such case, k-nearest neighbors and Delaunay triangulation are preferred.

- Trimmed Delaunay triangulation can also be used to ensure that neighbors lie inside a convex hull. Features outside the convex hull are not assigned to as neighbors. This method is suitable in the case of spatial outliers.

- A weight matrix can also be used to include user-defined weights that may also reflect time constraints. Still, the algorithm does not take into account the weights, as it needs a binary definition of contingency. If weights are 0 or 1, then the algorithm performs at the usual fashion. In cases where inverse distance is used without any cutoff point, all features will get some weight and treated as neighbors. As such, when using weights, there should be either a cutoff point or a binary representation.

- For additional temporal analysis, we can add variables containing time such as night, day or the day of the week. As such, the algorithm will be forced to include temporal distances.
- Similarly, we can add a spatial variable such as distance from the center of a town, distance from major roads, slope or land cover/land use type (Grekousis et al. 2015b). Including such variables will probably reinforce the spatial clustering process.
- The pseudo F-statistic can also be used as a way to assess the optimal number of clusters (as in the k-means algorithm). It can also be used to identify the most effective spatial constraint method as long as the variables analyzed are the same in each trial.
- Choosing the most appropriate variables to use is similar to k-means algorithm (see Section 5.5.2). Variables should be related to the problem in question, exhibiting also a high R^2, which reflects how much of the original variation of the variable (for the entire dataset) was retained after the clustering process. It is better to begin with a few variables and progressively increase them one by one to better understand how each variable contributes to the data separation. Moreover, variables selected should be standardized (see Section 2.4), as variables with large variance tend to have substantial influence on the clusters compared to variables with small variance (most of the times this is automatically done by the software used).
- When calculating the variables values in the new regions, we have to be cautious about how we treat percentages, average values and index values. In case we have the absolute value of a variable (e.g., human population), the new value of the region is just the sum of the variable's values within the aggregated polygons. In case of percentages though, we cannot just sum them or calculate an average. We should first transform percentages to their absolute values and then calculate the aggregated ones, reflecting the region. Likewise, for index or an average value (e.g., car per capita), we cannot sum the values. We should calculate a weighted average $\bar{X}$ as (Wang 2014 p. 207) (5.10):

$$\bar{X} = \frac{\sum_{i=1}^{n} w_i X_i}{\sum_{i=1}^{n} w_i} \tag{5.10}$$

where w_i is the weight (for example, the total population) in the i-th polygon, X_i, is the variable value (e.g., number of cars per capita) in the i-th polygon and n is the number of polygons that have been merged to create a new region. The final weighted average $\bar{X}$ reflects the value of the variable (car per capita) of the new region. This formula is similar with the one used for the weighted mean center presented in Section 3.1. Different weights can be used according to the problem in question.

- As the method calculates the probability of cluster membership, we should also set the number of permutations that defines the number of random spanning trees to be created. As this analysis takes a consider-able amount of time, it is advised to first define the optimal number of clusters and then perform permutations.

- Graphs and plots can be used to explore results further.

5.6.2   REDCAP Method

Definition
REDCAP (Regionalization with dynamically constrained agglomerative cluster-ing and partitioning) is a regionalization set of methods (Guo et al. 2008).

Why Use
REDCAP is used to perform spatially constrained agglomerative clustering and partitioning.

Interpretation
The features belonging to the same cluster are both contiguous and homogenous.

Discussion and Practical Guidelines
REDCAP creates homogeneous regions by aggregating contiguous areas based on attributes similarity. REDCAP consists of two levels (Wang 2014 p. 199). In the first level, hierarchical clustering is performed based on spatial constraints following a bottom-up approach. In the second level, the tree constructed in the first level (as a result of the hierarchical clustering) is partitioned following a top-down approach. The Rook's contiguity method (share edge) is used to define if any two features are contiguous or not. Like SKATER, REDCAP can also adopt attribute constraints as minimum or max-imum regional population, or the minimum number of events contained in each region (see Box 5.5). Similar guidelines referred for SKATER can be used to guide the process.

Box 5.5 REDCAP toolkit can be downloaded from www.spatialdatamining .org/software/redcap. ArcGIS offers a toolbox for district design (www .esri.com/software/arcgis/extensions/districting/download). This toolbox does not create regions automatically as SKATER and REDCAP, but it offers a rich variety of tools to experiment with different scenarios and plans that can be investigated based on available socioeconomic data. It is a spatial planning toolkit that can be used to evaluate the results of other automated methods and cross-compare with human-made scenarios. It also offers a sense of control by integrating human knowledge of planners and other stakeholders in the decision process.

5.7   Density-Based Clustering: DBSCAN, HDBSCAN, OPTICS

Definition
Density-based clustering algorithms regard clusters when points concentrate in a geographical region (high density), while they label as noise (low density) those points with no neighbors at a close distance (Halkidi et al. 2001).

Why Use
Density-based clustering algorithms are used to perform spatial clustering of point features and better handle spatial outliers (noise), especially when large amounts of point data are analyzed (e.g., big data).

Interpretation
Objects that lie inside a cluster are more similar than objects belonging in different clusters.

Discussion and Practical Guidelines
There are three main unsupervised machine-learning algorithms to perform density-based clustering named: DBSCAN, HDBSCAN, OPTICS. These algorithms are typically applied in the N-dimensional space, but from the geographical perspective are applied on a two-dimensional space for analyzing the spatial distribution of points.

DBSCAN (density-based spatial clustering of applications with noise) in the context of spatial analysis uses only two parameters namely the maximum radius and the minimum points to be included with this radius (Ester et al. 1996). It is based on the calculation of the geographical distances among points along with a threshold distance value to calculate the density. Each point within a cluster should contain a minimum number of points within a given radius (neighborhood). The density in the neighborhood of a point has to be larger than a threshold value. DBSCAN is a fast algorithm, but it is sensitive to the search distance. Still, if clusters have similar densities, then it performs well. DBSCAN does not require setting the number of clusters as the k-means does, and it can trace clusters of irregular shape.

HDBSCAN (hierarchical DBSCAN) applies incremental distances to partition data into meaningful clusters while removing noise (Campello 2015). It is an extension of DBSCAN, and it uses a hierarchical clustering algorithm. Compared to DBSCAN and OPTICS, it is the most data-driven algorithm, and in this sense, human interaction is kept to a minimum.

OPTICS (ordering points to identify the clustering structure) is another density-based algorithm that creates a reachability plot used to separate clusters from noise (Ankerst et al. 1999). This algorithm is an extension of DBSCAN, but it has been addressed to better identify clusters when density fluctuates. The reachability graph plots on the x-axis the points ordered as processed by

OPTICS and the reachability distance on the y-axis. Points belonging to the same cluster have a lower reachability distance. A sharp increase in reachability distance signifies a new cluster. Clusters form a series of low values with some high values in the beginning and the ending, creating a valley-like shape, and as such, the deeper a valley is, the denser the cluster.

All previously mentioned algorithms require (a) a minimum number of features (minPoints) to call a set of nearby points a cluster and (b) a distance threshold (radius) to define the neighborhood. Groups with less than the specified number of minimum features will either be considered as noise or will be merged with another cluster. A rule of thumb for selecting the minimum number of features can be derived from the number of dimensions N in the dataset as $minPoints \geq 2*N$, but it may be necessary to select even larger numbers in case of big data or noisy data (Sander et al. 1998). To select the distance threshold, one can calculate a k-distance graph and then plot the distance to the $k = minPoints - 1$ nearest neighbors ordered from the largest to the smallest (Schubert et al. 2017).

5.8   Similarity Analysis: Cosine Similarity

Definition
Similarity analysis utilizes a function to quantify the similarity among objects. One of the most widely used similarity metrics is the cosine similarity index defined as (5.11):

$$cosine\ similarity = \frac{\sum_{i=1}^{n} A_i B_i}{\sqrt{\sum_{i=1}^{n} A_i^2} \sqrt{\sum_{i=1}^{n} B_i^2}} \tag{5.11}$$

where

  A, B are attribute vectors
  i is a variable
  n is the total number of variables to be compared (number of columns containing the variables used for comparison)

By computing cosine similarity, we measure the direction-length resemblance between datasets represented as vectors (ESRI 2018b). In a spatial analysis context, suppose that A and B are two spatial features to be compared. A is the target object and stands for the attribute vector of this spatial feature (line in the attribute table). B is the candidate object and stands for the attribute vector of this object. The target object is the object that all other objects (candidates) will be compared to identify similarity. If more than one object is used as a target, then a composite target feature is created based on the average values of the target objects.

Why Use
Cosine similarity is used to identify if two objects have similar profiles, thus facilitating their comparison.

Interpretation
Objects closer in the ranking position are more similar in comparison to objects further away from each other. "Similarity" concept is used in spatial analysis as in the Euclidean geometry where two objects that are similar do not have to be of the same size. It is their shape that defines the similarity. Likewise, similarity in spatial analysis does not imply the same magnitude of values but a similar pattern. When using the cosine similarity, the outcome ranges between −1 to 1. A value toward of 1 indicates similarity while a value toward −1 reveals dissimilarity. A value of zero indicates orthogonality (decorrelation − reduction of cross-correlation).

Discussion and Practical Guidelines
Cosine similarity does not identify if two objects are identical but if they are similar. Similar objects are those having a similar pattern, but their scale or magnitude may significantly vary. Suppose we want to identify which island is more similar to Corfu (target) by using the data in Table 5.6 (we only compare three islands in this example). We observe a similar data pattern/profile between Corfu (target) and Santorini (candidate) (see Table 5.7). The cosine similarity index between these two islands is 0.99, indicating high similarity. Notice that the magnitude in values is quite different. On the other hand, Hydra has not as similar pattern with Corfu and gets a lower similarity index (0.85) (see Figure 5.10).

Standardized attribute values should be used for variables with large variances and different scales. Similarity analysis can be applied in many spatial analysis problems. For instance, in the previous example, we may attempt to identify those islands having similar potential of becoming top destinations like an existing one (target) but currently are not that developed or famous. This will probably offer better opportunities for investments as land price, or other related costs should be significantly lower. Another example might be the

Table 5.7 This table is the same with Table 5.6 but focuses on the specific features to be analyzed.

Object	ID	Island name	Average nights staying	Percentage of tourists coming from Italy	Percentage of tourists coming from the UK	Similarity coefficient
Candidate	3	Santorini	5	14	30	0.99
Target	4	Corfu	10	30	50	-
Candidate	6	Hydra	2	19	10	0.85

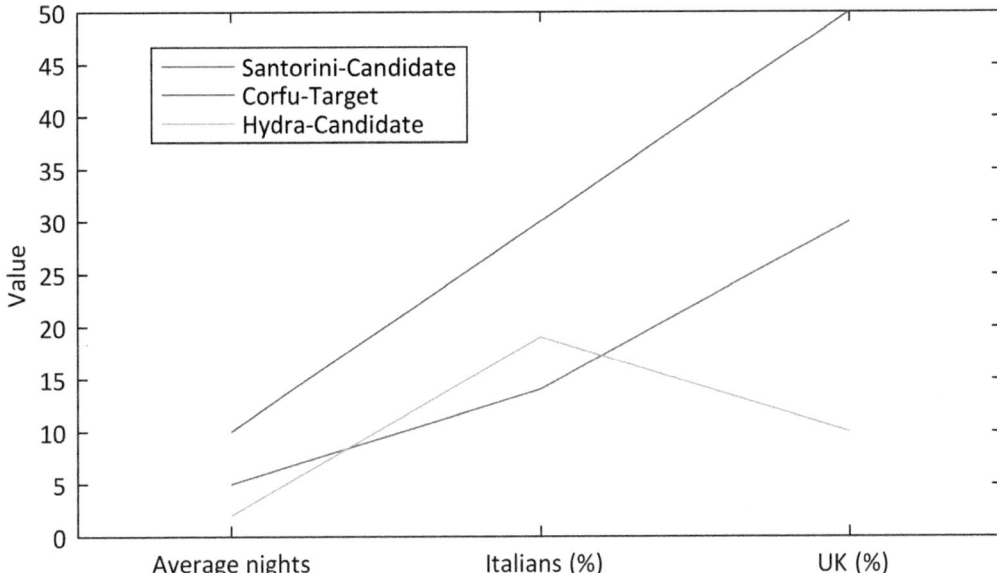

Figure 5.10 Graph depicting the values of three islands as presented in Table 5.7.

expansion of a retail franchise company. Finding new regions to expand with similar characteristics as those that the company exhibits substantial revenues will facilitate the location of new shops. According to the problem in question, spatial variables can be used, such as distance from point of interest, distance from competitors or distance from open space.

5.9   Chapter Concluding Remarks

- PCA, FA and MDS reduce the number of columns, while CA reduces the number of lines of a multivariate dataset.
- The main advantage of PCA and FA is that they may reveal components/latent variables that lead to uncovering hidden interactions.
- MDS great advantage is that it maps observations to two or three dimensions, providing a graphical representation of similarities for objects clustered and dissimilarities for objects lying apart.
- PCA performs a rotation of the original data so that variance is maximized in the new axes.
- PCA then projects original data to the new axes. We can visualize data in the new space (with lower dimensions) by keeping only two or three axes (the first two or three principal components).
- Usually, the first five components are enough to capture data information.
- The results of PCA depend on the scale of variables (e.g., different units).

- Geodemographic segmentation is a typical problem that PCA may be applied. By using PCA, data are condensed to meaningful groups of socioeconomic variables.
- Raw unstandardized data may be used only if variables are at the same scale and at relatively similar variance.
- In social and geographical analysis, it is rare not to standardize data.
- CA groups observations, which makes an enormous difference to the previous methods.
- In hierarchical clustering, when heights of a dendrogram are similar, clusters merged are not well separated.
- Cutting the tree at a certain height typically reveals the number of clusters to use for our analysis.
- A high inconsistency coefficient reveals that merging clusters are not homogenous, while a low inconsistency coefficient reveals clusters that can be merged.
- Three main topics should be defined prior to the k-means algorithm run: the initialization method, the number of clusters k and the variables to include.
- The maximum pseudo F-statistic value indicates the optimal number of clusters in k-means clustering.
- Regionalization methods main advantage is that they produce homogenous clusters of spatial features that are also adjacent.
- SKATER and REDCAP are two widely used methods for regionalization. The features belonging to the same cluster are both contiguous and homogenous.
- Density-based clustering methods like SCAN, DBSCAN and OPTICS are used to group points based on their concertation in a geographical region.
- Cosine similarity does not identify if two objects are identical but if they are similar.
- Similar objects are those having similar values pattern (also called profile), but the scale or magnitude may significantly vary.

Questions and Answers

The answers given here are brief. For more thorough answers, refer back to the relevant sections of this chapter.

Q1. What are multivariate data and multivariate analysis. Why are they used?

A1. Multivariate data are data with more than two values recorded for each observation. Multivariate statistical analysis is a collection of statistical methods to analyze multivariate data. Multivariate techniques dive inside the vast amount of data available through census bureaus, organizations and sensors to discover patterns and unexpected trends or behaviors and extract hidden knowledge valuable for spatial analysis and spatial planning.

Q2. Name the most commonly used multivariate methods for dimensionality reduction and data clustering.

A2. The most commonly used multivariate statistical analysis methods are
- Principal component analysis (PCA)
- Factor analysis (FA)
- Multidimensional scaling (MDS)
- Cluster analysis (CA, for classifying observations)
- Regionalization when clustering is made with spatial constraints
- Similarity analysis

Q3. How is PCA applied with raster data? Give a short example.

A3. In case of raster data, components are calculated based on measurements referring to each cell. This approach is common in remote sensing when many different raster datasets with the same spatial reference and extent are combined to produce composite indices. For example, we may combine a land use/land cover image, a raster for soil variables, a raster of temperature, a raster of CO_2 emissions and a raster of socio-economic and census data. By applying PCA, we detect the principal components, and we map each one of them as a new raster file. In this way, we depict the spatial distribution of each component's scores directly.

Q4. What are the main differences between PCA and factor analysis?

A4. PCA transforms the original observed variables, and thus, it can be considered as a mathematical transformation using linear combinations of the original variables. On the other hand, factor analysis attempts to capture the variations of the observed variables on the assumption of latent variables' existence, the factors, associated with error terms, and for this reason can be regarded a statistical process.

Q5. What is multidimensional scaling? What is ordination and what does it represent?

A5. Multidimensional scaling (MDS) is a technique that reduces the dimensionality of an N-dimensional dataset ($N > 2$) into two or three dimensions while preserving to some extent the relationships (similarities or dissimilarities) among the observations. Given pairwise dissimilarities, the algorithm reconstructs a map in the two- or three-dimensional space that is called ordination, which preserves as much as possible the original distances. The closer two objects lie in the ordination, the closer they lie to the N-dimensional space as well.

Q6. What is hierarchical clustering, what is a dendrogram and what are the main two approaches to partition data?

A6. Hierarchical clustering is an unsupervised method to group data by building a nested hierarchy of clusters. It is based on the creation of a tree-based representation of the data, which is called dendrogram. There are two approaches to perform hierarchical clustering, namely the agglomerative (bottom-up) and the divisive (top-down). An agglomerative hierarchical clustering starts from the bottom, assigning each object (also called leaf nodes) to a single separate cluster. Divisive hierarchical clustering operates in the opposite direction from what the agglomerative method does. It works in a top-bottom manner, considering in its first cycle all objects belonging at a single cluster.

Q7. What are the elbow method and the Calinski-Harabasz pseudo F-statistic in *k*-means algorithm?

A7. The elbow method plots the percentage of variance explained (ratio of the between-cluster variance to the total variance) as a function of increasing number of clusters. The point on the graph that adding an extra cluster does not explain much more of the data can be considered as the appropriate number of *k*. The Calinski–Harabasz pseudo F-statistic is another method to select the appropriate number of clusters. It is a measure of separation between the clusters, reflecting within-cluster similarity and between-cluster difference. The most appropriate *k* is the one with the highest pseudo F-statistic value.

Q8. What is cluster validity, and why we use validation measures?

A8. Cluster validity is the procedure for evaluating the results of a clustering algorithm. The purpose of calculating validation measures is twofold. First, validation measures evaluate the results of a clustering algorithm (shape/correctness of clusters) and, second, assist in assessing the optimal number of clusters. Validation measures can be also used to compare results across different clustering algorithms.

Q9. What is regionalization, and what are the main characteristics of the clusters created?

A9. Regionalization consists of a set of methods that cluster multivariate spatial data based on spatial constraints. Regionalization is a procedure of grouping a large number of spatial objects into a desirable smaller number of homogeneous clusters that also occupy contiguous regions in space. Regionalization methods produce clusters in which (a) features within clusters are as similar as possible, (b) between-cluster difference is as large as possible and (c) clusters are composed of contiguous spatial entities.

Q10. What is density-based clustering, and which are the best-known algorithms?

A10. Density-based clustering algorithms regard clusters when points concentrate in a geographical region (high density), while they label as noise (low density) those points with no neighbors at a close distance. They are used to perform spatial clustering and better handle spatial outliers (noise), especially when large amounts of point data are analyzed (e.g., big data). There are three main algorithms to perform density-based clustering named: DBSACN, HDBSCAN, OPTICS.

LAB 5
MULTIVARIATE STATISTICS: CLUSTERING

Overall Progress

Spatial Analysis/ Lab Workflow

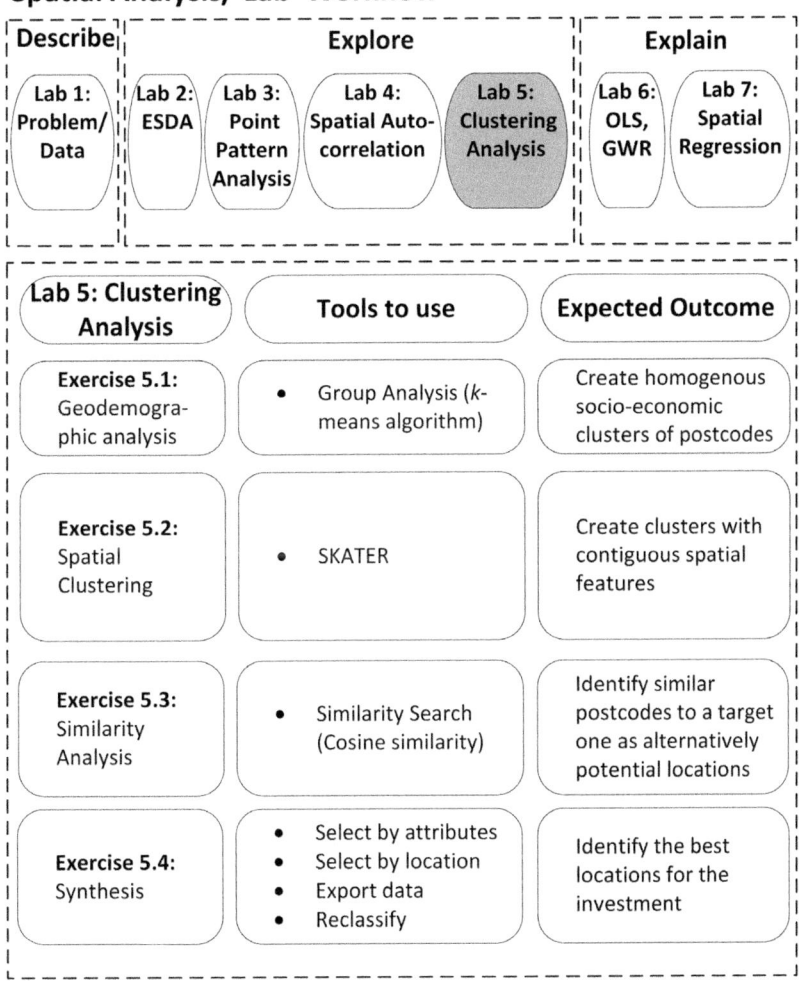

Figure 5.11 Lab 5 workflow and overall progress.

Scope of the Analysis

This lab deals with the following objectives:

- **Objective 3:** Clustering – High spenders (see Table 1.2)
- **Objective 1:** Locating high income areas (synthesis)
- **Objective 2:** Locating low crime areas (synthesis)

In this exercise we perform geodemographic analysis (see Figure 5.11). By utilizing the *k*-means algorithm, we group postcodes with similar demographical and socioeconomic characteristics. Pen portraits (short profile descriptions) are created for each cluster thereafter. Postcodes assigned to the cluster that better approximates the target group of the coffee shop will be preferred. We also conduct similarity search to identify which postcodes are more similar to a target postcode. We finally combine all results (synthesis) from the four previous labs to identify the best sites for the coffee shop.

Section A ArcGIS

Exercise 5.1 *k*-Means Clustering

In this exercise, we create homogeneous socioeconomic clusters of postcodes using the *k*-means algorithm. To run the *k*-means algorithm, we should define the initialization method, the number of clusters and the variables to include (see Figure 5.12).
   ArcGIS Tools to be used: Grouping Analysis

ACTION: *k*-means clustering for *k* = 4 (we select this randomly)

Navigate to the location you have stored the book dataset and click on Lab5_Clustering.mxd

Main Menu > File > Save As > My_Lab5_Clustering.mxd

In I:\BookLabs\Lab5\Output

ArcToolBox > Spatial Statistics Tools > Mapping Clusters > Grouping Analysis

Input Feature Class = City (see Figure 5.12)

Unique ID Field = PostCode

Output Feature Class = I:\BookLabs\Lab5\Output\Kmeans4.shp

Number of groups = 4 (We start by using an initial, arbitrarily chosen, number of clusters)

Analysis Fields = Check all except Population, Area, Regimes, PostCode.

Spatial Constraints = NO_SPATIAL_CONSTRAINT (*k*-means algorithm)

Exercise 5.1 (*cont.*)

⌛ Grouping Analysis

Input Features
| City | ▾ | 🖼 |

Unique ID Field
| PostCode | ⌄ |

Output Feature Class
| I:\BookLabs\Lab5\Output\Kmeans4.shp | 🖼 |

Number of Groups
| 4 |

Analysis Fields
☐ Population	⌃
☑ Density	
☑ Foreigners	
☑ Owners	
☑ SecondaryE	
☑ University	
☑ Phd_Master	
☑ Income	
☑ Insurance	⌄
‹	›

| Select All | Unselect All | Add Field |

Spatial Constraints
| NO_SPATIAL_CONSTRAINT | ⌄ |

Distance Method (optional)
| EUCLIDEAN | ⌄ |

Number of Neighbors (optional)
| |

Weights Matrix File (optional)
| | 🖼 |

Initialization Method (optional)
| FIND_SEED_LOCATIONS | ⌄ |

Initialization Field (optional)
| | ⌄ |

Output Report File (optional)
| I:\BookLabs\Lab5\Output\Kmeans4.pdf | 🖼 |

☑ Evaluate Optimal Number of Groups (optional)

| OK | Cancel | Environments... | << Hide Help |

Figure 5.12 *k*-means clustering dialog box for *k* = 4.

```
Distance Method = EUCLIDEAN

Initializing Method = FIND_SEED_LOCATIONS

Output Report File = I:\BookLabs\Lab5\Output\Kmeans4.pdf

Evaluate Optimal Number of Groups = Check

OK

Main Menu > Geoprocessing > Results > Current Session >
Grouping Analysis > DC on Output Report File: Kmeans4.pdf
```

Exercise 5.1 (*cont.*)

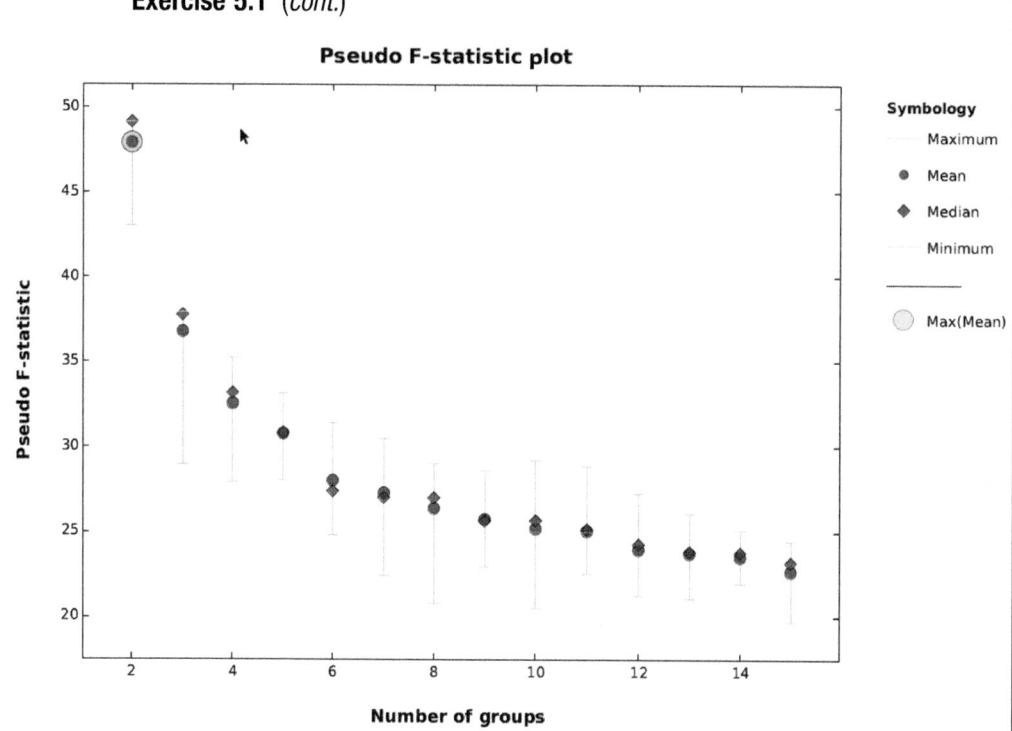

Figure 5.13 Pseudo F-statistic for defining the optimal number of clusters.

Interpreting results: Before we interpret results, we should first check the Calinski–Harabasz pseudo F-statistic (Figure 5.13; last two pages of the file Kmeans4.pdf).

The pseudo F-statistic plot indicates $k = 2$ as the optimal number of clusters (the maximum pseudo F-statistic value). In this respect, we should run the k-means algorithm again for $k = 2$.

ACTION: k-**means clustering** for $K = 2$

ArcToolBox > Spatial Statistics Tools > Mapping Clusters > Grouping Analysis

Input Feature Class = City

Unique ID Field = PostCode

Output Feature Class = I:\BookLabs\Lab5\Output\Kmeans2.shp

Number of groups = 2

Analysis Fields = Check all except Population, Area, Regimes, PostCode.

Exercise 5.1 (*cont.*)

Spatial Constraints = NO_SPATIAL_CONSTRAINT (*k*-means algorithm)

Distance Method = EUCLIDEAN

Initializing Method = FIND_SEED_LOCATIONS

Output Report File = I:\BookLabs\Lab5\Output\Kmeans2.pdf

Evaluate Optimal Number of Groups = Check

OK

Main Menu > Geoprocessing > Results > Current Session > Grouping Analysis > DC on Output Report File: Kmeans2.pdf

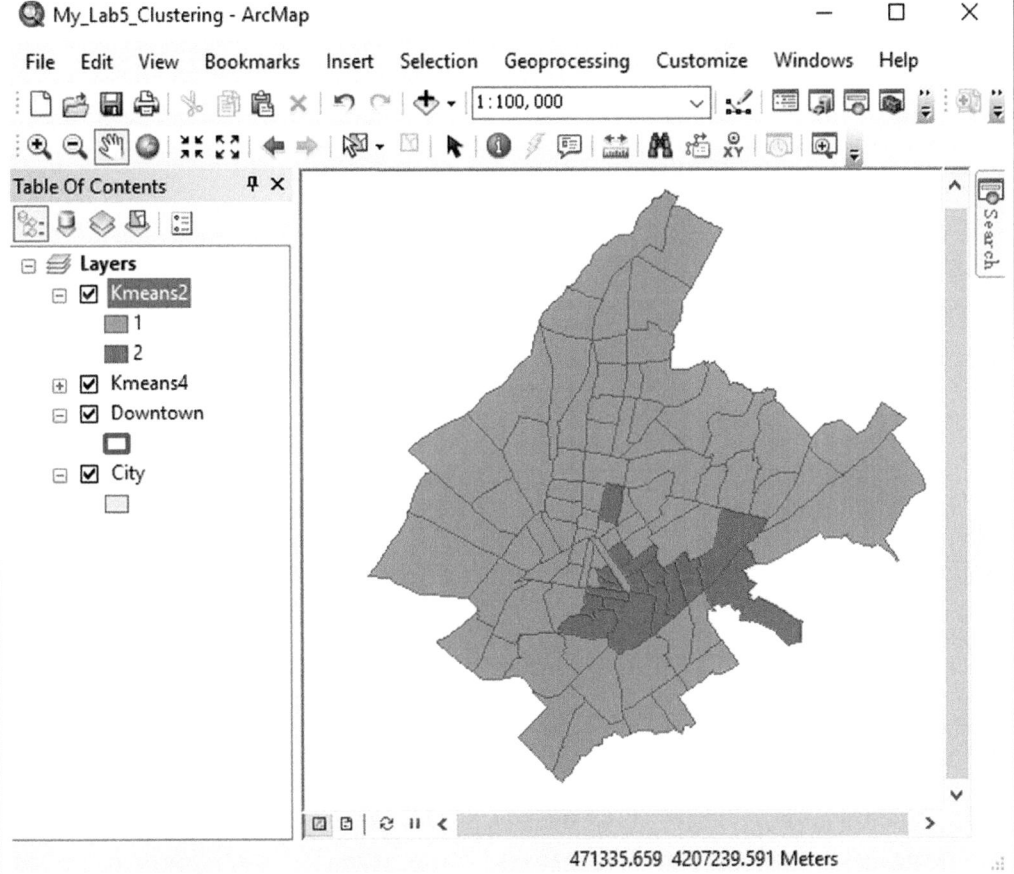

Figure 5.14 Postcodes grouped into two clusters using the *k*-means algorithm.

Exercise 5.1 (*cont.*)

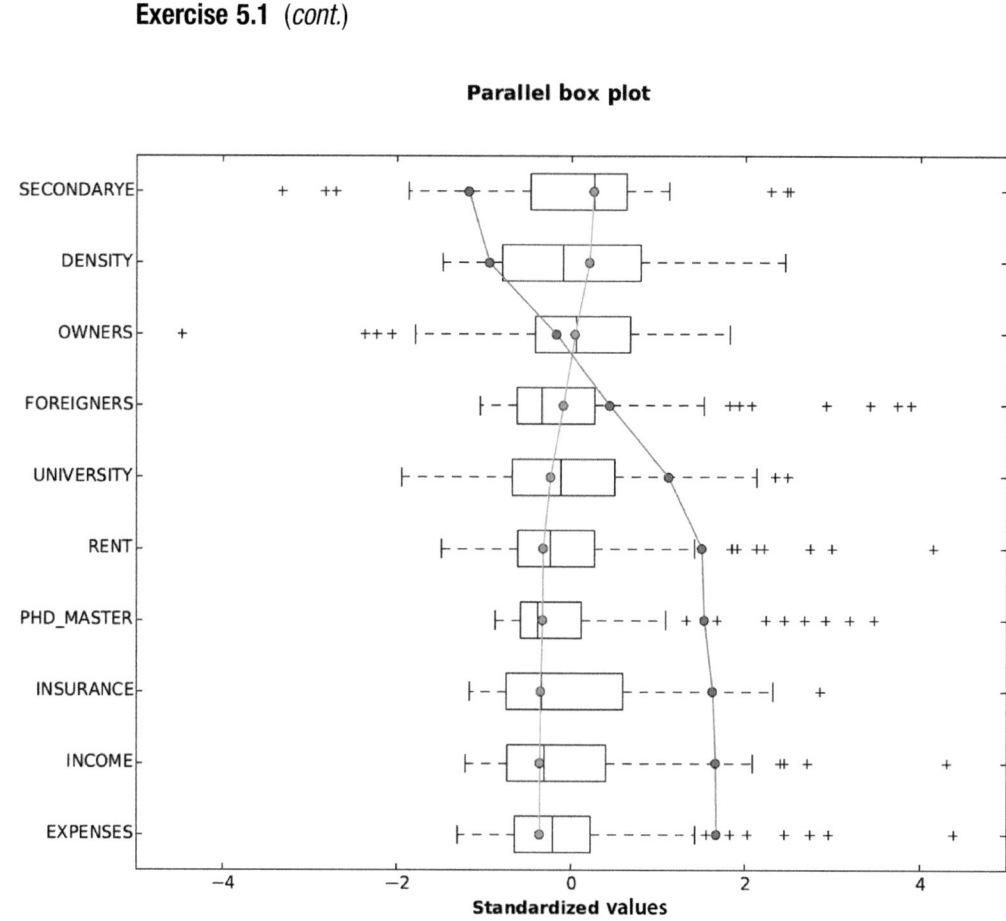

Figure 5.15 Clusters parallel boxplot.

Interpreting results: Before we interpret results (see Figure 5.14), we should first check the output file Kmeans2.pdf and check the R^2 of the variables used.

Variable-Wise Summary

Variable	Mean	Std. Dev.	Min	Max	R2
EXPENSES	179.921690	89.014105	64.621173	570.611969	0.603111
INCOME	16316.753639	4947.896105	10294.894892	37644.131352	0.596749
INSURANCE	239.645790	112.347156	108.397063	560.179919	0.571446
PHD_MASTER	2.366778	2.598303	0.100000	11.380000	0.506063
RENT	654.962744	142.125660	443.642480	1245.163541	0.487241

Exercise 5.1 (*cont.*)

SECONDARYE	45.496111	6.592007	23.690000	61.980000	0.303487
UNIVERSITY	18.477889	6.996596	4.870000	35.850000	0.271286
DENSITY	0.022990	0.015535	0.000051	0.061002	0.195142
FOREIGNERS	21.774222	15.418853	5.570000	81.760000	0.042156
OWNERS	37.258222	8.324889	0.000000	52.440000	0.006760

Results show that the variables FOREIGNERS and OWNERS have a very small R^2. This is also obvious from the parallel boxplot where FOREIGNERS and OWNERS converge to similar values (see Figure 5.15). Put simply, FOREIGNERS and OWNERS do not assist in creating well-separated clusters and should be removed.

We should run the *k*-means algorithm again with $k = 2$, removing FOR-EIGNERS and OWNERS from the Analysis Fields (see Box 5.6).

Box 5.6 In real case studies, it is likely to run the *k*-means algorithm several times with varying Analysis fields, Initialization methods, and Optimal number of clusters. In this exercise, we minimized the process to the most basic steps as it would be infeasible to present all combinations.

ACTION: ***k*-means clustering** for $k = 2$, FOREIGNERS and OWNERS removed

ArcToolBox > Spatial Statistics Tools > Mapping Clusters > Grouping Analysis

Input Feature Class = City

Unique ID Field = PostCode

Output Feature Class = I:\BookLabs\Lab5\Output\Kmeans2A.shp

Number of groups = 2

Analysis Fields = Check all except Population, Area, Regimes, PostCode, Foreigners, Owners

Spatial Constraints = NO_SPATIAL_CONSTRAINT (*k*-means algorithm)

Distance Method = EUCLIDEAN

Exercise 5.1 (*cont.*)

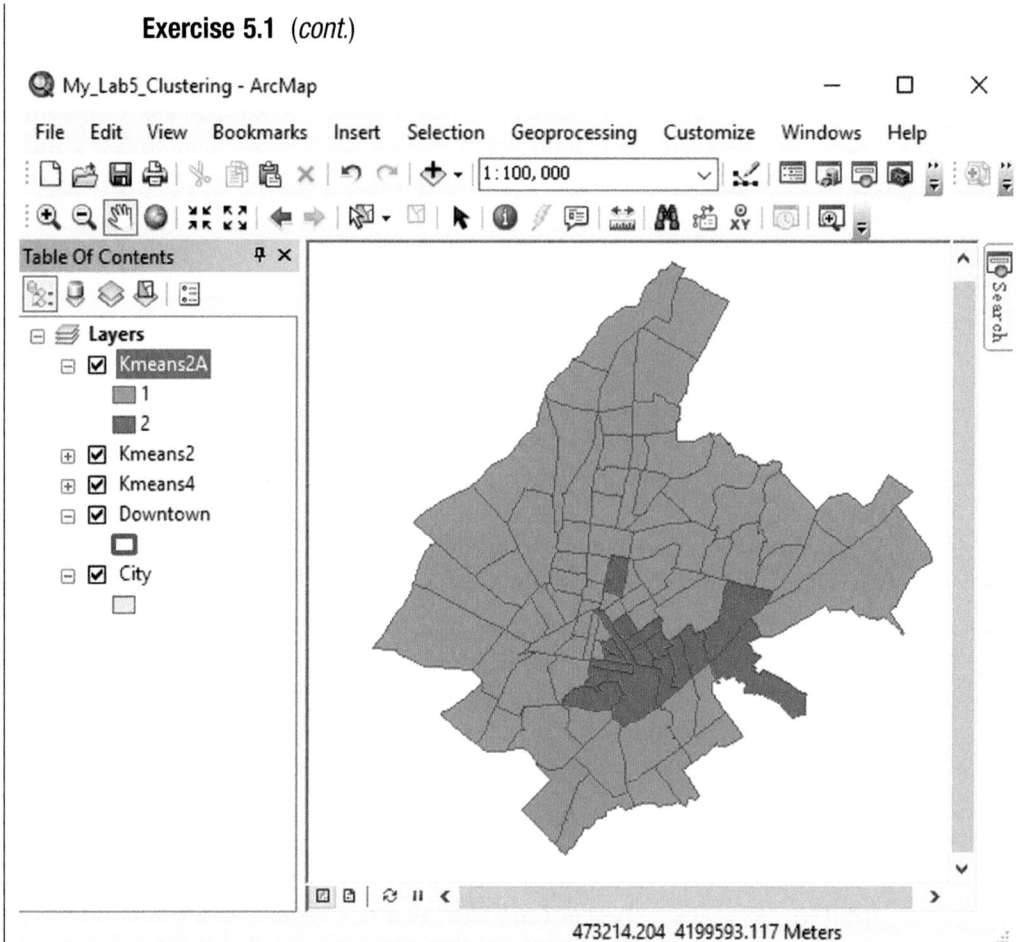

Figure 5.16 Postcodes grouped in two clusters using the *k*-means algorithm with variables "Foreigners" and "Owners" removed.

```
Initializing Method = FIND_SEED_LOCATIONS

Output Report File = I:\BookLabs\Lab5\Output\Kmeans2A.pdf

Evaluate Optimal Number of Groups = Check

OK
```

Main Menu > Geoprocessing > Results > Current Session > Grouping Analysis > DC on Output Report File: Kmeans2A.pdfMain Menu > File > Save

Interpreting results: Postcodes are clustered in identical groups, as clustered when FOREIGNERS and OWNERS were included (see Figures 5.14 and 5.16). Still, clusters are now better separated, as Figure 5.17 shows.

Exercise 5.1 *(cont.)*

Parallel box plot

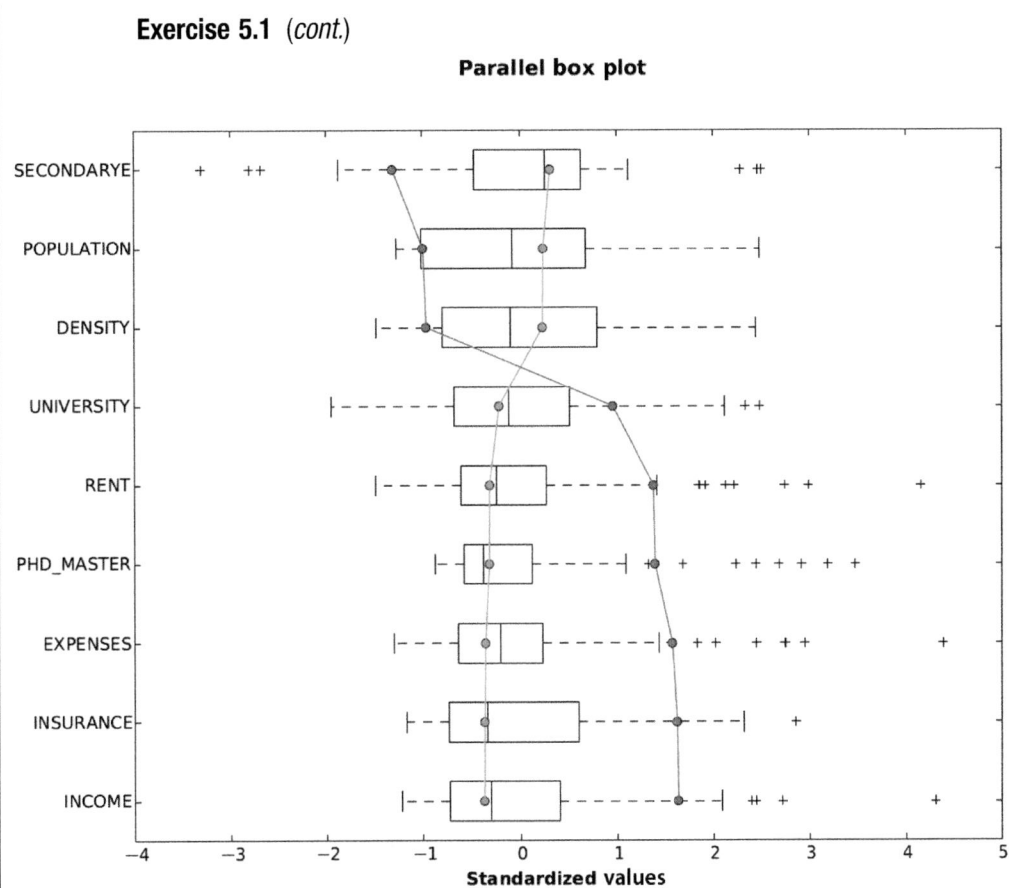

Figure 5.17 Clusters boxplot.

The parallel boxplot maps the standardized values of each variable for each group (see Figure 5.17). The colored dot refers to the standardized mean value of each variable per group plotted across the global variable (variable for the entire dataset denoted by the boxplot). The further apart the group standardized means lie, the more the separation among the clusters and the better the clustering scheme. We can further compare the variables between the two groups by using the following tables and box-plots that refer to the nonstandardized values (see Figure 5.18). Based on this type of analysis, pen portraits can be built for each group thereafter (short profile descriptions of each group).

Variable-Wise Summary

Using the output of Figure 5.18, each group can be compared to all other groups (here we have Group 1 and Group 2) and to the Global group (here marked as Total), which refers to the entire dataset. The boxplot on the right-hand side of each table refers to the distribution of the Global variable.

Exercise 5.1 (*cont.*)

Variable-wise summary

INCOME: R2 = 0.62

Group	Mean	Std. Dev.	Min	Max	Share	
1	14439.9561	2557.2007	10294.8949	21579.5650	0.4126	
2	24375.9430	4631.6794	18445.6591	37644.1314	0.7020	
Total	16316.7536	4947.8961	10294.8949	37644.1314	1.0000	

INSURANCE: R2 = 0.61

Group	Mean	Std. Dev.	Min	Max	Share	
1	197.3499	72.0736	108.3971	462.3114	0.7834	
2	421.2692	62.0204	314.5315	560.1799	0.5437	
Total	239.6458	112.3472	108.3971	560.1799	1.0000	

EXPENSES: R2 = 0.57

Group	Mean	Std. Dev.	Min	Max	Share	
1	147.3783	39.8644	64.6212	233.4759	0.3337	
2	319.6669	105.1066	164.7626	570.6120	0.8021	
Total	179.9217	89.0141	64.6212	570.6120	1.0000	

PHD_MASTER: R2 = 0.45

Group	Mean	Std. Dev.	Min	Max	Share	
1	1.5252	1.0583	0.1000	4.8700	0.4229	
2	5.9806	3.8512	0.1000	11.3800	1.0000	
Total	2.3668	2.5983	0.1000	11.3800	1.0000	

RENT: R2 = 0.44

Group	Mean	Std. Dev.	Min	Max	Share	
1	609.4267	76.8264	443.6425	842.2071	0.4973	
2	850.4998	185.6226	459.5455	1245.1635	0.9802	
Total	654.9627	142.1257	443.6425	1245.1635	1.0000	

SECONDARYE: R2 = 0.40

Group	Mean	Std. Dev.	Min	Max	Share	
1	47.5066	4.7662	34.5800	61.9800	0.7156	
2	36.8629	6.3730	23.6900	46.1300	0.5861	
Total	45.4961	6.5920	23.6900	61.9800	1.0000	

POPULATION: R2 = 0.23

Group	Mean	Std. Dev.	Min	Max	Share	
1	10386.6027	6651.9015	34.0000	25933.0000	0.9988	
2	1877.2353	2044.5310	3.0000	8014.0000	0.3089	
Total	8779.2778	6911.8216	3.0000	25933.0000	1.0000	

DENSITY: R2 = 0.22

Group	Mean	Std. Dev.	Min	Max	Share	
1	0.0265	0.0151	0.0002	0.0610	0.9969	
2	0.0080	0.0045	0.0001	0.0148	0.2418	
Total	0.0230	0.0155	0.0001	0.0610	1.0000	

UNIVERSITY: R2 = 0.21

Group	Mean	Std. Dev.	Min	Max	Share	
1	16.9236	4.9744	8.0300	27.9400	0.6427	
2	25.1524	9.8986	4.8700	35.8500	1.0000	
Total	18.4779	6.9966	4.8700	35.8500	1.0000	

Figure 5.18 Variable-wise summary per cluster.

Exercise 5.1 (*cont.*)

The colored dot refers to the mean value of the variable for the specific group and can be used for comparison with the Global mean value denoted with a black dot (line: Total). The color of the dot is the same as the color of the group (e.g., blue for Group 1 and red for Group 2). The colored vertical dashed lines reflect the minimum and maximum values (range) per group per variable. For example, the minimum and maximum values of INCOME for Group 1 are 10,294 euros and 21,579 euros respectively, denoted with the blue dashed lines vertically plotted along the boxplot.

The Share value is the ratio between the group range and the global range. For Group 1 and INCOME, the 41.26% share is obtained by dividing the group range (21,579.565 − 10,294.895 = 11,284.67) by the global range (37,644.31 − 10,294.89 = 27,349.19) (share = 11,284.67/27,349.19 = 0.4126). This share indicates that Group 1 contains 41.26% of the range of values of the global INCOME variable. In general, low shares reveal clusters where the range of values is not large compared to the global dataset. As shown before, by inspecting the boxplot we can also locate where this range lies. In the case of Group 1, income ranges between the two blue vertical lines. Finally, crosses represent outliers of the Global group. In our example, several postcodes with high income have been labeled as outliers.

People living in Group 2 have an annual mean income of 24,375 euros, which is significantly larger than both the Global mean income (16,316 euros) and the Group 1 mean income (see Figure 5.18 and boxplots on the right of each variable). People living in Group 1 have income that is on average nearly 2,000 less than the global mean income. People in Group 2 spend more than double for medical insurance (421 euros) than those living in Group 1 (197 euros). Furthermore, people in Group 2 spend on average 319 euros per person per month for everyday purchases (EXPENSES: grocery shops, coffee shops), which is twice as large as the expenses of Group 1 (147 euros). Group 2 has a high percentage of people with a PhD or master's degree (5.97%), which far exceeds the global average (2.36%), while Group 1 has significantly less (1.52%) than the global average. People in Group 2 renting a house pay, on average, 850 euros per month, nearly 40% more than people in Group 1 (609 euros per month). Group 2 has a lower percentage of people with secondary education (34.5%) than Group 1 (47.5%) but more people with a university degree (25.1%) compared to Group 1 (16.9%). Group 2 has areas with lower population density than Group 1.

Short Group's Profiling - Pen Portraits

Group 1: Low Spenders

People in this group have lower income than average. Their expenses for insurance, monthly costs and rent are below the global average. The majority of people in Group 1 have obtained lower education.

Exercise 5.1 (*cont.*)

Group 2: High Spenders

People in this group have significantly larger income compared to other groups. They spend twice as much for their medical insurance and monthly expenses compared to Group 1 and 40% more for their rent. Regarding education attainment, Group 2 has significantly higher percentages of people with higher education with one out of three having at least a bachelor's degree, master's degree or PhD.

Concluding, Group 2 seems to match the target group of the new coffee shop, as people residing there are of higher income and expenditure capacity. In addition, their higher educational level might also be a plus, as the coffee house can integrate some additional more "intellectual" services (see Box 5.7).

Box 5.7 Analysis Criterion C5 to Be Used in Synthesis Lab 5.4. The location of coffee shop should lie within an area that belongs to Group 2. [C5_HighSpenders.shp]

Main Menu > Selection > Select By Attributes

Layer = Kmeans2A

Method = Create a new selection

Double Click "SS_GROUP"

In the "SELECT*FROM Kmeans2A WHERE:" type after: "SS_GROUP" =2 (SS_GROUP is already written in the box)

OK

TIP: Have in mind that in case that Group 2 has been labeled as 1 in your output report, you should switch above to "SS_GROUP" = 1. This might happen because the algorithm runs on different computers.

TOC > RC Kmeans2A > Data > Export Data >

Export = Selected Features

Output feature class = I:\BookLabs\Lab5\OutputC5_HighSpenders. shp

Do you want to add the exported data the map as layer = Yes

Main Menu > Selection > Clear Selected Features

Main Menu > File > Save

Exercise 5.2 Spatial Clustering (Regionalization)

In this exercise we apply the SKATER algorithm to identify clusters where spatial features belonging in the same group are both contiguous and homogenous. Regionalization is not necessary in this project, as we are interested in selecting only one postcode that meets some predefined criteria. This exercise aims to showcase briefly how spatial clustering is conducted.

ArcGIS Tools to be used: Grouping Analysis

ACTION: Spatially constrained multivariate clustering

Navigate to the location you have stored the book dataset and click My_Lab5_Clustering.mxd

ArcToolBox > Spatial Statistics Tools > Mapping Clusters > Grouping Analysis (see also Figure 5.12)

Input Feature Class = City

Unique ID Field = PostCode

Output Feature Class = I:\BookLabs\Lab5\Output\SKATER2.shp

Number of groups = 2

Analysis Fields = Based on the results of Exercise 5.1, we select all variables except Population, Foreigners, Owners, Area, Regimes, Postcode

Spatial Constraints = CONTIGUITY_EDGES_CORNERS

Distance Method = EUCLIDEAN

Number of Neighbors = 0

Output Report File = I:\BookLabs\Lab5\Output\SKATER2.pdf

Evaluate Optimal Number of Groups = Check

OK

Before we interpret results, we should check the R^2 and the Calinski–Harabasz pseudo F-statistic in the related file: SKATER_2.pdf. Results indicate that SECONDARY and DENSITY have R^2 less than 0.1, and they can be omitted from the clustering process. Furthermore, the pseudo F-statistic indicates that either two or three clusters would be valid. Run again SKATER with the two above variables removed (see Figure 5.19). Results are presented in Figure 5.20.

Exercise 5.2 *(cont.)*

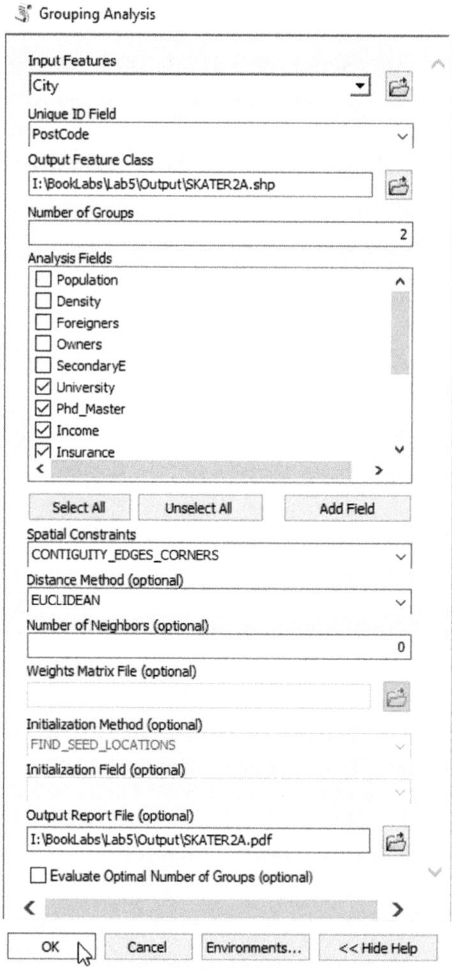

Figure 5.19 SKATER dialog box for two groups.

ArcToolBox > Spatial Statistics Tools > Mapping Clusters > Grouping Analysis

Input Feature Class = City

Unique ID Field = OBJECT ID

Output Feature Class = I:\BookLabs\Lab5\Output\SKATER_2A.shp

Number of groups = 2

Analysis Fields = Based on the results of exercise 5.1, we select University, Phd Master, Income, Insurance, Rent, Expenses and we remove Secondary and Density.

Exercise 5.2 *(cont.)*

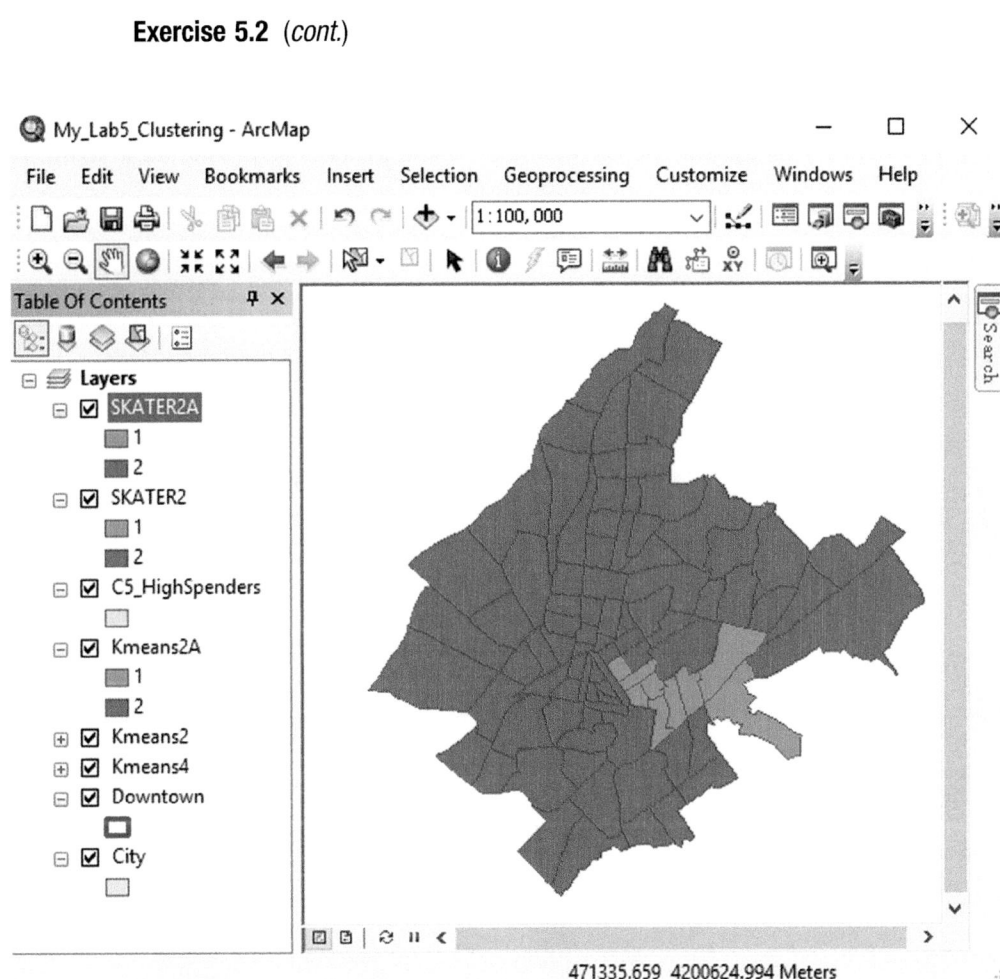

Figure 5.20 Postcodes grouped in two clusters using SKATER. Compared to Figure 5.16, we observe that features in the same group are contiguous, and that cluster 1 has fewer features compared to *k*-means output.

Spatial Constraints = CONTIGUITY_EDGES_CORNERS

Distance Method = EUCLIDEAN

Number of Neighbors = 0

Output Report File = I:\BookLabs\Lab5\Output\SKATER_2A.pdf

Evaluate Optimal Number of Groups = Do not check

OK

Main Menu > File > Save

Exercise 5.2 (*cont.*)

Interpreting results: For space limitation reasons, we will not repeat here the analysis related to statistics and group profiling. Based on `SKATER_2A.pdf`, you are advised to repeat the same process as presented in Exercise 5.1 and interpret results accordingly.

Exercise 5.3 Similarity Analysis

The scope of this analysis is to trace the two most similar postcodes (in terms of `University`, `Phd Master`, `Income`, `Insurance`, `Rent`, `Expenses`, as determined in Exercise 5.1) to the one (target group) that people living there are willing to spend more on coffee shop expenses. This analysis provides alternatives in locating the coffee shop in various similar areas with the target. We select as the target postcode to be the one with the highest value in daily expenses (belonging in Group 2) and should identify which other postcodes belonging to Group 2 have similar characteristics (across all variables). These characteristics are not necessarily of the same magnitude.

ArcGIS Tools to be used: `Select By Attributes`, `Similarity Search`

`ACTION:` **`Select the target object`** `(PostCode with highest expenses)`

`Main Menu > Selection > Select By Attributes`

`Layer = Kmeans2A`

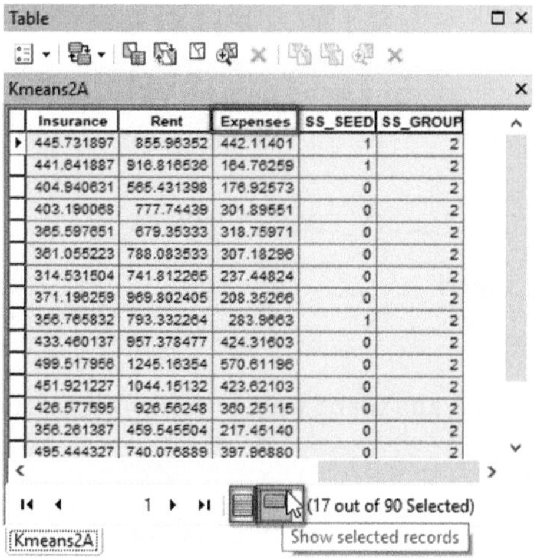

Insurance	Rent	Expenses	SS_SEED	SS_GROUP
445.731897	855.96352	442.11401	1	2
441.641887	916.816536	164.76259	1	2
404.940631	565.431398	176.92573	0	2
403.190068	777.74439	301.89551	0	2
365.597651	679.35333	318.75971	0	2
361.055223	788.083533	307.18296	0	2
314.531504	741.812265	237.44824	0	2
371.196259	969.802405	208.35266	0	2
356.765832	793.332264	283.9663	1	2
433.460137	957.378477	424.31603	0	2
499.517956	1245.16354	570.61196	0	2
451.921227	1044.15132	423.62103	0	2
426.577595	926.56248	360.25115	0	2
356.261387	459.545504	217.45140	0	2
495.444327	740.076889	397.96880	0	2

Figure 5.21 Show selected records.

Exercise 5.3 (*cont.*)

Method = Create a new selection

Double Click "SS_GROUP"

In the "SELECT*FROM Kmeans2A WHERE:" type after: "SS_GROUP" = 2
(SS_GROUP is already written in the box)

OK

TOC > RC Kmeans2A > Open Attribute Table > Click Show selected
records (in the bottom of the table; see Figure 5.21)

RC Expenses column > Sort Descending

Click the first row

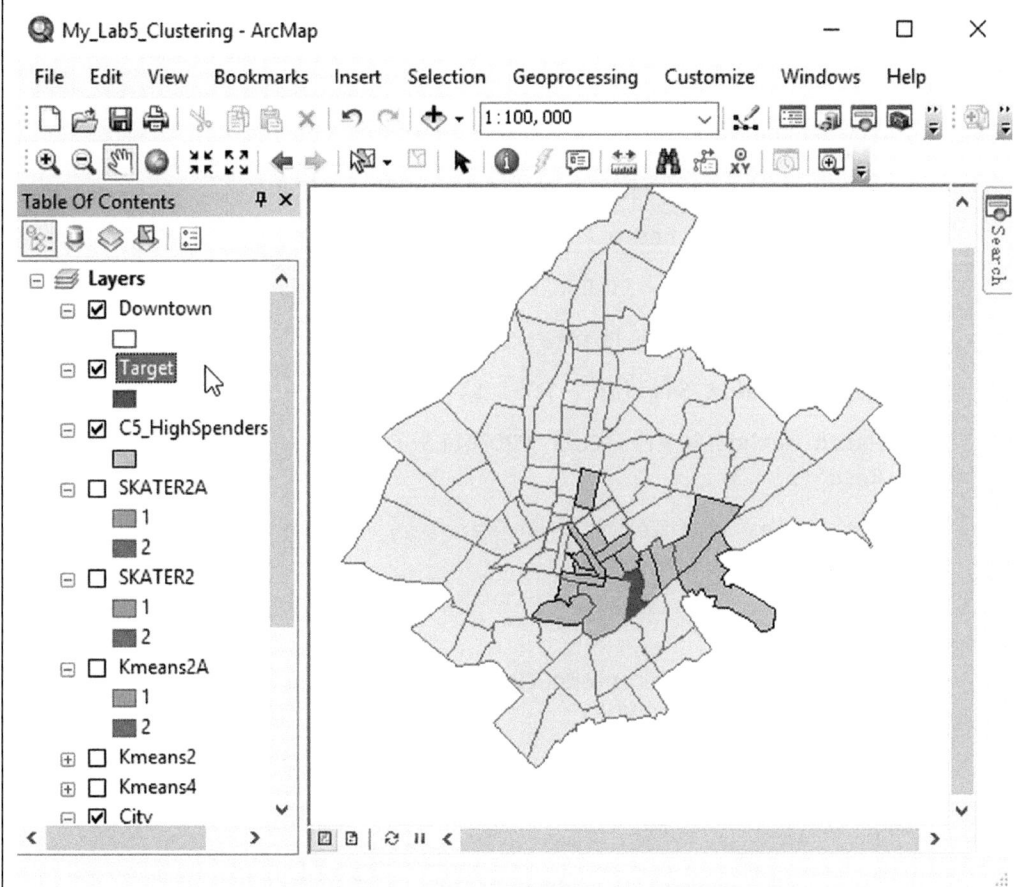

Figure 5.22 Target postcode 10674.

Exercise 5.3 (*cont.*)

This is the Target Group with PostCode = 10674

Click Show all records

Click Clear Selection

Select line with PostCode = 10674

TOC > RC Kmeans2A > Data > Export Data >

Export: Selected Features (only one feature is selected)

Save as: I:\BookLabs\Lab5\Output\Target.shp (save as shapefile)

Do you want to add the exported data the map as layer = Yes

Click Clear Selection

Close Table

ACTION: Similarity Search

ArcToolBox > Spatial Statistics Tools > Mapping Clusters > Similarity Search

Input Features To Match = Target (see Figure 5.23)

Candidate Features = C5_HighSpenders

Output Features = I:\BookLabs\Lab5\Output\Similarity.shp

Collapse Output To Points: Do not check

Most Or Least Similar = MOST_SIMILAR

Match Method = ATTRIBUTE PROFILES (Cosine similarity)Number of Results = 3

Attributes of Interest = University, Phd Master, Income, Insurance, Rent, Expenses

OK

3 Most Similar Locations (Values)
OID SIMRANK SIMINDEX
  10   1 0.0000
  11   2 6.3092
   9   3 7.7133

TOC > RC Similarity > Properties > TAB=Labels >

Exercise 5.3 (*cont.*)

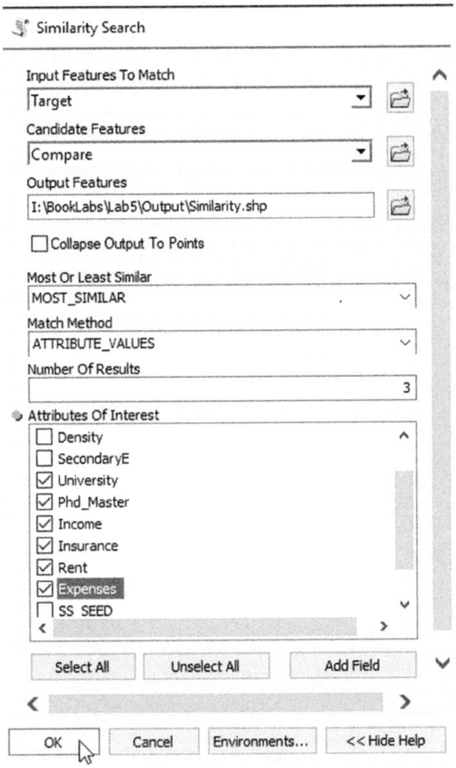

Figure 5.23 Similarity search dialog box.

```
Label features in this layer = Check

Label Filed = LABELRANK > OK

TOC > RC Similarity > Open Attribute Table

Main Menu > File > Save
```

Interpreting results: Figure 5.24 depicts the two most similar postcodes regarding characteristics and trends to the target postcode. The target post-code (PostCode = 10674 – red polygon) is characterized by high expenses, high annual income and high educational level. It gets Label = 0. Polygon with Label = 1 is identical to target (target postcode is included in the C5_HighSpenders as well). The first similar postcode to the target is PostCode = 10675 (Label = 2) and the second similar postcode is PostCode = 10673 (Label = 3) (Figure 5.24).

For the needs of this exercise, the target object can act as the most appropriate postcode to locate the new coffee shop because it meets the

Exercise 5.3 *(cont.)*

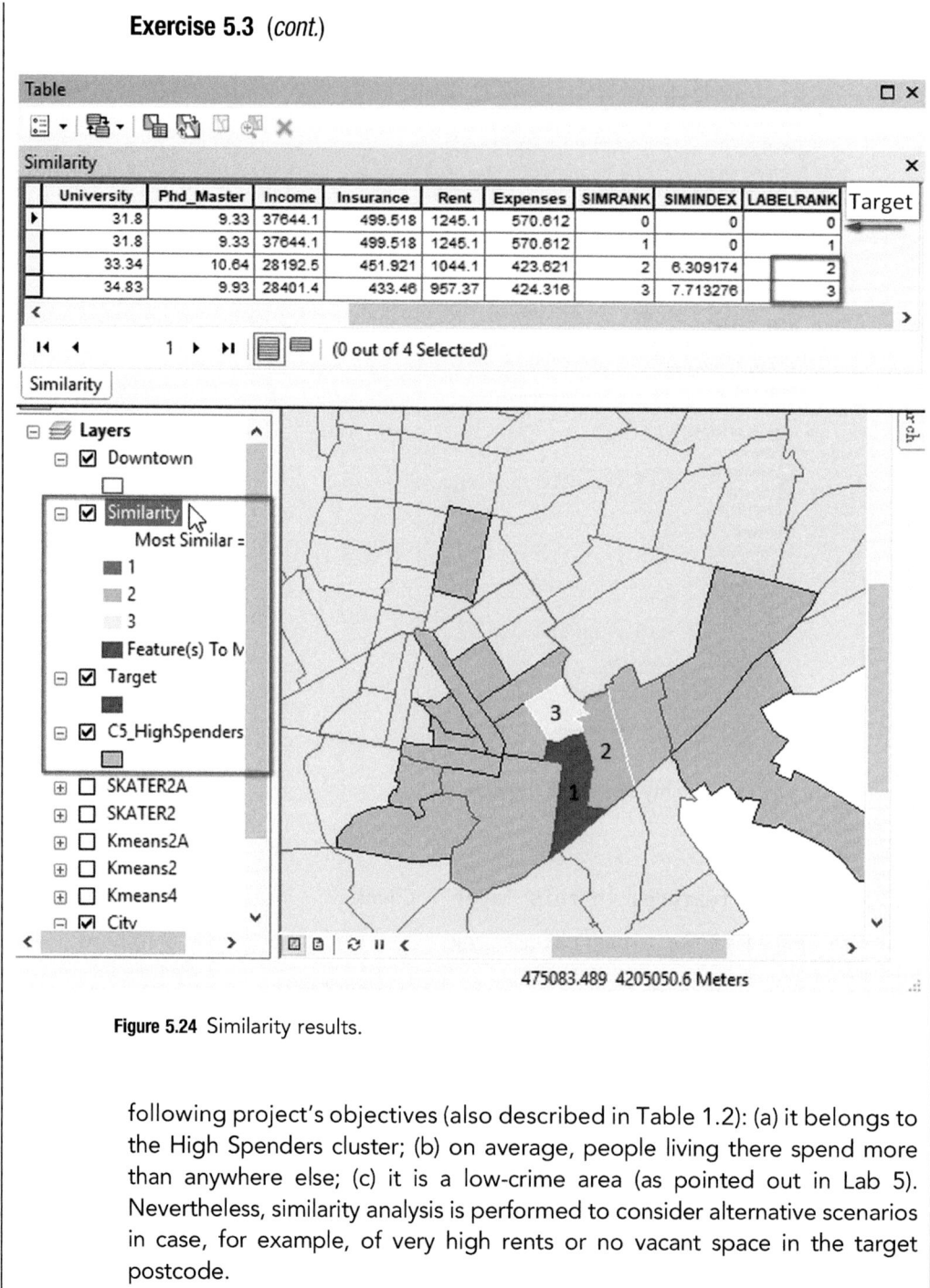

Figure 5.24 Similarity results.

following project's objectives (also described in Table 1.2): (a) it belongs to the High Spenders cluster; (b) on average, people living there spend more than anywhere else; (c) it is a low-crime area (as pointed out in Lab 5). Nevertheless, similarity analysis is performed to consider alternative scenarios in case, for example, of very high rents or no vacant space in the target postcode.

Exercise 5.4 Synthesis

In the four previous labs, various spatial analysis methods were applied to address the project's objectives to ultimately identify potential locations for a new coffee shop. The primary objective of these labs is to showcase how spatial analysis assists in different aspects of analyzing a problem. Not all techniques mentioned earlier are necessary in every analysis, nor should the proposed steps be followed in a strict way. It depends on the problem at hand and the expected outcomes. For example, regionalization methods may not be necessary if we only conduct spatiotemporal autocorrelation analysis. For the needs of this laboratory, we briefly synthesize the results extracted from each previous lab (analysis outcomes) to make our final decision: where the new coffee house should be located. To do so, we apply GIS techniques and analysis.

The criteria to be considered are shown in Table 5.8:

Table 5.8 Analysis criteria for identifying the proper location.

Criterion		Exercise	Objective	Further actions
C1	The location of the coffee shop should lie within areas of high annual average income.	2.1	1	Select areas with high income
C2	The location of the coffee shop should not lie within areas of high assault density.	3.3	2	Remove areas with high densities of assaults crime.
C3	The location of the coffee shop should lie within income hot spots.	4.4	1	Select income hot spot areas
C4	The location of the coffee shop should not lie within crime (other than assaults) hot spots.	4.5	2	Remove crime hot spot areas
C5	The location of the coffee shop should lie within an area that belongs to high spenders (Group 2).	5.1	3	Select areas of Group 2

The GIS analysis includes the following steps:

Step 1: Select polygons with annual average income exceeding 20,000 that are also hot spots of Income [Combine C1 and C3].

Step 2: From polygons selected, select those that also belong to the high-spenders group [C5].

Step 3: From those selected, exclude areas that are of high crime density [Exclude C2].

Step 4: From those selected, exclude areas that are crime hot spots [Exclude C4].

Exercise 5.4 (*cont.*)

Step 5: Assess the final potential sites using similarity analysis.

Step 1:   Combine C1 and C3

First, we select those postcodes with annual income higher than 20,000 (the average annual income is 16,317 euros, Exercise 2.1) that are also hot spots of income. The reason we focus on hot spot areas (in conjunction with absolute values of income) is that we do not just seek areas of high income but, rather, areas of high income that cluster, enlarging thus the catchment area.

ACTION: Select by attributes, Select by location

Navigate to the location you have stored the book dataset and click on Lab5_Synthesis.mxd

Main Menu > File > Save As > My_Lab5_Synthesis.mxd

In I:\BookLabs\Lab5\Output

Main Menu > Selection > Select By Attributes

Layer = C3_IncomeOptimizedHotSpot

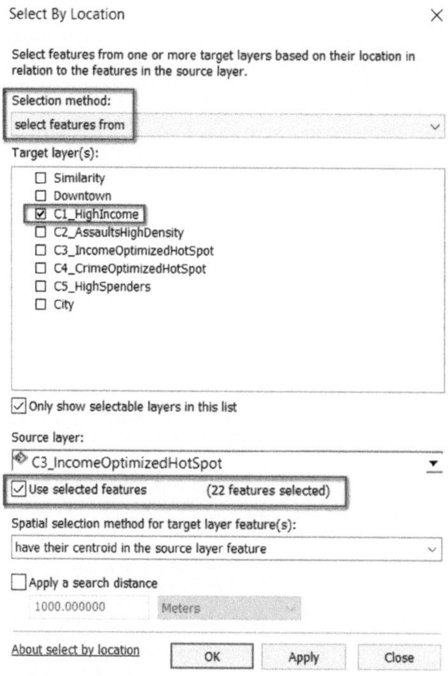

Figure 5.25 Select by location dialog box.

Exercise 5.4 (*cont.*)

SELECT * FROM C3_IncomeHotSpot Where: "Gi_Bin" >= 3

OK

Main Menu > Selection > Select By Location

Selection method: select features from (see Figure 5.25):

Target layer(s) = C1_HighIncome

Source layer: C3_IncomeOptimizedHotSpot

Use selected features = Check (22 features selected)

Spatial selection method for target layer feature(s): have their centroid in the source layer feature

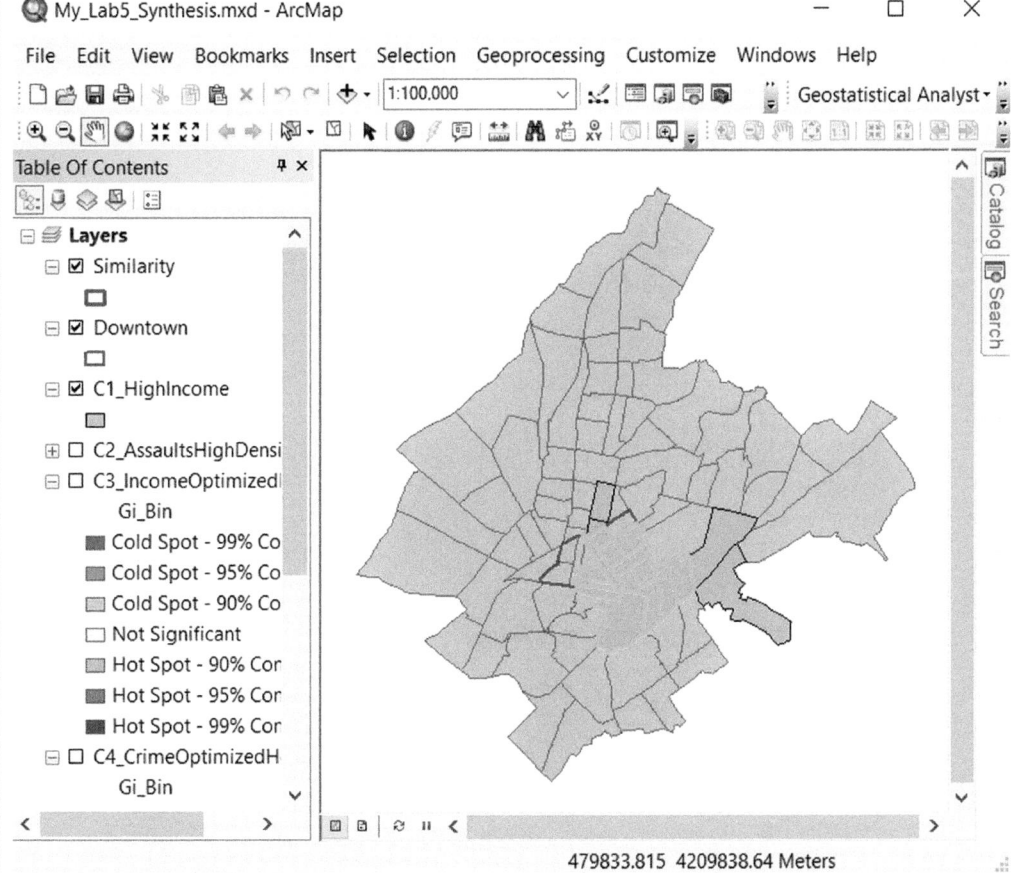

Figure 5.26 Fourteen postcodes selected, meeting C1 and C3 criteria (click on the C1_HighIncome layer).

Exercise 5.4 (*cont.*)

OK

TOC > RC C3_ IncomeOptimizedHotSpot > Open Attribute Table > Clear Selection > Close Table (see Figure 5.26)

Step 2: Select from High-Spenders [C5]

Main Menu > Selection > Select By Location

Selection method: select from the currently selected features in (see Figure 5.27)

Target layer(s) = C1_HighIncome

Source layer: C5_HighSpenders

Spatial selection method for target layer feature(s): have their centroid in the source layer feature

OK (see Figure 5.28)

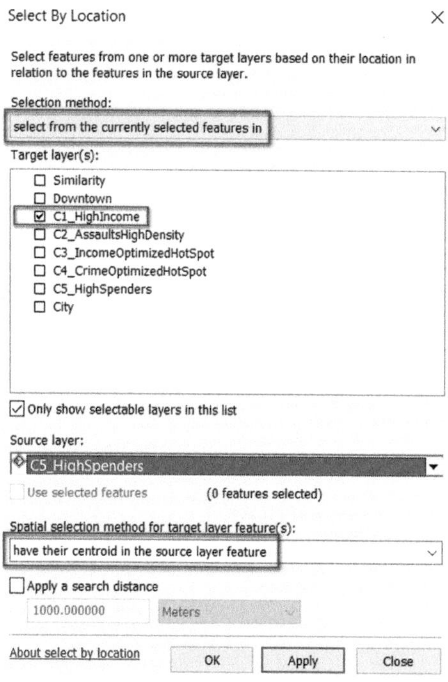

Figure 5.27 Select by location dialog box.

Exercise 5.4 *(cont.)*

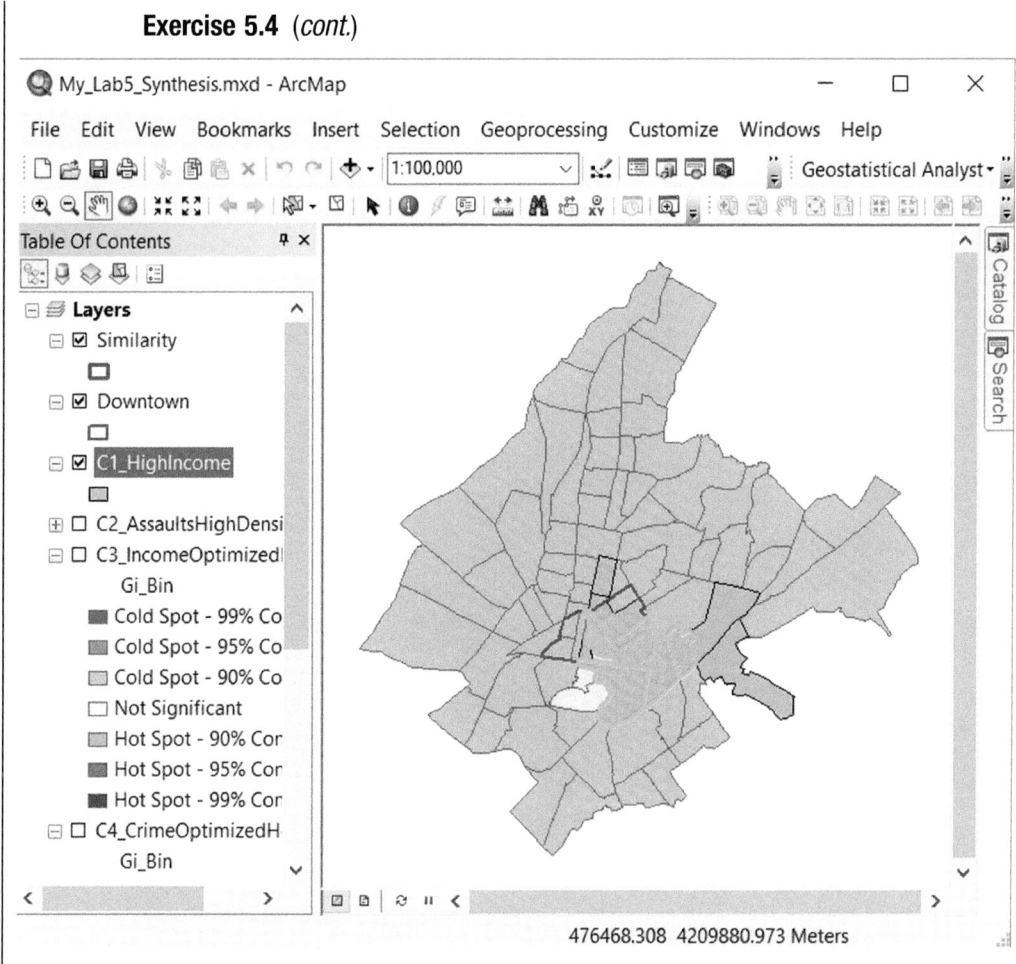

Figure 5.28 Eleven postcodes selected, meeting C1, C3 and C5 criteria.

Step 3: Exclude Areas with High Assault Density (C2)

ACTION: Select from high income (from the eleven selected previously) those not overlapping to areas with high density of assaults

Main Menu > Selection > Select By Location

Selection Method: remove from the currently selected features in

Target layer(s) = C1_HighIncome

Source layer: C2_AssaultsHighDensity

Spatial selection method for target layer feature(s): intersect (3d) the layer source feature

OK

Exercise 5.4 *(cont.)*

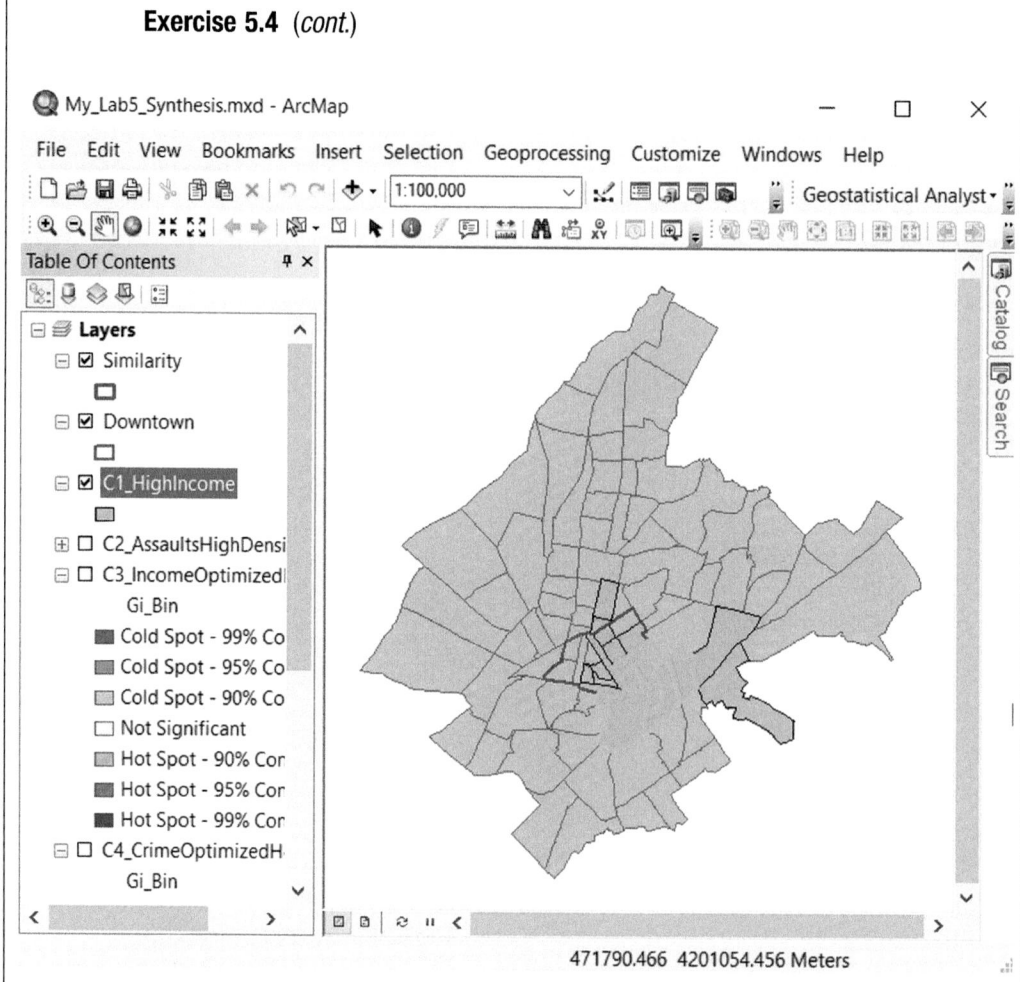

Figure 5.29 Seven polygons selected, meeting C1, C3, C5 and C2 criteria.

Step 4: Exclude Areas That Meet Criterion C4 (Not Crime Hot Spot)

ACTION: Select from C4_CrimeOptimizedHotSpot the polygons with GiBin 2 or 3

Main Menu > Selection > Select By Attributes

Layer = C4_CrimeHotSpot

SELECT * FROM C4_CrimeOptimizedHotSpot Where: "Gi_Bin" >=2

OK

In total, 48 squares are selected.

Exercise 5.4 *(cont.)*

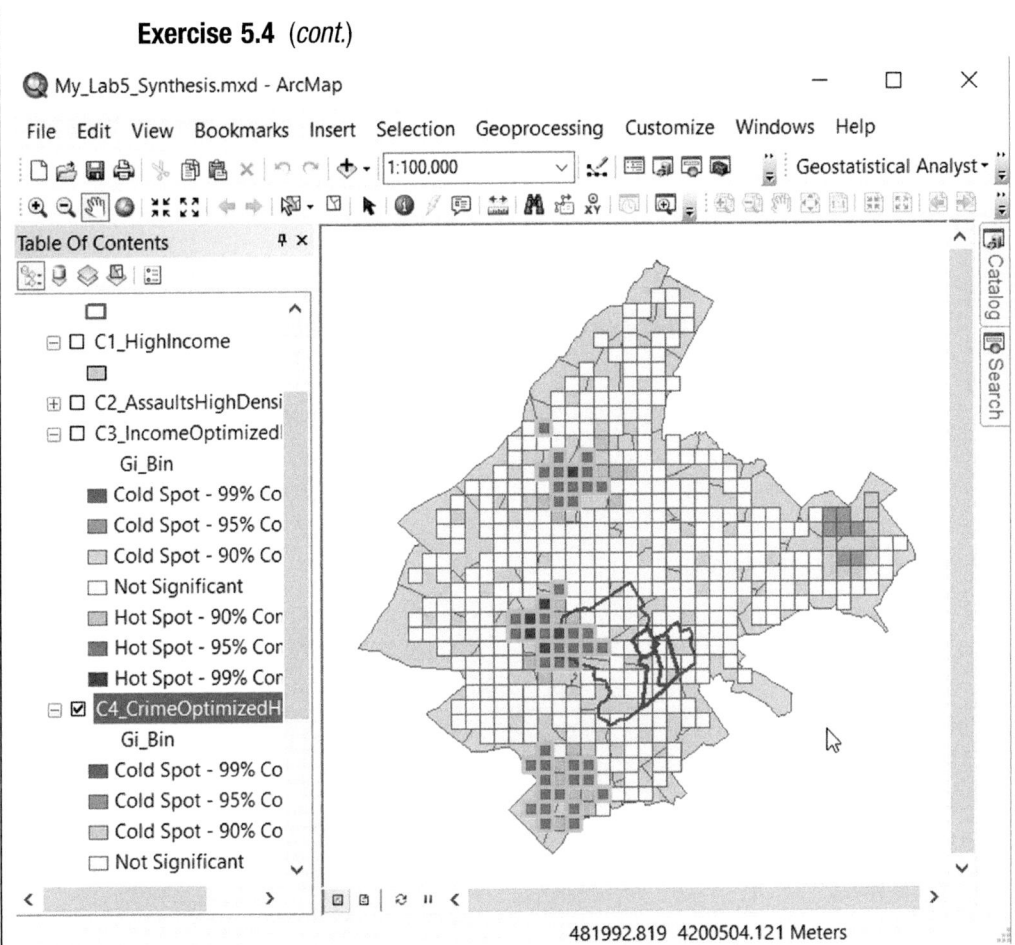

Figure 5.30 Select squares that are labeled as crime hot spots.

ACTION: Select from High Income (from the seven selected previously) those not overlapping with hot spots of crime

Main Menu > Selection > Select By Location

Selection Method: remove from the currently selected features in

Target layer(s) = C1_HighIncome

Source layer: C4_CrimeOptimizedHotSpot

Use selected features: Check

Spatial selection method for target layer feature(s): intersect (3d) the layer source feature

OK

Exercise 5.4 (cont.)

As there is no overlap between selected C1_HighIncome and selected C4_CrimeOptimizedHotSpot polygons the final High Income polygons remain seven.

Step 5: Find Potential Areas (Similarity Analysis)

RC> C1_HighIncome > Data > Export Data > Export: Selected Features > Output feature class: I:\BookLabs\Lab5\Output\SuitableLocations.shp

TOC > RC Similarity > Properties > TAB=Labels >

Label features in this layer = Check

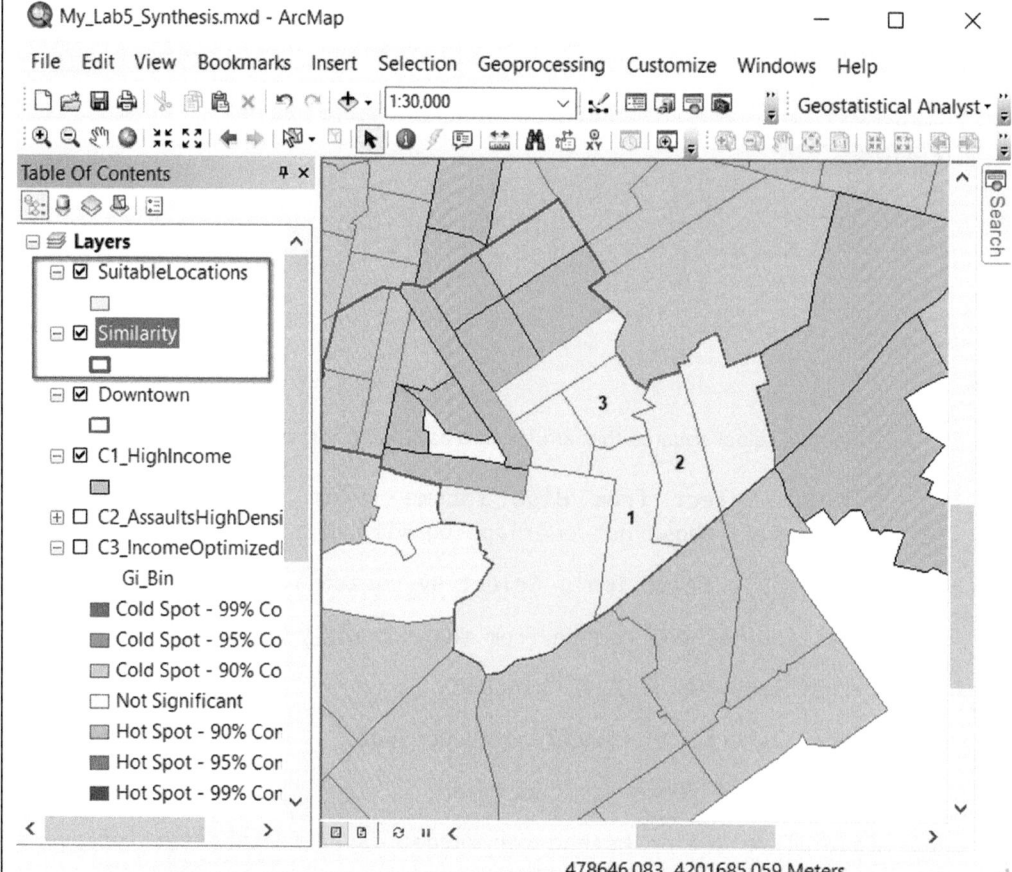

Figure 5.31 Seven suitable locations for the new coffee shop meeting all criteria. Ranking order for the first three most similar locations is also plotted.

Exercise 5.4 *(cont.)*

Label Filed = LABELRANK > OK

Main Menu > File > Save

Interpreting results: Seven areas meet all criteria (SuitableLocations. shp) and are potential sites for the new coffee shop (Figure 5.31). As shown in Exercise 5.3, the most appropriate postcode (among those suitable) is PostCode = 10674, as (a) it belongs to the High Spenders cluster, (b) on average, people living there spend more than anywhere else and (c) it is a low crime area. By using the results of Exercise 5.3, the two most similar postcodes to PostCode = 10674 are PostCode = 10675 and PostCode = 10673, also lying within the suitable locations (Figure 5.31). Similarity analysis allows for ranking sites when more than one location is provided as candidate – something that also facilitates testing alternative scenarios. In case, for example, that in PostCode = 10674, no vacant spaces exist or rents are beyond budget, we may consider alternative neighborhoods starting from PostCode = 10675 and so on, following the rank shown in Table 5.9.

Table 5.9 Site selection rank.

PostCode	Site selection rank
10674	1
10675	2
10673	3

Section B   GeoDa

Exercise 5.1 *k*-Means Clustering

In this exercise, we create homogeneous socioeconomic clusters of postcodes using the *k*-means algorithm. GeoDa does not evaluate results through reports as ArcGIS does. We either have to run multiple trials or use the settings directly produced from Exercise 5.1 in Section A.

GeoDa Tools to be used: k-Means

To run the *k*-means algorithm we should define first, the initialization method, the number of clusters and the variables to include (see Figure 5.32).

Exercise 5.1 (*cont.*)

ACTION: *k*-**means clustering** for k = 2 (we select based on Exercise 5.1, Section A)

Navigate to the location you have stored the book dataset and click on Lab5_Clustering_GeoDa.gda

Main Menu > Clusters > K Means

KMeans Dialog

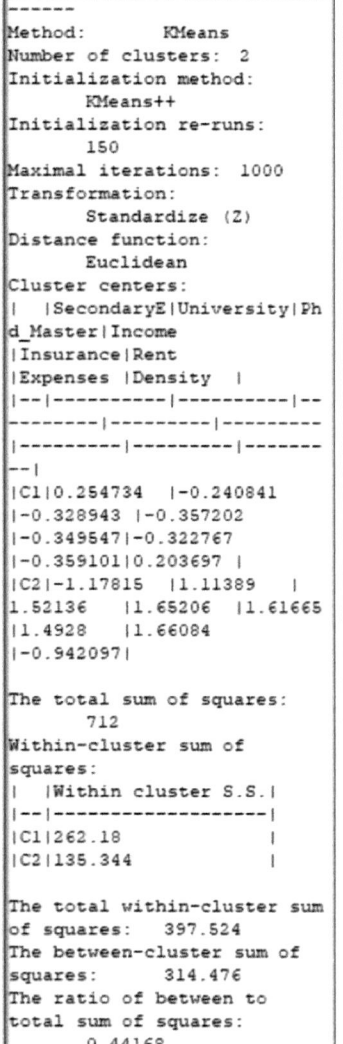

Input:

Select Variables

PostCode
Population
SecondaryE
University
Phd_Master
Income
Insurance
Rent
Expenses
Area
Regimes
Density
Foreigners
Owners
CLKMeans

☐ Use geometric centroids   Auto Weighting

Weighting: 0 ━━━━━━━━━━━━ 1 | 1

Parameters:

Number of Clusters: 2

Minimum Bound: ☐ ⌄ 1
━━━━━━━ 10%

Transformation: Standardize (Z)

Initialization Method: KMeans++

Initialization Re-runs: 150

Use specified seed: ☑ Change Seed

Maximal Iterations: 1000

Distance Function: Euclidean

```
------
Method:      KMeans
Number of clusters: 2
Initialization method:
     KMeans++
Initialization re-runs:
     150
Maximal iterations: 1000
Transformation:
     Standardize (Z)
Distance function:
     Euclidean
Cluster centers:
|  |SecondaryE|University|Ph
d_Master|Income
|Insurance|Rent
|Expenses |Density  |
|--|----------|----------|--
--------|---------|---------
|---------|---------|-------
--|
|C1|0.254734  |-0.240841
|-0.328943 |-0.357202
|-0.349547|-0.322767
|-0.359101|0.203697 |
|C2|-1.17815  |1.11389   |
1.52136   |1.65206  |1.61665
|1.4928    |1.66084
|-0.942097|

The total sum of squares:
     712
Within-cluster sum of
squares:
|  |Within cluster S.S.|
|--|-------------------|
|C1|262.18             |
|C2|135.344            |

The total within-cluster sum
of squares:   397.524
The between-cluster sum of
squares:      314.476
The ratio of between to
total sum of squares:
     0.44168
```

Figure 5.32 *k*-means dialog box and summary results.

Exercise 5.1 (*cont.*)

Input = Click the Shift key and then click: Density/SecondaryE/ University/Phd_Master/Income/Insurance//Rent/Expenses (see Exercise 5.1, Section A, on how these variables are selected).

Number of clusters = 2

Transformation = Standardize (Z)

Initialization Method = KMeans++

Initialization Re-runs=150

Use specified seed: Check

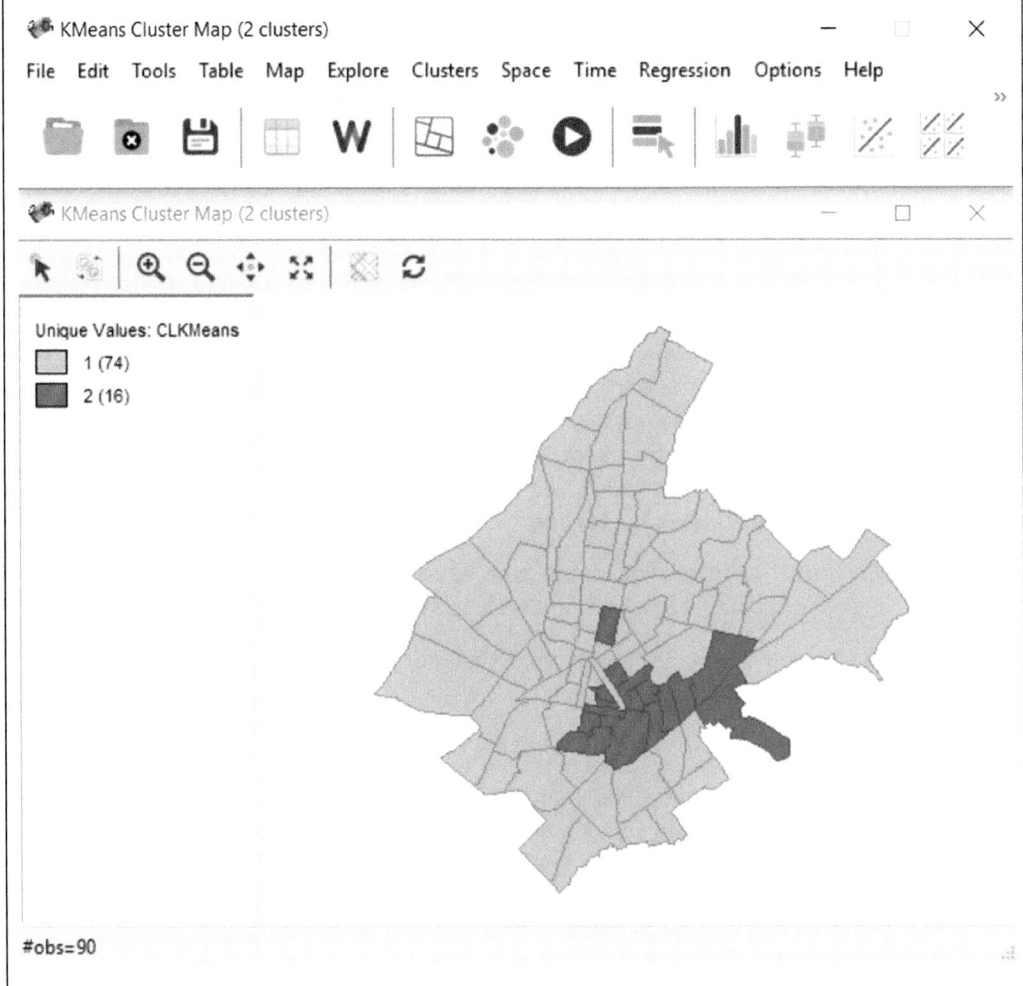

Figure 5.33 Postcodes grouped in two clusters using the *k*-means algorithm.

Exercise 5.1 (*cont.*)

Maximal Iterations:1000

Distance Function: Euclidean

Save Cluster in Field = CLKmeans

Run

Save

Interpreting results: The postcodes (see Figure 5.33) have been clustered almost identical with those presented in Figure 5.16. Slight differences in cluster centers and statistics are due to different initializing procedures and setting up of the *k*-means algorithm. See Section A, Exercise 5.1, interpretation section, for more analysis. To rerun the algorithm, delete the column CLKmeans from the attributes table first.

Exercise 5.2 Spatial Clustering

In this exercise, we apply the SKATER algorithm to identify clusters where spatial features belonging in the same group are both contiguous and homogenous. Regionalization is not necessary in this project, as we are interested in selecting only one postcode that meets certain predefined criteria. This exercise aims to showcase briefly how spatial clustering is conducted.
  GeoDa Tools to be used: SKATER

ACTION: Spatial clustering

Navigate to the location you have stored the book dataset and click on Lab5_Clustering_GeoDa.gda

Main Menu > Tools > Weights Manager > Load > CityGeoDa1150 (in Lab5/Geoda) > Close

Main Menu > Clusters > skater (see Figure 5.34)

Input = University/Phd_Master/Income/Insurance/Rent/Expenses (see Exercise 5.2, Section A, on how these variables are selected)

Number of clusters = 2

Weights = CityGeoDa1150

Exercise 5.2 (*cont.*)

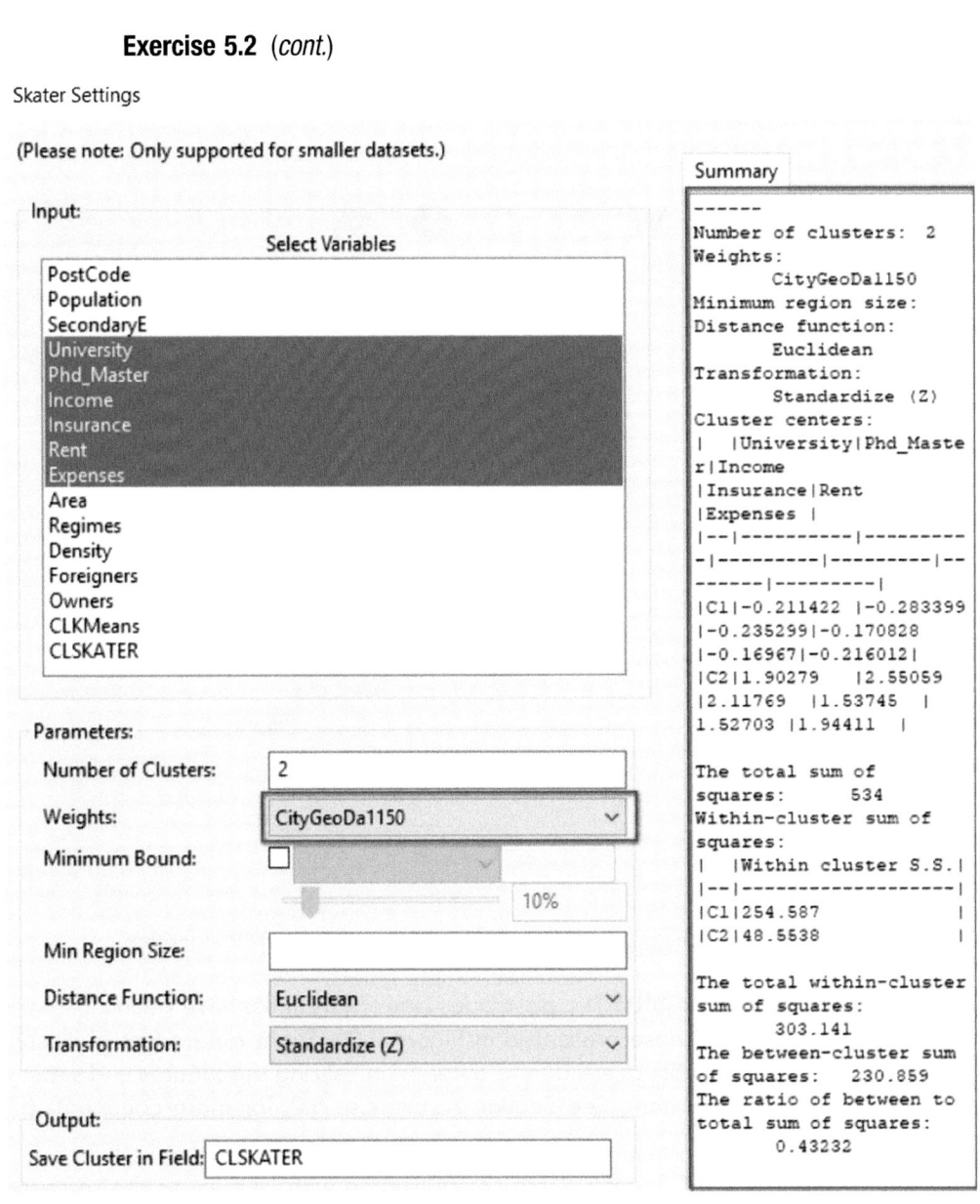

Skater Settings

(Please note: Only supported for smaller datasets.)

Figure 5.34 SKATER dialog box and summary results.

```
Distance Function = Euclidean

Transformation = Standardize (Z)

Save Cluster in Field = CLSKATER

RUN
```

Exercise 5.2 (*cont.*)

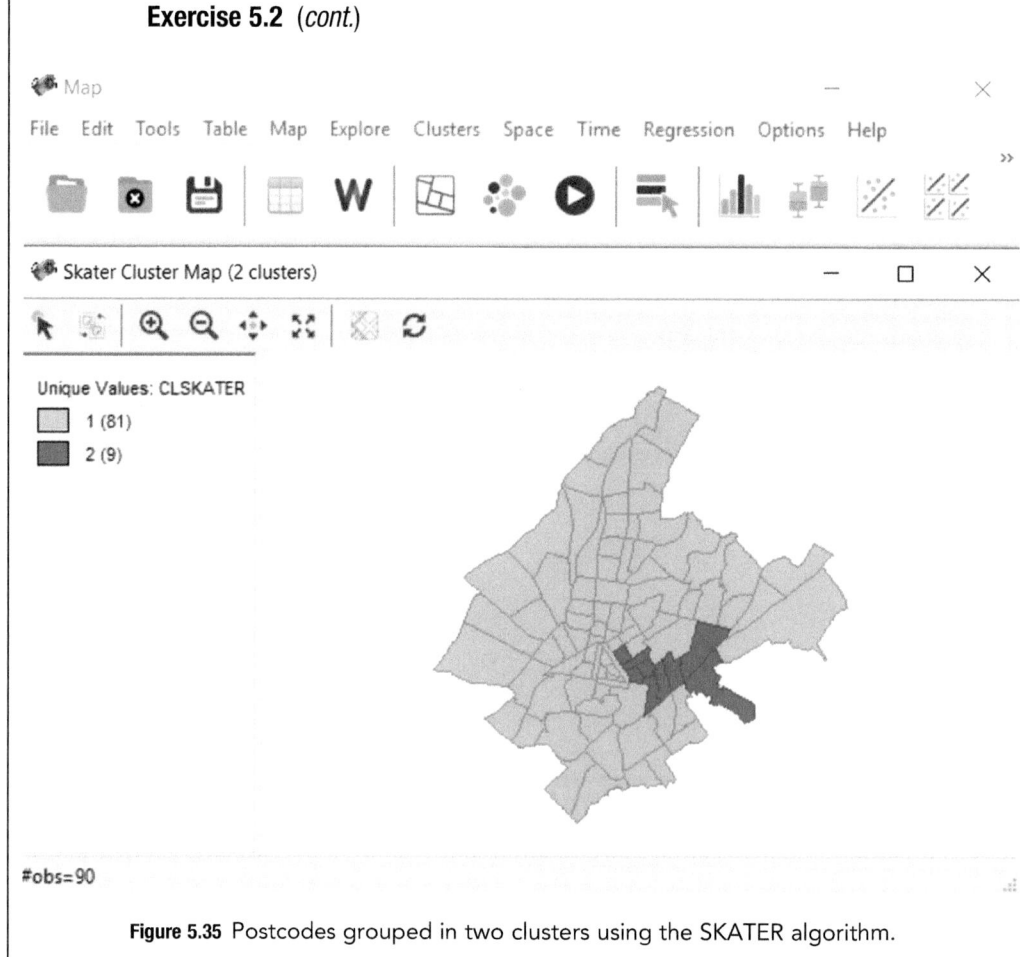

Figure 5.35 Postcodes grouped in two clusters using the SKATER algorithm.

Interpreting results: The postcodes (see Figure 5.35) have been clustered identical with those presented in Figure 5.20. Slight differences in cluster centers and statistics are due to different initializing procedures and setting up of the algorithm. See Section A, Exercise 5.2, for more comments on results.

Tip: Exercises 5.3 and 5.4 are not presented with GeoDa, as the software does not provide related functions and procedures.

6 Modeling Relationships
Regression and Geographically Weighted Regression

THEORY

Learning Objectives

This chapter deals with

- Linear regression notions and mathematical formations
- The assumptions of linear regression
- Identifying linear relationships between a dependent variable and independent variables
- Residual plots, leverage and influential points
- Evaluating the degree (effect) to which one variable influences the positive or negative change of another variable
- The geometrical interpretation of the intercept, the slope, and metrics and tests used to assess the quality of the results
- The interpretation of coefficients
- Studying casual relationships (under specific conditions)
- Building predictive models by fitting a regression line to the data
- Evaluating the correctness of the model
- Model overfitting
- Ordinary least squares (OLS)
- Exploratory OLS
- Geographically weighted regression (GWR)

After a thorough study of the theory and lab sections, you will be able to

- Distinguish which regression method is more appropriate for the problem at hand and the data available
- Interpret statistics and diagnostics used to evaluate a regression model
- Interpret the outcomes of a regression model such as the coefficient estimates, the scatter plots, the statistics and the maps created
- Identify if relationships, associations or linkages among dependent and independent variables exist
- Analyze data through regression analysis in Matlab

- Conduct exploratory regression analysis in ArcGIS
- Conduct geographically weighted regression in ArcGIS

6.1   Simple Linear Regression

Definition

Simple linear regression (also called bivariate regression) is a statistical analysis that identifies the linear relationships between a dependent variable, y, also called response, and a single independent variable, x, also called explanatory variable (for multiple independent variables, see Section 6.2). Independent x is any variable that is used to explain/predict a single variable that is called dependent y.

Linear regression analysis is based on the fitting of a straight line to the dataset in order to produce a single equation that describes the dataset. The equation of the regression line is (6.1) (see Figure 6.1A and B):

$$\hat{y} = a + bx \qquad (6.1)$$

where

$\hat{y}$ is the predicted value (also called fitted value) of the dependent variable (it is pronounced "y-hat"; the observed [real] value of the dependent variable is denoted as y – without a hat)
x is the independent variable
a is the intercept
b is the slope of the trend line

Slope (b) and intercept (a) are also called coefficients and are the parameters that have to be estimated. Methods to estimate the parameters of a linear regression include ordinary least squares (see Section 6.1.2), Bayesian inference, least absolute deviations and nonparametric regression.

The difference of the observed value with the predicted (fitted) value is the model error called residual e (6.2).

$$e = y - \hat{y} \qquad (6.2)$$

When residuals are positive, there is underprediction implying that the predicted (fitted) value is lower than the observed (see Figure 6.1.B). On the other hand, overprediction means that the fitted value is higher than the observed value. In this case, residuals are negative.

In regression analysis, we build a probabilistic model (6.3) consisting of a deterministic component ($\hat{y}$) and a random error component (e). In particular, each observation of the dependent variable is the sum of the predicted value with the residual (6.3) (see Figure 6.1B):

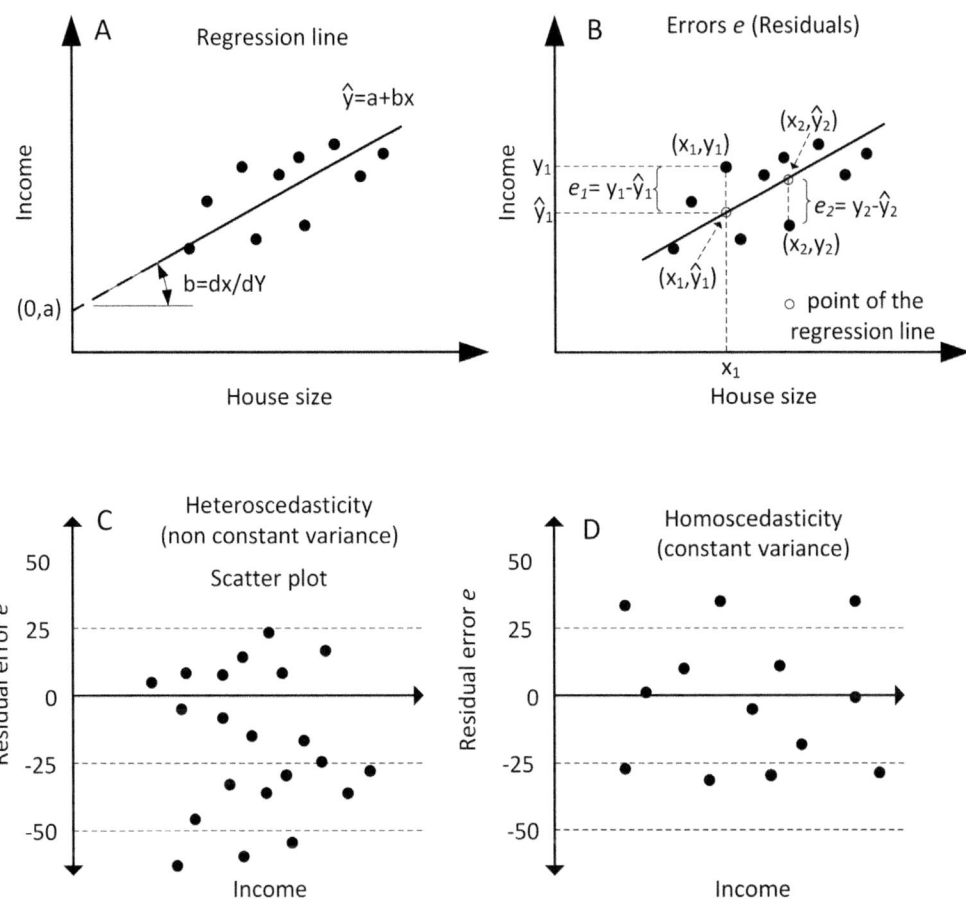

Figure 6.1 Regression basics. (A) A regression linear fit. (B) Calculating residuals. (C) Depicting heteroscedasticity with residual plot. (D) Depicting homoscedasticity with residual plot. These residual plots are not standardized. The y-axis has the same units as the dependent variable.

$$y = \hat{y} + e = a + bx + e \tag{6.3}$$

In deterministic models, outcomes (a) are precisely determined by well-defined equations, (b) no random variation exists and (c) a given input always produces the same output. In contrast, a probabilistic (stochastic) model handles complex problems where uncertainty exists. Errors are inevitable and should be included in the model. If under- or overpredictions occur in a systematic way, then the regression model is biased.

Interpretation
The intercept (a) is the value of the dependent variable when the independent variable is zero. The slope (b) is the expected change of the dependent variable for a one-unit change in the independent variable (Lee & Wong 2001).

Why Use

Simple linear regression is used to (Rogerson 2001 p. 104)

- Identify relationships between a dependent variable and an independent variable
- Evaluate the impact (importance) of the independent variable on the dependent variable
- Build a predictive model by fitting a regression line to the data
- Evaluate the correctness of the model

It should be emphasized that simple regression does not study casual relationships. Any hypothesized causal relationship can be tested under specific conditions within the multiple linear regression context presented in Section 6.2.

Discussion and Practical Guidelines

Suppose we study if "Income" is related to "House size." We can calculate the correlation coefficient between "Income" and "House size," which measures the direction and the strength of a relationship. For example, a strong positive correlation coefficient would reveal that when income increases, house size increases as well. What correlation cannot estimate is the change in "Income" for a given increase in "House size" or the expected (predicted) "Income" for a given "House size" value. By regression modeling, we can analyze more thoroughly the relationship between a dependent (Income) and an independent variable (House size) by fitting a straight line among the data points (see Figure 6.1B). The equation of the line reveals the trend and can be used for predicting purposes. For example, if we use satellite data to measure the size of a house, we can calculate the expected income of that specific household by plugging in the "House size" value to the regression function (which has been constructed already based on other data). Such models and approaches provide quick and economic ways to estimate socioeconomic data based on free, easily acquired imagery.

Regression works by fitting a line to the data points by determining parameters a and b. There are many ways to fit a line in a dataset. We can simply draw the mean value of y as a parallel line of the z-axis. We can connect the leftmost to the rightmost data point or draw a line that has the same number of points above and below the line. In fact, regression is much more than a fitting technique. It is a statistical inference approach, estimating the unknown parameters (coefficients).

The following subsections present the assumptions that a simple linear regression model is based on, the geometrical interpretation of the intercept and the slope and various metrics and tests used to assess the quality of the results.

6.1.1   Simple Linear Regression Assumptions

Simple linear regression builds a probabilistic model based on four assumptions (use the word LINE as mnemonic, standing for the first letter of each assumption):

1. **Linearity**. The relationship between *y* and *x* is linear. The linearity assumption can be tested by using scatter plots. Outliers and influential observations should be considered for exclusion, prior to any other analysis (more on this in Section 6.3).
2. **Independence – No autocorrelation**. The residuals are independent from each other, meaning that an error *e* in a data point *x* does not affect by any means another error *e* of some different value of the same variable *x*. The Durbin–Watson test statistic can be used to detect the presence of autocorrelation in the residuals of regression analysis.
3. **Normality.** The distribution of errors *e* at any particular *x* value is normal having zero mean value. This means that for fixed values of *x*, *y* has a normal distribution as well. Assumption can be checked by using a histogram and a fitted normal curve, a Q-Q plot, Jarque–Berra test, etc. (see multiple linear regression, Section 6.2).
4. **Equality – Homoscedasticity.** Errors *e* have a zero mean, equal variance and a constant standard deviation at any particular value of *x*. This is also called homoscedasticity and exists when the variance of residuals does not change (increase or decrease) with the fitted values of the dependent variable. Heteroscedasticity, on the other hand, exists when the variance is not constant. Residual plot inspection is a straightforward way to trace if heteroscedasticity exists (Figure 6.1C and D).

6.1.2   Ordinary Least Squares (Intercept and Slope by OLS)

Definition
Ordinary least squares (OLS) is a statistical method for estimating the unknown parameters (coefficients) of a linear regression model. The regression using this method is also called OLS linear regression, or just linear regression. Other methods used to estimate these parameters include the Bayesian inference, the least absolute deviations and the nonparametric regression. The parameters *a* and *b* are called intercept and slope, respectively (*b* is called the coefficient in multiple linear regression; see Section 6.2). OLS regression determines these values by minimizing the sum of the squared vertical distances from the observed points to the line (sum of the squared residuals – Eqs. [6.4,6.5,6.6], Figure 6.1B, Rogerson 2001 p. 107). The line produced is the least-squares line.

$$min_{a,b} = \sum_{i=1}^{n} (y - \hat{y})^2 \qquad (6.4)$$

$$b = \frac{\sum_{i=1}^{n}(x_i - \bar{x})(y_i - \bar{y})}{\sum_{i=1}^{n}(x_i - \bar{x})^2} \tag{6.5}$$

$$a = \bar{y} - b\bar{x} \tag{6.6}$$

where

$\bar{x}$ is the mean of the independent variable
$\bar{y}$ is the mean of the observed values of the dependent variable
n is the number of data points

Interpretation
The regression line is the line with a slope that equals b, passing through the point $(0,a)$ and point $(\bar{x}, \bar{y})$ (Figure 6.1B, Rogerson 2001 p. 109) The slope denotes that for each unit of change in the independent variable, the mean change in the dependent variable is b. We should highlight that the regression line of y on x should not be used to predict x on y. In other words, we cannot use this equation to predict x values if we have y values available because the regression line has not been constructed to minimize the sum of squared residuals in the x direction (Peck et al. 2012).

6.2   Multiple Linear Regression (MLR)

6.2.1   Multiple Regression Basics

Definition
Multiple linear regression (MLR) analysis identifies the linear relationships between a dependent variable y and a set of independent variables, also called explanatory variables x (Rogerson 2001 p. 124). Regression analysis fits a straight line to the dataset to produce a single equation that describes the data. For m independent variables and n observations, the regression equation (deterministic function) is

$$\hat{y}_i = b_0 + b_1 x_{i1} + b_2 x_{i2} + \cdots + b_m x_{im} \tag{6.7}$$

where

$\hat{y}_i$ is the fitted value for the i-th observation.
$i = 1, \ldots n.$
n is the total number of observations.
$b_0, b_1, b_2 \ldots b_m$ are the coefficients.
b_0 is the intercept. It is the value of the equation if all variables or coefficients have zero value.
m is the total number of the independent variables.
x_{im} is the i-th observation on the m-th independent variable.

The regression probabilistic model is (6.8) (O'Sullivan & Unwin 2010 p. 226):

$$y_i = \hat{y}_i + e = b_0 + b_1 x_{i1} + b_2 x_{i2} + \cdots + b_m x_{im} + e_i \tag{6.8}$$

where

> y_i is the i-th value of the dependent variable
>
> e_i is the i-th error of the model (random deviation from the deterministic function, Eq. 6.7).

In matrix terms, the model can also be expressed as (6.9, 6.10) (O'Sullivan & Unwin 2010 p. 226):

$$\begin{bmatrix} y_1 \\ \vdots \\ y_n \end{bmatrix} = \begin{bmatrix} 1 & \cdots & x_{1m} \\ \vdots & \ddots & \vdots \\ 1 & \cdots & x_{nm} \end{bmatrix} \begin{bmatrix} b_0 \\ \vdots \\ b_m \end{bmatrix} + \begin{bmatrix} e_1 \\ \vdots \\ e_n \end{bmatrix} \tag{6.9}$$

$$Y = Xb + e \tag{6.10}$$

where

> Y is the vector containing the n observations for the dependent variable
> b is the vector of the estimated regression model coefficients
> X is the data matrix containing the n observations for the m independent variables (along with a column of ones)
> e is the error term

Interpretation

Each coefficient b_i can be interpreted as the change (impact) in mean y, for a one unit change of x_i, holding the remaining independent variables constant. Regression coefficients express how much impact an independent variable has on the dependent variable (de Vaus 2002 p. 279). The sign of the coefficient indicates the direction of the impact.

Why Use

MLR is used to (Rogerson 2001 p. 104)

- Identify linear relationships between a dependent variable and independent variables
- Evaluate the degree (effect) to which one variable influences the positive or negative change of another variable
- Build a predictive model by fitting a regression line to the data
- Study casual relationships (under specific conditions)

Discussion and Practical Guidelines

For most real-world problems, a plethora of variables are available, and simple linear regression is inadequate. Most of the concepts and terms used in simple linear regression can be applied in MLR with no or only a slight modification. The calculations though in MLR are extremely tedious in comparison to simple linear regression. MLR is not preferred solely due to variables' abundance. Explaining a dependent variable by a single independent variable, most of the time, is insufficient, as their relationship may be weak. In such case, we should add extra independent variables to increase the explained variation of the dependent variable. For example, estimating the value of a house is not just a matter of its size, although size is strongly related to price. Other variables that influence the value of a house include the year of construction (if it's new or old), the type of house (apartment, detached house, etc.) and other spatial variables such as the distance from city center, the distance from transportation network (e.g., subway, highway) or the location (i.e., neighborhood, city, county, country in which the house lies). Still, integrating many variables does not necessarily lead to a better model. On the contrary, including many independent variables is likely to lead to model overfitting, which is undesirable (see Section 6.2.2).

MLR can be also used to identify causal relationships (causes and effects) among the dependent and the independent variables. Still, not all types of MLR can establish causal relationships. Identifying causes and effects is a wider analysis lying in the explanatory analysis filed. Explanatory analysis attempts to identify the factors or mechanisms that produce a certain state of some phenomenon (Blaikie 2003). From the social analysis perspective, explanatory analysis attempts to trace the causes that create an effect, by carrying out controlled experiments. Such experiments establish a linking relationship between a variable (cause) and one or more variables (effects) by controlling the order in which causes and effects (and related population's groups) are analyzed (Blaikie 2003). Regression analysis may be used for identifying causes and effects, but not all regression models can be used for this purpose (see Section 6.2.6). In other words, although some regression models may be robust and capable to identify linear relationships among variables, this does not straightforwardly lead to the conclusion of a cause-and-effect relationship (see also Section 2.3.4). In addition, it has to be emphasized that any **hypothesized causal relationship** is one way, meaning that the dependent variable is responsive to the independent variable and not the other way around (Longley et al. 2011 p. 357). That is why the dependent variable is also called the response variable, and the independent variable is called the predictor. It should also be noted that in case of spatial analysis and due to spatial heterogeneity and spatial dependence, the discovered casual relationships through regression modeling might change from place to place, and that is why a spatial aspect of regression should be considered. For more details on spatial regression, see Section 6.5 and Chapter 7.

Finally, prior to any regression analysis, there is a set of decisions that should be determined (explained in the following subsections):

- Choose the variables to be used to avoid overfitting (Section 6.2.2)
- Check the dataset for missing values (Section 6.2.3)
- Check the dataset for outliers and leverage points (Section 6.2.4)
- Check if any dummy variable is needed (Section 6.2.5)
- Select the method of entering variables in the model (Section 6.2.6)

6.2.2   Model Overfit: Selecting the Number of Variables by Defining a Functional Relationship

Definition
An **overfit model** is a model that performs well on a specific dataset but cannot generalize new solutions (to newly presented data) with accuracy.

Discussion and Practical Guidelines
In general, overfitting occurs when a model has too many parameters relatively to the number of observations. The number of variables to be included in the model depends on the problem. The first step is to identify if any functional relationship (or else a meaningful relationship or causation) among the variables under consideration exists through an established theoretical background. Solid comprehension of such background leads to the identification of potential independent variables. The regression model will then carry out the burden to convert this functional relationship into a meaningful mathematical equation.

We should avoid the temptation to use as many variables as possible. The more variables included, the more likely for an overfitted model. Model overfitting is not desirable, as it leads to a model with poor predictive performance. One way to avoid overfitting is to have considerably more observations than the number of variables. The following rules of thumb can be used for selecting the number of variables in relation to the number of observations available (de Vaus 2002 p. 357):

- When using standard MLR the ratio of observations to variables should be larger than 20/1 (Tabachnick et al. 2012).
- The sample size should be at least $100 + k$, where k is the number of independent variables (Newton & Rudestam 1999)
- In case of a non-normal distributed dependent variable, samples should be increased to more than 100.

These rules of thumb are for general guidance. The final decision on the choice of variables should be made in relation to the problem at hand and the data available. A smooth approach is to initially build a model that is simple, explaining a large proportion of variation, that is also backed up by a solid theoretical background. By using the *R*-squared adjusted, we can determine which variables are useful and which are not by observing the fluctuations in the *R*-squared adjusted value (see Section 6.3.3). Exploratory regression explained in Section 6.7 provides such approach. Section 6.2.6 also presents the effect that the order of variables' entry infers to the model.

6.2.3   Missing Values

Before we begin our analysis, we should inspect if missing values exist. Three of the most common ways to confront missing values are

* Deleting observations that contain missing values.
* Replacing the missing values of a variable with the mean value of this specific variable. Although in this approach, we retain all observations, it is not guaranteed that the final results will be meaningful. For instance, if we miss age values and we use the mean value of age for the entire variable for those missing, then we will probably get wrong estimations.
* Estimate the missing values by using simple OLS, or MLR , of other variables with no missing values. For example, if the "House size" variable has no missing values and some respondents refused to reveal their income, we can build a simple linear regression of "Income" and "House size." If the model is statistically significant and all related diagnostics are checked, we can estimate/predict the income of the missing observations by using the known size of a house.

6.2.4   Outliers and Leverage Points

Outliers and high **leverage points** change a lot the results of any analysis (see Section 6.3.11). There are mainly three approaches to deal with such observations in regression analysis.

* First, trace and remove outliers before regression analysis takes place. To do so, one should typically follow the procedures for tracing outliers as explained in Section 2.2.7. As long as outliers are removed, then regression analysis is carried out along with the calculation of high leverage values (see Section 6.3.11).
* Second, perform MLR including all data, and create the standardized residual plots where both outliers and high leverage points can be traced. After we trace influential points, we may remove them and test the model again. If the model is significantly improved, we can drop these influential observations (see how this is done in detail in Section 6.3.11). Any observation that leads to a large residual should be scrutinized.
* Apply robust regression that is specifically designed to handle outliers and heteroscedasticity (not further explained in this book).

In general, we should check if an outlier reflects an error or if it is an actual value revealing unusual behavior/status. Although outliers are usually removed, it depends on the studied problem and the researcher's choice of how outliers will be handled.

6.2.5 Dummy Variables

Definition
A **dummy variable** is a binary variable getting values either 1 or 0, indicating the presence (1) or absence (0) of some categorical effect (in the case of a dependent binary variable, we consider logistic regression). If a categorical variable is decomposed to its categories, then each single category can be a new variable, named a dummy variable.

Why Use
Apart from ratio variables, categorical independent variables may be integrated to a regression model. For example, *Income* is often classified into a category (i.e., *Low*, *Average* or *High*). A person with average income is assigned a value of 1 for the category Average and 0 for the two others. *Place of living* may also be reported as a nominal variable such as *Urban*, *Suburban* and *Rural*. A person living in rural area is assigned a value of 1 for *Rural* and 0 for the other two categories (*Urban* and *Suburban*). Other examples include *Gender* (i.e., *Male* or *Female*) and *Land Use* (i.e., *Forest*, *Water*, *Urban* and *Shrubland*; Grekousis et al. 2015a).

Discussion and Practical Guidelines
To handle dummy variables with k categories, we create $k - 1$ variables by omitting one category to avoid multicollinearity (Rogerson 2001 p. 128). For example, for the variable *Place of living*, suppose that X_1 is the *Urban* category, X_2 is the *Suburban* category and X_3 is the *Rural* category. The following regression equation would be inappropriate, violating the non-multicollinearity assumption for MLR (see Section 6.4), since the sum of all k columns should always equal 1 (Rogerson 2001 p. 129) (6.11):

$$Y = b_0 + b_1X_1 + b_2X_2 + b_3X_3. \tag{6.11}$$

We can arbitrarily omit the category "Rural" (6.12):

$$Y = b_0 + b_1X_1 + b_2X_2. \tag{6.12}$$

A person living in a rural area gets 1 for *Rural* and 0 for *Urban* and *Suburban*. In other words, we can determine from only two out of three variables where a person lives. Once dummy variables are defined, then regression analysis proceeds at the usual fashion.

Dummy variables can be combined to ratio variables within the same regression model. For example, to model $Y = Income$, the ratio independent variables $X_1 = Years\ of\ education$ and $X_2 = House\ size$ can be combined to the categorical variable *Place of living* (three categories: *Urban*, *Suburban* and *Rural*). From the three categories, we create two dummy variables, $X_3 = Urban$ and $X_4 = Suburban$ by arbitrarily omitting *Rural*.

The regression equation is (6.13):

$$Y = b_0 + b_1X_1 + b_2X_2 + b_3X_3 + b_4X_4 \tag{6.13}$$

$$Income = b_0 + b_1YearsEdu + b_2HouseSize + b_3Urban + b_4Suburban \tag{6.14}$$

The model would lead to different coefficients but precisely the same conclusions if we dropped any other category.

Interpreting Coefficients of Dummy Variables

The regression coefficients for the dummy variables can be interpreted in relation to the omitted category (Rogerson 2001 p. 129). For Eq. (6.14), if a person lives in an urban area, then X_3 gets 1 and X_4 gets 0. Keeping constant the other variables, then *Income* is, by b_3 units, higher for a person living in an urban area, relatively to the omitted category (it would be less if b_3 were negative).

Being located in *Suburban* means that X_4 gets 1, X_3 gets 0 and *Income* is by b_4 units different, relatively to *Rural*, keeping constant the other variables. Finally, being located in *Rural* implies that both X_3 and X_4 are 0, and by keeping constant the other variable, *Income* is b_0, which is the intercept.

Suppose we add two more categorical variables, namely *Gender* and *Age*. *Gender* consists of two categories, *Male* and *Female*, and *Age* consists of three, *Young*, *Middle* and *Old*. We can add these variables in the previous model to test whether *Income* is further explained by these variables (6.15). We choose to omit the *Male* and *Old* categories:

$$Income = b_0 + b_1YearsEdu + b_2HouseSize + b_3Urban \\ + b_4Suburban + b_5Female + b_6Young + b_7Middle \tag{6.15}$$

Suppose that the coefficients of the model are (6.16):

$$Income = 2{,}500 + 10YearsEdu + 0.5HouseSize + 1{,}500Urban \\ + 750Suburban - 500Female - 1{,}000Young + 2{,}200Middle \tag{6.16}$$

The coefficients will be interpreted relatively to the omitted categories. Thus, a person living in a city earns on average $1,500 more than a person living in rural areas. A woman earns $500 less than a man. A young person earns $1,000 less than an old person.

From the policy perspective, this type of analysis is appropriate for evaluating various scenarios. Suppose we build a regression model for land use changes, based on population, income, GDP and other socioeconomic variables. Based on the coefficients produced from the model, various scenarios can be tested on projected pressures on land use changes – for example, of a 20% urban population increase or a 10% GDP increase within the next decade (Grekousis et al. 2016).

6.2.6 Methods for Entering Variables in MLR: Explanatory Analysis; Identifying Causes and Effects

Although for some regression models, controlling the order in which variables are entered does not matter, for others, it offers the choice to conduct statistical experiments and identify causes and effects through explanatory analysis (de Vaus 2002 p. 363). There are four main methods of MLR based on the sequence that the independent variables are entered in the regression analysis:

- **Standard or single-step regression**, when all variables are analyzed in a single step. This method (a) identifies which independent variables are linked to the dependent variable and (b) quantifies how much of the total variation of the dependent variable is explained (through R-square adjusted). In this method, we explore linkages, not causal relationships (de Vaus 2002 p. 360).

  When to use: When we are dealing with a small set of variables and we do not have a clear view of which independent variables are appropriate, when the main task is the identification of linkage and not causation.

- **Stepwise regression** is used when we have more than one independent variable and we want to use an automated procedure to select only those significant to the model (those that explain as much variation as possible of the dependent variable). Instead of calculating statistics and diagnostics for the whole dataset in a single step (as in the standard method), variables are added or removed to the model sequentially. When data are added to the model, the method is called forward stepwise regression, starting from a simple regression model having a few variables or just the constant. On the other hand, when variables are removed, the method is called backward stepwise, and the initial model contains all variables. The order of entry plays a crucial role in stepwise regression. In cases where variables are uncorrelated, the order is not an issue and does not affect results. Still, it is hard to find completely uncorrelated variables in a dataset. In cases of correlated variables, order matters. Although the final adjusted R-squared value will be the same, no matter what the order is, the relative contribution of each variable to the model will change, resulting in different coefficients. Getting different models from the same set of potential terms is a disadvantage of the method; by selecting different criteria/order for entering or removing variables, stepwise yields different results.

When to use: When we want to use an automated method to select the appropriate variables, when we want to know how much additional variance each variable explains, when we want to maximize the prediction capabilities of the model with as few variables as possible (de Vaus 2002 p. 365).

- The **hierarchical method** is used when the order of entering or removing the independent variables is determined by the researcher (also called hierarchical regression). Controlling the order of entry allows for statistical experimentation and the identification of casual relationships by testing how much the adjusted R-squared increases each time a new variable is added/removed. By applying hierarchical regression, we use the same statistics and diagnostics as in MLR. The difference lies in the way we built the sequence of the independent variables. The focus is on the change in the predictability of the model associated with independent variables entered later in the analysis in comparison to those entered earlier. The order is based on the scope of the analysis, the associated theory and the intuition of the analyst.

More specifically, applying the hierarchical method allows for

(a) Testing theories and assumptions by conducting experiments and controlling the order of entry.

(b) Calculating the extra exploratory power of each variable or block of variables.

(c) Controlling for confounding variables. There are cases in which two variables (i.e., dependent and independent) exhibit correlation but with no causal relationship between them. A confounding variable is a third variable that is associated with both the dependent and the independent variables and causes extraneous changes. It is the confounding variable's impact that explains the variance on both the dependent and independent variables. In this sense, hierarchical regression allows for the identification of causal relationships (de Vaus 2002 p. 365).

When to use: To have more control over the analysis as well as to test causal relationships.

- **Exploratory regression** runs all models' combinations to find the optimal one. It applies OLS regression to many different models to select those that pass all necessary OLS diagnostic tests. A close inspection of the detailed results and models can lead to a robust analysis similar to hierarchical regression. In other words, exploratory regression, if used wisely, allows for assessing the exploratory power of each variable as well for controlling of confounding variables by cross-comparing the models that better reflect the tested theories and assumptions. Still, exploratory regression is a data-mining approach, more similar to stepwise regression and not to hierarchical regression, as it does not allow for much experimentation (ESRI 2016a).

When to use: When we want to test all combinations in an automated way, when we want to identify causal relationships.

6.3   Evaluating Linear Regression Results: Metrics, Tests and Plots

There are various metrics, tests and graphs to evaluate the performance of a regression model. In the next subsections, the following methods are presented:

- Multiple *R*
- Coefficient of determination *R*-squared
- Adjusted *R*-squared
- Predicted *R*-squared
- F-test
- T-statistic
- Wald test
- Standardized coefficients (beta)
- Residual plots and standardized residual plots
- Influential points (outliers and leverages)

6.3.1   Multiple *r*

Definition
Multiple *r* is the absolute value of the correlation between the observed *y* (response) and the estimated *ŷ* predicted by the regression equation.

Interpretation
Multiple *r* measures correlation and can be interpreted accordingly. In simple linear regression, the sign of *b* coefficient (slope) reveals the positive or negative correlation. Multiple *r* is calculated by the same way for multiple regression as well; that is why it is called "multiple." Still, multiple *r* is not very indicative of assessing the results of regression analysis, and it is not commonly analyzed further.

6.3.2   Variation and Coefficient of Determination *R*-Squared

Definition
Before we define *R*-squared, let's define the following quantities (see Figure 6.2):

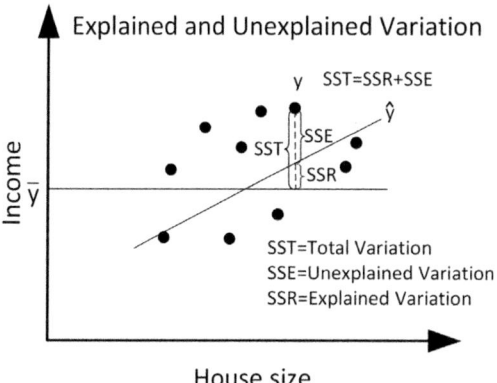

Figure 6.2 Partitioning the variation in *y*. The distance of the total variation, SST, is the summation of SSR and SSE distances.

Sum of Squared Regression (6.17), also called explained variation (of the model):

$$SSR = \sum_{i=1}^{n} (\hat{y}_i - \bar{y})^2 \tag{6.17}$$

It is a measure of the explained variation or the amount of variation in *y* that can be modeled by a linear relationship to *x*.

Sum of Squared Error (6.18), also called unexplained variation, or residual sum of squares:

$$SSE = \sum_{i=1}^{n} (y_i - \hat{y})^2 \tag{6.18}$$

It is a measure of the unexplained variation or the amount of variation in *y* that cannot be modeled by a linear relationship to *x*.

Sum of Squared Total (6.19) is a measure of the total variation of *y*:

$$SST = \sum_{i=1}^{n} (y_i - \bar{y})^2 \tag{6.19}$$

The total variation in *y* can be calculated as:
  Total variation = Explained Variation by Regression + Unexplained variation,

which is equivalent to $SST = SSR + SSE$ (6.20)

In practice, the variation of *y* can be considered as the result of two quantities: (a) the variation explained by the linear model assumed to describe better the data and (b) the unexplained variation that the linear model did not achieve to explain (Linneman 2011 p. 226).

Coefficient of determination denoted as R^2 (*R*-squared) is the *Percent* of the variation explained by the model (Figure 6.2; de Vaus 2002, p. 354). It is calculated as the ratio between the variation of the predicted values of the dependent variable (explained variation *SSM*) to the variation of the observed values of the dependent variable (total variation SST; Eq. 6.21). It equals the squared correlation coefficient (see Figures 6.1D and 6.2):

$$R^2 = \frac{SSR}{SST} = \frac{\sum_{i=1}^{n}(\hat{y}_i - \bar{y})^2}{\sum_{i=1}^{n}(y_i - \bar{y})^2} \tag{6.21}$$

Put simply, it is the squared sum of the difference of predicted values $\hat{y}_i$ with the mean value of the observed values $\bar{y}$ (squared sum of regression), divided by the squared sum of the difference of the observed values y_i, with the mean value of the observed values $\bar{y}$. The coefficient of determination is also calculated by the following formula (6.22):

$$R^2 = 1 - \frac{SSE}{SST} = 1 - \frac{\sum_{i=1}^{n}(y_i - \hat{y})^2}{\sum_{i=1}^{n}(y_i - \bar{y})^2} \tag{6.22}$$

As the residual sum of squares is minimized in the ordinary least squares regression, the ratio SSE/SST is always smaller than 1.

Why Use
R-squared is used to assess if a linear regression model fits well to the data points. It provides information about the goodness of fit of the model.

Interpretation
The coefficient of determination ranges from 0 to 1 and expresses the percentage of the variation explained by the model (see Figure 6.3). A zero value indicates that no variation is explained and, thus, the model cannot be used, whereas a value of 1 explains all data variability perfectly fitting a line on the data points. For example, if R-squared is 82%, this typically means that 82% of the variation in Y is due to the changes in values of X, meaning that X is a good predictor for Y having a probabilistic linear relationship. The smaller the unexplained variation, the smaller the ratio (in Eq. 6.22) and the larger the coefficient of determination or the larger the explained variation. The larger the determination of coefficient is, the more robust the model. As the linear regression model fits a line to minimize residuals, an ideal fit would result in zero residuals. When residuals are small, we have a *good fit* of the line. As residuals get larger, the less robust the model becomes; it produces predicted values that deviate a lot from the observed ones.

Let's see how R-squared is graphically explained. In general, when variables are not correlated, each one is expected to explain a different amount of variation of the dependent variable. Still, variables are rarely uncorrelated, and to some extent, they explain the same percentage of the dependent variable's variation. In other words, the variation they explain partially overlaps. For example, in Figure 6.3A, X_1 variable explains portion a (the shaded overlapping part) of Y variation through simple linear regression. In Figure 6.3B, two independent variables are used, each one explaining a different amount of

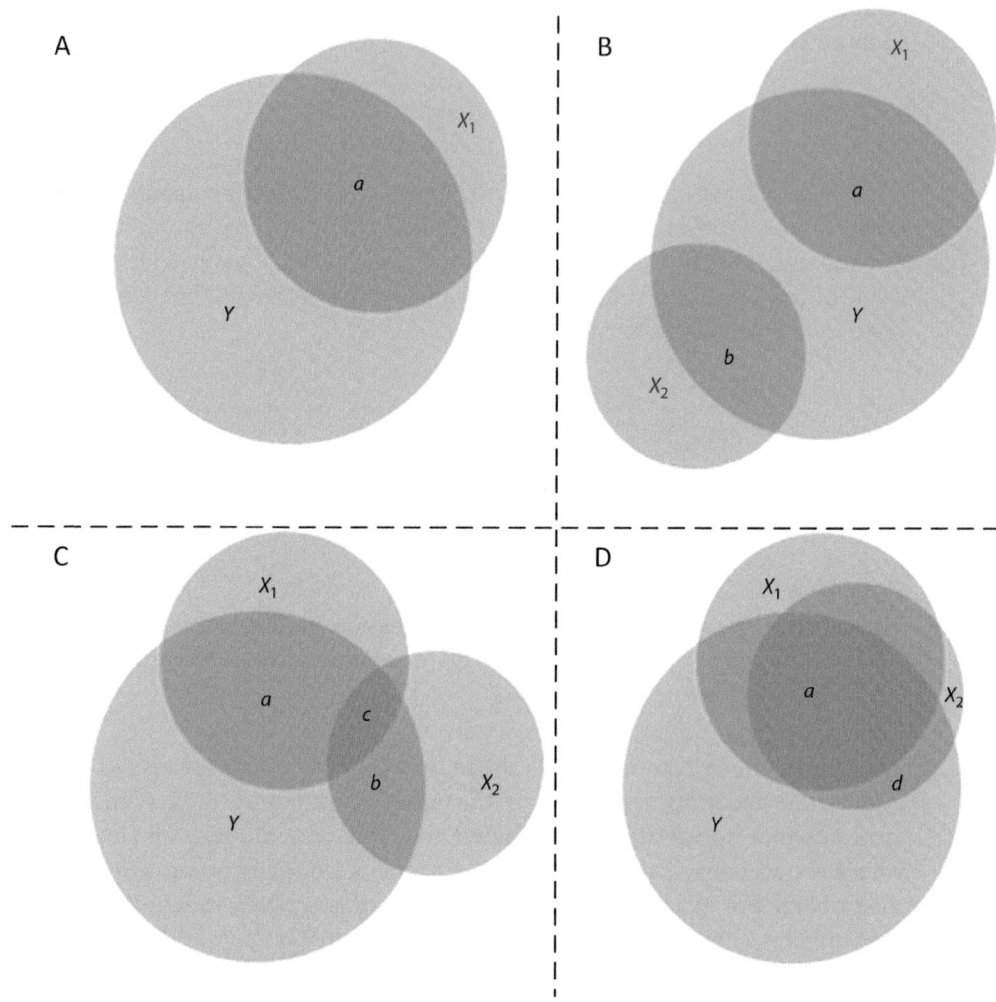

Figure 6.3 Variation explained in a graphical representation. (A) In cases of simple linear regression, variable X_1 explains portion a of independent's variable Y variation. (B) In MLR, an inclusion of a second variable leads to additional explanation b of variation. (C) These variables explain a common (overlapping) variation c. (D) X_1 and X_2 are highly correlated and explain almost the same proportion of variation.

variation. The total explained variation R-squared is $a + b$. There is no overlap between X_1 and X_2, as these two variables are uncorrelated. In Figure 6.3C, X_1 and X_2 are moderately correlated. There is an overlapping portion c of Y variation that is explained by both independent variables, so it should be included only once in the total variation explained. It is only the extra non-overlapping proportion $(b - c)$ of variation explained through X_2, which is of great importance. The final R-squared is $a + (b - c)$. In Figure 6.3D, X_1 and X_2 are strongly correlated and the extra variation explained, when including X_2 in the model, is portion d. The final R-squared is $a + d$, where a is the variation explained by X_1 and d is the non-overlapping variation between X_1 and X_2. If $X1$ explains 80% of the variation and by adding X_2 the explained variation increases to 82%. This typically means that X_2 assists in explaining an extra 2% of the explained variation and not that X_2 would explain 2% if X_1 was not included in the model. The more correlated the variables are, the less the extra variation explained when adding a new (correlated) variable. The less correlated the variables are, the more extra variation explained when adding a new (uncorrelated) variable.

Discussion and Practical Guidelines

A small R-squared is not necessarily a sign of an insufficient regression model. How we interpret R-squared depends on the scopes of the study and the research questions. For example, for identifying the relationships between dependent and independent variables, even a low R-squared can still be useful. Although there is not much of variation explained (small R-squared), the model might still have statistically significant independent variables (low p-values). Having statistically significant coefficients means that we cannot reject the null hypothesis that these coefficients are zero (do not affect the dependent variable), thus revealing valid relationships among variables. In other words, even with low variation explained, relationships might still exist, and as such, the model will be beneficial in identifying trends (i.e., increasing X will increase Y, according to coefficients). For this reason, R-squared should always be evaluated under the prism of the residual plot (see example in the following sections).

On the other hand, it would not be wise to use a model with low R-squared for predictive purposes, as the large unexplained variation will lead to large predictive intervals for the Y values and thus to a model with poor predictive quality. Prediction should be considered when R-squared is high. Furthermore, the regression line should not be used for predicting values far out of the range of initial Y and X values, as we cannot directly assume that relationships extracted inside a specific range stand well outside this range. This is called the danger of extrapolation (Peck et al. 2012).

R-squared provides a meassure of the strength of the relationships among dependent and independent variables. Still, it is not so helpful in cases of multiple regression because for every new variable added, R-squared increases. This reveals potential model overfitting, which is undesirable. To overcome this problem, we introduce the adjusted R-squared (see Section 6.3.3). Finally, R-squared does not provide a measure of how statistically significant the hypothetical relationship is. To do so, we use the F-test of overall significance (Section 6.3.6).

6.3.3 Adjusted R-Squared

Definition

Adjusted R-squared is the coefficient of determination, adjusted for the number of coefficients. The formula to calculate adjusted R-squared is (6.23):

$$R^2_{adj} = 1 - \left(\frac{n-1}{n-p}\right)\frac{SSE}{SST}$$

(6.23)

where

> SSE is the sum of the squared error (residuals; see Eq. 6.18)
> SST is the sum of the squared total (see Eq. 6.19)
> n is the number of observations
> p is the number of parameters — regression coefficients (with intercept included)

Why Use

In MLR, for every new variable added, R-squared increases, revealing potential model overfitting. To account for this problem, we use adjusted R-squared, which does not necessarily increase with the addition of extra variables. In MLR, adjusted R-squared should be used instead of R-squared, which is mainly used in simple linear regression.

Interpretation

Adjusted R-squared reveals the additional variation explained if an extra independent variable is included in the model. If we add a new variable and R-squared adjusted increases, then this is a useful variable. If R-squared adjusted decreases, then this variable might not be useful for the model.

Discussion and Practical Guidelines

Both R-squared and adjusted R-squared are used to evaluate the percentage of total variation explained. Still, there is a significant difference between them. While R-squared assumes that every single variable explains the variation of Y, adjusted R-squared explains the percentage of variation of only those independent variables that have an impact on Y. R-squared tends to increase with each additional variable, leading us to the wrong conclusion that we create a better model. In fact, by adding more and more variables, we confront the problem of overfitting data. If we use the adjusted R-squared instead, we can determine which variables are useful and which are not by inspecting the adjusted R-squared values. We may consider keeping only those variables that increase the adjusted R-squared. The adjusted R-squared is a downward adjustment of R-squared and, as such, is always smaller from the R-squared value. Adjusted R-squared is also used to compare models with different number of predictors (see also predicted R-squared in Section 6.3.4).

6.3.4   Predicted R-Squared

Definition

Predicted R-squared is a statistic that measures the ability of a regression model to predict a set of newly presented data (Ogee et al. 2013). Predicted

R-squared is an iterative process and is calculated by removing each observation successively from the dataset and then estimating the regression equation again. The new model is used to calculate the observation removed, and the result is compared with the actual value.

Why Use
R-squared and adjusted *R*-squared explain how well the dependent variable is explained by the independent variables available. Still, they do not directly provide an estimation of the predictive quality of the model. When there is a big difference between *R*-squared and adjusted *R*-squared, it is a sign that the model contains more predictors than necessary, leading to low predictive quality. As such, the closer to 1 the ratio of adjusted *R*-squared to *R*-squared is, the better the model fits the existing data. A better way to assess the predictive quality of the model is by using the predicted *R*-squared.

Interpretation
Predicted R-squared reveals if overfitting of the original model exists. The higher the predicted *R*-squared is, the better the original model is. The lower the value is, the more unsuitable the model is, indicating overfitting.

Discussion and Practical Guidelines
Adjusted *R*-squared is used to test whether additional independent variables affect the explained variation of the dependent variable. On the other hand, predicted *R*-squared is more valuable than adjusted *R*-squared when it comes to evaluating the predictive capabilities of the model. Predicted *R*-squared is calculated using existing observations with known *Y* values that are not included in the initial model creation (out of sample). For example, suppose we get adjusted *R*-squared = 0.89 and predicted *R*-squared = 0.61. The high adjusted *R*-squared reveals that the original model fits existing data well, but when the model is fed with unseen data, its predictive quality decreases sharply to the value of 0.61 (predicted *R*-squared). A low predicted *R*-squared might be a sign of an overfitted model. A remedy would be to remove independent variables, add observations, or both. Predicted *R*-squared is not widely available in statistical software packages, but it can be calculated if one has basic scripting knowledge following the preceding steps/procedures.

6.3.5   Standard Error (Deviation) of Regression (or Standard Error of the Estimate)

Definition
The **standard error (deviation) of regression** for samples is the square root of the sum of squared errors divided by the degrees of freedom (6.24) (de Smith 2018 p. 507):

$$s_e = \sqrt{\frac{SSE}{DFE}} = \sqrt{\frac{SSE}{n-p}} \qquad (6.24)$$

where

SSE is the sum of squared errors (errors; see Eq. 6.18)
DFE = $n - p$ is the degrees of freedom for error
n is the number of observations
p is the number of parameters (coefficients) of the model (including slope)

The quantity $\frac{SSE}{n-p}$ is also known as the mean-squared error (MSE). As a result, the estimate of the standard error s_e is the square root of the MSE (Lacey 1997).

In cases of simple linear regression, the standard error is calculated as (6.25):

$$s_e = \sqrt{\frac{SSE}{n-2}} \qquad (6.25)$$

where $p = 2$, as two parameters are used, the slope and the intercept.

Why Use

To calculate the extent (average distance) of the observations' deviation from the regression line. The standard error of regression is also used to compute the confidence intervals for the regression line (de Smith 2018 p. 507).

Interpretation

The magnitude of the standard deviation reveals how close or far, on average, observations lie from the least squares line (average distance that a data point deviates from the fitted line). Approximately 95% of the observations should range within two standard errors from the regression line (Scibila 2017). In this respect, we can use the standard deviation of the regression as a rough estimate of the 95% prediction interval. The standard error of regression has the same units with the dependent variable. Consequently, it gives an estimation of the prediction's precision in the same units to the dependent variable. Lower values of standard errors are preferred, indicating smaller distances between the data points and the fitted values.

Discussion and Practical Guidelines

With the standard error of regression, we assess the precision of prediction measured in the units of Y, which is more straightforward compared to the R-squared – a percentage. By combining R-squared and standard regression error, we better evaluate the validity of our model. The standard error of regression can sometimes assist in defining how high R-squared should be in order to consider the model valid. For example, a not very high R-squared value with a small standard error of regression might be accepted according to the study and the research questions.

6.3.6   F-Test of the Overall Significance

Definition

The **F-test** statistic of the overall significance of the model, evaluates whether the coefficients in MLR are statistically significant (de Smith 2018 p. 508) (6.26):

$$F = \frac{SSR/DFM}{SSE/DFE} \tag{6.26}$$

where

> SSR is the sum of squared regression (see Eq. 6.17)
> SSE is the sum of squared errors (see Eq. 6.18)
> $DFM = p - 1$ is the degrees of freedom for the model
> p is the number of model parameters
> $DFE = n - p$ is the degrees of freedom for error
> n is the number of observations

The hypotheses for the F-test of the overall significance are as follows:

- **Null hypothesis:** coefficients' values are zero H_0: $b = 0$.
- **Alternative hypothesis:** coefficients' values are not zero H_1: $b \neq 0$.

The F-test is a test on the joint significance of all coefficients except the constant term and is also called joint F-statistic. F-test also produces a p-value, which is called F-significance. The F-value of a regression model is compared to the F-critical value resulting from the F-tables and the corresponding degrees of freedom. It is not essential to showcase how these values are extracted, but for those interested more explanation can be found in Peck et al. 2012 (p. 840).

Why Use

F-test is used to test if the regression fit is statistically significant, or else, if the regression has been successful at explaining a significant amount of the variation of Y (Rogerson 2001 p. 110).

Interpretation

In the context of regression analysis, the F-statistic value is not useful by itself. It is used for comparison with the F-critical value. As such, large or low values do not indicate any significance of the model. Instead, we should be directed to the F-significance value (p-value) to draw our conclusions. If the F-significance is small (e.g., less than 0.05), then we can reject the null hypothesis and accept that the regression model is statistically significant. In this case, we state that the observed trend is not due to random sampling of the population. If the F-significance is high (e.g., larger than 0.05), then we cannot reject the null hypothesis. In this case, we have to be cautious on how to interpret results, as we cannot definitely state that there is no relationship between Y and X. Possible interpretations are the following:

- There is not a linear relationship between X and Y. Still, there might exist a nonlinear relationship such as exponential or logarithmic. A scatter plot of X over Y is the first step to identify potential curvature.
- The X variable might explain a small portion of the variation of Y, but it is not sufficient enough to consider the model statistically significant by using only X. We might need to add some more variables to explain Y. By adding extra variables, we might discover a linear relationship that was not unraveled initially. Additional variables are likely to increase the percentage of variation explained as well.
- There might exist a linear trend, but if our sample size is small, then we are not able to detect any linearity.

Discussion and Practical Guidelines

In general, F-statistic is not very useful as it's quite common that the output is statistically significant (Anselin 2014). Other statistics such as the Wald test and the adjusted R-squared can be used to assess the overall performance of the model.

6.3.7   t-Statistic (Coefficients' Test)

Definition

The **t-statistic** in regression analysis is a statistic used to test if coefficient b is statistically significant. Coefficients are calculated as point estimates through the n data points available. In any point estimate, an indication of the accuracy is needed (Peck et al. 2012). The hypotheses to be tested are

- **Null hypothesis:** The coefficient value is zero H_0: $b = 0$.
- **Alternative hypothesis:** The coefficient value is not zero H_1: $b \neq 0$.

The t-statistic is calculated as (6.27):

$$t = \frac{b}{SE_b} \tag{6.27}$$

where SE_b is the standard error of the estimated coefficient b.

Why Use

To assess if the coefficients values are statistically significant and decide which independent variables to include in the model.

Interpretation

For a small p-value, we reject the null hypothesis that coefficient b is zero, and we accept its value as statistically significant. In this case, the independent variable X that the coefficient is assigned is important for the calculation of the dependent Y. For a large p-value, we cannot accept b as statistically significant, and we should consider removing the corresponding independent variable from the model.

Discussion and Practical Guidelines

It is important to highlight that each measurement is not error-free. The ability of each independent variable to contribute to increased explained variation is also relative to the measurement error. When we interpret the t-statistic, we should have in mind that a potential rejection of a predictor in our model might be just due to measurement error and not due to a nonexisting relationship. It is likely (but still not that common) that with another, more accurate dataset, this same variable turns out to be significant.

Before we evaluate the t-statistic, we should check the Koenker (BP) statistic (Koenker & Hallock 2001). When the Koenker (BP) statistic is statistically significant, heteroscedasticity exists, and the relationships of the model are not reliable (see Section 6.4). In this case, a robust t-statistic and robust probabilities (calculated automatically) should be used instead of the t-statistic p-value. Robust probabilities can be interpreted as the p-values, with values smaller than 0.01 for example, being statistically significant.

6.3.8 Wald Test (Coefficients' Test)

Definition

The **Wald test** is a statistical test that evaluates the significance of the coefficients.

The hypotheses for the Wald test of coefficients' significance are as follows:

- **Null hypothesis:** The coefficient value is zero H_0: $b = 0$.
- **Alternative hypothesis:** The coefficient value is not zero H_1: $b \neq 0$.

Interpretation

If the test rejects the null hypothesis (p-value smaller than the significance level), then we accept that the coefficient b is not zero, and thus, we can include the related variable in the model. If the null hypothesis is not rejected, then removing the respective variable will not substantially harm the fit of the model.

Discussion and Practical Guidelines

The Wald test may additionally be applied for testing the joint significance of several coefficients, and that is why it is called joint Wald test as well. In this case, it can be used as a test for the general performance of the model, or else as a test for model misspecification. If the p-value is small, it is an indication of robust overall model performance. The joint Wald test should be checked instead of the F-significance when the Koenker (BP) statistic is statistically significant.

6.3.9   Standardized Coefficients (Beta)

Definition
Standardized coefficients are the coefficients resulting when variables in the regression model (both dependent and independent) are converted to their z-scores before running the fitting model. The standardized regression coefficients – also called beta coefficients, beta weights or just beta – are expressed in standard deviation units, thus removing the measurement units of the variables. Coefficients that are not standardized are also called unstandardized coefficients.

An alternative way to compute the beta coefficients, is to use the regression coefficients b_i estimated when variables are not standardized, multiplied by the ratio of the standard deviation of the independent variable X_i, to the standard deviation of the dependent variable Y (6.28):

$$Standardized\ b_i = b_{i*} \frac{Standard\ Deviation\ (X_i)}{Stdandard\ Deviation\ (Y)} \tag{6.28}$$

where

b_i is the unstandardized coefficient of X_i

Why Use
To decide which independent variable X is more important to model dependent Y. To compare the effect of each independent variable to the model, especially when the units or the scale among the independent variables is different. In fact, in case of unstandardized coefficients, we cannot state which independent variable is more important in determining the value of Y since the value of the regression coefficients depend on the units with which we measure X.

Interpretation
In cases of beta coefficients, a change in X_i by 1 standard deviation of X_i leads, on average, to beta coefficient (b_i) standard deviations (s_y) change, in the dependent variable Y or else: an average change in Y of $Standardized\ b_i \times s_y$.

Discussion and Practical Guidelines
Coefficients can be defined as the numbers by which the variables in the regression equation are multiplied. Unstandardized coefficients reveal the

effect on the dependent variable of a one-unit change (increase or decrease depending sign) of an independent variable, if all other independent variables remain stable.

For example, consider the following model for estimating the monthly income of individuals

$$Income = 2,000 + 50(Years\ of\ education) + 10(Size\ of\ house)$$

This model suggests that for every additional year of education, holding constant the *Size of house*, income increases by 50 units. Still, as coefficients are estimated from variables measured in different units and scales, their importance in determining the dependent variable cannot be compared. In other words, we cannot infer that *Years of education* is more important because its coefficient value (b_1 = 50) is larger than the coefficient of *Size of house* (b_2 = 10). The significance level is sometimes used to compare coefficients but that is not what significant levels are meant to be used for (Linneman 2011). A better way to compare coefficients is to standardize them.

Suppose that for the model above, the beta coefficients are

$$Income = 0.32 + 0.18(Years\ of\ education) + 0.23(Size\ of\ house)$$

As *Years of education* increase by one standard deviation, income increases by 0.18 standard deviations. Having the largest effect on income is *Size of house*, as it has a larger beta value (0.23) compared to the beta of *Years of education* (0.18). By standardizing the coefficients, we may assess the importance of each variable better compared to the unstandardized. For example, it is now revealed that the *Size of house* is more important predictor for income, compared to *Years of education*, having a higher beta coefficient. By estimating beta coefficients, we calculate how important each statistically significant variable is to the model. The higher the beta value (ignoring the sign), the more critical the variable is, compared to the other variables of the model.

Standardized coefficients have been criticized, as they express standard deviations (thus removing the scale of the actual unit of measurement from each variable), and as such, they become less meaningful and are not easily interpreted, especially when distributions are skewed. Furthermore, comparisons across groups are difficult as the standardization is different for each group. In this respect, as beta coefficients most of the time are automatically calculated by statistical software and are readily available, they can be used as a way to assess which variable is more important. Still, it would be more reasonable to use unstandardized coefficients to interpret what is the effect of one unit change of *X* to *Y*.

To answer the question of which variable is more important in a regression model, strongly depends on the specific context in which any analysis is carried out. Statistical analysis provides measures to evaluate any model, but the researcher should consider the meaning of a one-unit change in an independent variable in relation to the dependent variable in a real-world context. From the policy perspective, some variables might not be feasible to significantly change. If, for example, we build a model where the *Number of cars* has a strong effect on *Pollution*, then applied policies can hardly interfere with *Number of cars*, as the number of cars is unlikely to decrease worldwide. It may be more rational to switch from petrol to another type of fuel that might result in a positive effect on mitigating pollution. From the cost perspective, one might also consider whether a change in one independent variable through applied policies is preferred to a change in another independent variable to obtain an equivalent change in the dependent variable. Depending on the problem studied, it might be more cost-effective to prefer a small change to a variable through specific policy actions instead of a larger change to another variable that is more rigid.

6.3.10   Residuals, Residual Plots and Standardized Residuals

Definition

The **residual** (as shown in Eq. 6.2) is the difference of the observed value with the predicted (fitted) value (Figure 6.1B).

    Residual plots are a set of plots used to verify the accuracy of the regression model in relation to the residuals' errors and the regression's assumptions. The most commonly used residual plot is the scatter plot that plots the residuals e of a regression in the y-axis for each fitted (predicted) value $\hat{Y}$ in the x-axis ($\hat{Y}$, e) (Figure 6.4A and B). It is used to test if the residuals' errors have a constant variance (homoscedasticity). Additional residual plot include the normal probability plot of residuals that determine whether they are normally distributed and the histogram of residuals to test if the residuals are skewed or if outliers exist.

    Standardized residuals are residuals divided by their estimated standard deviation. The standardized residual for the *i*-th observation is (6.29):

$$standardized\ residual_i$$

$$= \frac{residual(i)}{estimated\ standard\ deviation\ of\ residual\ (i)} = \frac{e_i}{\sqrt{MSE(1 - h_{ii})}} \tag{6.29}$$

where

    MSE is the mean squared error
    h_{ii} is the leverage value for observation i (see more in Section 6.3.11)
    e_i is the residual of the *i*-th observation

Standardized residuals have a mean of zero and a standard deviation of 1.

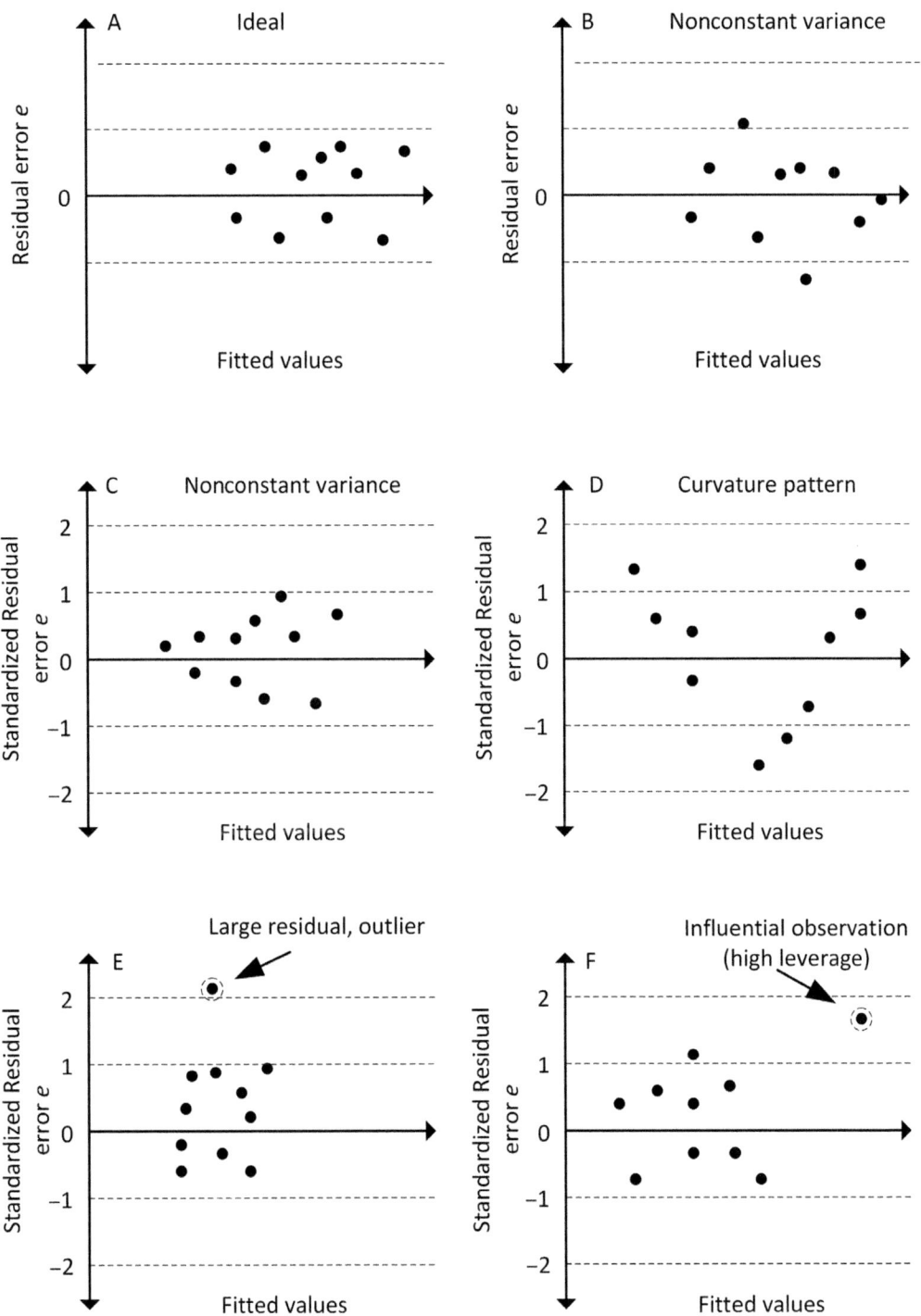

Figure 6.4 (A) Residual plot with randomly dispersed residuals (no pattern). (B) Residual plot with nonconstant variance. (C) Standardized residual plot with nonconstant variance. (D) Standardized residual plot. A curvature pattern in residuals is detected, indicating that a better fit would be a nonlinear model. (E) Standardized residual plot with a large residual (outlier). (F) Standardized residual plot with influential observation.

The **standardized residual plot** is the plot of the fitted value against the standardized residuals (y-axis depicts the number of standard deviations, and the x-axis the fitted values; see Figure 6.4C and D).

Why Use

The residual plot and standardized residual plot are ways to assess the quality of the regression model and identify if the normality or the homoscedasticity assumptions are violated. A standardized residual plot offers an easy way to locate large discrepancies of the dataset. For example, data points that lie further away in the y-axis direction create large residuals that may be potential outliers (see Figure 6.5D). Similarly, data points lying further away from the average fitted values are potential influential points (see Section 6.3.11 and Figure 6.5D).

Interpretation

When residuals are positive, there is underprediction. Underprediction means that the fitted (predicted) value is lower than the observed. On the other hand, overprediction means that the fitted value is higher than the observed. In this case, residuals are negative. If under- or overpredictions occur systematically, then there is bias in the model. To check if bias exists, we use the residual plot.

Ideally, the residual plot should exhibit no particular pattern, and data points should be randomly distributed, have constant variance and be as close to the x-axis as possible (see Figure 6.4A). These conditions satisfy the third (normal distribution of errors) and the fourth (homoscedasticity) assumptions of linear regression (see Section 6.4). If clusters, outliers, or trends exist in the data points pattern, it is an indication of one or more assumptions' violation (see Figure 6.4B–E). If assumptions are severely violated, then the model is unreliable and cannot be used although some statistics may be statistically significant.

By standardizing residuals, we calculate how many standard deviations a residual lies from the mean (mean, in our case, is 0; see Chapter 2 for standardization methods). Standardized residual plots are interpreted like the nonstandardized plots in regards to their shape and variance. Additionally, observations that lie vertically more than ± 2 standard deviations from the x-axis (in the direction of y-axis) might be regarded as outliers (see Figure 6.4E). If we locate an outlier, we should check if there is an error in the dataset or if it is a value revealing unusual behavior.

Discussion and Practical Guidelines

Residuals analysis deals with analyzing the residuals (or the standardized residuals) of a regression model by plotting them against the fitted values. Residual plots, standardized residual plots, scatter plots and other graphs are as important as the outputs of the statistical tests and metrics reported in regression analysis. These graphs provide a graphical representation of inspecting data and regression results concurrently, thus increasing our perception about the model performance. For example, inspecting Figure 6.4, an ideal pattern of residuals would look similar to the one in subplot A. No trends, clusters or outliers are observed. The observed pattern resembles a random point pattern. Additionally, the closer to the x-axis the points lie, the less magnitude of errors e. Subplot B reveals some trend at a diagonal direction (upper left to bottom right). This arrangement indicates a nonconstant variance of errors e, violating the homoscedasticity assumption of regression. Subplots C to E depict the standardized residuals. Subplot C indicates an increasing trend in errors. A linear pattern might still be valid, but we may also test a weighted least squares method instead of OLS. By this method, we assign more weights to the observations with high variability and less to those observations of low variability. An opposite pattern (reflection) would be treated similarly. We can also transform data and then apply OLS again. The pattern in subplot D indicates a curvature pattern. Either we use a nonlinear model or we transform data.

6.3.11 Influential Points: Outliers and High-Leverage Observations

Definition

An **influential point** is any data point that substantially influences the geometry of the regression line (Figure 6.5D).

Outliers are data points for which their response Y lie far away from the other response values (see Figure 6.5A). In cases of regression analysis, we may use the standardized residuals to trace potential outliers, as those data points exceeding more than 2 (sometimes more than 2.5) standard deviations away from the mean (vertical distance of a point from the x-axis; see Figure 6.4E).

Leverage of a data point is a measure of the effect of this point, in respect to its X values, on the regression predictions (see Figure 6.5B and D). It is the value of the i-th diagonal element of the hat matrix H (6.30), where

$$H = X(X^TX)^{-1}X^T \tag{6.30}$$

The diagonal elements satisfy

$$0 \leq h_{ii} \leq 1 \tag{6.31}$$

$$\sum_{i=1}^{n} h_{ii} = p \tag{6.32}$$

where

h_{ii} is the leverage value for observation i
p is the number of parameters (coefficients) in the regression model and
n is the number of observations.

Why Use

Outliers are used to identify extreme Y values. Leverage is used to identify outliers with regard to their X values. An influential point should be identified and potentially removed, as it significantly distorts the geometry of the regression line, leading to inaccurate models.

Interpretation

Those data points that exceed vertically more than 2 (or 2.5) standard deviations away from the x-axis of the standardized residual plot can be considered outliers (see Figure 6.4E). Data points with extreme values in the x-axis direction (or else those lying far away from the center of the input data space) have greater leverage than those close to the center of the input space (see Figure 6.5B).

As a rule of thumb, leverage is large when it exceeds (6.33):

$$2(p + 1)/n \qquad (6.33)$$

where p is the number of parameters (intercept including) and n is the number of observations.

Discussion and Practical Guidelines

An influential point might be both an outlier point (at the y-axis direction of the residual plot or far away from the regression line) and a point with high leverage (extreme value at the x-axis direction; see Figures 6.4F and 6.5D). It is quite common that influential points are those with high leverage. The typical meaning of leverage is the exertion of force using a lever. In cases of simple linear regression, a high-leverage point tends to rotate the regression line as a lever unless it lies very close to the regression line (Figure 6.5B). An outlier lying around the mean of the x-axis values does not create large changes in the geometry of the regression line (see Figure 6.5C).

Similarly, having a point with high leverage that is close to the regression line infers small changes in the slope. The more it deviates from the regression line, the sharper the changes in the regression are. For example, a point lying far right will produce a lower slope if it also lies away from the regression line and, therefore, will be an influential point (see Figure 6.5D). A point with high leverage that lies on the regression line infers no change in the geometry of the line (see Figure 6.5B). In general, we can remove influential objects and test if the model is improved. If the model is considerably improved, then we have to consider dropping these observations. Another method we may apply to control for outliers is the robust regression.

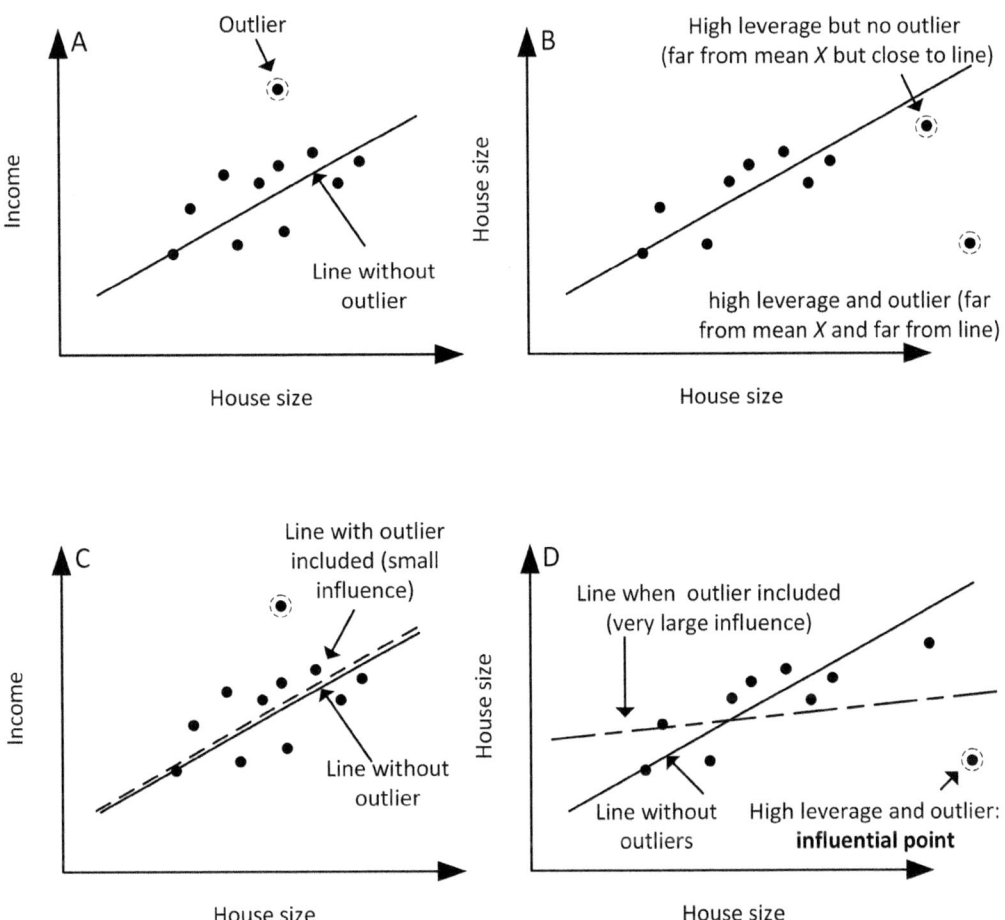

Figure 6.5 Outliers, high-leverage points and influential points. (A) A point far away from the regression line might be an outlier. (B) A point that can change the slope of the regression line is a leverage point. (C) An outlier with small influence on the regression line. (D) A high-leverage point that influences the slope of the regression line.

6.4 Multiple Linear Regression Assumptions: Diagnose and Fix

MLR builds a probabilistic model based on five assumptions (tip to remember: LINE-M):

1. **Linearity**. The relationships between Y and X_i are linear. Regression analysis is based on Pearson's r correlation coefficient, which is appropriate for linear relationships. As such, outliers and influential observations should be considered for exclusion, prior to any other analysis.
    The effect if the linearity assumption is violated (see Table 6.1):

- A large portion of dependent *Y* variation will not be explained (de Vaus 2002 p. 343)
- Important relationships will not be detected

How to diagnose if linearity exists: Linearity assumption can be tested using

- Scatter plots
- Residual plots

What to do if the linearity assumption is violated (fixing nonlinearity):

- Apply transformations
- Use another type of regression not sensitive to linear relationships (e.g., logistic)

2. **Independence – No spatial autocorrelation**. The residuals are independent of each other, meaning that an error *e* in a data point *X* does not affect by any means another error *e* for some different value of the same variable *X*. In other words, errors should be randomly scattered.

   The effect if the independence assumption is violated (see Table 6.1):

   - Inefficient coefficient estimates
   - Biased standard errors
   - Unreliable hypothesis testing
   - Unreliable predictions

   How to diagnose if independence exists:

   - For non–time series data or nonspatial data, residual plots may be used to inspect if errors are randomly scattered.
   - For spatial data, various spatial autocorrelation indices can be used, including Moran's I index, Geary's C index, General G-Statistic (see Chapter 4).

   What to do if independence is violated (fixing error non-independence):

   - For non–time series data or nonspatial data, apply transformations.
   - For spatial data, if residuals are spatially autocorrelated, then we may improve the model by adding or removing variables. We can also use a spatial regression model that controls for spatial autocorrelation – e.g., spatial filtering with eigenvectors (Thayn & Simanis 2013). Geographically weighted regression (GWR; see Section 6.8) can be also tested.

3. **Normality.** The distribution of residual errors *e* at any particular *X* value is normal, having zero mean value. This means that for fixed values of X_i, *Y* has a normal distribution as well.

   The effect if the normality assumption is violated (see Table 6.1):

   - Inefficient coefficient estimates
   - Biased standard errors
   - Unreliable hypothesis testing
   - Unreliable predictions

How to diagnose if normality exists: Assumption can be checked by using

- A histogram and a fitted normal curve of residuals.
- Q-Q plot of standardized residuals.
- Standardized predicted values plotted against standardized residual values. No pattern reveals normality.
- Histogram of standardized residuals. For normality, residuals should approximate a normal distribution.
- Jarque–Berra test. If the value of the test is statistically significant (*p*-value smaller than the significance level), then there is an indication for non-normality. The hypotheses for the Jarque–Berra test are

  Null hypothesis: normal distribution of regression errors
  Alternative hypothesis: non-normal distribution of regression errors (when *p*-value is smaller than the significance level, then we reject the null hypothesis).

What to do if the normality assumption is violated (fixing non-normality):

- Apply transformations

4. **Equality – Homoscedasticity.** Residuals (errors *e*) have a zero mean, equal variance and constant standard deviation at any particular value of *X*. This is also called homoscedasticity and exists when the variance of residuals does not change (increase or decrease) with the fitted values of the dependent variable. Heteroscedasticity, on the other hand, exists when the variance is not constant.

   The effect if the homoscedasticity assumption is violated (see Table 6.1):

   - Inefficient coefficient estimates
   - Biased standard errors
   - Unreliable hypothesis testing
   - Unreliable predictions

   How to diagnose if heteroscedasticity exists: Assumption can be checked through

   - Residual plot inspection. It is a straightforward way to diagnose if heteroscedasticity exists.
   - Statistical tests such as the White test (White 1980), Breusch–Pagan test (Breusch & Pagan 1979) and Koenker (KB) test. If the value of the test is statistically significant (*p*-value smaller than the significance level), then we reject the null hypothesis, and there is an indication for heteroscedasticity. The hypotheses for the aforementioned tests for diagnosing heteroscedasticity are:

     Null hypothesis: constant variance of regression error (homoscedasticity)

Alternative hypothesis: nonconstant variance of regression error (heteroscedasticity)

- If the difference between the Breusch–Pagan test and the Koenker (KB) test values (not the p-values) is large, it is an indication of potential non-normality of the error, as, in the case of normality, the values should be similar. For the implementation of these and additional diagnostic tests, consider LeSage et al. (1998).

What to do if heteroscedasticity exists (fixing non-homoscedasticity):

- Heteroscedasticity often exists due to the skewness of one or more variables. Apply transformations to fix normality, and heteroscedasticity will be fixed to some extent. Often a log transformation is appropriate.
- Use robust standard errors, also called Huber–White sandwich estimators. It is a regression with robust standard errors that estimates the standard errors using the Huber–White sandwich estimators (and is preferred over weighted least squares regression). Regression with robust standard errors is one of the available approaches to perform robust regression. In most statistical software, robust standard errors are automatically calculated. Although coefficients remain the same (with non-robust standard errors), p-values change as standard errors change. In this respect, we should consider the new p-values, as some variables might not be statistically significant anymore.
- Apply a weighted least squares regression (if we know the form of heteroscedasticity).
- Use quantile regression.
- Consider using geographically weighted regression in cases of spatial data.

Although heteroscedasticity is undesirable for regression analysis leading to unreliable models, it reveals interesting patterns for the data. Heteroscedasticity usually occurs when subpopulations exhibit substantial differences in their values. For example, suppose we study *Income* versus *Cost of house*. To some extent, it is expected that people with lower incomes will buy houses with relatively less variability in their price, as their budgets are more rigid. On the other hand, for people with considerably higher income, the price of a house may vary a lot depending on extra criteria (extra variables to the model). To exaggerate a little, a preference (variable) might be the size of the indoor pool; that would never trouble the low-income group. A regression fit line based only on income to model the house cost would most likely yield a residual plot similar to the one in Figure 6.4C. Although it might have a good fit for the low-income values, as income increases, residual errors are expected to increase as well, simply because extra variables should be included to reflect reality better.

Residuals' heteroscedasticity is a typical problem of model misspecification. In practice, we should eliminate heteroscedasticity by adding variables, using robust standard errors or applying more sophisticated methods, such as geographical regression. Still, we have to point out that in a geographical context, heterogeneity reveals valuable information for the data. The variation of a phenomenon from place to place indicates that there are underlying processes at play, and geographical analysis should trace and explain them – not hiding them just because a statistical model should work. Instead of using robust standard errors, which is another mathematical operation applied on data (by weighting them with weights not easily interpreted), it is worth considering clustering data to homogeneous groups or using local regression models. As a general advice, it is better to select methods that are easier to interpret when dealing with geographical analysis instead of using statistical tests that are not easily comprehended in relation to the distortion they infer to the dataset.

5. **Multicollinearity** absence. Variables X_i should be independent of each other, exhibiting no multicollinearity (see Section 6.5).

The preceding assumptions reveal that a model should be appropriately designed to avoid errors. Errors occurring due to poor model design are called misspecification errors, leading to **model misspecification** and to a weak unreliable model. Misspecification may arise when

- Proper variables are omitted
- Wrong variables are included (irrelevant variables' presence)
- Data from included proper variables are not well selected (samples not appropriate or some measurements are wrong)
- Considering a relationship linear when it is not
- Violating the assumptions of linear regression (LINE-M)
- Independent variable is a function of the dependent (e.g., using population as the dependent variable and population density as one of the independent variables)

Table 6.1 presents a list with the assumptions violated, their definition, the effects of these violations, the diagnostic tests and the remedies.

6.5   Multicollinearity

Definition
Multicollinearity exists among two or more variables X_i when they are highly or moderately correlated. No multicollinearity exists when their correlation is absent or very small.

Table 6.1 Assumptions violations, diagnostics and remedies.

Violations	Definition	Effect	Diagnose	Fixing violations
Nonlinearity	The relationships between Y (dependent) and X_i (independent variable) are not linear	• A large portion of the variation of dependent Y will not be explained • Significant relationships remain undetected	• Scatter plots • Residual plots	• Apply transformations • Other types of regression, not sensitive to linear relationships, may be used (e.g., logistic) • Outliers and influential data should be removed
Non independence	Residuals are not independent of one other	• Inefficient coefficient estimates • Biased standard errors • Unreliable hypothesis testing • Unreliable predictions	• Residual plots • Standardized residual plots • For spatial data, various spatial autocorrelation indices such as Moran's I index, Geary's C index, General G-Statistic	• Apply transformations • Improve the model by adding or removing variables • Use a spatial regression model that controls for spatial autocorrelation – e.g., spatial filtering with eigenvectors
Non-normality	The distribution of errors e at any particular X value does not follow the normal distribution	• Inefficient coefficient estimates • Biased standard errors • Unreliable hypothesis testing • Unreliable predictions	• Histogram and a fitted normal curve • Q-Q plot • Jarque-Bera test • Standardized predicted values plotted against standardized residual values • Histogram of standardized residuals	• Apply transformations
Heteroscedasticity	Nonconstant errors variance	• Inefficient coefficient	• Residual plots • White test	• Apply transformations

Issue	Description	Effects	Diagnostic tests	Remedies
		estimates Biased standard errors Unreliable hypothesis testing • Unreliable predictions	• Breusch–Pagan test • Koenker (KB) test	• Weighted least squares regression • Robust standard errors (Huber-White sandwich estimators) • Use quantile regression • Use geographically weighted regression when dealing with spatial data
Multicollinearity	When variables X_i in a dataset have high correlations among them	• Sensitive coefficient estimates • Coefficients are inflated • Not reliable confidence intervals and predictions • Insignificant independent variables might appear as significant	• Bivariate or pairwise correlation • Diagnostic statistics such as singular value decomposition, condition index (CI) • Variance decomposition proportions, variance inflator factor (VIF)	• Remove one of the variables that are highly correlated • Reduce the total number of variables by using principal component analysis • Use stepwise regression or exploratory regression
Model misspecification/ Overall performance low	Misspecification arises when the model is not well designed	• Not accurate model	Adjusted R-squared F-statisticJoint Wald's test	• Add omitted variables • Remove redundant variables • Apply transformations • Change model

Why Use

The absence of multicollinearity among independent variables is one of the five assumptions that should not be violated in order for the regression model to be reliable. If multicollinearity exists, then (Rogerson 2001 p. 126, de Vaus 2002 p. 343):

- Estimates of the coefficients are sensitive to individual observations, meaning that if we add or delete some observations, coefficients' values may change significantly.
- The variance of the coefficients estimates is inflated (increases).
- Significance levels, confidence intervals and prediction intervals are not reliable and are also wider.
- Insignificant independent variables might appear to be significant due to the large variability in coefficients.

Eliminating or reducing the multicollinearity in a dataset yields more accurate results, and more reliable (and less wide) confidence intervals for the coefficients.

Interpretation

Diagnostic statistics for detecting multicollinearity include the variance inflator factor (VIF), the condition index (CI) and the variance decomposition proportions (Belsley et al. 1980). Table 6.2 presents the values that multicollinearity is evident based on these diagnostics.

Discussion and Practical Guidelines

- **Variance inflation factor (VIF)** is a measure that estimates how much the variance of a coefficient is inflated due to multicollinearity existence (Rawlings et al. 1998). VIF values between 4–10 indicate increased multicollinearity, while high values VIF ($>$10) are a sign of severe collinearity (Table 6.2). For example, if VIF for an independent variable is 7.2, it means that the variance of the estimated coefficient is inflated by a factor of 7.2 as this independent variable is highly correlated with at least one of the other independent variables in the model.
- **Condition index (CI) and variance decomposition proportions** are examined based on the following conditions (Belsley et al. 1980):
  1. A large condition index associated with

Table 6.2 Multicollinearity diagnostics.

VIF	Collinearity	Condition index (CI)	Collinearity	Variance decomposition	Collinearity
1–4	No	$5 < CI < 30$	Weak	<0.5	Weak
4–10	Further investigation needed	$30 < CI < 100$	Moderate	>0.5	Severe
>10	Severe	$CI > 100$	Severe		

2.   A large variance decomposition proportion for two or more covariates

   If both conditions are true, then multicollinearity might be present in the dataset. The variance decomposition index is checked when the condition index is large. Put simply, when the condition index is larger than 30 (condition 1), then there may be multicollinearity issues. In this case, we inspect the variance decomposition index. If a large CI is associated with two or more variables with a large variance decomposition index (over 0.5) (condition 2), then it is a sign of multicollinearity in the corresponding variables.

- **Bivariate or pairwise correlation** between the independent variables. Very high correlations (larger than 0.90) will produce collinearity problems. Correlation analysis has been presented in Section 2.3.

In case we detect multicollinearity, we may consider (see Table 6.1):

- Removing one of the highly correlated variables. Deciding which variable to delete strongly depends on the problem at hand. We should keep the variable that is conceptually more evidently linked to the dependent variable. In this respect, the choice should be based primarily on the assumed underlying processes and less on the magnitude of the multicollinearity metric.
- Reducing the total number of variables by using principal component analysis or by creating a composite index (variables combined in a new single variable).
- Applying exploratory regression, stepwise regression or hierarchical regression, which allows for controlling the order of variables entry.

6.6 Worked Example: Simple and Multiple Linear Regression

Suppose we want to model Income (dependent) to Medical expenses (independent) for a set of 64 postcodes (see Box 6.1). Both variables are measured in euros, and values refer to the average income and the average medical expenses of the people living in each postcode. The results of the OLS simple linear regression are:

Metrics and Statistics for Simple Linear Regression

Box 6.1 Matlab. You can find data, Matalab commands and related files to reproduce the following graphs and results in I:\BookLabs\Lab6\Matlab\

Run SimpleOLS.m

```
*******************************RESULTS*****************************
Linear regression model:

    Income ~ 1 + MedExpenses
Estimated Coefficients:
                     Estimate        SE       tStat         pValue

                     _____      _____     _____      _____

    (Intercept)      7446.8        365.84     20.355      3.823e-29
    MedExpenses      46.569        3.0092     15.476      5.7356e-23
Number of observations: 64, Error degrees of freedom: 62
Standard Error: Root Mean Squared Error: 897
```

Tip: Calculated based on eq.6.25 as: Square Root
(SumSqResidual/(n-2))=>(4.9866e+07/(64-2))^-0.5=896.82   {(n-2)
is the Degrees of freedom (DF)}

```
R-squared: 0.794, Adjusted R-Squared 0.791
F-statistic vs. constant model: 239, p-value = 5.74e-23
Linear regression model:
    Income = 7446.8 + 46.569*MedExpenses
ANOVA Results
                SumSq          DF       MeanSq        F         pValue

                _____     __       _____    _____     _____

    Total       2.4249e+08     63       3.8491e+06
    Model       1.9263e+08      1       1.9263e+08    239.5     5.7356e-23
    Residual    4.9866e+07     62       8.0429e+05

*******************************************************************
```

Interpreting Results

The adjusted R-squared is high, indicating goodness of fit (0.791) and that 79% of Income's variation is explained by MedExpenses variable (Figure 6.6). As the F-significance is less than 0.05 (F-statistic p-value = 5.74e-23), we reject the null hypothesis and accept that the regression model is statistically significant, and the identified trend is a real effect, not due to random sampling of the population. In addition, MedExpenses coefficient's p-value, calculated through t-statistic, is 5.7356e-23 (less than 0.05). As such, we accept the coefficient value as statistically significant for the model. In other words, MedExpenses independent variable is statistically significant and can be included in the model as a good predictor of Income. The regression of income on medical expenses is Income = 7446.8 + 46.569*MedExpenses (Figure 6.6). It practically means that for a one-unit increase in medical expenses (i.e., 1 euro), income increases by b = 46.569 euros. The standard error of the regression is 897 euros, expressing the typical distance that the data points fall from the

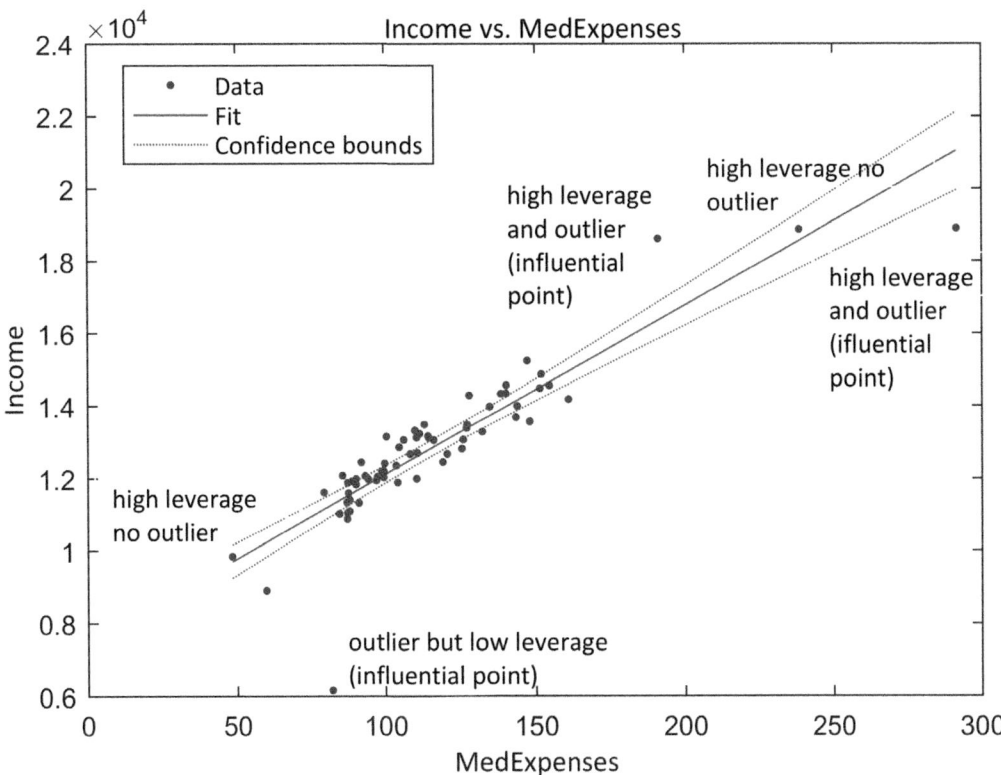

Figure 6.6 Plot observations, regression line and confidence interval for mean (95%). Leverage points and outliers are also mapped.

regression line on average. It also expresses how precise the model's predictions are using the units of the dependent variable. As the income values in the dataset range between 10,000 and 22,000 (Figure 6.6), we consider this error relatively low (although it could be lower; potential outliers might have a negative effect). We should also test (later) our model for homoscedasticity, errors normality and influential points.

Standardized Residual Plots
The standardized residual plot is depicted in Figure 6.7.

Interpretation
The residual plot (Figure 6.7) shows constant variance in errors indicating that there is not a serious heteroscedasticity issue. High leverage points (points with leverage value larger than 0.0625; see Eq. 6.33), along with outliers and influential points, are depicted in Figures 6.6 and 6.7. We should eliminate those points one by one and run the model again. If the model is improved, we may consider dropping these observations and keeping the new model.

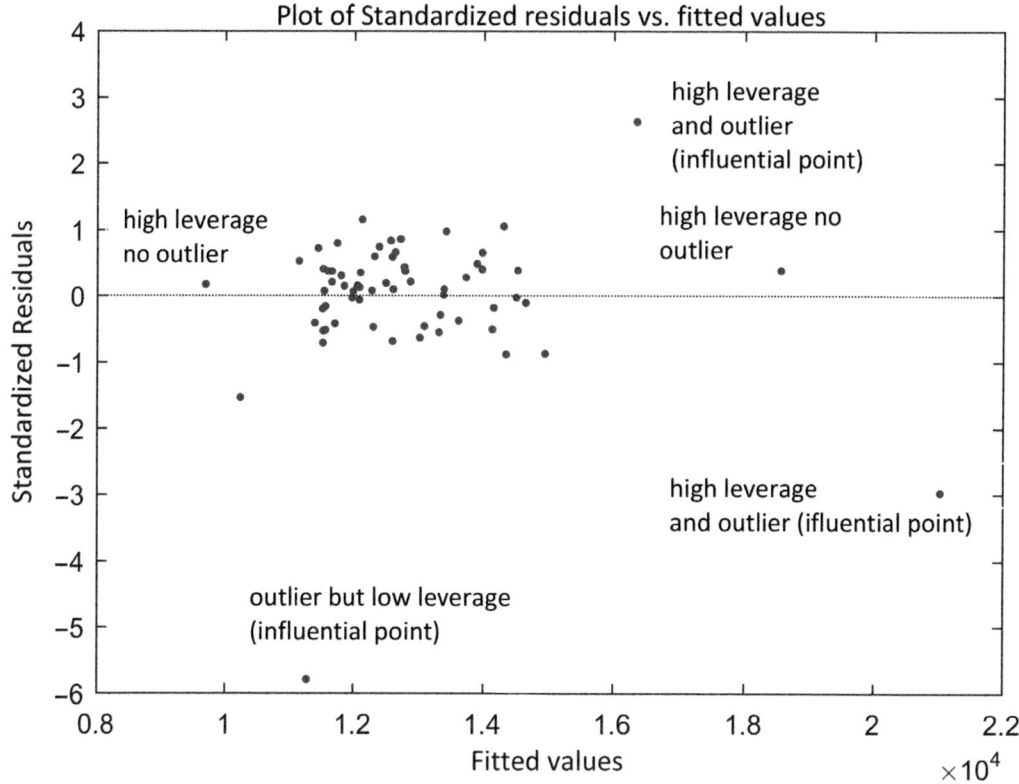

Figure 6.7 Standardized residuals vs. fitted values along with influential points. The lower-left point has low leverage because it is relatively close to the mean fitted value. Still, it is an outlier as it lies vertically from the x-axis more than 2.5 standard deviations. The lower-right point, on the other hand, is both outlier and high leverage point, as it is away from the mean fitted value and lies vertically from the x-axis more than 2.5 standard deviations. The upper-right point is for the same reasons both outlier and high leverage. Finally, the upper-left point is of high leverage but not an outlier.

Multiple Linear Regression

Continuing on the same example, suppose we want to model Income (dependent variable) as a function of the following independent variables (see Box 6.2):

- Sec: percentage of people obtained secondary education
- Unv: percentage of people that graduated from university
- Med: medical expenses per month in euros
- Ins: money spent on monthly insurance in euros
- Ren: monthly rent in euros

Multicollinearity

Before we apply regression modeling, we test if multicollinearity exists using the Belsley collinearity diagnostics.

> **Box 6.2 Matlab.** You can find data, Matalab commands and related files to reproduce the following graphs and results in I:\BookLabs\Lab6\Matlab\
>
> Run MLR.m

```
*****************************RESULTS*****************************

                    |      Variance Decomposition          |
 sValue    condIdx     Sec     Unv      Med      Ins      Ren
---------------------------------------------------------------
 2.2000        1     0.0013  0.0029   0.0004   0.0003   0.0014
 0.3118    7.0561    0.0186  0.4484   0.0144   0.0101   0.0253
 0.1886   11.6618    0.4080  0.2966   0.0666   0.0078   0.1281
 0.1510   14.5684    0.3764  0.1439   0.0174   0.0232   0.8339
 0.0677   32.4967    0.1956  0.1083   0.9012   0.9586   0.0113

***************************************************************
```

Results show that only the last row has condition index larger than the default tolerance value 30. Inspecting this row, the third (Med) and fourth (Ins) independent variables have variance-decomposition proportions exceeding the default tolerance 0.5, suggesting that these variables exhibit multicollinearity.

The pairwise correlation for all variables and the related correlation matrix are as follows (see Figure 6.8):

```
*****************************RESULTS*****************************

 R =     Sec       Unv       Med       Ins       Ren
 Sec   1.0000    0.7630    0.3655    0.3560    0.4371
 Unv   0.7630    1.0000    0.3281    0.2563    0.2609
 Med   0.3655    0.3281    1.0000    0.9372    0.7151
 Ins   0.3560    0.2563    0.9372    1.0000    0.7399
 Ren   0.4371    0.2609    0.7151    0.7399    1.0000
 Pvalue =
         1.0000    0.0000    0.0030    0.0039    0.0003
         0.0000    1.0000    0.0081    0.0409    0.0373
         0.0030    0.0081    1.0000    0.0000    0.0000
         0.0039    0.0409    0.0000    1.0000    0.0000
         0.0003    0.0373    0.0000    0.0000    1.0000

***************************************************************
```

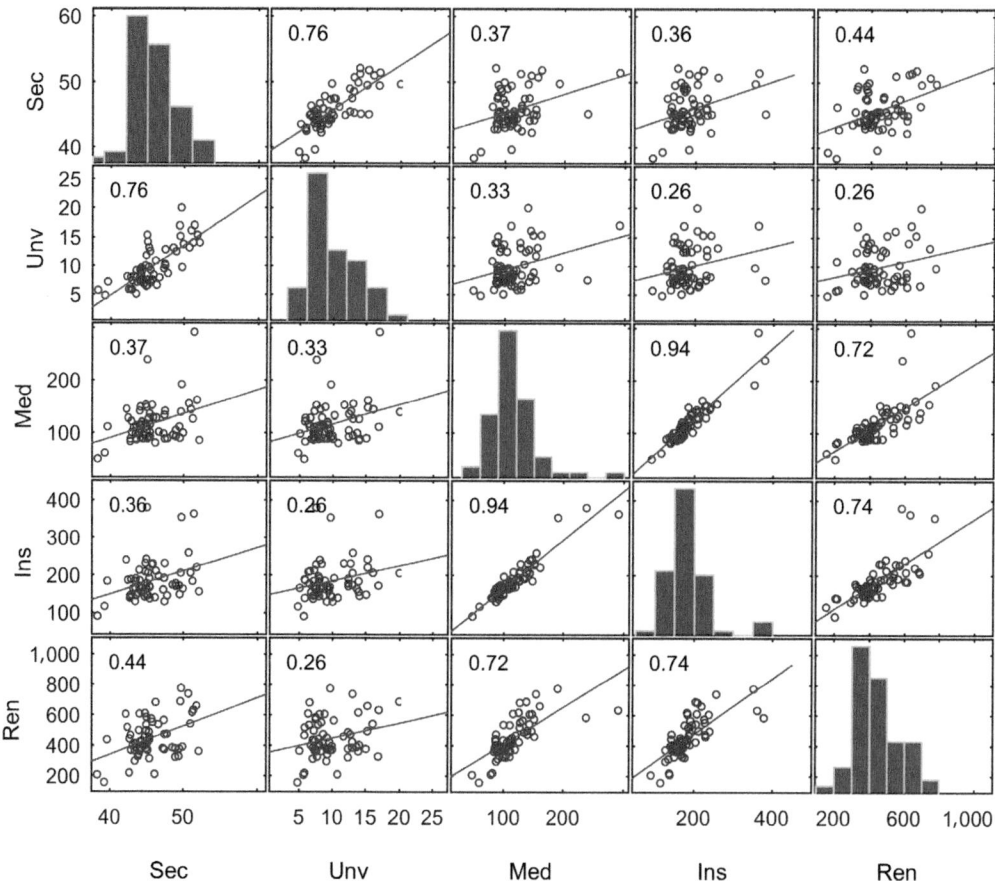

Figure 6.8 Correlation matrix.

Although correlations around 0.70 exist, we consider variables exhibiting serious collinearity issues as those exceeding correlation of 0.90. Med and Ins have a high pairwise correlation (0.94), indicating multicollinearity. We also calculate the variance inflation factor.

```
*****************************RESULTS*****************************

        VIF =
            2.8929

            2.6859

            9.1455

            9.4201

            2.4803

*****************************************************************
```

A high VIF value (close to 10) for the third and fourth variable (Med, Ins) is identified, a sign of multicollinearity existence. VIF shows how much the variance of the coefficient of an independent variable is inflated, but it does not point directly to which variable it exhibits multicollinearity. We should turn to the pairwise correlation and the variance decomposition index calculated earlier, to conclude that Med and Ins exhibit multicollinearity. We should continue by running the MLR but without including Ins.

```
*******************************RESULTS***************************
Linear regression model:
    Income ~ 1 + Sec + Unv + Med + Ren
Estimated Coefficients:
                  Estimate        SE          tStat          pValue

                  _____      _____     _____      _____

    (Intercept)    9034.2       2299.7        3.9284       0.00022677
    Sec            -56.92       60.785       -0.93642      0.35287
    Unv             34.14       50.989        0.66957      0.50575
    Med            38.072       4.2493        8.9594       1.3349e-12
    Ren            3.7421       1.2923        2.8958       0.0052968

Number of observations: 64, Error degrees of freedom: 59
Root Mean Squared Error: 860
```

Tip: Calculated based on eq.6.24 as: Square Root (SumSqResidual/ (n-p))=> (4.3653e+07/(64-5))^-0.5=860.16   {(n-p) is the Degrees of freedom (DF)}

```
R-squared: 0.82,  Adjusted R-Squared 0.808
F-statistic vs. constant model: 67.2, p-value = 2.71e-21
                  SumSq         DF       MeanSq          F          pValue

                _____        __     _____     _____      _____

    Total       2.4249e+08      63     3.8491e+06
    Model       1.9884e+08      4      4.971e+07      67.187      2.709e-21
    Residual    4.3653e+07      59     7.3988e+05
*****************************************************************
```

Coefficient's p-values for variables Sec (p-value = 0.35287) and Unv (p-value = 0.50575) are not statistically significant, indicating that we should remove these variables from the model. Only Ren and Med are statistically significant and can be included in the model, with p-values 0.0052968 and 1.3349e-12, respectively.

We run the model again with only Ren and Med, and we calculate the unstandardized and the standardized (beta) coefficients.

```
******************************RESULTS******************************
Linear regression model:
    Income ~ 1 + Med + Ren
Estimated Coefficients:
                      Estimate        SE        tStat          pValue

                      _____      _____     _____       _____

      (Intercept)      6912.6       397.63     17.385       2.5899e-25
      Med              38.473       4.0907      9.405       1.7656e-13
      Ren               3.312       1.1967      2.7676       0.0074669

Number of observations: 64, Error degrees of freedom: 61
Root Mean Squared Error: 852
R-squared: 0.817, Adjusted R-Squared 0.811
F-statistic vs. constant model: 136, p-value = 3.04e-23


Standardized coefficients (beta coefficients):
Linear regression model:
    Income ~ 1 + Med + Ren

Estimated Standardized Coefficients:
                    Estimate          SE          tStat          pValue

                   _____     _____    _____     _____

(Intercept)     4.6221e-16      0.054298     8.5124e-15              1
Med              0.73632        0.078291         9.405       1.7656e-13
Ren              0.21668        0.078291         2.7676       0.0074669

Number of observations: 64, Error degrees of freedom: 61
Root Mean Squared Error: 0.434
R-squared: 0.817, Adjusted R-Squared 0.811
F-statistic vs. constant model: 136, p-value = 3.04e-23
******************************************************************
```

Interpretation

Adjusted R-squared is 0.811, meaning that 81.1% of the total variation of Income is explained by Med and Ren independent variables. It slightly improves the simple OLS model with MedExpenses (adjusted R-squared = 0.791). The higher adjusted R-squared typically means that adding Ren to the model (on top of Med) contributed by 2% in explaining the total variation if Income.

As the F-significance is small (3.04e-23 less than 0.05), we reject the null hypothesis. We accept that the regression model is statistically significant and that the trend is a real effect, not due to random sampling of the population. Coefficient's p-values are statistical significant (Med:1.7656e-13

Ren:0.007466). The Medical expenses variable is more important (has more effect on income) compared to Ren as its beta (b_1 = 0.73632) is more than triple than the Ren beta (b_2 = 0.21668).

The purpose of the preceding model is to identify which variables are linked to income and which are not. Education attainment seems to be irrelevant to the income, as neither related variables (Sec, Unv) are statistically significant. We also quantified how much of the total variation of income is explained by including medical expenses and rent along with the effect of each statistically significant variable to the model. The higher the beta value, the more critical the variable is. With this single-step approach (all data analyzed in one step), we explore linkages, not causal relationships. The main conclusion is that a large portion of income's variation (81.1%) is linked to/explained by/dependent on first medical expenses (higher beta) and then monthly rent expenses. Looking at the unstandardized coefficients, keeping rent stable, a 100-unit increase in medical expenses leads to 3,847.3 euros extra annual income. Be careful here. As previously mentioned, MLR at one step does not explore causations. The preceding finding does not mean that by spending more money on medical expenses, one will increase his income (i.e., get a raise in salary). It is not a cause-and-effect relationship. It just links medical expenses to income. It is more rational to assume that because one has more money, he/she can afford to pay more for medical expenses.

Such type of models can be used when we need to estimate a variable, but we cannot accurately measure it. In a survey, for example, asking about personal income might be inappropriate. The question is unlikely to receive an honest answer, as people seldomly reveal their actual income. Having the preceding model at hand, it would be more rational to ask about medical expenses, as we anticipate to get an accurate response. Plugging this value into the model would reveal a good estimate of the actual income. Another example is establishing the relation of the house size to the owner's income through the linear regression model. By using satellite images, we can efficiently estimate the size of a house and then the potential income of its owner without having to conduct a costly survey. Similar regression models are also used to assess the population in areas that censuses are rare or not accurately updated. By utilizing satellite images to measure the size of settlements, estimations can be made for the total living population.

We should also mention that the preceding findings are statistically significant only for the specific dataset. Conclusions drawn are valid only for the specific case study and do not necessarily hold true elsewhere. In this respect, when we report results, we should avoid generalization as well as the temptation to refer to causations when we have not carried out such analysis. It is always advisable to seek other similar studies for the same case study to explore if common findings exist. Comparisons with other geographical regions can be made mostly for discussing the outcomes but not for approving or rejecting the results of these studies.

6.7   Exploratory Regression

Definition
Exploratory regression is a process assisting in the appropriate selection of an OLS model when the number of independent variables is large.

Why Use
Exploratory regression is used to automatically test all possible combinations of different OLS models when many independent variables are available. In the previous sections, we handled a handful of variables to showcase how regression works. As the number of available independent variables increases, the process of selecting the most appropriate model becomes tedious. For example, for 10 independent variables to run all possible combinations (all models with just one variable, all models with two variables and so on up to a final single model that will include all variables), it involves creating 1,023 different regression models. It is a combination problem of selecting r by n with final variables' order not important but with repetition not allowed (a variable cannot participate more than once in the same model). An additional independent variable would increase the models to be tested to 2,047. Exploratory regression is used to take this burden and run all models' combinations to find the most appropriate one.

Interpretation
For a thorough analysis of how exploratory regression outputs are interpreted, see Exercise 6.1.

Discussion and Practical Guidelines
Exploratory regression is a data-mining tool and not a new regression technique, as it just applies OLS regression to many different models, finally to select those that pass all necessary OLS diagnostic tests. By testing all available models:

- Chances of finding the optimal regression modal are increased.
- Hidden independent variables that were not theoretically associated with dependent variable might be traced.
- Independent variables that are statistically significant in the majority of the tested models are more likley to be associated with the dependent variable.

To evaluate the results of exploratory regression and test whether the regression assumptions are violated, we apply the following tests (see Table 6.3):

- The Jarque–Bera test for residuals normality as proof of not biased residuals.
- The Breusch–Pagan test for checking residual errors' heteroscedasticity.
- The Moran's I test to identify if residuals errors are spatially autocorrelated.

Table 6.3 Diagnostic tests used in exploratory regression. Keep in mind that in these tests the desired result is not to reject the null hypothesis. In other words, we prefer p-values larger than the significance levels. For example, if the significance level is 0.05, and the result for the Jarque–Bera test is 0.01, then we reject the null hypothesis and we accept that the residual errors do not follow a normal distribution.

Name of diagnostic test	Detects	Hypothesis tested (When p-value is smaller than the significance level, we reject the null hypothesis.)
Jarque-Bera test	Non normality	**Null hypothesis:** normal distribution of residuals errors **Alternative hypothesis:** non-normal distribution of regression errors
Breusch-Pagan test	Heteroscedasticity	**Null hypothesis:** constant variance of residuals error (homoscedasticity) **Alternative hypothesis:** nonconstant variance of regression error (heteroscedasticity)
Moran's I	Spatial autocorrelation	**Null hypothesis:** there is no spatial autocorrelation in residuals errors **Alternative hypothesis:** there is spatial autocorrelation in residual errors

Even with modern-day excessive computer power, it is still not wise to run all possible combinations. Not only does computational time increase but results are hard to interpret. For example, models with more than 10 variables would be too complex to interpret. A function of many variables is inappropriate because it is difficult to comprehend how they are meaningfully interconnected with the dependent variable. For this reason, we can set the maximum number of independent variables, no matter how many variables are available, in order to expedite the overall process and reach more straightforward conclusions.

Additionally, we may set the following parameters:

- The threshold of adjusted R-squared accepted, as proof of goodness of fit.
- The p-values that coefficients are statistically significant as proof of variables' significance (variables with a justifiable relationship with the dependent variable).
- The VIF threshold value to account for multicollinearity as proof of non-redundant variables (no multicollinearity).
- The p-value for the Jarque–Bera test for residuals' normality, as proof of unbiased residuals.
- The p-value for the Global Moran's I tool on model's residuals, as a proof of no spatial autocorrelation.

From the practical perceptive, exploratory regression is similar to stepwise regression, but its evaluation is not only based on the change of the adjusted R-squared. In exploratory regression, all common tests and diagnostics are used to evaluate the performance of the model, adding more reliability and also offering better insight into the model structure and the associated data.

Moreover, a stepwise regression or a regression defined by the researcher is biased on the order that variables are entered or removed from the model. In contrast, exploratory regression is not based on an initial model specification alerting it by each run, but it creates new models that are self-evaluated each time. In this respect, exploratory regression is more powerful than stepwise regression.

Still, there is some controversy in using the exploratory regression, depending on the way we approach a scientific (geographical in our case) problem. There are roughly two approaches, namely the theoretically based approach and the data-mining approach:

- Theoretically based approach: In this approach, researchers build their assumptions based on theoretical evidence before exploring available data. They conceptually form a model by identifying the variables, backed up theoretically, and then adopt a model to solve the problem. The extent of researching additional variables is small. For example, when we model the variable *House price*, the theoretical approach would mainly seek relationships among variables – such as location, size or income – that have been reported to have links to a house's value. In the case of exploratory regression analysis, though, models may be over-fitted, as the more variables included, the more unstable the model becomes. A second aspect that the theoretically based approach, as opposed to the data-mining approach, is that regression statistics are heavily based on probabilities expressed by the resulting p-values. A 95% confidence interval means, for example, that a coefficient may be statistically significant, but there is a 5% chance that it is not significant. In other words, we may reject the null hypothesis when, in fact, we shouldn't (Type I error, see Section 2.5.5). In case of exploratory regression, where we test hundred or even thousand models, Type I error is more common. A smaller p-value (e.g., 0.01) is a potential solution, but, again, the problem remains.
- Data-mining approach: Those adhering the data-mining approach advocate that in many complex problems, theories do not provide all necessary background to shape a rigid model. In this respect, experimenting with additional variables might lead to hidden patterns and relations in the dataset that should be further investigated.

Although the pitfalls of exploratory regression are acknowledged, a fusion of both approaches is suggested in this book. It is more rational to move on solid theoretical foundations. Still, analysts/researchers should think outside the box, as doubting is the key to discovering knowledge. Exploratory regression is a good starting point to analyze data and discover potentially hidden patterns. With the knowledge gained regarding variables' associations, we may move forward to deploy other, more advanced models if necessary, including spatial econometric models, spatial regime models and geographically weighted regression.

It is always advised to select variables that are supported by theory, experts' knowledge or common sense. Statistically significant variables that are not supported by any of these approaches are either useless or a significant discovery. However, if it is a significant discovery, it has to be excessively studied before findings announced. For example, suppose we model human health as a function of variables related to monthly groceries purchases. Through exploratory regression analysis, we might discover that the *variable Canned dog food* (i.e., number of cans containing food for our dog) is statistically significant to the model with a positive coefficient. A first interpretation of this model is, "The more canned dog food you buy, the better your health is." A more exaggerated statement would be, "Eating dog food improves your health." Although this is obviously misleading, it's quite common that research results are misinterpreted or lack some theoretical grounds. Common sense suggests that you buy this type of food if you have a dog. If you have a dog, you probably walk the dog daily. You exercise a lot, so it makes more sense why your health is better (see more on cause and effects in Sections 2.3.4 and 6.2.1).

In more complex problems, where common sense or knowledge cannot straightforwardly guide us on whether to reject an "astonishing" finding or not, we might go completely out of the way. It's common for posts on websites and social media or in the news to be quite startling (e.g., "People that read more live longer," "People who eat burgers are wiser," or "People who eat canned dog food are healthier."), as their primary objective is to gain attention. But be aware of the underlying assumptions, methods and data used, as well as the study's overall approach.

6.8   Geographically Weighted Regression

Definition
Geographically weighted regression (GWR) is a local form of linear regression model used to detect heterogeneity and analyze spatially varying relationships (Fotheringham et al. 2002). GWR provides a local model of the dependent variable allowing for coefficients to vary across space by fitting a weighted linear regression model to every spatial object in the dataset (O'Sullivan & Unwin 2010 p. 228). GWR should obey the same assumptions as multiple linear regression (LINE-M).

The model can be expressed as (6.34) (Wang 2014 p. 178):

$$y_i = \beta_{0i} + \beta_{1i}x_{1i} + \beta_{2i}x_{2i} + \cdots + \beta_{mi}x_{mi} + \varepsilon_i \tag{6.34}$$

where $i = 1, ..n$ with n as the total number of locations/observations

y_i is the value of the dependent variable Y at location i

x_{mi} is the observation of the variable X_m at location i

β_{0i} is the intercept of the model at location i

$\beta_{1i}, \beta_{2i}, \ldots \beta_{mi}$ is the coefficient set at location i

m is the total number of independent variables

ε_i is the error at location i

GWR estimates the coefficients β_i at each location i such as (Wheeler & Páez 2010 p. 462) (6.35):

$$\beta_i = (X^T W_i X)^{-1} X^T W_i Y \tag{6.35}$$

where

Y is the n-by-1 vector of the dependent variable

X is the design matrix of the independent variables (containing a leading column of ones for the intercept)

W_i is the n-by-n diagonal matrix with diagonal elements being the weights w_{ij} at location i (6.36).

$$W_i = \begin{bmatrix} w_{i1} & 0 & \ldots & 0 \\ 0 & w_{i2} & \ldots & 0 \\ \ldots & \ldots & w_{ij} & \ldots \\ 0 & \ldots & \ldots & w_{in} \end{bmatrix} \tag{6.36}$$

w_{ij} is the weight between the regression location i and the location j. w_{ij} takes a value between [0, 1]. The weight matrix W_i has to be calculated prior to local regression coefficients' estimation. There are numerous weighting schemes that can be used, such as the binary scheme (Brunsdon et al. 1996) (6.37):

$$w_{ij} = \begin{cases} 1 \ if \ d_{ij} \leq d \\ 0 \ if \ d_{ij} > d \end{cases} \tag{6.37}$$

In this case, if an object j lies less than a distance threshold value d from i, it gets a weight of 1. If it lies further away than this threshold value, it gets a zero weight. Spatial kernels are most commonly used on the calculation of weights, as shown in the next section.

Why Use

GWR is used to better handle spatial heterogeneity by allowing the regression coefficients (and, consequently, the relationships among the variables), to vary from place to place, in a similar way that that variables' values vary in space (Wheeler et el. 2010 p. 461). Instead of having a global intercept and global coefficients, local values are estimated to better reflect the spatial heterogeneity of the phenomenon studied.

6.8.1   Spatial Kernel Types

In practice, a spatial kernel is used to provide the geographic weighting for the GWR method and should be selected before calibrating a GWR model. Objects closer to the location i are weighted more than those further away, based on the assumption of spatial autocorrelation. There are two kernel types (functions) that are most commonly used in the context of GWR, namely fixed and adaptive.

- Fixed: The Gaussian kernel function (a monotonically decreasing function) is used to calculate local weights and is based on a fixed non-varying bandwidth (h) distance (6.38):

$$w_{ij} = \exp\left(-\frac{1}{2}\left(\frac{d_{ij}}{h}\right)^2 \right) \tag{6.38}$$

where d_{ij} is the distance between locations i and j, and h is a kernel bandwidth parameter that controls the decay. The weighting between locations i and j will decrease according to the Gaussian curve, as the distance between the two points increases (Brunsdon et al. 1996). While the bandwidth is fixed, a different number of spatial objects might be used for calculating the weights for different locations. For large-bandwidth values the gradient of the kernel becomes less steep, and as such, more locations/observations are included in the local estimations. Bandwidth h has to be optimally selected for better GWR performance (as explained in next section).

- Adaptive: The bi-square kernel function used to calculate local weights is based on the number N of nearest neighbors. In this case, the same number of spatial objects is used to calculate the weights for each location i (6.39) (Wheeler et al. 2010 p. 464).

$$w_{ij} = \begin{cases} \left(1 - \dfrac{d_{ij}^2}{d_{iN}^2}\right)^2 & \text{if } j \text{ is one of the } N\text{-th nearest points of } i \\ 0 & \text{otherwise} \end{cases} \tag{6.39}$$

Where d_{iN} is the distance of location i from the N-th nearest neighbor (for example, if we select five nearest neighbors, then $N = 5$). In this case, we have to find the optimal number N of nearest neighbors

The choice of the kernel type depends on the problem and on the spatial distribution of the objects. In cases where objects are evenly distributed following a dispersed spatial point pattern, a fixed kernel type is a reasonable selection. When spatial objects are clustered (also revealing spatial autocorrelation), an adaptive kernel type is more appropriate. Adaptive kernel is most widely used, as spatial autocorrelation is evident in most geographical problems.

6.8.2   Bandwidth

The bandwidth of the kernel type – that is, the unknown parameter h in the fixed kernel type – and the number of nearest neighbors N in the adoptive kernel type define the size of each kernel. There are three main methods for bandwidth selection/estimation:

- **AICc method.** The corrected AIC estimates the bandwidth, which minimizes the AICc value.
- **Cross-validation (CV) method.** The CV method calculates the bandwidth that minimizes the cross-validation score.
- **User-defined method.** The distance or the number of neighbors that will be used to define the kernels are selected by the researcher.

From the three methods, the last one is not automated and does not guarantee any minimization. As GWR heavily relies on the weights' matrix, it is crucial to define this matrix most appropriately. The optimal bandwidth is calculated to keep a balance between bias and variance. A small bandwidth leads to considerable variance in local estimates, while a large bandwidth leads to a significant bias in the local estimates (Fotheringham et al. 2002). The first two methods automatically esimate the bandwidth value based on well-defined criteria and are preferred to the user-defined bandwidth method. Cross-validation method and AICc method are commonly used (Fotheringham et al. 2002, Wheeler et al. 2010 p. 466).

6.8.3   Interpreting GWR Results and Practical Guidelines

GWR results can be evaluated and interpreted in three ways: (A) by using common statistical diagnostics, (B) by mapping standardized residuals and (C) by mapping local coefficients.

(A)   Common Diagnostics

A GWR model is firstly compared to an OLS model. The OLS is also called the global model, as it provides a single model for the entire study area. In this respect, GWR is referred to as the local model. Typically, we start by comparing adjusted R-squared, AICc and other typical metrics between the local and the global model. Issues of multicollinearity should be mentioned in cases of condition index values larger than 30 (see Lab Exercise 6.3). The main advantage of GWR is that it provides local estimates that can be mapped and further analyzed through spatial statistics.

(B)   Mapping Standardized Residuals (Output Feature Class)
Mapping the standardized residuals allows tracing the following:

1. **High or low residuals**. Residuals reveal how much spatial data variations are not explained by the independent variables (Krivoruchko 2011 p. 485). In case that standardized residuals are relatively high, then we may locate under- or overpredictions. Underprediction means that the fitted value is lower than the observed. In this case, residuals are positive. Overprediction means that the fitted value is higher than the observed. In this case, residuals are negative. If under- or overpredictions occur systematically, then there is bias in the model. From the geographical analysis perspective, it is necessary to delve deeper if we identify extremely low or high residuals, as underlying processes at play might explain their presence. For example, suppose we study humans' respiratory health, and after using GWR, we locate extremely large residuals in a region. These extreme values may be an indication of a pollution source (e.g., factory, or industry not conforming with regulations).

2. **Spatial autocorrelation.** We should run a spatial autocorrelation test to identify if spatial autocorrelation exists in residuals. If there is not, then GWR removed the potential spatial autocorrelation existing in the global model. If spatial autocorrelation remains and clusters of over- and/or underpredicted values are formed, this is a sign of an ineffective GWR model, possibly due to omitting one or more independent variables (model misspecification). On the other hand, if we locate clusters of residuals, we should analyze why this clustering occurs. For example, there might be some specific regulations that apply to these areas and not to the others, and as such, local regression models are misspecified. Keep in mind that our ultimate scope is to perform geographical analysis and trace trends and spatial processes that lead to meaningful results and not always to create models that perform at a near-optimally predictive way. Although a model with high predictive performance is sometimes desired, problem complexity, model misspecification and wrong model adoption might drive us away from building a robust model. Still, conclusions can always be drawn if we interpret findings based on the identified relationships, even if the predictive model is weak.

(C)   Mapping Local Coefficients (Coefficient Raster Surfaces)
Mapping local coefficients (through a raster surface for points or by polygon rendering, in cases of polygons) shows the variation in the coefficient estimation for each independent variable. Moreover, it shows how much each independent variable impacts the dependent variable locally. For example, suppose we model *House price*, and one of the independent variables is *House size*. Areas with high coefficients in *House size* are areas where the estimation of *House price* is more influenced by this variable compared to other areas with

lower coefficients (of the same variable). Directional trends or spatial clusters might also be spotted when mapping coefficients, leaving room for additional geographical analysis and unraveling hidden spatial processes. The range of the local coefficients might be high or low with the same sign. Still, there are rare cases where the majority of coefficients have a specific sign, and relatively few observations have the opposite one. In other words, a coefficient might have both negative and positive values. Although this seems not accurate, it typically suggests that some coefficient values will be zero or very close to zero. In OLS, the t-statistic is used to test if a coefficient is statistically significant. In GWR, there is no such test carried out, and those coefficients near zero may be regarded insignificant (Fotheringham et al. 2002). From the policy perspective, the coefficient analysis may help to identify the following types of variables:

- Statistically significant global variables that do not vary significantly from place to place. This type of variables can be used for analysis at the regional level. For example, the percentage of illiteracy may not vary significantly among postcodes for a cluster of cities in a remote prefecture of a developing country. A regional policy, such as offering free language lessons in each municipality of the region, might be appropriate to decrease rates of illiteracy.
- Statistically significant global variables that vary significantly from place to place. This type of variables can be used for analysis at the local level. For the illiteracy example, by using GWR coefficient mapping, we may identify specific variables that are related to high illiteracy rates in specific areas – e.g., lack of easy access to a learning center. By applying a local policy through better spatial planning, we can select only those areas scattered in the entire region that lack learning centers and create new ones.
- Statistically significant variables that are spatially autocorrelated. Although we cannot include these variables in the GWR model, still we can use them to make conclusions for applying local policies. For example, high values of illiteracy rates may cluster to a specific part of the study area (e.g., the most deprived one). Local policies can be applied again to those areas suffering from high illiteracy rates.

Practical Guidelines:

- GWR is a linear model and must follow the same assumptions the linear regression model does (LINE-M).
- GWR performs better when applied to large datasets with hundreds of features. As a rule of thumb, spatial objects should exceed 160. In cases of smaller datasets, results might be unreliable (ESRI 2016b).
- Data should be projected using a projected coordinate system rather than a geographic coordinate system, as distance calculations are essential to the creation of the spatial weights matrix.

- Begin with an OLS model to specify a reasonably good model. Test VIF values for multicollinearity. VIF values larger than 7.5 reveal global multicollinearity and will prevent GWR from producing a good solution. Use the exploratory regression tool also if the dataset is large.
- Map each independent variable separately and look if clustering emerges. Keep in mind that as spatial autocorrelation exists in most real-world problems, we expect a degree of spatial clustering. If spatial clustering (in the values of a specific independent variable) is very pronounced, then we should consider removing this variable.
- The dependent variable in GWR cannot be binary.
- Remove any dummy variables representing different regimes (geographical areas).
- If multicollinearity is detected, we can remove variables or combine them to increase value variation. For example, if we analyze *Expenses for supermarket* and *Expenses for groceries* and we identify multicollinearity between them, we can just add these values and create a new variable labeled *Expenses*.
- Problems of local multicollinearity might exist when the values of an independent variable cluster geographically. The condition number reveals if local multicollinearity exists. As a rule of thumb, condition values larger than 30, equal to null or negative reveal multicollinearity, and results should be examined with caution. When categorical variables have few categories or regime variables exist, we run the risk of spatial clustering and, thus, local multicollinearity.
- Mapping coefficients offers a comprehensive view of the spatial variability of coefficient values. By applying a rendering scheme (e.g., cold to hot) to the raster surface of the coefficients of a specific variable, we trace fluctuations across the study area, locate large or small values and potentially identify spatially clustering.
- Run spatial autocorrelation tests in residuals. If the test is statistically significant, then the model may be misspecified.
- Map residuals to identify any heteroscedasticity. If there is heteroscedasticity, the model may be misspecified.
- In cases of misspecification, variables might have been omitted from the model. Use exploratory regression to identify any missing variables.

6.9   Chapter Concluding Remarks

- Before we begin our analysis, we should inspect if missing values exist.
- Underprediction means that the fitted (predicted) value is lower than the observed.
- Overprediction means that the fitted value is higher than the observed.

- If under or overpredictions occur systematically, then there is bias in our model.
- By regression, we built a probabilistic model that is related to a random amount of error (e).
- 82% R-squared typically means that 82% of the variation in Y is due to the changes in values of X, meaning that X is a good predictor for Y having a probabilistic linear relationship.
- In case of a low R-squared, we may add more independent variables or more observations (if available) to check if the R-squared will increase.
- Adjusted R-squared reveals if an extra variable is useful or not.
- Predicted R-squared is more valuable than adjusted R-squared because it is calculated based on existing observations with known Y values that are not included in the model creation (out of sample).
- By standardizing residuals, we calculate how many standard deviations a residual lies from the mean.
- If a residual plot indicates that some assumptions are severely violated, then we cannot use the model although some statistics may be statistically significant.
- Overfitting occurs when a model has too many parameters relative to the number of observations.
- To diagnose multicollinearity, the following metrics can be used: Singular value decomposition, condition index (CI), variance decomposition proportions, and variance inflator factor (VIF).
- Before we evaluate the t-statistic, we should check the Koenker (BP) statistic.
- When the Koenker (BP) statistic is statistically significant, then heteroscedasticity exists, and the relationships of the model are not reliable. In this case, robust t-statistic and robust probabilities should be used instead of the t-statistic p-value.
- The joint Wald test should be checked when the Koenker (BP) statistic is statistically significant.
- In case that the Koenker (BP) statistic is statistically significant, the joint Wald test should be inspected instead of the F-significance.
- An influential point might be both an outlier point (at the Y direction of the residual plot or far away from the regression line) and a point with high leverage (far away from the X or fitted values mean).
- Beta coefficients are used to compare the importance of each independent variable to the model especially when the units or the scale among the independent variables are different.
- Spatial kernel is used to provide the geographic weighting for the GWR method.
- In case of multicollinearity, AIC and CV methods might not accurately calculate the optimal distance or the optimal number of neighbors for the

bandwidth parameter selection. Detect and remove multicollinearity issues first, and then apply the previous methods.

Questions and Answers

The answers given here are brief. For more thorough answers, refer back to the relevant sections of this chapter.

Q1. What is ordinary least squares regression? How does OLS work? What is a residual?

A1. Ordinary least squares (OLS) is a statistical method for estimating the unknown parameters (coefficients) of a linear regression model. OLS regression works by fitting a line to the data points by determining the parameters of the model that minimize the sum of the squared vertical distances from the observed points to the line (sum of the squared residuals). The difference of the observed value with the predicted (fitted) value is the model error called residual (e).

Q2. What is the coefficient of determination R-squared, and why it is used?

A2. The coefficient of determination denoted as R^2 (R-squared) is the percentage of the variation explained by the model. It is calculated as the ratio between the variation of the predicted values of the dependent variable (explained variation SSR) to the variation of the observed values of the dependent variable (total variation SST). Its values range from 0 to 1.

Q3. What is the F-test and the F significance level, and why it is used in comparison to R-squared?

A3. The F-test and the F-significance level are used to test if the regression line fit is statistically significant, or else if the regression has been successful at explaining a significant amount of the variation of Y. R-squared assesses the strength of the relationship among predicted and independent variables. Still, it does not provide a measure of how statistically significant the hypothetical relationship is. To do so, we use the F-test of overall significance.

Q4. What is a standardized residual plot, and what is used for?

A4. Standardized residual plot is the plot of the fitted values against the standardized residuals (the y-axis depicts the number of standard deviations, and the x-axis depicts the fitted values). Using a standardized residual plot assists in assessing the quality of the regression model and identifying if normality or homoscedasticity assumptions are violated. Ideally, the standardized residual plot should exhibit no particular pattern; and data points should be randomly distributed, have constant variance and be as close to the x-axis as possible. These conditions satisfy the third (normal distribution of e) and the fourth (homoscedasticity) assumptions of linear regression (see Section 6.4). If clusters, outliers or trends exist in the data point pattern, it is an indication of one or more assumptions' violation.

Q5. What are the standardized (beta) coefficients?

A5. Standardized coefficients are the coefficients resulting when variables in the regression model (both dependent and independent) are converted to their z-scores before running the fitting model. The standardized regression coefficients – also called beta coefficients, beta weights or just beta – are expressed in standard deviation units thus removing the measurement units of the variables. They are used to decide which independent variable X is more important to model Y.

Q6. What is influential point, and why tracing it is important?

A6. Influential point is any data point that substantially influences the geometry of the regression line. Influential points should be identified and potentially removed as they significantly distort the geometry of the regression line leading to inaccurate models.

Q7. Which are the assumptions of multiple linear regression?

A7. MLR builds a probabilistic model based on five assumptions. (1) Linearity: The relationships between dependent and independent variables are linear. (2) Independence: The residuals are independent of each other, meaning that an error e in a data point X does not affect by any means another error e for some different value of the same variable X. (3) Normality: The distribution of errors e at any particular X value is normal. 4) Homoscedasticity: Errors e have a zero mean equal variance and a constant standard deviation at any particular value of X. (5) Multicollinearity absence: Variables X should be independent of each other exhibiting no multicollinearity.

Q8. What is exploratory regression, and why is it used?

A8. Exploratory regression is a process that facilitates the appropriate selection of an OLS model when the number of independent variables is large. Exploratory regression is a data-mining tool and not a new regression technique, as it just applies OLS regression to many different models to select those that pass all necessary OLS diagnostic tests. It is used to automatically test all possible combinations of different OLS models when many independent variables are available.

Q9. What is geographically weighted regression (GWR)? Which assumptions should obey?

A9. GWR is a local form of linear regression model used to detect heterogeneity and analyze spatially varying relationships. GWR provides a local model of the dependent variable allowing for coefficients to vary across space, by fitting a weighted linear regression model to every spatial object in the dataset. GWR should obey the same assumptions as multiple linear regression (LINE-M).

Q10. Why spatial kernels and bandwidth are used in GWR?

A10. A spatial kernel is used to provide the geographic weighting for the GWR method and should be selected before calibrating a GWR model. Objects closer to the location i are weighted more than those further away, based on the assumption of spatial autocorrelation. The bandwidth of the kernel type – that is, the unknown parameter h in the fixed kernel type – and the number of nearest neighbors N in the adoptive kernel type define the size of each kernel. There are three main methods for bandwidth selection/estimation: AICc method, cross-validation (CV) method and the bandwidth method.

LAB 6
OLS, EXPLANATORY REGRESSION, GWR

Overall Progress

Spatial Analysis/Lab Workflow

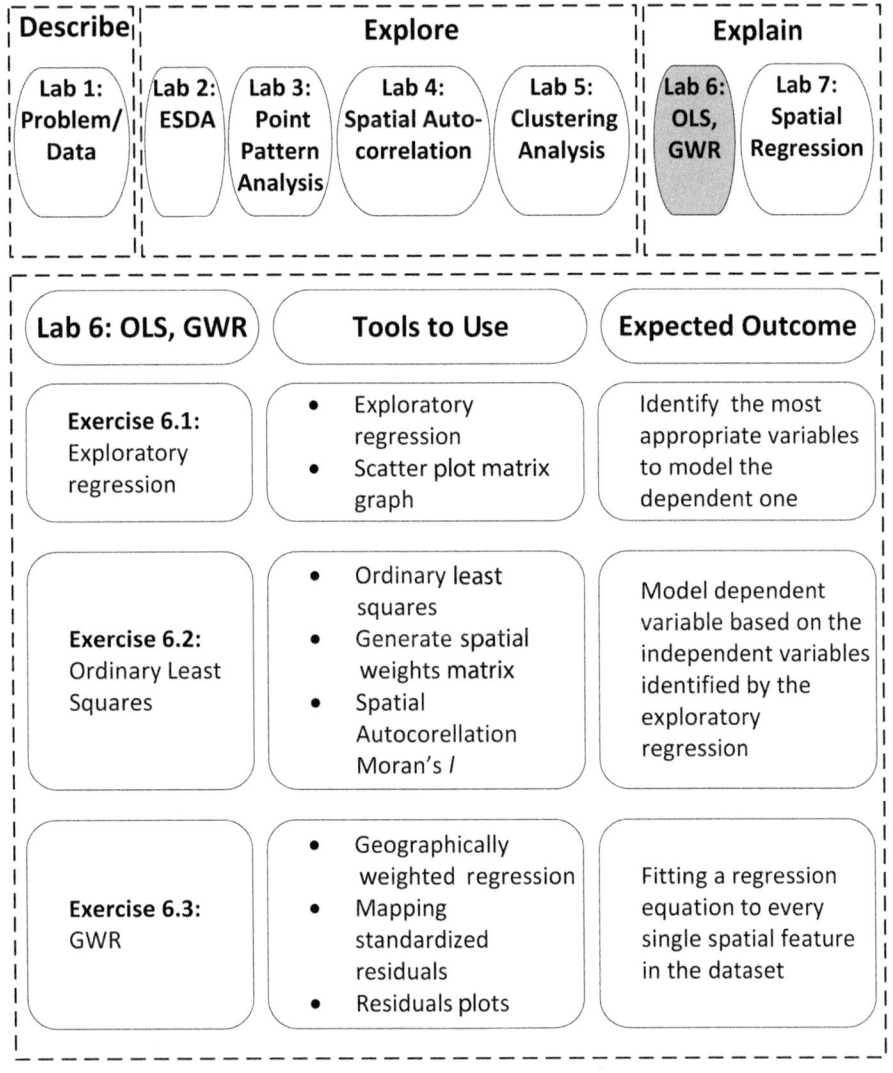

Figure 6.9 Lab 6 workflow and overall progress.

Scope of the Analysis

This lab deals with

- **Objective 4:** Modeling. Identifying socioeconomic drivers behind people's monthly expenses (including those for coffee-related services) (see Table 1.2).

In this lab, we model Expenses for those people living in the city, based on various socioeconomic variables (see Figure 6.9). The overall scope of the analysis is to identify the socioeconomic drivers behind monthly expenses (including coffee-related expenses). From the investor perspective, this type of analysis can be used, first, to better trace areas with the higher potential of expenses and, second, to identify what makes people spend more in coffee-related purchases. This type of analysis is not linked directly to finding an optimal location (as we did in previous labs), but it focuses on modeling relationships that can be used for better market penetration and customer analysis. A complete analysis should include much more detailed variables such as consumer preferences, everyday habits, type of job and money spent on coffee shops. For educational reasons and to keep analysis short, we only focus on 10 basic socioeconomic variables. We start with exploratory regression first, and then we analyze the most appropriate model resulted, with additional OLS diagnostics. Finally, we apply GWR to handle spatial heterogeneity.

Exercise 6.1 Exploratory Regression

Before we run the tool, we show how ArcGIS presents the results of exploratory regression analysis, and we also highlight how we should interpret outputs (for more details on this section, you can visit ESRI's website, ESRI 2016a). The primary output is a report file summarizing the most successful regressions. Keep in mind that there might be cases where no model passes all tests. The report can still be valuable, as we can analyze which coefficients (and thus associated variables) are consistently statistically significant and which diagnostics most commonly yield non-statistical significant results. In cases of no model passing the tests, we may relax the parameter values and rerun exploratory regression. Results can be summarized in the following five sections:

- **Section 1: Best models summary.** It provides the three best models with the highest adjusted R-squared. It also provides the following diagnostics:
  1. AICc (corrected Akaike Information Criterion), as a general measure of goodness of fit, and as a measure to compare different models. The one with the smallest value provides best goodness of fit among those tested (for more on AICc, see Section 6.8.3).
  2. JB (Jarque–Bera) p-value, as test of residuals normality.
  3. K(BP) (Koenker's studentized Breusch–Pagan) p-value, as a measure of heteroscedasticity.
  4. VIF (variance inflation factor). In order to assess multicollinearity, the largest variance inflation factor is reported.
  5. SA is the p-value of the Global Moran's I test for detecting spatial autocorrelation.

Exercise 6.1 (*cont.*)

The models (if any) passing all tests are labeled as "Passing Models." The results are reported as many times as the *Maximum Number of Explanatory Variables* parameter has been set. As mentioned before, we might have many variables, but for simplicity, we consider only those models having no more than a specific number of independent variables. For instance, for twenty independent variables if we set the *Maximum Number of Explanatory Variables* parameter to 5, then the exploratory regression will test the combinations of all models having from one independent variable up to a maximum of five. The output includes the best three models (along with the five diagnostics) with one independent variable, the three best models with two independent variables and so on up to the three best models with five independent variables. Furthermore, all passing models will be reported. In a nutshell, five different subsections will be included in this section, summarizing the best models and the passing ones. If there are no passing models, then the sections that follow reveal a lot regarding the data, assisting in determining which direction the analysis should follow.

- **Section 2: Global summary.** This section lists the five diagnostics and the percentage (as well as the absolute values) of the models that passed each of these tests. In case that no model passes, this section assists in the identification of potential reasons that models do not perform well. For example, if the normality test is not statistically significant for 90% of the models, then the normality assumption is at stake. We might consider transforming some of the variables, as nonlinear relationships might exist in the data (to detect nonlinear relationships we can create a scatter plot matrix, as shown earlier). Potential outliers' existence might also be a cause of non-normality of errors. Furthermore, we might encounter spatially autocorrelated residuals for all models. Spatially autocorrelated residuals is an indication of a missing variable. It is quite common for this omitted variable to be a geographical variable such as the distance from a landmark (e.g., distance from city center). Spatial regimes may also assist on the elimination of spatially autocorrelated residuals. Another approach is to select a good model that passes all or the most of the rest diagnostics and run OLS model to inspect potential problems in the residuals scatter plot.

- **Section 3: Variable significance summary.** Each variable is listed along with three percentage values related to a specific coefficient:
  1. The percentage that a coefficient (and thus the associated variable) is significant. It is the ratio of the number of models that a specific variable is statistically significant to the number of all models tested.

Exercise 6.1 (*cont.*)

2.   The percentage that a coefficient is positive.
3.   The percentage that a coefficient is negative.

Variables with high percentage of being significant are strong predictors and are expected to have consistently the same sign, either positive or negative. Variables with small percentage of being significant are not explaining the dependent variable. In case we want to drop variables from our model, these variables are the ones we should drop first. In case no passing models emerge, we can use the variables with the highest percentages of being significant and add omitted variables, in anticipation of building models that pass the tests.

- **Section 4: Multicollinearity summary.** It reports how many times each variable with a high VIF (multicollinearity) is included in a model along with all independent variables (in the same model). When two or more variables coexist in many models, with high multicollinearity, they practically explain the same percentage of variation. If we keep only one, we avoid multicollinearity and build a better model. Among those variables that are highly correlated, we may keep the one with the highest percentage of being significant, reported in the previous section, or the one that is more appropriate conceptually.

- **Section 5: Diagnostics summary.** The three models with the highest Jarque–Berra p-values (errors' normality) and the three models with the highest Moran's I p-values (spatial autocorrelation) are reported in this section along with all other associated diagnostics and independent variables. Models reported here do not necessarily pass all diagnostics. This section is quite helpful when no model passes the tests. By inspecting the p-values of the Jarque–Berra test and the Moran's I test, we estimate how far the model lies from fulfilling the assumptions of having normally distributed residuals and nonspatially autocorrelated residuals. For example, if the maximum Moran's I p-value is less than 0.001, then spatial autocorrelation of residuals is a major issue. However, if this value is 0.095 with a significance level of 0.10, then we are close to considering that residuals are not spatially autocorrelated. We can also check which independent variables are those that result in higher values (preferred) on these tests, leading to passing or nearly passing models.

The second output of the tool is a table summarizing all models that meet the maximum coefficient cutoff value and the maximum VIF cutoff value. Each row in the table describes a model that meets these two criteria regardless of if it passes the tests overall. All additional diagnostics and independent variables are also included in the same line. This table is supportive when we consider using a model although one or more

Exercise 6.1 (*cont.*)

assumptions are not fulfilled. For example, the most common violated assumption is the errors normality. In such cases, we can sort the table by the AICc values and use the model with the minimum AICc value that also meets as many of the diagnostics as possible except for Jarque–Berra.

Let's proceed to our exercise. In total, 10 independent variables are tested to model the dependent variable (Expenses). We do not define which of the variables will be finally included in the final model. Instead, we lean on the data-mining capabilities that the exploratory regression offers. The 10 independent variables are as follows:

- Population: population per postcode
- Density: people per square meter
- Foreigners: percentage of people of different nationality
- SecondaryE: percentage of people obtained secondary education
- University: percentage of people that graduated from university
- PhD Master: percentage of people obtained a master's degree or PhD
- Income: annual income per capita (in euros)
- Insurance: annual money spent on insurance policy (in euros)
- Rent: monthly rent (in euros)
- Owners: percentage of people living in their own house

If we attempt to test every possible combination of the independent variables, we should build 1,023 different models. Exploratory regression builds these models automatically. Before we perform exploratory regression, we should create a scatter plot matrix to inspect the presence of potential multicollinearity among the independent variables.

ArcGIS Tools to be used: Scatter Plot Matrix Graph, Exploratory Regression

ACTION: Scatter Plot Matrix Graph
We create this plot to trace multicollinearity, outliers or trends among the variables.

Navigate to the location you have stored the book dataset and click Lab6_Regression.mxd

Main Menu > File > Save As > My_Lab6_Regression.mxd

In I:\BookLabs\Lab6\Output

Main Menu > View> Graphs > Create Scatter Plot Matrix >

Layer/Table: City

Fields > Click in line 1, under the Field name. Select variables one by one, starting with Expenses and adding the 10 independent variables and the dependent variable, as shown in Figure 6.10.

Exercise 6.1 (*cont.*)

Check Show Histograms

Number of bins = 3

Apply > Next > Finish

RC on graph > Add to layout > Return to Data View > Close graph > Save

You should inspect one by one these plots for potential multicollinearity issues or outliers presence (see Figure 6.11).

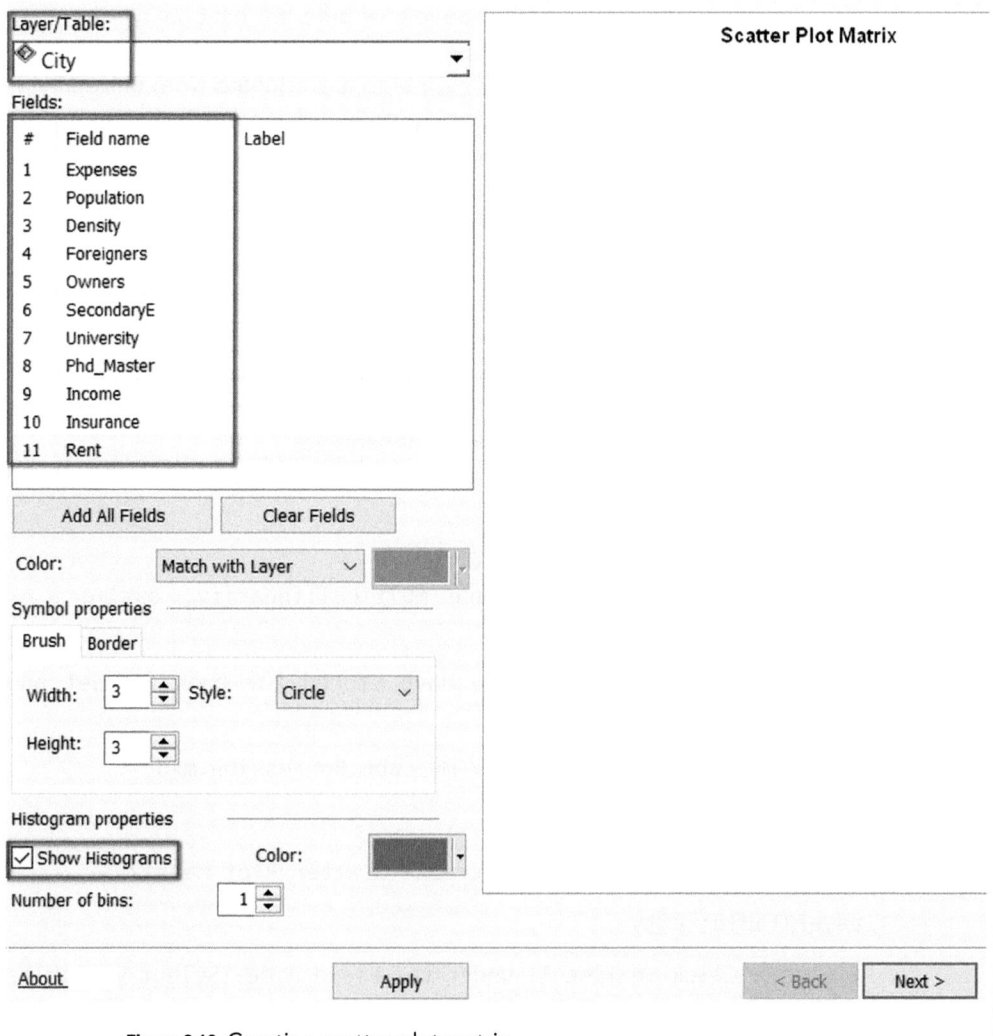

Figure 6.10 Creating scatter plot matrix.

Exercise 6.1 *(cont.)*

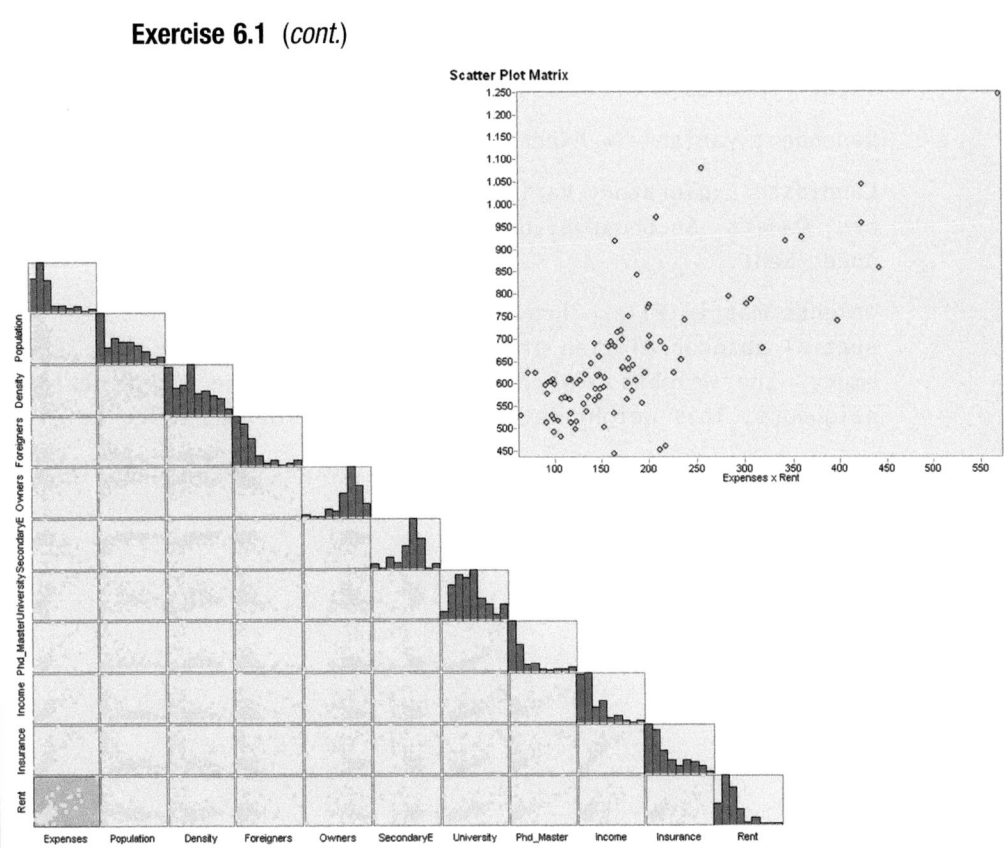

Figure 6.11 Scatter plot matrix of all variables. When clicking a scatter plot in the matrix, a larger version is presented in the upper-right corner of the graph.

ACTION: Exploratory Regression

ArcToolbox > Spatial Statistic Tools > Modeling Spatial Relationships > Exploratory Regression (see Figure 6.12)

☐ 🜨 Spatial Statistics Tools
   ⊞ 🜨 Analyzing Patterns
   ⊞ 🜨 Mapping Clusters
   ⊞ 🜨 Measuring Geographic Distributions
   ☐ 🜨 Modeling Spatial Relationships
      🝰 Exploratory Regression
      🝰 Generate Network Spatial Weights
      🝰 Generate Spatial Weights Matrix
      🔨 Geographically Weighted Regression
      🝰 Ordinary Least Squares

Figure 6.12 Exploratory regression tool.

Exercise 6.1 (*cont.*)

Input Features = City (see Figure 6.13)

Dependent Variable = Expenses

Candidate Exploratory Variables = Population, Density, Foreigners, Owners, SecondaryE, University, PhD Master, Income, Insurance, Rent,

Weights Matrix File = Leave empty. It is used to calculate the spatial autocorrelation of the residuals. By leaving this field empty, the weights are calculated based on the eight nearest neighbors. This weight matrix is not used for OLS calculations.

Output Report File = I:\BookLabs\Lab6\Output\Exploratory.txt

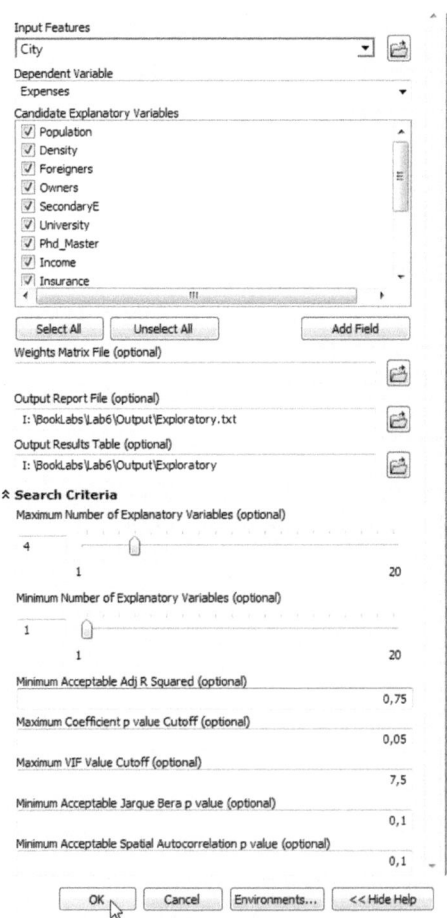

Figure 6.13 Exploratory regression dialog box.

Exercise 6.1 *(cont.)*

Output Results Table = I:\BookLabs\Lab6\Output\Exploratory

Maximum Number of Explanatory Variable = 4 (We select to build models with a maximum of four independent variables)

Minimum Number of Explanatory Variable = 1

Minimum Acceptable Adj R Squared = 0.75

Maximum Coefficient p value Cutoff = 0.05

Maximum VIF Cutoff = 7.5

Minimum Acceptable Jargue Bera p value = 0.1

Minimum Acceptable Spatial Autocorrelation p value = 0.1

OK

Main Menu > Geoprocessing > Results > Current Session > Exploratory Regression > DC on Output Report File: Exploratory.txt

The output report (Exploratory.txt) is now presented in detail.

- **Section 1: Best models summary**

```
****************************************************************
Choose 1 of 10 Summary
    Highest Adjusted R-Squared Results
AdjR2    AICc   JB K(BP)  VIF   SA    Model
 0.81   918.02 0.02  0.00 1.00 0.00  +INCOME***
 0.60   985.34 0.00  0.00 1.00 0.00  +INSURANCE***
 0.57   992.01 0.01  0.00 1.00 0.00  +RENT***
            Passing Models
AdjR2 AICc JB K(BP) VIF SA Model
****************************************************************
Choose 2 of 10 Summary
          Highest Adjusted R-Squared Results
AdjR2    AICc  JB K(BP)  VIF   SA    Model
0.85   900.89 0.00  0.01 1.91 0.00  +INCOME***       +RENT***
0.84   905.78 0.00  0.00 1.66 0.00  +UNIVERSITY**  +INCOME***
0.83   912.56 0.00  0.00 1.47 0.00  +POPULATION*** +INCOME***
        Passing Models
   AdjR2 AICc JB K(BP) VIF SA    Model
****************************************************************
```

Exercise 6.1 *(cont.)*

```
Choose 3 of 10 Summary
                    Highest Adjusted R-Squared Results
AdjR2   AICc   JB K(BP)  VIF   SA   Model
0.87 884.45 0.00  0.00 2.59  0.26  +UNIVERSITY***  +INCOME***  +RENT***
0.86 895.71 0.00  0.01 2.49  0.60  +POPULATION***  +INCOME***  +RENT***
0.86 896.43 0.00  0.01 1.92  0.11  +OWNERS**       +INCOME***  +RENT***
                    Passing Models
        AdjR2 AICc JB K(BP) VIF SA    Model
        **************************************************************
        Choose 4 of 10 Summary
                         Highest Adjusted R-Squared Results
AdjR2    AICc   JB K(BP)   VIF    SA   Model
0.88 882.52 0.00   0.01 2.64 0.56  +OWNERS**     +UNIVERSITY*** +INCOME***  +RENT***
0.88 883.87 0.00   0.01 3.70 0.86  +POPULATION*  +UNIVERSITY**  +INCOME***  +RENT***
0.88 884.32 0.00   0.00 3.81 0.18  +UNIVERSITY** -PHD_MASTER    +INCOME***  +RENT***
                    Passing Models
        AdjR2 AICc JB K(BP) VIF SA    Model
        **************************************************************
```

Interpreting results: This section provides the three best models that have between one and four independent variables (we set this as one of our parameters). Although many models have high adjusted R-squared (>0.80), there is no passing model (meeting all criteria). Multicollinearity is not an issue in the aforementioned models, as VIF is smaller than 4 in all cases. A closer inspection of the table reveals that models do not pass errors normality (JB) and heteroscedasticity K(BP) tests, but some of them pass the spatial autocorrelation (SA) test. The fact the heteroscedasticity in residuals exists is a potential indication that GWR should be applied. We have to go through the following sections to better understand why these assumptions are violated.

- **Section 2: Global summary**

```
********** Exploratory Regression Global Summary (EXPENSES) ************
            Percentage of Search Criteria Passed
          Search Criterion Cutoff Trials # Passed % Passed
          Min Adjusted R-Squared      > 0.75     385      180     46.75
          Max Coefficient p-value     < 0.05     385       60     15.58
                 Max VIF Value        < 7.50     385      367     95.32
          Min Jarque-Bera p-value     > 0.10     385       16      4.16
  Min Spatial Autocorrelation p-value > 0.10      15        6     40.00
  _____
```

Exercise 6.1 (*cont.*)

Interpreting results: From the results, we identify one major problem. Only 4.16% of the models passed the errors normality test (Jarque–Bera). Regarding the violation of normality assumption, we might consider transforming some of the variables, as nonlinear relationships might exist in the data. Potential outliers' existence might also be a cause of non-normality of errors. To detect nonlinear relationships or outliers, we can check the scatter plot matrix as produced in Figure 6.11. For example, comparing Expenses to all other variables indicates that there might be a need for transformation in Population or PhD_Master (see Figure 6.11). In addition, by close inspection of the scatter plot matrix, we locate outliers – for example, in SecondaryE with Expenses or Insurance with Expenses. Removing independent variables is an alternative way that may improve normality issues.

We will not deal with outliers or transformations, as this example is intended to showcase the exploratory regression method and not to produce a complete analysis for the city. In real case studies, though, we should test potential transformations like logarithmic or Box–Cox and evaluate if results are improved. In fact, applying spatial regression methods may mitigate the previous problems by including spatial variables and handling spatial autocorrelation.

Another problem we encounter in this example is that residuals are spatially autocorrelated for nearly 60% of the models tested. Spatial autocorrelation of residuals mostly occurs when a variable is missing from the model. Potentially, one or more geographical variables are omitted, such as the distance from a point (e.g., distance from the city center). To identify which variable is missing, we can also select a good model that passes most of the diagnostics and run the OLS model to map residuals. Inspecting residuals' distribution along with knowledge of the study area assists in identifying potential missing variables. The missing variable might also be the spatially lagged Income, but this is a topic of spatial econometrics covered in Chapter 7 (a spatial lag or spatial error model might be the solution to removing spatial autocorrelation).

• **Section 3: Variable significance summary**

```
Summary of Variable Significance
Variable   % Significant % Negative % Positive
INCOME             100.00       0.00      100.00
RENT               100.00       0.00      100.00
UNIVERSITY          84.62       0.00      100.00
POPULATION          74.62      49.23       50.77
INSURANCE           71.54      26.92       73.08
DENSITY             60.77      50.77       49.23
PHD_MASTER          52.31       6.15       93.85
SECONDARYE          18.46      72.31       27.69
FOREIGNERS          14.62      32.31       67.69
OWNERS              12.31      12.31       87.69
```

Exercise 6.1 (*cont.*)

Interpreting results: Coefficients of Income and Rent are consistently significant and positive and are regarded as strong predictors of Expenses. University is by 84.62% significant and can be considered to be included in the models. In all tests (100%), the coefficient of University has a positive sign. On the other hand, Population and Insurance, which have significant coefficients for 74.62% and 71.54% of the models tested, respectively, do not have a constant sign and, as such, are not reliable predictors of expense. The rest of the variables are not stable and should not be included in the final model. In conclusion, in our model, we should use Income, Rent and University

- **Section 4: Multicollinearity summary**

```
----------------------------------------------------------------
Summary of Multicollinearity
Variable      VIF Violations Covariates
POPULATION   2.49      0      -------
DENSITY      1.89      0      -------
FOREIGNERS   2.08      0      -------
OWNERS       1.92      0      -------
SECONDARYE   1.72      0      -------
UNIVERSITY   3.55      0      -------
PHD_MASTER   4.49      0      -------
INCOME      10.97     16      INSURANCE (10.53)
INSURANCE    8.60      6      INCOME (10.53)
RENT         2.09      0      -------
```

Interpreting results: Income exhibits multicollinearity with Insurance. These variables should not be included concurrently in the same model. We keep Income, as it is significant in more cases (100%) and also has constant sign, compared to Insurance (71.54%). Income conceptually fits better the scopes of this analysis.

- **Section 5: Diagnostic Summary**

```
----------------------------------------------------------------

                    Summary of Residual Normality (JB)
    JB     AdjR2     AICc      K(BP)      VIF      SA     Model
0.996766 0.833190 911.084266 0.000000 2.139714 0.000005 +POPULATION*** +FOREIGNERS  +OWNERS  +INCOME***
0.964232 0.821283 917.289778 0.000000 1.730483 0.000000 +FOREIGNERS  +OWNERS* +SECONDARYE  +INCOME***
0.879930 0.822030 915.667491 0.000000 1.562164 0.000000 +FOREIGNERS  +OWNERS* +INCOME***
----------------------------------------------------------------

                Summary of Residual Spatial Autocorrelation (SA)
    JB     AdjR2     AICc      K(BP)      VIF      SA     Model
0.855801 0.876721 883.867623 0.000000 0.011252 3.695209 +POPULATION* +UNIVERSITY** +INCOME*** +RENT***
```

Exercise 6.1 *(cont.)*

```
0.604094 0.857425 895.710602 0.000000 0.012039 2.488589 +POPULATION*** +INCOME*** +RENT***
0.558420 0.878560 882.515318 0.000000 0.008369 2.643798 +OWNERS** +UNIVERSITY*** +INCOME*** +RENT***
-------------------------------------------------------------------------------------------------

Table Abbreviations
AdjR2 Adjusted R-Squared
AICc  Akaike's Information Criterion
JB    Jarque-Bera p-value
K(BP) Koenker (BP) Statistic p-value
VIF   Max Variance Inflation Factor
SA    Global Moran's I p-value
Model Variable sign (+/-)
Model Variable significance (* = 0.10, ** = 0.05, *** = 0.01)
```

Interpreting results: The model with the highest *p*-value for errors normality (Jarque–Berra) includes the variables `+POPULATION*** +FOREIGNERS +OWNERS +INCOME***`. This model is unreliable, as most of the variables included are not statistically significant (Section 3).

The model with the highest *p*-value for spatial autocorrelation includes the variables `+POPULATION* +UNIVERSITY** +INCOME*** +RENT***`. From the geographical analysis perspective, spatial autocorrelation is not evident in this model. Still, it includes `Population` – which, as shown in Section 3, has unstable sign and is not advised to be included in the model.

Finally, an additional output is a table summarizing the models that meet the maximum coefficient cutoff value and the maximum VIF cutoff value (see Figure 6.14). In conjunction with the previous findings, we use this table to select the appropriate model for OLS, even if some tests are rejected.

ACTION: Open Exploratory Table (List by Source)

RC Exploratory > Open > RC AdjR2 > Sort Descending

Sorting the table by adjusted *R*-squared, we get the best models relatively to this metric, no matter if they meet all passing criteria.

We only present the first five out 60 models included in the table (see Figure 6.14).

Table

Exploratory

AdjR2	AICc	JB	K_BP	MaxVIF	SA	NumVars	X1	X2	X3	X4
0.87856	882.515318	0	0.008369	2.643798	0.55842	4	OWNERS	UNIVERSITY	INCOME	RENT
0.874191	884.451425	0	0.003895	2.590905	0.264513	3	UNIVERSITY	INCOME	RENT	
0.857425	895.710602	0	0.012039	2.488589	0.604094	3	POPULATION	INCOME	RENT	
0.856285	896.427437	0	0.005268	1.92393	0.105628	3	OWNERS	INCOME	RENT	
0.851471	900.637445	0.002872	0.000014	1.812094	-1.797693e+308	4	FOREIGNERS	OWNERS	UNIVERSITY	INCOME

0 (1 out of 60 Selected)

Exploratory

Figure 6.14 Models' diagnostics.

Exercise 6.1 (*cont.*)

Interpreting results: The maximum adjusted *R*-squared value is 0.878, with four independent variables: Owners, University, Income and Rent. Still, as Owners is not a stable predictor (see table in Section 3), it should not be included in the model. The model with the second-highest adjusted *R*-squared includes University, Income and Rent. These three variables, embedded in many different models, seem to be the best available predictors of Expenses and will be further analyzed by OLS regression in the next exercise. This OLS model does not produce different results than those presented in the preceding table, but it provides us with more statistics, graphs and capabilities for further analysis.

Exercise 6.2 OLS Regression

Before we proceed to GWR analysis we run OLS to the three variables (University, Income, Rent) selected by the exploratory regression analysis in Exercise 6.1.

ArcGIS Tools to be used: Ordinary Least Squares, Generate Spatial Weights Matrix, Spatial Autocorrelation Moran's I

ACTION: OLS regression

Navigate to the location you have stored the book dataset and click on My_Lab6_Regression.mxd

ArcToolbox > Spatial Statistic Tools > Modeling Spatial Relationships > Ordinary Least Squares

Input Feature Class = City (see Figure 6.15)

Unique ID Field = PosctCode

Output Feature Class = I:\BookLabs\Lab6\Output\OLS.shp

Dependent Variable = Expenses

Explanatory Variable = University, Income, Rent

Output Report File = I:\BookLabs\Lab6\Output\OLSReport.pdf

Coefficient Output Table = I:\BookLabs\Lab6\Output\OLSCoefficientsOutput

Diagnostic Output Table = I:\BookLabs\Lab6\Output\OLSDiagnostics

OK

Exercise 6.2 (*cont.*)

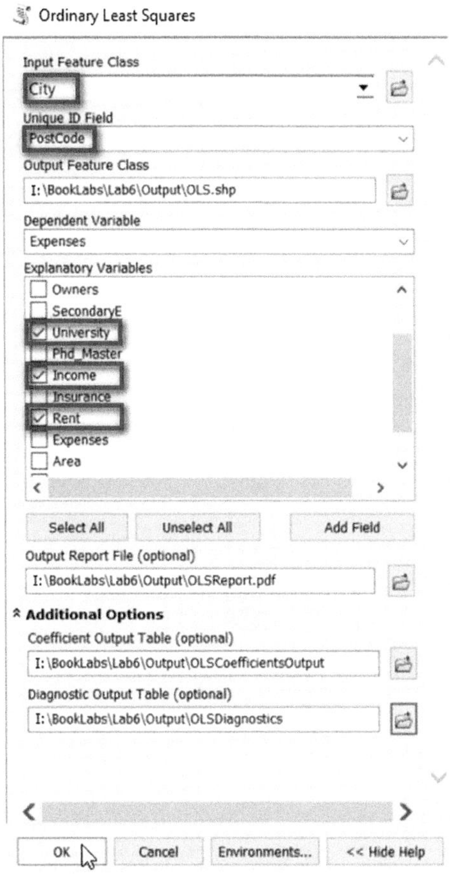

Figure 6.15 OLS settings.

The standardized residuals are mapped automatically (Figure 6.16).

The software also provides a very informative report in PDF format, including statistics, coefficients and diagnostics tables. We briefly present the report next.

```
Main Menu > Geoprocessing > Results > Current Session >
Ordinary Least Squares > DC on Output Report File: OLSReport.pdf
```

Variable	Coefficient	StdError	t-Statistic	Probability	Robust_SE	Robust_t	Robust_Pr	VIF
Intercept	-150.468653	15.988300	-9.411173	0.000000*	16.417585	-9.165090	0.000000*	—
UNIVERSITY	2.748526	0.617052	4.454288	0.000027*	0.990945	2.773640	0.006796*	1.664086
INCOME	0.010503	0.001089	9.646756	0.000000*	0.001486	7.067846	0.000000*	2.590905
RENT	0.165248	0.032568	5.073918	0.000003*	0.034200	4.831819	0.000007*	1.912884

Exercise 6.2 (*cont.*)

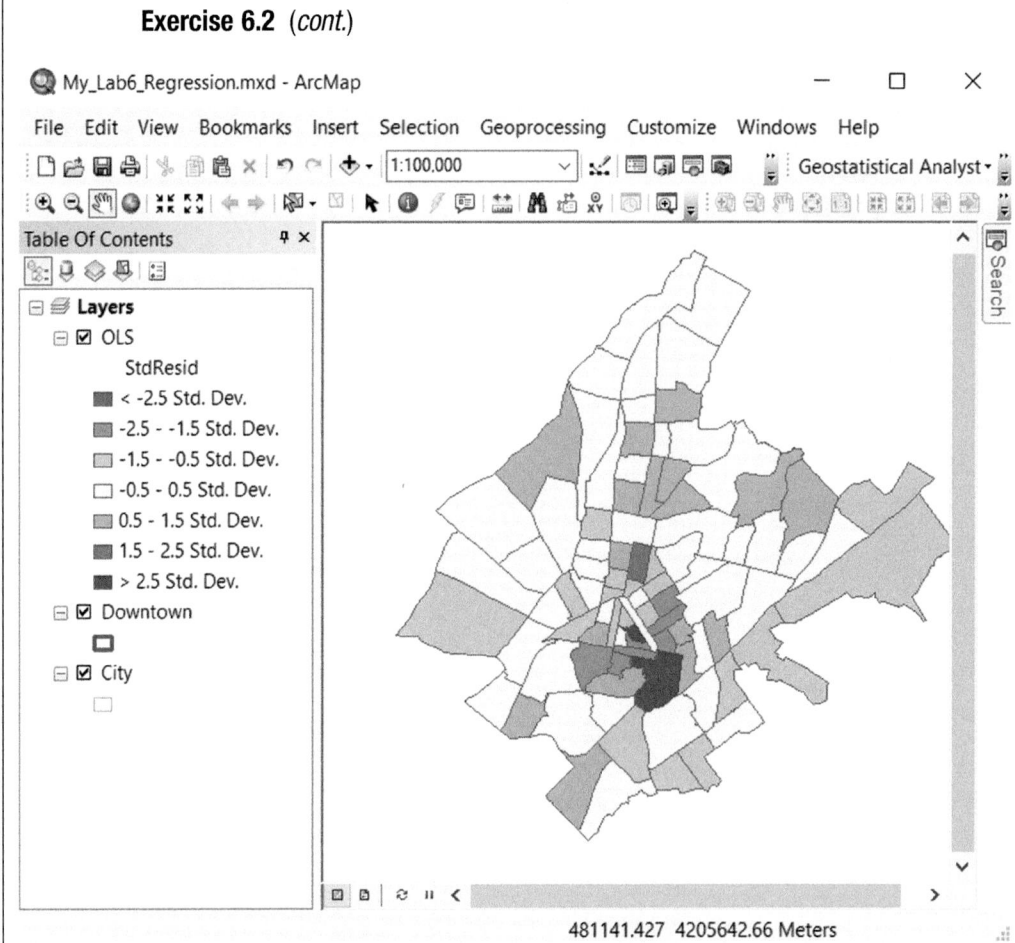

Figure 6.16 Standardized residuals.

This report presents the robust standard errors and VIF as well. Robust errors should be used if the Koenker (BP) test is statistically significant. Results are the same with the model from the exploratory regression (see Figure 6.14, second line).

```
Input Features: City
Dependent Variable: EXPENSES
Number of Observations: 90
Sum of Residual Squares: 86692.533
Akaike's Information Criterion (AICc) [d]: 884.451425
Multiple R-Squared [d]: 0.878431
Adjusted R-Squared [d]: 0.874191
```

Exercise 6.2 (cont.)

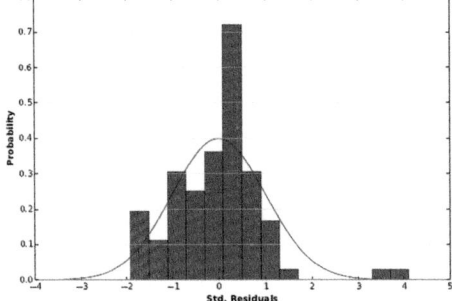

Figure 6.17 Histogram of standardized residuals.

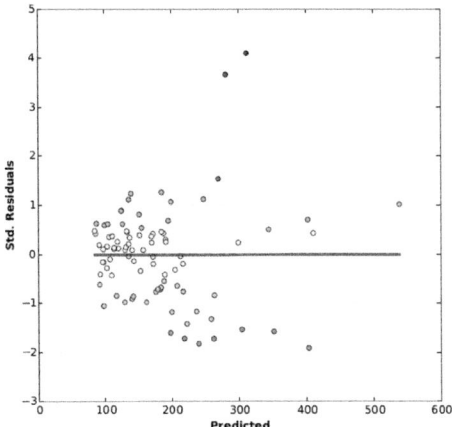

Figure 6.18 Residuals vs. predicted plot.

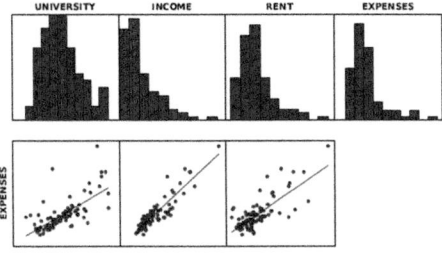

Figure 6.19 Variable distributions and relationships.

Exercise 6.2 (cont.)

```
Joint F-Statistic [e]: 207.139777 Prob(>F), (3,86) degrees of freedom: 0.000000*
Joint Wald Statistic [e]: 415.966846 Prob(>chi-squared), (3) degrees of freedom: 0.000000*
Koenker (BP) Statistic [f]: 13.373289 Prob(>chi-squared), (3) degrees of freedom: 0.003895*
Jarque-Bera Statistic [g]: 70.548247 Prob(>chi-squared), (2) degrees of freedom: 0.000000*
```

Interpreting results:

- First, we should check the Koenker (BP) statistic. When it is statistically significant, we should check the joint Wald statistic to determine the overall model significance and the robust probabilities to determine the coefficients significance. The Koenker (BP) statistic is statistically significant (0.003895*), indicating the existence of heterogeneity. As such, we should check the joint Wald statistic (instead of the F-significance) and the robust probabilities.
- The joint Wald statistic is statistically significant (0.000000*), and we accept the overall significance of the model with a high adjusted R-squared (0.874).
- In addition, all robust probabilities are statistically significant, and we accept the importance of each independent variable to the model.
- As such, the model can be written as:

Expenses = -150.46 + 2.75University + 0.01Income + 0.16 Rent

- These coefficients are not the beta coefficients, so we cannot infer which variable is more important for the model. Still, from the conceptual perspective, the model states that the more people with a university degree, of higher income and higher rent, the higher expenses for daily purchases. Following the typical analysis of the coefficients, if Income (annual income) increases by 2,000 euros and all other variables remain constant, then the mean change of Expenses (monthly expenses) will increase by 2,000X0.01 =+20 euros.
- Still, the Jarque–Berra statistic reveals violation of the normality assumption, as it is statistically significant (Jarque-Bera: 0.000000*), something that is also evident in the histogram of the standardized residuals (the blue line indicates a normal curve; see Figure 6.17).
- Moreover, the standardized residuals vs. predicted values scatter plot reveals heteroscedasticity, which increases especially for predicted values larger than 300 euros (see Figure 6.18). Ideally, residuals should exhibit no structure, shaping a random point pattern. Two outliers (more than three standard deviations) are also spotted (see the red dots in Figure 6.18). The postcodes linked to these outliers are rendered in red in Figure 6.16, indicating that the predicted value

Exercise 6.2 *(cont.)*

(fitted value) for Expenses deviated a lot from the actual one. Inspecting the robust standard errors reveals that all coefficients are robust to heteroscedasticity, as they are statistically significant.

- Heteroscedasticity and nonlinearity infer bias to the model when used for predictions. To deal with heteroscedasticity and nonlinearity, we can inspect the distributions of the independent to the dependent variables (see Figure 6.19).
- The histograms and scatter plots among Expenses and the other independent variables show in general linear patterns. Still, in cases of strongly skewed variables, we may apply a transformation. Due to the significant extent of such analysis, we do not apply any transformation in this example, but you are encouraged to transform income and investigate the outcomes (keep in mind that transformations most of the time are not easily interpreted and, for this reason, many researchers avoid them).
- Finally, we should check if residual errors are spatially autocorrelated.

ACTION: Spatial Autocorrelation Moran's I

OLS does not automatically calculate spatial autocorrelation for residuals. Before applying Global Moran's I, we have to create the weights matrix. The spatial autocorrelation test used in the exploratory regression analysis in Exercise 6.1 chose by default the eight nearest neighbors to build the weights matrix. Still, this option is not included in the Moran's I tool, so we have to build the weights matrix first (see Box 6.3).

ArcToolBox > Spatial Statistics Tools > Modeling Spatial Relationships > Generate Spatial Weights Matrix

Input Feature Class = I:\BookLabs\Lab6\Output\OLS.shp

Unique ID Field = PostCode (see Figure 6.20)

Output Spatial Weights Matrix File =

I:\BookLabs\Lab6\Output\SW8K.swm

Conceptualization of Spatial Relationships =

K_NEAREST_NEIGHBORS

Distance Method = EUCLIDEAN

Number of Neighbors = 8

Row Standardization = Check

OK

Exercise 6.2 (*cont.*)

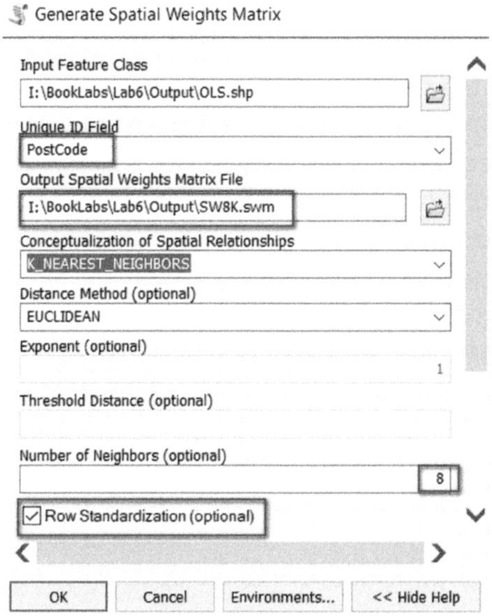

Figure 6.20 Spatial weights matrix dialog box.

Continue by calculating Moran's I.

ArcToolbox > Spatial Statistic Tools > Analyzing Patterns > Spatial Autocorrelation (Morans I)

Input Feature Class = OLS (see Figure 6.21)

Input Field = Residual

Generate Report = Check

Conceptualization of Spatial Relationships =

GET_SPATIAL_WEIGHTS_FROM_FILE

Weights Matrix File = I:\BookLabs\Lab6\Output\SW8K.swm

OK

Main Menu > Geoprocessing > Results > Current Session > Spatial Autocorrelation > DC Report File: MoransI_Results.html

Main Menu > File > Save

Exercise 6.2 (*cont.*)

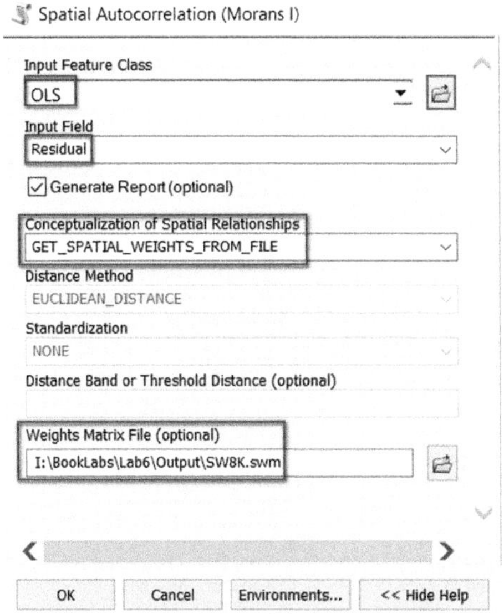

Figure 6.21 Spatial autocorrelation (Moran's I).

The *p*-value is 0.26453, indicating no spatial autocorrelation (same as the results of exploratory regression).

Box 6.3 Keep in mind that for the same problem, a different conceptualization method (e.g., Queen's contiguity) may lead to contradicting results (e.g., spatial autocorrelation existence instead of nonexistence), showcasing how a single choice might change results significantly. Each parameter choice should be made after thorough investigation. Conceptualization of spatial relationships presented in Chapter 1 is very important in the output of many spatial statistics tools. We can run exploratory regression with various conceptualization methods to inspect how results are differentiated. Analysis is not a straight road, and we may have to take loops and start once again when new evidence comes to light. For practice, you are encouraged to run exploratory regression again with the Queen's contiguity method for calculating the weights matrix file.

Interpreting results: The overall interpretation for OLS model is presented in Table 6.4:

Expenses = -150.46 + 2.75University + 0.01Income + 0.16 Rent

Exercise 6.2 (*cont.*)

Table 6.4 Regression summary results.

Value		Significant	Interpretation	Remarks	Evaluation
Model Overall significance/ performance					
R-Squared	0.878	Not applicable (N/A)	The model explains 87.8% of Expenses variation.	Adjusted R-squared should also be checked to avoid model overfitting. Adjusted R-squared is preferred, especially when their difference is large.	☑
Adjusted R-Squared	0.874	(N/A)	Adjusted R-squared is 0.874, indicating high goodness of fit.	In cases of low goodness of fit, we should add extra variables and/or test other regression methods (e.g., spatial lag model).	☑
F-Statistic (Also called Joint F-Statistic	207.139	Significant at the 0.01 significance level	We reject the null hypothesis and accept that the regression model is statistically significant. The trend is a real effect, not due to random sampling of the population.	F-statistic is of minor importance as most of the times is significant. It should be cross-checked with joint Wald statistic if Koenker (KB) test is statistically significant.	☑
Joint Wald Statistic	415.966	Significant at the 0.01 significance level	As the joint Wald test is statistically significant, it is an indication of overall model significance.	Joint Wald statistic is used to analyze the overall performance of the model especially when Koenker BP statistic is statistically significant ($p < 0.05$).	☑
Akaike Info Criterion corrected (AICc)	884.451	N/A	It is used for comparison with the spatial model. AICc indicates better fit when values decrease.	AICc is a measure of the quality of models for the same set of data. It is used to compare different models and select the most appropriate.	
Coefficients					
t-statistic	Uni/sity: 0.000027* Income: 0.000000* Rent: 0.000003*	Variables significant at the 0.01 significance level	All variables significant	As Koenker BP statistic is statistically significant, robust probabilities should be checked instead.	☑
White standard errors (Robust probabilites)	Uni/sity: 0.006796* Income: 0.000000* Rent: 0.000007*	Variables significant at the 0.01 significance level	All variables significant	Regression coefficients remain the same as in the OLS model. Standard errors and p-values change. Coefficients' values significant in OLS might be different after the white standard calculation. The new coefficients that are significant are robust to heteroscedasticity.	☑
Multicollinearity					
Variance Inflator Factor	Uni/sity: 1.664086 Income: 2.590905 Rent: 1.912884	VIF = 1-4 No collinearity VIF = 4-10 Further analysis needed VIF > 10 Severe collinearity	All variables have values smaller than 4. No multicollinearity issues.		☑

Exercise 6.2 *(cont.)*

Table 6.4 *(cont.)*

Value		Significant	Interpretation	Remarks	Evaluation
Normality of residual errors					
Jarque–Bera test	70.548	Significant at the 0.01 significance level	We reject the null hypothesis (errors normality) as the value of the test is statistically significant.	Biased model	☒ Violating errors normality assumption
Heteroscedasticity					
Koenker–Bassett test	13.373	Significant at the 0.01 significance level	We reject the null hypothesis (homoscedasticity) as the value of the test is statistically significant.	When the test is significant, relationships modeled through coefficients are not consistent. We should look at standard errors probabilities (White, HAC) to determine coefficients' significance. Large difference between the values of Breusch–Pagan test and Koenker (KB) test also indicates violation of the errors normality assumption.	☒ Violating the assumption of constant variance of errors. Still, standard errors are statistically significant.
Spatial Dependence					
Moran's I	0.26	Not significant at the 0.01 significance level	There is not sufficient evidence to reject the null hypothesis (no spatial autocorrelation).		☑
Overall conclusion		Model suffers from non-errors normality and heteroscedasticity			
Next step		Try a GWR model			

Although the OLS performs moderately well, it would be reasonable to attempt improving the reliability of the predictions from the models by using GWR. GWR would also allow us to map the coefficient estimates and to examine further whether the process is spatially heterogeneous.

Exercise 6.3 GWR

In this exercise, we model Expenses as a function of University, Income and Rent using GWR. With GWR, a local model is produced by fitting a regression equation to every single spatial feature in the dataset.

Before we run the tool, we show how ArcGIS presents the results of GWR. For more details on this section, you can visit ESRI's website (ESRI 2016b).

Exercise 6.3 (*cont.*)

GWR output: GWR as generated in ArcGIS yields three main outputs: (A) common diagnostics, (B) output feature class (shapefile) containing all local estimates and (C) coefficient raster surfaces mapping each coefficient separately. In more detail,

(A) **Common diagnostics:** This table describes the model variables and the diagnostic results. It is used for comparison reasons with the global OLS model (GWR is the local model). It contains the following:

- **Bandwidth or the number of neighbors:** This is the distance or the number of neighbors used in the spatial kernel adopted for each local estimation. It controls the degree of smoothing. The larger the values and the smoother the model, the more global the results are. The smaller the number, the more local the results are.
- **Residual squares:** It is the sum of the squared residuals of the model. The smaller the value, the better the fit of the GWR model.
- **Effective number of parameters:** It is a measure of the complexity of the model and is also used in the calculation of other diagnostics. It is equivalent to the number of parameters (intercept and coefficients) in OLS. The value of the effective number of parameters is a trade-off between the variance of the model and the bias in the coefficient estimates, and is heavily dependent on the bandwidth. Suppose that the global model (OLS) has k parameters for n data points (observations/locations). If the bandwidth value tends to infinity, then the local model tends to be global and the number of parameters tend to k. If the bandwidth value tends to zero, then the local models are based only on the regression points, as weights tend to zero as well. In this case, the parameters are n (as many as the data points). The effective number of parameters is a number ranging from k to n and might not be an integer. Additionally:
  - ➢ When the effective number of parameters tends to k (global model), then the weights tend to 1 and the local coefficient estimates have small variance but are biased.
  - ➢ When the effective number of parameters tends to n (data points), then the weights tend to zero and the local coefficient estimates have large variance but are not that biased.
- **Sigma** is the square root of the normalized residual sum of squares. It is the estimated standard deviation of the residuals. As in the case of residuals, the smaller the value, the better the model. As sigma is used to calculate AICc, we may only analyze AICc in our results.

Exercise 6.3 (*cont.*)

- **AIC corrected (AICc)** is a measure of the quality of models for the same set of data based on the AIC criterion. It is used to compare different models (i.e. GWR vs. OLS) in order to select the most appropriate one. The comparison is relative to those models available, which are not necessarily the most accurate. As a rule of thumb, if the AICc value of two models differs by more than 3, then the one with the smallest AICc is better. For small samples or for samples with a relatively large number of parameters, AIC corrected is advised to be used instead of uncorrected AIC.
- ***R*-squared** is a measure of goodness of fit indicating the explained variation of the dependent variable.
- **Adjusted *R*-square:** is a measure of goodness of fit that is more reliable than *R*-squared.

(B) **Output feature class** (shapefile): It is used primarily for mapping standardized residuals. It contains:

- **Condition number:** The condition number (field COND) reveals if local multicollinearity exists. As a rule of thumb, condition values larger than 30, equal to null or negative reveal multicollinearity, and results should be examined with caution. A high condition number reveals high sensitivity in the regression equation for small changes in the variables' coefficients. Problems of local multicollinearity for independent variables might exist when values (whether ratio, nominal or categorical) cluster geographically. For this reason, categorical variables with few categories, dummy or regime variables should not be used.
- **Local *R*-squared:** It is the *R*-squared value for the local model and is interpreted as a measure of goodness of fit like the global *R*-squared. By mapping *R*-squared, we observe where GWR model predicts well and where it predicts poorly. Local *R*-squared can be used to identify potential model misspecification.
- **Predicted values:** The estimated fitted values of the dependent variable.
- **Coefficients:** Coefficients and intercepts for each location.
- **Residual values:** The subtraction of the fitted values from the observed values (dependent variable).
- **Coefficient standard errors:** When standard errors are small in relation to absolute values, then coefficient values are more reliable. Large standard errors may reveal local multicollinearity issues.

Exercise 6.3 *(cont.)*

(C) **Coefficient raster surfaces:** A raster surface per coefficient is created. It shows the variation in the coefficient estimation for each independent variable.

ArcGIS Tools to be used: Geographically Weighted Regression, Spatial Autocorrelation Moran's I, Generate Spatial Weights Matrix, Create Graph

ACTION: GWR regression
Navigate to the location you have stored the book dataset and click on My_Lab6_Regression.mxd.

ArcToolbox > Spatial Statistic Tools > Modeling Spatial Relationships > Geographically Weighted Regression

Input Feature Class = City (see Figure 6.22)

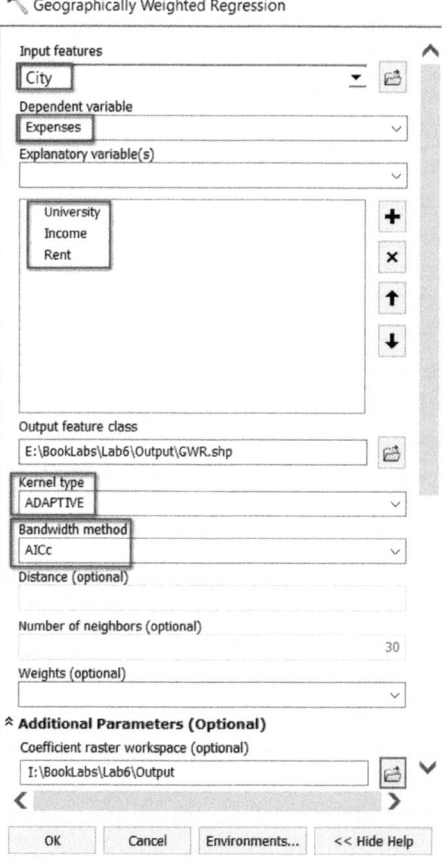

Figure 6.22 GWR dialog box.

Exercise 6.3 *(cont.)*

Dependent Variable = Expenses

Explanatory Variable = University, Income, Rent

Output Feature Class = I:\BookLabs\Lab6\Output\GWR.shp

Kernel Type = ADAPTIVE

Bandwidth method = AICc

Weights = Leave blank

Coefficient raster workspace = I:\BookLabs\Lab6\Output\ > Add

Leave all other fields blank or as filled by default

OK

(A) Common Diagnostics (Supplementary Table)

ACTION: Add GWR_supp to the TOC

ArcCatalog > Navigate to I:\BookLabs\Lab6\Output\ > Drag and drop GWR_supp.dbf to the TOC

TOC > RC GWR_supp > Open

We begin with analyzing the supplementary table GWR_supp.dbf, which presents the primary results of the GWR model (see Figure 6.23).

Interpreting results: The base for analyzing GWR output (see Figure 6.23) is the OLS results, which are referred to as the global model as it provides a single model for the whole study area. GWR results along with a comparison with the global OLS model presented in Exercise 6.2 are listed and interpreted in Table 6.5.

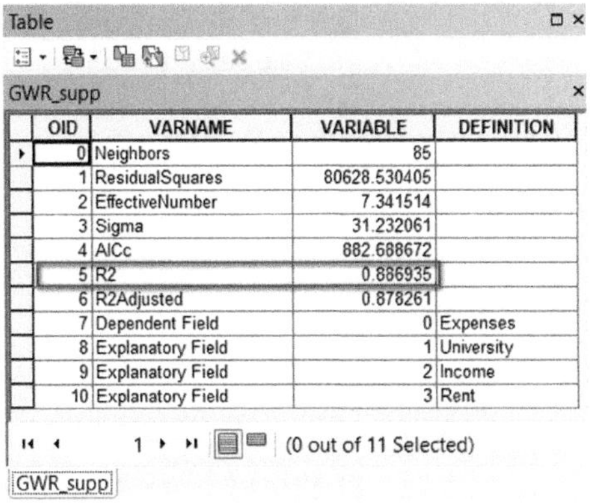

Table

GWR_supp

OID	VARNAME	VARIABLE	DEFINITION
0	Neighbors	85	
1	ResidualSquares	80628.530405	
2	EffectiveNumber	7.341514	
3	Sigma	31.232061	
4	AICc	882.688672	
5	R2	0.886935	
6	R2Adjusted	0.878261	
7	Dependent Field	0	Expenses
8	Explanatory Field	1	University
9	Explanatory Field	2	Income
10	Explanatory Field	3	Rent

(0 out of 11 Selected)

GWR_supp

Figure 6.23 Common diagnostics of GWR.

Exercise 6.3 *(cont.)*

Table 6.5 Model overall significance/performance for the local GWR and OLS model

	GWR (local) Value	OLS (global) Value	Interpretation	Evaluation
Bandwidth or number of neighbors	85 nearest neighbors	No direct comparison. OLS uses all spatial objects (90).	85 out of 90 spatial objects are used for each local estimation. Under each kernel, about 94% of the data are used, which is quite large. The model tends rather to a global one.	
Sum of residual squares	80628.530	86692.533	The local model has a smaller sum of residual squares than the global model, indicating better performance.	☑
Effective Number	7.34	4	In this dataset, the effective value could vary between 4 and 90. The value of 7.34 is very close to 4, indicating that as the bandwidth value tends to 90, the local model tends to be global, and the number of parameters tends to 4.	
Akaike Info Criterion corrected (AICc)	882.688	884.451	Their value difference is smaller than 3, indicating that there is not any substantial improvement in the GWR model.	☒
R-Squared	0.886	0.878	Value is practically the same with the global model. The	☑

Exercise 6.3 (*cont.*)

Table 6.5 (*cont.*)

	GWR (local) Value	OLS (global) Value	Interpretation	Evaluation
Adjusted R-Squared	0.878	0.874	model explains 88.6% of expenses' variation. Value is practically the same with the global model. There is high goodness of fit.	☑
Overall conclusion		Local GWR model did not significantly improve the global OLS model. GWR is better applied when hundreds of spatial units exist. For the educational needs of this example, we use 90, which probably leads to a poor GWR model.		
Next Step		Continue by mapping residuals and coefficients.		

(B)   Mapping Standardized Residuals (Output Feature Class) and Calculating Spatial Autocorrelation: Over- and Underpredictions
Interpreting results: The standardized residuals reveal how much of the spatial data variation is not explained by the independent variables. From the map inspection, we observe that overpredictions (blue) and underpredictions (red) are located around the downtown area, engulfing the historic and financial center (see Figure 6.24). To better analyze residuals, we plot them using a scatter plot. (see Figure 6.26).

```
ACTION: Create residual plot

Main Menu > View > Graphs > Create Graphs

Graph type = Scatter Plot (see Figure 6.25)

Layer/Table = GWR

Y field = StdResid

X field = Predicted

Uncheck Add to legend > Next > Title = Standardized Residual
plot > Finish > RC on graph > Add to layout
```

Exercise 6.3 (*cont.*)

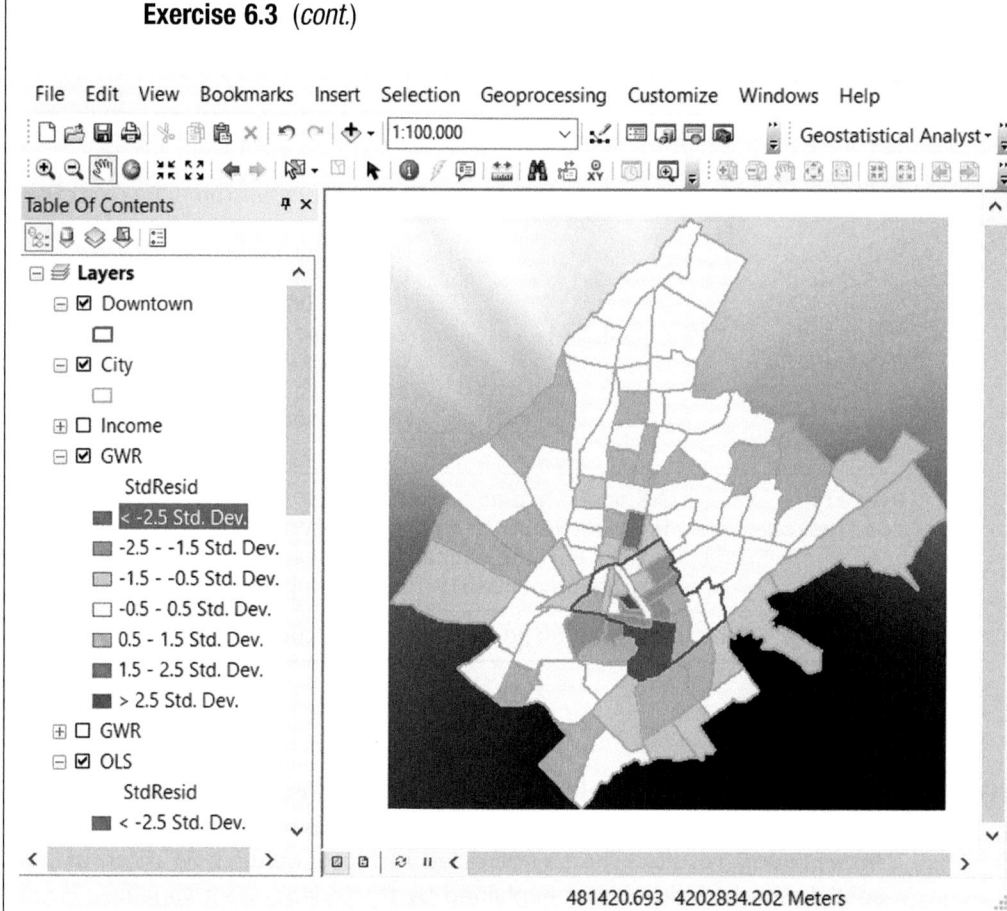

Figure 6.24 Standardized residuals mapping.

Get back to the Data View > Brush the dots in the upper-right corner

Main Menu > Save

Interpreting results: In general, the scatter plot reveals heteroscedasticit,y although the majority of residuals seem to have a random pattern (see Figure 6.26). There are two residuals (center up – underprediction), and one residual (right down – overprediction) that deviate a lot from the general pattern (brush to see the related postcodes). These extreme residuals lie inside the historic center and the financial district (downtown) and should be closely inspected to discover the potential reasons for these significant differences from the observed values.

Exercise 6.3 (*cont.*)

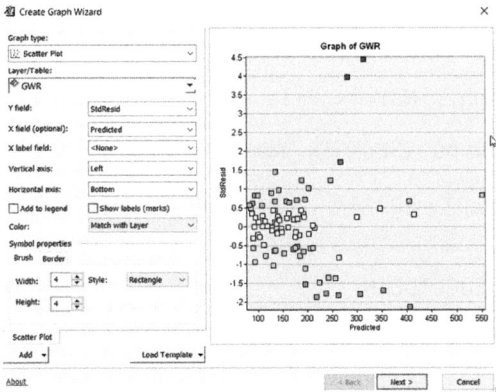

Figure 6.25 Creating scatter plot.

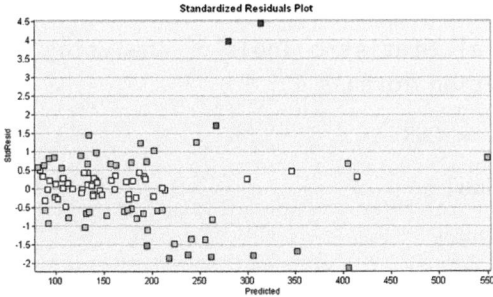

Figure 6.26 Standardized residual plot.

Spatial Autocorrelation Global Moran's I

Continue by calculating Moran's I (see Box 6.4).

ACTION: Spatial Autocorrelation (Morans I)

Box 6.4 Before running the tool, calculate again the weights matrix to ensure its smooth operation.

ArcToolBox > Spatial Statistics Tools > Modeling Spatial

Relationships > Generate Spatial Weights

Matrix Input Feature Class = GWR.shp

Unique ID Field = Source_ID

Exercise 6.3 *(cont.)*

> **Box 6.4** *(cont.)*
>
> Output Spatial Weights Matrix File =
>
> I:\BookLabs\Lab6\Output\SW8KGWR.swm
>
> Conceptualization of Spatial Relationships = K_NEAREST_ NEIGHBORS
>
> Distance Method = EUCLIDEAN
>
> Number of Neighbors = 8
>
> Row Standardization = Check
>
> OK

ArcToolbox > Spatial Statistic Tools > Analyzing Patterns > Spatial Autocorrelation Moran's I

Input Feature Class = GWR

Input Field = Residual

Generate Report = Check

Conceptualization of Spatial Relationships =

GET_SPATIAL_WEIGHTS_FROM_FILE

Weights Matrix File = I:\BookLabs\Lab6\Output\SW8KGWR.swm

OK

   To view the results of the test:

Main Menu > Geoprocessing > Results > Current Session > Spatial Autocorrelation > DC Report File: MoransI_Results0.html

Global Moran's I Summary

Moran's I Index:-0.011492

Expected Index:-0.011236

z-score:-005704

p-value:0.995449

Save
TOC > RC GWR > Open Attribute Table > RC Cond > Sort Descending
Close table

Exercise 6.3 *(cont.)*

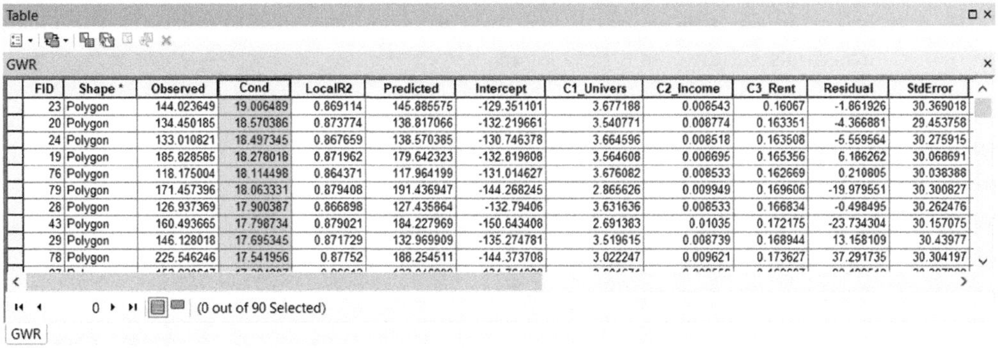

	FID	Shape *	Observed	Cond	LocalR2	Predicted	Intercept	C1_Univers	C2_Income	C3_Rent	Residual	StdError	
	23	Polygon	144.023649	19.006489	0.869114	145.885575	-129.351101	3.677188	0.008543	0.16067	-1.861926	30.369018	
	20	Polygon	134.450185	18.570386	0.873774	138.817066	-132.219661	3.540771	0.008774	0.163351	-4.366881	29.453758	
	24	Polygon	133.010821	18.497345	0.867659	138.570385	-130.746378	3.664596	0.008518	0.163508	-5.559564	30.275915	
	19	Polygon	185.828585	18.278018	0.871962	179.642323	-132.819808	3.564608	0.008695	0.165356	6.186262	30.068691	
	76	Polygon	118.175004	18.114498	0.864371	117.964199	-131.014627	3.676082	0.008533	0.162669	0.210805	30.038388	
	79	Polygon	171.457396	18.063331	0.879408	191.436947	-144.268245	2.865626	0.009949	0.169606	-19.979551	30.300827	
	28	Polygon	126.937369	17.900387	0.866898	127.435864	-132.79406	3.631636	0.008533	0.166834	-0.498495	30.262476	
	43	Polygon	160.493665	17.798734	0.879021	184.227969	-150.643408	2.691383	0.01035	0.172175	-23.734304	30.157075	
	29	Polygon	146.128018	17.695345	0.871729	132.969909	-135.274781	3.519615	0.008739	0.168944	13.158109	30.43977	
	78	Polygon	225.546246	17.541956	0.87752	188.254511	-144.373708	3.022247	0.009621	0.173627	37.291735	30.304197	

(0 out of 90 Selected)

GWR

Figure 6.27 Inspecting Cond field.

Interpreting results: Given the z-score of -0.00570, the pattern does not appear to be significantly different than random. This is a sign of an effective GWR model (OLS lead to the same conclusion as well). In cases where Moran's I does not run due to an error in weights matrix, save the project, exit and run again.

By opening the attribute table of the output feature class, we observe no issues of local multicollinearity, as no value (Cond field in GWR.shp attribute table) is larger than 30, a threshold value to consider that multicollinearity exists (current values are less than 19; see Figure 6.27). Standard errors of coefficients are relatively small, something expected when effective number is small (close to OLS parameters). Local *R*-squared is high in every single model, with the lowest being 0.85.

(C)   Mapping Local Coefficients (Coefficient Raster Surfaces)

We do not analyze every single coefficient, as this would require much space. We mostly focus on how to assess raster coefficients to trace potentially hidden underlying processes. In cases where GWR is not effective due to spatial autocorrelation, mapping coefficients remains informative. For example, deviations or trends of coefficients might reveal valuable information that will help us to build a better model in a later stage.

ACTION: Raster surface

```
TOC > RC Income (raster layer) > Properties > TAB = Symbology
Color Ramp = Blue to Red
OK
TOC > Move Income on the top before Downtown and City layers
(List By Drawing Order)
```

Exercise 6.3 (*cont.*)

We can also map coefficients per postcode by just using the symbology menu under the properties of the GWR.shp. We should first save the GWR residuals scheme as a new layer.

```
TOC > RC GWR > Save As Layer File >
I:\BookLabs\Lab6\Output\GWRStResid.lyr
Add layer to the TOC
TOC > RC GWR > Properties > TAB = Symbology > Quantities >
Graduated colors
Value = C2_Income
Color Ramp = Blue to Red > OK
Main Menu > File > Save
```

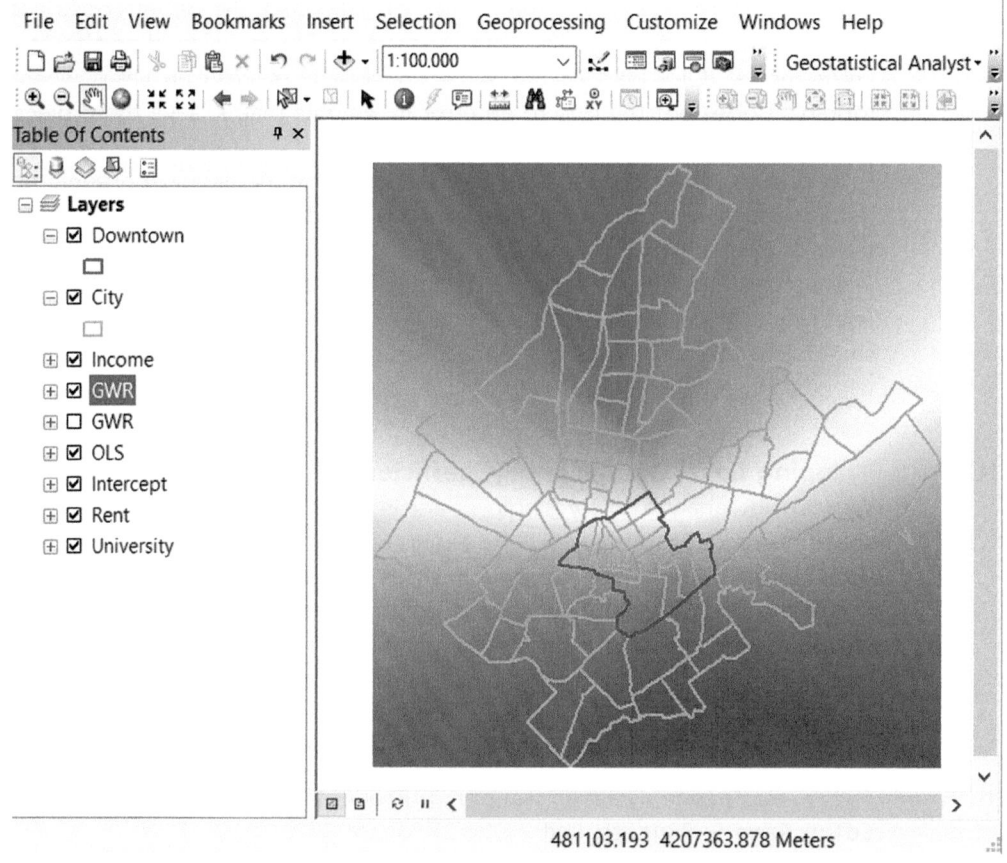

Figure 6.28 Mapping local coefficients of Income as a raster surface.

Exercise 6.3 (*cont.*)

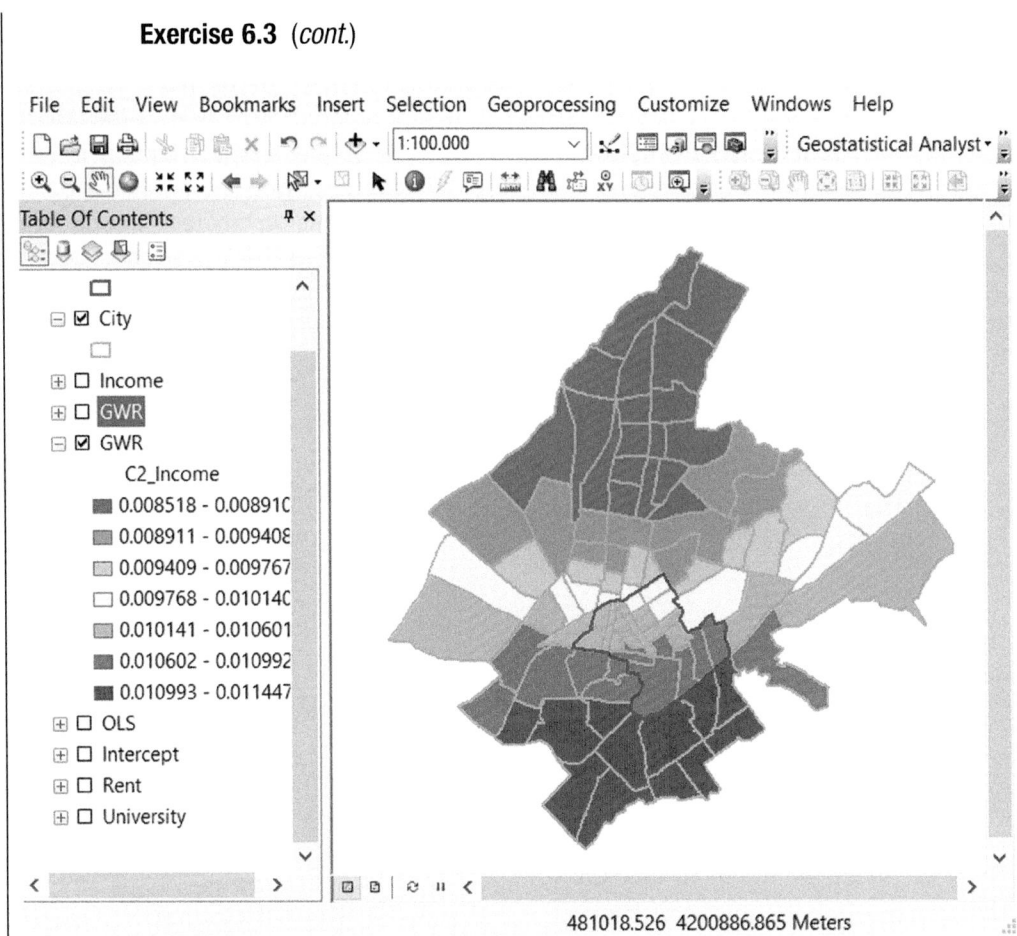

Figure 6.29 Mapping local coefficients of Income per spatial unit.

Interpreting results: Mapping Income coefficient estimates of the local model reveals a trend from north to south, with intensity of coefficients particularly starting to increase at the historical and financial center (see Figures 6.28 and 6.29). Moreover, coefficient values vary significantly from place to place (0.008 to 0.01). From the policy perspective, as the map for the local coefficients reveals that the influence of Income varies considerably across the study area (spatial heterogeneity presence), local policies seem more appropriate than regional policies. The global coefficient and the local coefficients are all positive for Income, so there is an agreement between the local and the global model on how this independent variable influences the dependent Expenses.

Exercise 6.3 *(cont.)*

As an overall comment, GWR performs relatively better than the global OLS model, providing us with rich mapping capabilities that better assist in identifying varying spatial relationships. Spatial autocorrelation is not present in residuals, and the model exhibits high goodness of fit and no multicollinearity issues. By mapping coefficients estimates, trends are traced, and valuable information may be extracted. GWR performs better with larger datasets (more than 160 spatial units) than those in the current layer.

7 Spatial Econometrics

THEORY

Learning Objectives

This chapter deals with

- Spatial econometrics notions
- Handling spatial dependence and spatial heterogeneity through spatial econometric models
- Diagnostics for spatial dependence (Lagrange multiplier tests)
- Spatial autoregressive model (SAR)
- Spatial error model
- Spatial filtering
- Spatial regression with regimes
- Spatial two-stage least squares (S2SLS)
- Maximum likelihood
- Generalized method of moments (GMM)

After a thorough study of the theory and lab sections, you will be able to

- Distinguish which spatial regression method to use in order to account for spatial dependence and spatial heterogeneity
- Interpret spatial diagnostics used in OLS regression that can guide you to the appropriate spatial regression model
- Build spatial regression models and interpret results
- Identify if spatial dependence or spatial heterogeneity is accounted for by the adopted method
- Conduct spatial econometric analysis through GeoDa space.

7.1 Spatial Econometrics

Definition
Spatial econometrics is a set of methods dealing with spatial dependence (spatial autocorrelation) and spatial heterogeneity (spatial structure) in

regression modeling for cross-sectional and panel data (Anselin 2003). Such regression models are also called spatial regression models.

Spatial regression models are regression models that take into account geographical concepts (i.e., spatial dependence, spatial heterogeneity) to identify relationships among dependent and independent variables.

Why Use

We may use regression models and spatial econometrics (a) to identify relationships among dependent and independent variables of sample observations with locational reference and (b) to handle potential spatial dependence and spatial heterogeneity of spatial data.

Discussion

Econometrics is a set of statistical methods originally introduced to analyze economic data. These methods focus on models in which parameters are estimated using regression analysis. Spatial econometrics overlaps with spatial analysis and econometrics and is applied mostly in regional studies.

As already mentioned in the beginning of this book, two problems arise when sample data observations have a locational component, namely (a) spatial dependence (spatial autocorrelation among observations) and (b) spatial heterogeneity (in the sense that the relationships to be modeled are not stationary throughout the study area; LeSage et al. 1998). Traditional econometric models ignore to a large extent these issues and for this reason are not reliable when used for spatial data. For example, in OLS regression (discussed in the previous chapter), the observed samples are assumed to be independent, and the residual errors to be normally distributed. In cases where spatial autocorrelation occurs, this assumption is violated, as spatial dependence exists and, thus, the values of a variable are not independent. Observations are now spatially clustered or dispersed. Moreover, due to potential non-stationarity, geographical trends may exist in the data that also violate the independence assumption. Spatial heterogeneity also violates the assumption that a single linear relationship exists for the entire dataset as the relationship changes according to location (LeSage et al. 1998).

To sum up, when spatial dependence (a functional relationship between what is happening in one location and what is happening in its vicinity) and spatial heterogeneity (lack of space uniformity) exist, residuals are not independent from each other, errors are not normally distributed and a single model cannot accurately describe the existing relationships (Murray 2010). In such a case, OLS regression as described in Chapter 6 is no longer valid (Wang 2014 p. 176), and spatial regression models should be tested.

The most widely used models taking into consideration spatial dependence are

- Spatial autoregressive models model (spatial lag)
- Spatial error models
- Spatial filter model (spatial filtering)

For handling spatial heterogeneity, we may use

- Spatial regression with regimes
- GWR (presented in Chapter 6).

7.2 Spatial Dependence: Spatial Regression Models and Diagnostics

When spatial dependence/autocorrelation exists and we use an OLS model,

- Correlation coefficients are larger than they really are (biased upward).
- The coefficient of determination (R-squared) is larger than it should be. This inflation misleads to the wrong conclusion that relationships are stronger than they are in reality.
- Standard errors appear smaller than in reality, and variability seems smaller due to spatial autocorrelation.
- p-values and F-significance might be found statistically significant just because errors seem to be small (precision is exaggerated).
- We might conclude that a relationship exists although it does not.
- Samples might not be appropriately selected. Spatial autocorrelation infers redundancy in the dataset. If we select samples, then each new observation is expected to provide less new information affecting the calculation of confidence intervals (O'Sullivan & Unwin 2010).

To deal with spatial dependence in regression analysis, several models have been proposed as (Anselin 2003):

- **Spatial autoregressive model (SAR).** An additional variable to the already existing independent variables, named spatially lagged variable, is added. It is calculated based on the dependent variable. The model is called **spatial lag model** (Ord 1975). This model is appropriate when we want to discover spatial interaction and its strength. The lag model controls spatial autocorrelation in the dependent variable.
- **Spatial error models.** The spatial error model controls spatial auto-correlation in the residuals, so it controls dependent and independent variables. This model is appropriate when we want to correct the bias of spatial autocorrelation due to the existence of spatial data (no matter whether the model of interest is spatial or not) (Anselin 2003). For this reason, it is preferred occasionally as it is more robust (Briggs 2012).
- **Spatial filter model** (spatial filtering; Getis 1995). Spatial filtering uses a different approach than spatial autoregressive model and spatial error model. This model separates the spatial effects from the nonspatial effects of the independent variables. Then any standard regression analysis can be used.

7.2.1   Diagnostics for Spatial Dependence

Diagnostics

To detect if spatial dependence exists, we use Global Moran's I and a collection of diagnostic tools called Lagrange multiplier test statistics (or Rao Score test statistics; Anselin 2014 p. 104). These tests are applied while we perform an OLS model to guide us on whether we should conduct spatial regression or just keep the OLS results (see Table 7.1). These tests are reported here from their implementation perspective (for further mathematical details, refer to Anselin 2014 p. 104). Table 7.1 presents both spatial and nonspatial diagnostics for the OLS regression, the assumption violation each test detects and the null hypothesis tested.

The **Lagrange multiplier collection** consists of four tests (Anselin 2014 p. 104):

- Lagrange multiplier (lag; or Lagrange multiplier test against spatial lag): Used to detect if spatial lag autocorrelation exists and if a lagged variable should be included in the regression. If the *p*-value is smaller than the significance level, then there is an indication that we should add the lagged variable in the model.
- Lagrange multiplier (error; or Lagrange multiplier test against error): Used to detect if spatial error autocorrelation exists and if a spatial error model should be used instead of the OLS. If the *p*-value is smaller than the significance level, then there is an indication that we should use the spatial error model.
- Robust Lagrange multiplier (lag): The robust statistic value is always smaller than the Lagrange multiplier (lag). If the difference (from the non-robust value) is small and the *p*-value is small, then it is an indication that spatial lag autocorrelation exists and we should apply a spatial lag model. If the difference is large, then typically *p*-value is also large, indicating that we should consider the spatial error model instead of a spatial lag model.
- Robust Lagrange multiplier (error): The robust statistic value is always smaller than the Lagrange multiplier (error). If the difference is small and the *p*-value is small, then it is an indication that spatial error autocorrelation exists and we should apply a spatial error model. If the difference is large, then typically the *p*-value is also large, indicating that we should consider the spatial lag model instead of a spatial error model.

Robust Lagrange multiplier tests are only used when both Lagrange multiplier lag and error are statistically significant.

Lagrange multiplier-SARMA test pertains to the higher-order alternative of the model with both spatial lag and spatial error terms existing in the model (Anselin 2005 p. 197, Anselin 2014 p. 197). Still, this test is not very informative, and it is not very useful in practice, as it tends to be significant when either the lag

Table 7.1 Statistical tests applied to OLS to detect assumptions violations and potential spatial effects. For further reading on the nonspatial tests, refer back to Chapter 6 and Table 6.3.

Name of diagnostic test	Detects	Hypothesis tested (When the p-value is smaller than the significance level, we reject the null hypothesis.)
		Spatial diagnostics
Lagrange Multiplier (lag)	Spatial lag autocorrelation	**Null hypothesis:** spatial autoregressive parameter ρ is zero H_0: $\rho = 0$ **Alternative hypothesis:** spatial autoregressive parameter ρ is **not** zero H_1: $\rho \neq 0$
Lagrange Multiplier (error)	Spatial error autocorrelation	**Null hypothesis:** spatial autoregressive parameter λ is zero H_0: $\lambda = 0$ **Alternative hypothesis:** spatial autoregressive parameter λ is **not** zero H_1: $\lambda \neq 0$
Robust Lagrange Multiplier (lag)	Spatial lag autocorrelation	**Null hypothesis:** spatial autoregressive parameter ρ is zero H_0: $\rho = 0$ **Alternative hypothesis:** spatial autoregressive parameter ρ is **not** zero H_1: $\rho \neq 0$
Robust Lagrange Multiplier (error)	Spatial error autocorrelation	**Null hypothesis:** spatial autoregressive parameter λ is zero zero H_0: $\lambda = 0$ **Alternative hypothesis:** spatial autoregressive parameter λ is **not** zero H_1: $\lambda \neq 0$
Lagrange Multiplier (SARMA)	Spatial lag and error autocorrelation	**Null hypothesis:** spatial autoregressive parameters ρ and λ are zero H_0: $\rho = 0$, $\lambda = 0$ **Alternative hypothesis:** spatial autoregressive parameters ρ and λ are not zero H_1: $\rho \neq 0$, $\lambda \neq 0$
Moran's I	Spatial autocorrelation	**Null hypothesis:** there is no spatial autocorrelation in residual error **Alternative hypothesis:** there is spatial autocorrelation in residual error
Anselin and Kelejian test	Spatial lag autocorrelation (applied only for the spatial lag model)	**Null hypothesis:** there is no spatial autocorrelation remaining in the model **Alternative hypothesis:** spatial autocorrelation remains in the model
		Nonspatial diagnostics
Jarque–Bera test	Non normality	**Null hypothesis:** normal distribution of regression errors **Alternative hypothesis:** non-normal distribution of regression errors
White test	Heteroscedasticity	**Null hypothesis:** constant variance of regression error (homoscedasticity)

Table 7.1 (*cont.*)

Name of diagnostic test	Detects	Hypothesis tested (When the *p*-value is smaller than the significance level, we reject the null hypothesis.)
Breusch–Pagan test	Heteroscedasticity	**Alternative hypothesis:** nonconstant variance of regression error (heteroscedasticity) **Null hypothesis:** constant variance of regression error (homoscedasticity)
Koenker (KB) test	Heteroscedasticity	**Alternative hypothesis:** nonconstant variance of regression error (heteroscedasticity) **Null hypothesis:** constant variance of regression error (homoscedasticity) **Alternative hypothesis:** nonconstant variance of regression error (heteroscedasticity)
Chow test	Difference between coefficients when we use regimes	**Null hypothesis**: Intercept and coefficients are equal across different regimes. (Anselin 2014 p. 295) **Alternative hypothesis**: Intercept and coefficients are not equal across regimes.
Joint Wald Test	Overall performance / Misspecification	**Null hypothesis:** coefficients' values are zero H_0: $b = 0$ **Alternative hypothesis:** coefficients' values are not zero H_1: $b \neq 0$
t-Statistic	Coefficients performance	**Null hypothesis:** coefficients' values are zero H_0: $b = 0$ **Alternative hypothesis:** coefficients' values are not zero H_1: $b \neq 0$
Joint F-Statistic (or else just F-statistic)	Overall performance of all coefficients except the constant term	**Null hypothesis:** coefficients' values are zero H_0: $b = 0$ **Alternative hypothesis:** coefficients' values are not zero H_1: $b \neq 0$

or the error model is also statistically significant (Anselin 2014 p. 197). Even if the Lagrange multiplier-SARMA test is statistically significant, it is advised to consider the second-order alternative (both spatial lag and spatial error) only when a single spatial parameter model (either lag or error) has first been implemented and has not achieved in removing spatial autocorrelation (Anselin 2014 p. 121).

Moran's I test has been excessively presented in Chapter 4. When Moran's I test null hypothesis of no autocorrelation is rejected, we consider that spatial autocorrelation does exist. Still, Moran's I test does not provide any guidance about which direction we should search for autocorrelation, namely lag or error autocorrelation. As a result, there is no indication of using a spatial lag model instead of a spatial error model. In this respect, Moran's I test can be seen more as a misspecification test.

Anselin and Kelejian test (Anselin & Kelejian 1997) is a test for the remaining spatial autocorrelation of residuals after the spatial lag model has been applied. In this respect, it is calculated only when we apply a spatial lag model and not for an OLS model. If the p-value is smaller than the significance level, then there is an indication that spatial autocorrelation remains in the spatial lag model (Anselin 2014 p. 142).

7.2.2   Selecting between Spatial Lag or Spatial Error Model

To decide which model (spatial lag or spatial error) is more appropriate we should first run an OLS model along with the related spatial diagnostics as presented in Table 7.1. The diagram in Figure 7.1 can guide us through the appropriate choice between the two models. This process has been successfully applied in many experiments (Anselin 2014 p. 109). To effectively use spatial diagnostics, the original OLS model should provide a reasonable fit to the data. If it does not, it is better to reconsider the structure (e.g., variables) of the OLS model to better fit existing data and then proceed to the application of a spatial lag or error model. As such, it is important to have a solid knowledge of multiple linear regression theory and practice (see Chapter 6).

As the figure suggests:

- **If one of the tests (LM lag, or LM error) is statistically significant,** then we run the corresponding (statistically significant) model.
- **If no test is significant,** we keep the OLS results, as there is no obvious need for spatial error or spatial lag model and no indication of spatial autocorrelation.
- **In cases where both tests are significant,** then we should check the corresponding robust tests. In general, the Lagrange multiplier test is sensitive to the other alternative. For example, the Lagrange multiplier (lag) is sensitive to the presence of spatial error autocorrelation. Just calculating Lagrange Multiplier (lag) as statistically significant is not enough, as we should consider the other alternative (spatial error autocorrelation presence) as well. Similarly, just calculating Lagrange Multiplier (error) as statistically significant is not enough. The reason is that we might get both tests as statistically significant for the same dataset. To correct for this test sensitivity, the robust Lagrange multiplier is introduced. In fact, we use the robust forms of Lagrange multipliers only in the cases where both Lagrange multiplier errors – lag and Lagrange multiplier – are statistically significant.

Figure 7.1 Select spatial lag or spatial error model according to Lagrange multiplier tests (see also figure 5.1 in Anselin 2014 p. 110).

When calculating robust tests:

- **If only one robust test** is **statistically significant,** we keep the statistically significant corresponding model.
- **If both robust tests are statistically significant,** (in cases with a large number of observations), we keep the model with the larger value of the robust statistic. Still, this choice should be made with extra consideration. It is probably better to check if there is any model misspecification (i.e., missing variables). We should probably reconsider the spatial conceptualization method and the related spatial weights matrix.

7.2.3   Estimation Methods

Once we decide on the model to account for spatial dependence, estimation methods should be used to produce the coefficients, the parameters of the model as well as the relevant statistics. The most widely used estimation methods in spatial econometrics also presented in this book are the following (Anselin 2014 p. 2 and 4):

- The spatial two-stage least squares (S2SLS) for the spatial lag model
- Maximum likelihood estimation (ML) for the spatial lag and the spatial error model
- The general method of moments (GMM) for models containing both a spatial lag and spatial error term (Kelejian & Prucha 2010)

The spatial two-stage least squares estimation method and the general method of moments are also referred to as modern methods of spatial econometrics.

7.3   Spatial Lag Model

Definition
A **spatial lag model** is a spatial regression method to account for spatial autocorrelation of the dependent variable and includes a new variable (on the right-hand side of the equation) called spatially lagged dependent variable (*Wy*) (Baller et al. 2001). A spatial lag model is also called *spatially autoregressive model* (SAR model; Fotheringham et al. 2000) or *mixed*[1]

---

[1] If we do not include any independent variables, then *Xb* is not included in the model. This is the pure spatially lagged autoregression model (SAR). If we add independent variables *X*, then we have a *mixed* (that is where mixed comes from) *regressive spatial autoregressive* model (MRSA; de Smith 2018 p. 547)

regressive spatial autoregressive model (SAR model), (Darmofai 2015 p. 97, Ord 1975). The model is expressed according to Anselin's (2014 p. 159) notation as (7.1):

$$y = \rho W y + X b + u \tag{7.1}$$

where

> y is the $n \times 1$ vector of observations of the dependent variable Y. n is the number of observations
> W is the $n \times n$ spatial weights matrix (also called spatial lag operator)
> Wy is a product called the spatial lag term and is the extra variable added; it is called spatially lagged dependent variable
> ρ is a spatial autoregressive coefficient
> X is the $n \times k$ matrix of independent variables, where k the number of independent variables
> b is the $k \times 1$ coefficient vector
> u is the $n \times 1$ vector of random errors of the model

The spatially lagged variable is the weighted sum of the neighboring values of the dependent variable Y for each location in its neighborhood (Anselin 2014 p. 40). For the observation i, the spatial lag of y_i is (7.2):

$$[Wy]_i = \sum\nolimits_{j=1}^{n} w_{i,j} y_j = w_{i,1} y_1 + w_{i,2} y_2 + \cdots + w_{i,n} y_n \tag{7.2}$$

where $w_{i,j}$ is the weight between the i-th object and the j-th object stored on spatial weights matrix W.

The spatial lag model in Equation (7.1) can be rearranged (in a reduced form) as (7.3):

$$y - \rho W y = X b + u \tag{7.3}$$

If ρ was known then, b could be estimated by OLS.

The equation can also be written as (7.4):

$$y = (I - \rho W)^{-1} X b + (I - \rho W)^{-1} u \tag{7.4}$$

This reduced form indicates that the value of y at each location i is not only determined by the values of X_i but also from the X values of the neighbors through the spatial multiplier expressed as (7.5):

$$Spatial\ Multiplier = (I - \rho W)^{-1} \tag{7.5}$$

where I is the identity matrix (Wang 2014 p.177, Baller et al. 2001)

There are two methods to estimate the parameters of the spatial lag model, namely

- Spatial two-stage least squares (S2SLS)
- Maximum likelihood (ML)

In Sections 7.3.1 and 7.3.2, we will present their primitives, their implementation and the analysis of the results. We will not delve deeper into technical details of the statistical background of these methods, as this is beyond the scope of this book (for a more thorough analysis read, Anselin 2014).

Why Use

To build a regression model that handles spatial autocorrelation. A spatial lag model gives the ability to assess (Yrigoyen 2007)

- The degree of spatial dependence when controlling for the effects of other variables
- The significance of the other independent variables after the spatial dependence is controlled

Discussion and Practical Guidelines

The concept behind the spatial lag model is that the outcome (dependent variable) is not only influenced by a set of independent variables but from the neighboring values of the same dependent variable as well. In other words, the outcome of Y (e.g., house price) in one location depends on the Y values (house price) of nearby locations (spatial autocorrelation) along with other variables (e.g., size, floors). For this reason, we should also include this influence in the equation by setting up a neighborhood and the related spatial weights matrix.

As described in Chapter 1, spatial weights are numbers that reflect some sort of distance, time or cost between a target spatial object and every other within a specified neighborhood. If we use the standardized spatial weights matrix, where all row weights add up to 1 with $\sum_{j=1}^{n} w_{i,j} = 1$, then the spatial lagged variable becomes a weighted average of the dependent values at the neighboring observations of I (Anselin 2014 p. 40).

Suppose we have the spatial objects (postcodes) depicted in Figure 7.2. The weights matrix W based on the Rook's case adjacency (only sharing edges), considering those polygons that are adjacent as neighbors, is (7.6):

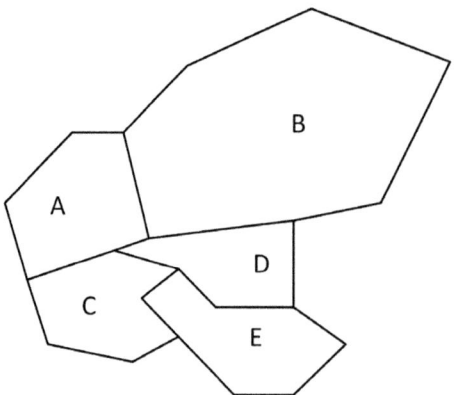

Figure 7.2 Example for adjacency matrix.

$$
Adj = \begin{bmatrix}
 & A & B & C & D & E & SUM \\
A & * & 1 & 1 & 1 & 0 & 3 \\
B & 1 & * & 0 & 1 & 0 & 2 \\
C & 1 & 0 & * & 1 & 1 & 3 \\
D & 1 & 1 & 1 & * & 1 & 4 \\
E & 0 & 0 & 1 & 1 & * & 2 \\
SUM & 3 & 2 & 3 & 4 & 2 &
\end{bmatrix}
\tag{7.6}
$$

Suppose that the dependent variable Y is the average *Income* of people living in each postcode. The value of the spatially lagged variable to be used in addition to other dependent variables is given by (7.2) as

$$
[Wy]_i = w_{i,1}y_1 + w_{i,2}y_2 + \cdots + w_{i,n}y_n
$$

In this example y_i refers to the income of the i-th postcode. For instance, the value for postcode E is

$$
[Wy]_E = w_{E,C}y_C + w_{E,D}y_D
$$

as C and D are the postcodes with nonzero weights.

If we use a different conceptualization method, we will get a different spatial weights matrix. A spatially lagged variable is usually automatically created (i.e., GeoDa software). Still, the user has to decide which type of conceptualization method is appropriate (see Chapter 1).

7.3.1   Spatial Two-Stage Least Squares (S2SLS)

Definition
Spatial two-stage least squares (S2SLS) is a method that can be used for the estimation of the parameters in the spatial lag model.

Endogenous variables are those independent variables in a linear regression model that are correlated with the error term. Endogenous variables' existence in a regression model violates the independence assumption of linear regression (the error term should be independent from the predictor variables).

An **instrumental variable** (or just instrument) is a variable that is highly correlated to the endogenous variable but not to the error term.

Why Use

Because of the presence of spatial dependence and the existence of endogenous variables, the OLS estimators cannot be used for a spatial model. As such, alternative estimators should be applied – such as the S2SLS,or the maximum likelihood, explained in the next section.

In the case where endogenous variables do exist, the regression model can be expressed as (7.7):

$$y = Xb + Y\gamma + u \tag{7.7}$$

where

Y denotes the endogenous variables. It is a $n \times s$ matrix, where s is the number of endogenous variables.

γ is the $s \times 1$ vector of coefficients of the endogenous variables.

y, X, b, u are explained in equation 7.1.

The main idea/assumption to deal with an endogenous variable is to replace the endogenous variable with another one (called instrument) that is

(a)   highly correlated with the endogenous variable
(b)   not correlated to the error term

In other words, if we find/build a variable that is highly correlated to the endogenous variable and not to the error term, then we can follow a method called two-stage least squares (2SLS) approach to replace the original with a new one.

As the name says, 2SLS is an estimation taking place in two stages (Anselin 2014 p. 140):

1.   At the first stage, each of the endogenous variables is regressed with all exogenous variables and instruments (matrix of instruments).
2.   The predicted values of the preceding regression are used as independent variables, replacing the endogenous variables.

This method as presented up to now is for linear regression and not for spatial regression, as we have not included any spatial weights matrix. In cases of a spatial lag model, the spatially lagged variable is also endogenous. The model is then expressed as (7.8) (Anselin 2014 p. 159):

$$y = \rho Wy + Xb + Y\gamma + u \tag{7.8}$$

where

Y is an $n \times s$ matrix of observations of s endogenous variables (excluding the spatially lagged dependent variable).

In other words, the preceding model has

(a)   the spatially lagged dependent variable as endogenous
(b)   potentially additionally independent endogenous variables (Y)

Discussion and Practical Guidelines

The choice of instrumental variables for the endogenous independent variables is not easy. One approach is to set as instrumental variables those that have been successfully utilized in similar studies. Another method is to make assumptions and test if an instrumental variable removes correlation with the error term. Instruments might also be the exogenous variables (as in the case of the spatially lagged dependent variable). As soon as instrumental variables are defined, exogenous variables are then replaced by the predicted values from a regression of the endogenous variables and the instruments (stage 2; Anselin 2014 p. 161). S2SLS is a robust estimation method since it does not require the assumption of normality in the errors (Yrigoyen 2007). Still, the selection of additional instruments is sometimes subjective. Finally, some general discussion and guidelines are provided next for (i) goodness of fit, (ii) robust coefficient variance estimation and (iii) coefficient interpretation.

(i)   *Interpreting Goodness of Fit*

To measure the goodness of fit of the model, we use the pseudo R-squared and the spatial pseudo R-squared, which are calculated as the squared correlation between the observed dependent variable and the corresponding predicted value (see more in Anselin 2014 p. 193). Both metrics are not true R-squared metrics, as they do not reveal the variation explained by the model. In fact, R-squared is not valid when spatial dependence exists and, as such, is not applicable in spatial lag models. The reason is that the predicted values and residuals are calculated based on equations that ignore the error that follows from having to predict Y, and also include Y in the right-hand side (of the equations). As such, the predicted values give an overly optimistic assessment of the model fit (Anselin 2014 pp. 166, 171). In this sense, these metrics can be used as rough indicators of relative goodness of fit. In addition, these metrics are neither comparable to OLS R-squared nor to maximum likelihood spatial lag model pseudo R-squared (next section; Yrigoyen 2007). More appropriate and comparable fits are the log-likelihood, the Akaike information criterion (AIC) and the Schwarz criterion (SW). An increased log-likelihood value is an indication of better fit. On the other hand, AIC and SW indicate better fit when their values decrease.

(ii)   *Robust Coefficient Variance Estimation*
To test if the adopted spatial lag model has accounted for the effects of spatial autocorrelation and spatial heterogeneity, we compute the standard errors to be robust to the presence of unspecified heteroscedasticity, or robust to both heteroscedasticity and unspecified spatial autocorrelation. To do so, we calculate the White standard errors, along with the heteroscedasticity and autocorrelation consistent (HAC) standard errors (Anslein 2014 p. 125).
   Robust coefficient estimates (White and HAC) can also be calculated when we perform OLS regression solely. In such cases, results provide information pertaining to potential heterogeneity or spatial autocorrelation existing in the residual errors. We can cross-check these results with the relevant statistical tests as White test, Jarque–Bera test and Koenker (KB) test.

(iii)   *Coefficient Interpretations*
For the spatial lag model, as shown from the Equation (7.4), the value of y at any location i is the function of the value of x at this location as well as the value of x at neighboring locations (Wang 2014 p. 177). Coefficients are not interpreted as straightforwardly as in the nonspatial regression, where a one-unit change in x induces a b change in y. In spatial lag regression, the effect of a change in one unit of x is just a part of the change of y. The neighboring values of x cause an extra change in y. Specifically, in cases of spatial lag regression, a one-unit change in variable x_h (that is, the h column vector of X change by one unit for all locations) leads to a total effect on y by $\frac{b_h}{1-\rho}$ (Kim et al. 2003).

In essence, the total effect of a one-unit change in variable x_h is composed from a direct effect b_h of the variable x_h at location i and an indirect effect $b_h\rho/(1-\rho)$ driven by the neighboring values of x_h (7.9). The indirect effect is called spatial multiplier (Anselin 2014 p. 165).

$$\text{Total effect on } y = (\text{direct effect of } x_h \text{ in location } i) \tag{7.9}$$
$$+ (\text{indirect effect of } x_h \text{ in neighborhood of } i)$$
$$\text{Total effect on } y = (\text{direct effect of } x_h \text{ in location } i) + (\text{Spatial Multiplier})$$
$$\frac{b_h}{1-\rho} = b_h + (\text{Spatial Multiplier}) \tag{7.10}$$
$$\text{Inderect effect} = \text{Spatial Multipler} = \frac{b_h}{1-\rho} - b_h = b_h\rho/(1-\rho)$$

b_h is the coefficient of variable x_h calculated by the spatial lag model.

Furthermore, a change in only one location spreads the change in all other locations like a wave. If, for example, a variable x_i changes in one location , then by using the fitted spatial model, we can predict the value of the dependent variable y_i for this location.

As y changes in one location, it will also affect the value of y in its neighborhood, as this is how the spatial lag model is defined. But the changes do not stop in the neighborhood. The neighbors of the neighbors of the original location will also be affected, and so on, until all locations reached.

From the policy perspective, the splitting of the effect of a change in a one-unit change of x at a location i into two parts gives the ability to account (a) for the effect of space and (b) the effect of the variable. By such approach, one can study the following:

(a)   The effect of equal change of a variable for the entire study area (e.g., how house price will change – dependent variable – if inflation – one of the independent variables – increases by 1% in the entire study area

(b)   How localized changes of a specific independent variable (that only take place in a specific spatial entity) induce changes to the dependent variable for the whole study area. For example, we can analyze the changes in house prices (dependent variable) for the entire study area if we select to change an independent's variable value (i.e., increase money spent for public safety or for urban renewal program) in a specific spatial unit (i.e., postcode)

Apart from predicting y values for different x values for spatial entities already included in the model, the prediction may also be used to estimate the y values of spatial entities not included in the original model. A very good presentation about predictions in SAR models is given in Goulard et al. (2017 p. 2). In general, using the spatial autoregressive models for prediction purposes is not widely applied within the context of spatial econometrics, in contrast to geostatistics. This is a research area that should be further developed.

7.3.2   Maximum Likelihood

Definition
Maximum likelihood (ML) estimation is a parametric statistical approach used to estimate the parameters of a statistical model.

Why Use
In the context of spatial econometrics, ML can be used to estimate the parameters of a spatial lag model and a spatial error model.

Discussion and Practical Guidelines
A parametric approach is based on strong statistical assumptions. When the assumptions adopted are correct, parametric methods yield more accurate

estimates than the nonparametric methods. ML assumes that the error term is normally distributed. Moreover, the spatial weights used in ML estimation for the spatial log model must refer to symmetric contiguity relation. For this reason, ML estimation of this model can be used only in conjunction with Rook's contiguity, Queen's contiguity and distance band, and not with *k*-nearest neighbors (Anselin 2005 p. 202).

As a measure of fit, the pseudo *R*-squared, log-likelihood, Akaike information criterion (AIC) and Schwarz criterion (SW) are used. As mentioned in Section 7.3.1, pseudo *R*-squared can be used as rough indicator of relative goodness of fit and is not as useful as the log-likelihood, AIC or SW. An increased log-likelihood value is an indication of an improved fit. On the other hand, AIC and SW indicate better fit when their values decrease (Anselin 2005 p. 175). In addition to the Lagrange multiplier tests referred to previously, the likelihood-ratio test is used to test the significance of the autoregressive parameter ρ on the null hypothesis that $\rho = 0$ (Anselin 2014 p. 198, Anselin 2005 p. 209). If the *p*-value is smaller than the significance level, then we reject the null hypothesis that $\rho = 0$. In such cases, we accept the result of the test as statistically significant and adopt a spatial lag model (refer to Figure 7.1 and recall that spatial error might also be significant, so there is an extra step to take). To test heteroscedasticity, we use the Breusch–Pagan test (Table 7.1).

Comparison of ML and S2SLS

Results of ML method might be better than S2SLS method. This does not necessarily indicate a more precise model. We should keep in mind that the ML estimator is based on an idealized asymptotic setting and does not take into account potential non-normality of the errors or heteroscedasticity (Anselin 2014 p. 202). S2SLS estimates, on the other hand, remain valid in the presence of the preceding problems, as the assumption of normality in the errors is not required (Yrigoyen 2007 slide 16). Finally, White and HAC standard errors are not calculated for the ML method. In this respect, using the S2SLS method is more appropriate for addressing non-normality and heterogeneity issues.

7.4   Spatial Error Model

Definition

The **spatial error model** is a form of regression model that handles spatial dependence by the inclusion of a spatial autoregressive error term (de Smith et al. 2018 p.548, Wang 2014 p. 177).

The model is expressed according to Anselin's (2014 p. 207) notation as (7.11):

$$y = Xb + u \tag{7.11}$$

where u is the $n \times 1$ vector of errors following a spatial autoregressive process (7.12):

$$u = \lambda W u + \varepsilon \qquad (7.12)$$

and

y is the $n \times 1$ vector of observations of the dependent variable Y. n is the number of observations

X is the $n \times k$ matrix of independent variables, and k the number of independent variables

b is the $k \times 1$ coefficient vector

W is the $n \times n$ spatial weights matrix

λ is the spatial autoregressive parameter

ε is the $n \times 1$ vector of idiosyncratic errors; these errors are uncorrelated but may exhibit heteroscedasticity

The preceding equations yield the reduced form (7.13):

$$y = Xb + (I - \lambda W)^{-1}\varepsilon \qquad (7.13)$$

This equation shows that the value of the dependent variable y_i, at each location i, is affected by the stochastic error ε at all nearby locations, according to the spatial filter $(I - \lambda W)^{-1}$.

Why Use

Instead of handling the dependent variable as autoregressive as in the case of a spatial lag model, the error of the model is now considered as spatially autoregressive (Wang 2014 p. 177).

Discussion and Practical Guidelines

Estimation of the spatial error model is implemented mainly by two methods, namely the generalized method of moments (GMM) and the maximum likelihood method (ML).

- **Generalized method of moments (GMM).** The generalized method of moments (GMM) is used to estimate the autoregressive coefficient λ. The GMM method is robust to heteroscedasticity, and no assumption on normality is needed. Pseudo R-squared is used as a measure of goodness of fit.
- **Maximum likelihood.** Similarly to the spatial lag model, the pseudo R-squared, log-likelihood, Akaike information criterion (AIC) and Schwarz criterion (SW) are used as measures of fit. An increased log-likelihood value is an indication of a better fit. On the other hand, AIC and SW indicate better fit when their values decrease. The likelihood-ratio test is used to test the

significance of the autoregressive parameter λ. If the p-value is smaller than the significance level, then we reject the null hypothesis that $\lambda = 0$. In such cases, the result of the test is statistically significant a spatial error model is adopted. To test heteroscedasticity, we use the Breusch–Pagan test.

Comparing GMM to ML, we would argue that GMM is more robust, as estimates remain valid in the presence of heterogeneity. On the other hand, even if ML might produce statistically significant results, there might be strong evidence (through the Breusch–Pagan test or likelihood-ratio test) of remaining heteroscedasticity.

7.5   Spatial Filtering

Definition
Spatial filtering (based on the Getis approach; Getis 1990, Getis 1995) is a method to handle spatial autocorrelation in spatial regression by separating spatial effects from the variables' total effects. It partitions each original spatially autocorrelated variable (either dependent or independent) into a filtered nonspatial variable (spatially independent) and a spatial residual variable (Getis 1995 p. 192, Getis & Griffith 2002, de Smith 2018 p. 555). As soon as the spatially autocorrelated variables have been filtered, and each variable on the right-hand side of the equation is not spatially autocorrelated, a typical OLS model is applied in the usual fashion (Getis 1995 p. 192). The filtered observation x_i in the Getis filtering approach is defined as (7.14) (Getis & Griffith 2002):

$$x_i^* = \frac{W_i / (n-1)}{G_i(d)} x_i \tag{7.14}$$

where

x_i is the original observation
$G_i(d)$ is the local G_i statistic (see Equation [4.8])
n the number of observations
$W_i / (n-1)$ is the expected value for $G_i(d)$

The difference between the observed x_i and the expected (filtered) value stands for a new variable L that represents the spatial effects of the variable X at location i (7.15):

$$L_{xi} = x_i - x^* \tag{7.15}$$

In practice, Equation (7.14) compares the observed value of $G_i(d)$ with the expected one. If there is no spatial autocorrelation, then the observed x_i and the expected (filtered) value x_i^* will be the same (see Eq. 7.15). When the difference in Equation (7.15) is positive, then spatial autocorrelation exists among high values of the variable X. On the contrary, a negative difference indicates spatial autocorrelation among low values of the variable X (Getis & Griffith 2002).

By this filtering, the original spatial autocorrelated variable X is now compounded by a spatial component $L_{xi} = x_i - x^*$ (not correlated to X) and a nonspatial component X^* (7.16):

$$Original\ Variable = Spatial\ Component + non\ Spatial\ Component$$

$$X = L_x + X^* \qquad (7.16)$$

Why Use

Spatial filtering is used to remove the spatial dependence within each dependent or independent variable that is spatially autocorrelated and then proceed with a typical OLS model (Getis 1995 p. 192).

Discussion and Practical Guidelines

The Getis spatial filtering approach provides the following advantages (Getis 1995 p. 194)

- There is no spatial autocorrelation in the X^* variable
- The difference L (7.15) in cases that X is spatially dependent is a spatially autocorrelated variable
- If all variables that exhibit spatial autocorrelation are spatially filtered, then the residuals are not spatially autocorrelated
- After spatial filtering, variables should be statistically significant. Still, the appropriateness of such variables should be theoretically backed up to ensure a rational interpretation.

7.6   Spatial Heterogeneity: Spatial Regression Models

Spatial heterogeneity has potentially the most damaging effects on the results of regression analysis (nonspatial; Griffith et al. 1997), as it provokes

- Misleading significance levels (Paez & Scott 2004)
- Biased parameter estimations
- Suboptimal predictions (Anselin 1988)
- Accepting that a relationship exists, although it does not
- False perception that results apply for the entire dataset (global), although the model might be accurate only for specific geographic regions

Spatial heterogeneity can be classified into (Anselin 2010 p. 4)

- Discrete heterogeneity when there are predefined spatial zones, also called spatial regimes, among which heterogeneity exists and model specifications, parameters and coefficients are free to vary
- Continuous heterogeneity when regression coefficients are free to change over space (no predefined zones needed)

The most widely used methods to deal with spatial heterogeneity in spatial regression analysis are

- Spatial regimes (discrete heterogeneity)
- Geographically weighted regression (continuous heterogeneity, presented in Chapter 6).

7.7   Spatial Regimes

Definition

The **spatial regimes** model (Anselin 1988) is an approach to handle discrete heterogeneity where different coefficients are computed for different subsets of the dataset based on various regression models.

The basic spatial regime model is (7.17):

$$y_{ij} = a_j + b_j x'_{ij} + \varepsilon_{ij} \tag{7.17}$$

where

> $i = 1, \ldots n$ with n the total number of observations
> $j = 1, \ldots J$ with J the total number of spatial regimes
> x'_{ij} is the observation of the variable X at observation i, belonging to spatial regime j
> a_j is the intercept of the model for the spatial regime j
> b_j is the coefficients of the model for the spatial regime j
> ε_{ij} is the error at observation i, belonging to spatial regime j

The preceding model specification states that each regime j has a distinct regression model with a different intercept a_j and different coefficients b_j.

For example, for two regimes, the models to be used will be

$$y_{i1} = a_1 + b_1 x'_{i1} + \varepsilon_{i1}$$

and

$$y_{i2} = a_2 + b_1 x'_{i2} + \varepsilon_{i2}$$

Why Use

The model is used when a different error variance is assumed to hold for each spatial regime.

Interpretation

When using spatial regimes and coefficients that are different in each regime, we have an indication of heterogeneity. In other words, the dependent variable varies differently across space. The spatial regimes method does not identify the factors in which heterogeneity exists, but it diagnoses its presence.

Discussion and Practical Guidelines

The spatial regime model is spatial in its nature when regimes have been constructed based on a form of spatial structure. For example, regimes such as urban and non-urban counties, Metropolitan Statistical Areas and non–Metropolitan Statistical Areas, or forest and non-forest areas have a clear spatial structure (Grekousis & Mountrakis 2015). If we create regimes whose structures are based on another type (e.g., counties starting with the letter A), then this is more a type of classification rather than a regime. In such cases, we can apply regression methods at each class in the usual fashion (nonspatial regression), although we always should look for potential spatial heterogeneity/autocorrelation existence, as our data might be spatially dependent for other reasons than the regime definition.

Methods for Estimates

An OLS, S2SLS or ML approach can be used for the estimates. Still, as ML is based on assumptions of normality and homoscedasticity, it is advised not to be preferred, as in most real case studies, these assumptions are unrealistic (Anselin 2014 p. 330). By using diagnostics for spatial autocorrelation (e.g., Lagrange multiplier), we might also consider other spatial econometric models to mix in, such as spatial lag or spatial error model for each regime. Furthermore, by comparing the results of heteroscedasticity tests when regimes are not used with the results of heteroscedasticity tests when regimes are adopted, we can quantify if heteroscedasticity was accounted for or if it remains in the model.

Chow Test

The Chow test is used to identify whether the coefficients resulting in each regime are statistically different among these regimes. The null hypothesis is that the intercept and the coefficients are equal. If the p-value is smaller than the adopted significance level, then we can reject the null hypothesis and accept that coefficients are statistically different. The Chow test is calculated for the coefficients globally and for each set of coefficients at the regime level. We might get a low p-value at the global level, but this does not mean that each specific set of coefficients is statistically different as well. We have to examine each set of coefficients one by one and based on the significance of each coefficient to the regime it belongs to decide whether different coefficients are rational. In general, if the Chow test is statistically significant, it is an indication that different spatial processes are at work in each regime, revealing heterogeneity. If the Chow test is not statistically significant, then there is not an indication of heterogeneity, and a global model for the entire dataset may be used.

Spatial Weights with Spatial Regimes

The spatial weights structure that is used to identify spatial effects has to be additionally considered, as neighbors might not belong to the same regime. In the general case, we use the weights matrix W where a spatial unit might have a neighbor belonging to another regime. In other words, spatial relationships are built not with a regime restriction but among all observations. For example, consider the US counties in different states (regimes). Although counties across borders belong in different regimes, we might want to consider them neighbors, as interactions may be strong. Still, in many cases, it might be more appropriate to construct a weights matrix in which spatial relationships (e.g., being neighbors or not) are not allowed to exist across regimes borders (e.g., prefectures in neighboring counties). In such cases, the regimes are considered to be isolated, and no interactions are assumed among them.

A different weights matrix is introduced for this case, called regimes weights W_R where no interactions exist among regimes (Anselin 2014 p. 289). The regime weights matrix is a diagonal matrix where each one of the J diagonal elements is a subset of the original matrix W of the whole dataset (7.18):

$$W_R = \begin{bmatrix} W_1 & 0 & \ldots & 0 \\ 0 & W_2 & \ldots & 0 \\ \ldots & \ldots & \ldots & \ldots \\ 0 & \ldots & \ldots & W_j \end{bmatrix} \tag{7.18}$$

Each element is initially constructed from the W matrix. For example, for a spatial object i belonging in j regime, the true k-nearest neighbors (if we select this type spatial conceptualization) are selected from the W matrix, no matter if they belong to a different regime. Then those of the k-nearest neighbors belonging in different regimes are removed. Consequently, the i spatial object does not always have k-nearest neighbors and sometimes might even have no neighbors, if all of them belong to a different regime. In this case, this spatial object is completely isolated. Similar to the previous notion, W_j is also different in case we construct a weights matrix per regime. Although the k-nearest neighbors belong to the same regime, they may not be the true neighbors as other spatial objects belonging to other regimes might be the nearest ones. In conclusion, the type of spatial weights matrix to be used depends on the problem and the conceptualization method adopted.

7.8   Chapter Concluding Remarks

- Two issues arise when sample data observations have a locational component: (a) spatial dependence and (b) spatial heterogeneity.
- Traditional econometric models ignore to a large extent these issues and for this reason are not reliable when used for spatial data.
- When spatial autocorrelation exists, the coefficient of determination (*R*-squared) is larger than it should be (in OLS), and we might wrongly conclude that a relationship is significant although it is not.
- Moran's I test does not provide any guidance for the direction we should look for autocorrelation, namely lag or error autocorrelation.
- As a result, there is no indication of using a spatial lag model instead of a spatial error model. In this respect, Moran's I test can be seen more as a misspecification test.
- The choice between spatial lag or spatial error model can be made with the use of the Lagrange multiplier tests.
- Spatial filtering uses a different approach than spatial error and spatial lag models (Getis 1995). Spatial filtering separates the spatial effects from the nonspatial effects of the variables. Then any standard regression analysis can be used.
- The spatial regime is a method where different coefficients are computed for different subsets of the dataset based on various regression models to handle discrete heterogeneity.
- The Chow test is used to identify whether the coefficients resulting in each regime are statistically different among these regimes.

Questions and Answers

The answers given here are brief. For more thorough answers, refer back to the relevant sections of this chapter.

Q1. What is spatial econometrics?

A1. Spatial econometrics is a set of methods dealing with spatial dependence (spatial autocorrelation) and spatial heterogeneity (spatial structure) in regression modeling for cross-sectional and panel data. Such regression models are also called spatial regression models.

Q2. Which are the most widely used spatial econometric models?

A2. Modeling spatial dependence: Spatial autoregressive models model (SAR, spatial lag), spatial error models, spatial filter models (spatial filtering). Modeling spatial heterogeneity: spatial regression with regimes, geographically weighted regression.

Q3. What is a spatial lag model?

A3. A spatial lag model is a spatial regression method to account for spatial autocorrelation of the dependent variable and includes a new variable (on the right-hand side of the equation) called the spatially lagged dependent variable. This model is appropriate when we want to discover spatial interaction and its strength. The lag model controls spatial autocorrelation in the dependent variable.

Q4. What is a spatial error model?

A4. A spatial error model is a model that controls the spatial autocorrelation in the residuals to control dependent and independent variables. This model is appropriate when we want to correct the bias of spatial autocorrelation due to the existence of spatial data (no matter if the model of interest is spatial or not). For this reason, it is preferred sometimes, as it is more robust.

Q5. What are the problems in OLS estimations when spatial autocorrelation exists?

A5. Correlation coefficients and coefficient of determination (*R*-squared) are likely to appear to be larger than they are (biased upward). This inflation misleads to the wrong conclusion that relationships are stronger than they actually are. Standard errors appear smaller than in reality, and variability seems smaller due to spatial autocorrelation. The *p*-values and the F-significance might be found to be statistically significant just because errors seem to be small. Samples might not be appropriately selected, as spatial autocorrelation infers redundancy in the dataset. If we select samples, then each new observation is expected to provide less new information, affecting the calculation of confidence intervals.

Q6. What are the Lagrange multiplier tests?

A6. It is a set of tests applied while we perform an OLS model to guide us on whether we should conduct spatial regression or just keep the OLS results.

Q7. How can we estimate the parameters of the spatial lag model?

A7. There are two main methods to estimate the parameters of the spatial lag model, namely the spatial two-stage least squares (S2SLS) method and the maximum likelihood (ML) method.

Q8. What is spatial filtering?

A8. Spatial filtering is a method of handling spatial autocorrelation in spatial regression by separating spatial effects from the variables' total effects. It partitions each original spatially autocorrelated variable (either dependent or independent) into a filtered nonspatial variable (spatially independent) and a spatial residual variable. As soon as the spatially autocorrelated variables have been filtered and each variable on the right-hand side of the equation is not spatially autocorrelated, a typical OLS model is applied in the usual fashion.

Q9. What problems arise in nonspatial regression analysis when spatial heterogeneity exists?

A9. Spatial heterogeneity existence leads to misleading significance tests; suboptimal predictions; accepting that a relationship exists, although it does not; and the false perception that results apply to the entire dataset (global), although the model might be accurate only for specific geographic regions.

Q10. What is the spatial regimes method?

A10. The spatial regimes method is an approach to handle discrete heterogeneity where different coefficients are calculated for different subsets of the dataset based on various regression models. When using spatial regimes and coefficients are different per regime, we have an indication of heterogeneity. In other words, the dependent variable varies differently across space. The spatial regimes method does not identify the factors in which heterogeneity exists, but it diagnoses its presence.

LAB 7
SPATIAL ECONOMETRICS

Overall Progress

Spatial Analysis/Lab Workflow

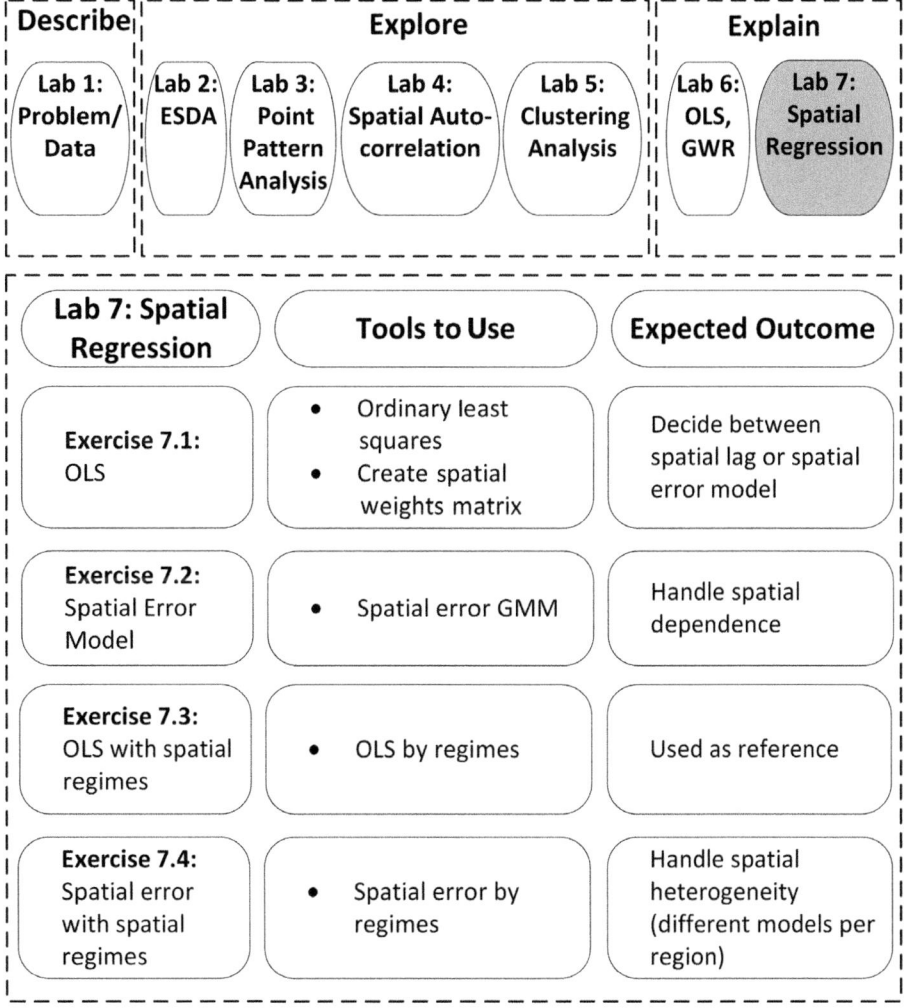

Figure 7.3 Lab 7 workflow and overall progress.

Scope of the Analysis

In this lab, we apply various spatial econometric models to account for potential spatial dependence and spatial heterogeneity in modeling Income (dependent variable; see Figure 7.3). To keep the analysis simple and to better showcase the effects of space in regression analysis, we only use one independent variable (Insurance). We use GeoDa open-source spatial analysis software, to experiment with many different models and also assess the results through various statistics and metrics. We begin with a typical OLS model and related tests and diagnostics to select between a spatial lag or spatial error model. In Exercise 7.2, we account for spatial dependence by testing the spatial error model. Finally, in Exercises 7.3 and 7.4, we study spatial heterogeneity using the spatial regimes method.

Exercise 7.1 OLS

In this exercise, we follow the process described in Figure 7.1 to identify which model (either the spatial lag or spatial error) is more appropriate.
GeoDa and GeoDa Space Tools to be used: Create Weights, OLS

ACTION: Create spatial weights matrix (GeoDa)

Navigate to the location you have stored the book dataset and click on Lab7_SpatialEconometrics.gda

Main Menu > Tools > Weight Manager > Create > Select ID Variable = PostCode

TAB = Distance Weight (see Figure 7.4)

TAB = K-Nearest neighbors

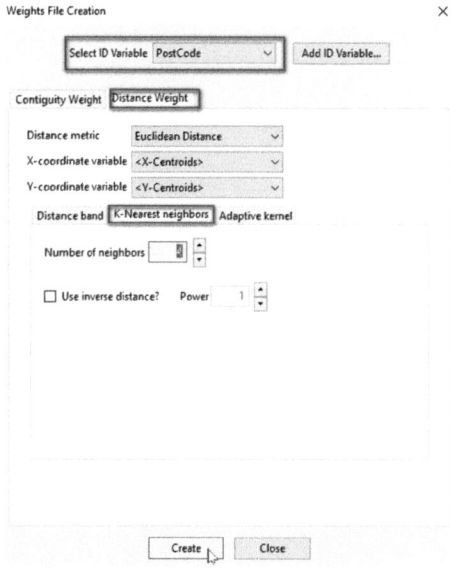

Figure 7.4 Weights file creation.

Exercise 7.1 (*cont.*)

TAB = K-Nearest neighbors

Number of neighbors = 8 > Create

Save as I:\BookLabs\Lab7\Output\City_k8.gwt

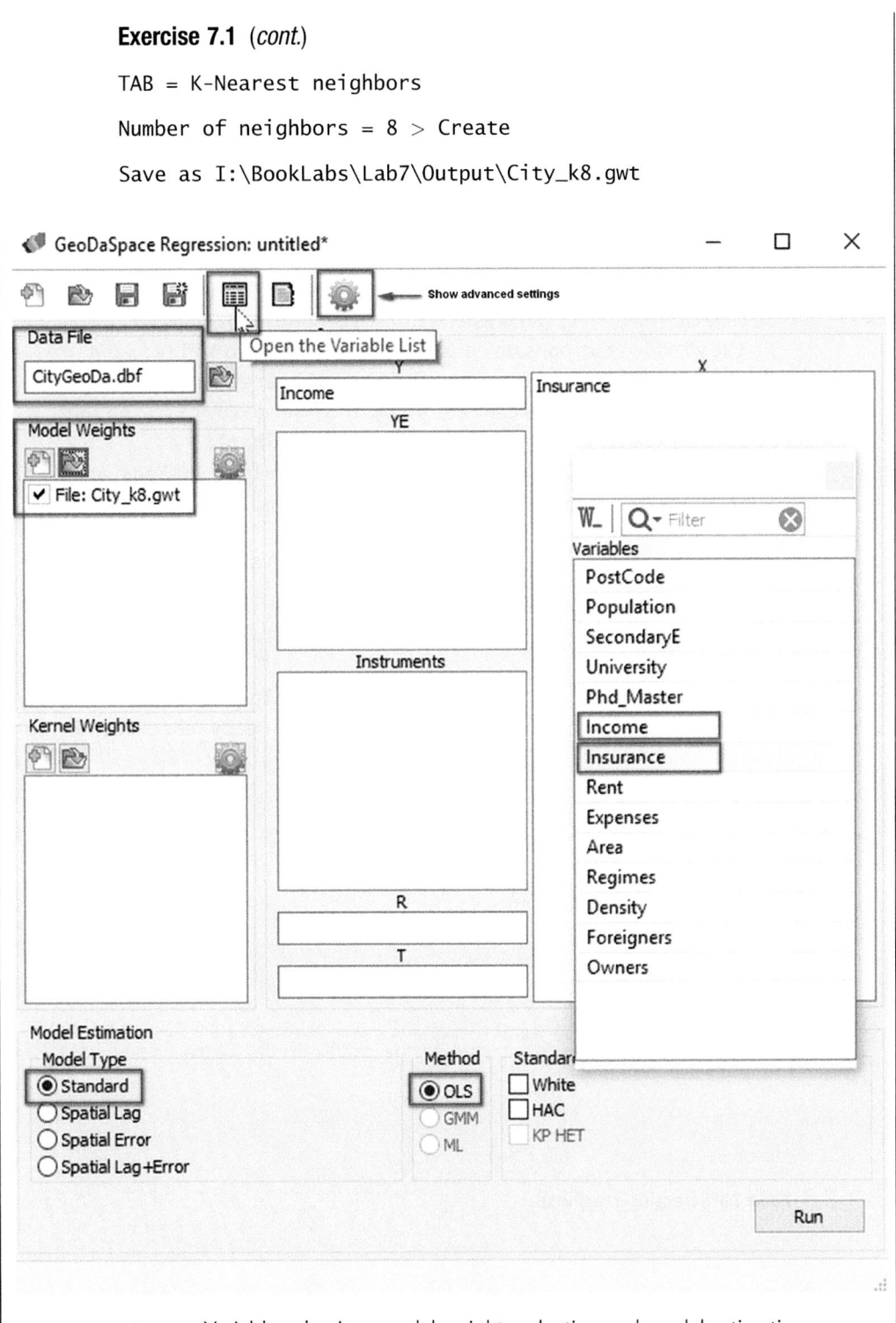

Figure 7.5 Variable selection, model weights selection and model estimation.

Exercise 7.1 (*cont.*)

Close

Save

ACTION: OLS (GeoDa Space)

Start GeoDaSpace by double clicking on GeoDaSpace icon at your desktop >
Click the open file button next to Data File (see Figure 7.5).
Set Data File = I:\BookLabs\Lab7\CityGeoDa.dbf > Open
In the window that pops up, drag and drop Income to Y filed, and
Insurance to X (You can experiment with more variables).
Close this window. (You can open it again if needed by clicking the
"Open Variable List" tool in the main toolbar).
Model Weights > Click the open button > Select I:\BookLabs\Lab7
\Output\City_k8.gwt

Main Tool Bar> Show Advanced Settings (icon) (see Figure 7.5)

TAB = Other (see Figure 7.6)

Check OLS Diagnostics

Check Moran's I of the residuals (to calculate spatial autocorrelation of residuals). If they are already checked, do not
check again.

Save

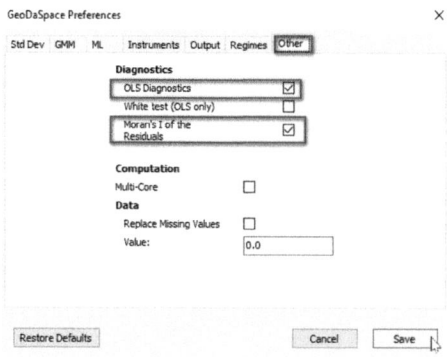

Figure 7.6 Selecting diagnostics.

Exercise 7.1 (*cont.*)

Model Estimation = Standard (see Figure 7.5)

Method = OLS

Do not check Standard Errors

Run

By clicking Run, you will be asked to save the predicted values and residuals in csv format

Save in > I:\BookLabs\Lab7\Output\OLSCityGeoDa_predY_resid.csv > Save

Save Results File As > I:\BookLabs\Lab7\Output\OLSResults.txt

Main Toolbar > Save Model As > I:\BookLabs\Lab7\Output\OLSMo-del.mdl

Results of OLSREsults.txt are:

REGRESSION

SUMMARY OF OUTPUT: ORDINARY LEAST SQUARES

Dataset	:CityGeoDa.dbf		
Weights matrix	:File: City_k8.gwt		
Dependent Variable :	Income	Number of Observations:	90
Mean dependent var :	16316.7534	Number of Variables :	2
S.D. dependent var :	4975.6157	Degrees of Freedom :	88
R-squared :	0.8104		
Adjusted R-squared :	0.8083		
Sum squared residual:	417719419.599	F-statistic :	376.1749
Sigma-square :	4746811.586	Prob(F-statistic) :	1.571e-33
S.E. of regression :	2178.718	Log likelihood :	-818.477
Sigma-square ML :	4641326.884	Akaike info criterion :	1640.955
S.E of regression ML:	2154.3739	Schwarz criterion :	1645.955

Exercise 7.1 (*cont.*)

Variable	Coefficient	Std.Error	t-Statistic	Probability
CONSTANT	6815.4555302	541.0387214	12.5969829	0.0000000
Insurance	39.6472210	2.0441739	19.3952292	0.0000000

REGRESSION DIAGNOSTICS
MULTICOLLINEARITY CONDITION NUMBER   4.489

TEST ON NORMALITY OF ERRORS

TEST	DF	VALUE	PROB
Jarque-Bera	2	216.887	0.0000

DIAGNOSTICS FOR HETEROSCEDASTICITY
RANDOM COEFFICIENTSTEST

	DF	VALUE	PROB
Breusch-Pagan test	1	170.257	0.0000
Koenker-Bassett test	1	36.189	0.0000

DIAGNOSTICS FOR SPATIAL DEPENDENCE

TEST	MI/DF	VALUE	PROB
Moran's I (error)	0.5676	8.927	0.0000
Lagrange Multiplier (lag)	1	40.638	0.0000
Robust LM (lag)	1	2.450	0.1175
Lagrange Multiplier (error)	1	67.845	0.0000
Robust LM (error)	1	29.656	0.0000
Lagrange Multiplier (SARMA)	2	70.294	0.0000

================================ END OF REPORT ==

Interpreting results: Detailed results and interpretation are presented in Table 7.2.

Exercise 7.1 (*cont.*)

Table 7.2 Interpretation of basic regression diagnostics for OLS model with additional spatial diagnostics.

	Value	Significance	Interpretation	Remarks	Evaluation
Model's overall significance/performance					
R-Squared	0.8104	Not applicable (N/A)	The model explains 81.04% of income variation.	Adjusted R-squared should also be checked to avoid model overfitting. Adjusted R-squared is preferred, especially when their difference is large.	☑
Adjusted R-Squared	0.8083	(N/A)	Adjusted R-squared is 0.8083, indicating high goodness of fit.	In cases of low goodness of fit, we should add extra variables and/or test other regression methods (e.g., spatial lag model).	☑
Pseudo R-Squared	Not calculated	N/A		Used in spatial econometric models. Pseudo R-squared is not a true R-squared, as it does not reveal the variation explained by the model. It cannot be compared to R-squared.	
Spatial Pseudo R-Squared	Not calculated	N/A		Used in spatial econometric models. Spatial pseudo R-squared is not a true R-squared, as it does not reveal the variation explained by the model. It cannot be compared to R-squared.	
F-Statistic (Also called Joint F-Statistic	376.175	Significant at the 0.01 significance level	We reject the null hypothesis and accept that the regression model is statistically significant. The trend is a real effect, not due to random sampling of the population.	F-statistic is of minor importance, as usually it is significant. It should be cross-checked with joint Wald statistic if Koenker (KB) test is statistically significant.	☑
Joint Wald Statistic	Not calculated	N/A	If the joint Wald test is statistically significant, it is an indication of overall model significance.	Joint Wald statistic is used to analyze the overall performance of the model, especially when Koenker (KB) statistic is statistical significant ($p < 0.05$).	

Exercise 7.1 (cont.)

Table 7.2 (cont.)

	Value	Significance	Interpretation	Remarks	Evaluation
Akaike Info Criterion (AIC)	1,640.955	N/A	It is used for comparison with the spatial model. AICc indicates better fit when values decrease.	AICc is a measure of the quality of models for the same set of data. It is used to compare different models and select the most appropriate one.	
Log-Likelihood	-818.477	N/A	It is used for comparison with the spatial model. An increased log-likelihood value is an indication of better fit.	Log-likelihood is a measure of the quality of models for the same set of data. It is used to compare different models and select the most appropriate one.	
Schwarz Criterion (SW)	1,645.95	N/A	It is used for comparison with the spatial model. SW indicates a better fit when values decrease.	SW is a measure of the quality of models for the same set of data. It is used to compare different models and select the most appropriate one.	
Coefficients					
t-Statistic	19.395	Variable significant at the 0.01 significance level: Insurance	Independent variable (Insurance) is statistically significant		☑
White standard errors	Not calculated		Text here refers to which variables are statistical significant and which are not.	Regression coefficients remain the same as in the OLS model. Standard errors and p-values change. Coefficients' values significant in OLS might be different after the White standard calculation. The new coefficients that are significant are robust to heteroscedasticity.	
HAC Standard Error	Not calculated		Text here refers to which variables are statistical significant and which are not.	Regression coefficients remain the same as in the OLS model. Standard errors and p-values change. Coefficients' values significant in OLS might be different after the HAC calculation. The new coefficients that are significant are robust to heteroscedasticity and autocorrelation.	

Exercise 7.1 (*cont.*)

Table 7.2 (*cont.*)

	Value	Significance	Interpretation	Remarks	Evaluation
Multicollinearity					
Condition Number	4.489	Rule of thumb: values larger than 30 indicate multicollinearity	Multicollinearity condition number is smaller than 30. No serious multicollinearity issues.	By only inspecting the condition number, we cannot varify which variables should be removed in case of multicollinearity. VIF and variance decomposition should be also used.	☑
Variance Inflator Factor	Not calculated	VIF = 1–4 No collinearity VIF = 4–10 Further analysis needed VIF > 10 Sever collinearity			
Variance Decomposition	Not calculated	<0.5 weak >0.5 severe			
Normality of Residual Errors					
Jarque–Bera test	216.887	Significant at the 0.01 significance level	We reject the null hypothesis (errors normality) as the value of the test is statistically significant.		☒ Violating errors normality assumption
Heteroscedasticity					
Breusch–Pagan test	170.257	Significant at the 0.01 significance level	We reject the null hypothesis (homoscedasticity), as the value of the test is statistically significant, indicating high heteroscedasticity.		☒ Violating assumption of constant variance of errors
Koenker–Bassett test	36.189	Significant at the 0.01 significance level	We reject the null hypothesis (homoscedasticity), as the value of the test is statistically significant.	When the test is significant, relationships modeled through coefficients are not consistent. We should look at standard errors' probabilities (White, HAC) to determine coefficients' significance. A large difference between the values of Breusch–Pagan test and Koenker (KB) statistic also indicates violation of the errors' normality assumption.	☒ Violating assumption of constant variance of errors
White test	Not calculated				

Exercise 7.1 (*cont.*)

Table 7.2 (*cont.*)

	Value	Significance	Interpretation	Remarks	Evaluation
Spatial Dependence					
Moran's I	8.927	Significant at the 0.01 significance level	We reject the null hypothesis (no spatial autocorrelation), as the value of the test is statistically significant. Spatial autocorrelation exists in residuals.	Moran's I does not indicate whether the autocorrelation is due to a spatially lagged dependent variable or due to spatially autoregressive error term.	☒ Violating assumption of residuals independence
Anselin and Kelejian Test	Not calculated		If the null hypothesis is rejected, then spatial autocorrelation remains in the model.	It is used only when we adopt a spatial lag autocorrelation model. Similarly to Moran's I, the rejection of the null hypothesis of no spatial autocorrelation does not point to either a lag or error model to use as a preferable alternative. If this test is statistically significant, the HAC standard errors calculation is necessary, as spatial autocorrelation remains in the model.	
Lagrange Multiplier Lag/Robust	40.638/ 2.450	LM lag significant/ Robust LM lag not significant	Not an indication for utilizing the spatial lag model	Used in OLS to guide us in the selection between the spatial lag or the spatial error model. See Figure 7.1.	☑
Lagrange Multiplier Error/Robust	67.845/ 29.656	LM Error significant/ Robust LM error significant	Due to spatial error autocorrelation presence, we should consider the spatial error model as an alternative, to account for spatial dependency	In this case, both LM lag and LM error are significant, so we should check their robust forms. Only the robust LM error is significant, so we should test a spatial error model.	☒

Overall Conclusion	Model suffers from heteroscedasticity and spatial error autocorrelation issues.
Next Step	Test a spatial error model with GMM estimation method.

Exercise 7.2 Spatial Error Model

GeoDa Space Tools to be used: Spatial Error GMM

ACTION: Spatial Error GMM

Open GeoDaSpace > Open I:\BookLabs\Lab7\Output\OLSModel.mdl

Before we run the new model, we should point where the predicted values and the residuals will be saved. We will use them later in GeoDa to plot Morans's I plot and residuals plot.

Main Tool Bar > Show Advanced Settings (icon) >

TAB = Output (see Figure 7.7)

Check only Save Predicted Values and Residuals. If it is already checked, do not check it again.

Save

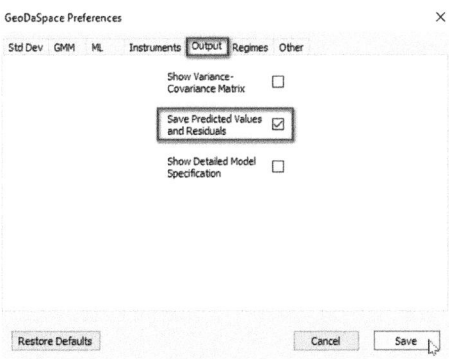

Figure 7.7 GeoDa Space preferences.

Model Weights = City_k8.gwt (see Figure 7.8)

Model Estimation = Spatial Error

Method = GMM

Run

As soon as you click Run, you will be asked to save the predicted values and residuals in csv format

Save in > I:\BookLabs\Lab7\Output\SpatialErrorCityGeoDa_predY_resid.csv > Save

Exercise 7.2 (*cont.*)

```
Save Results File As >
I:\BookLabs\Lab7\Output\SpatialErrorResults.txt
```

```
Main Toolbar > Save Model As >
I:\BookLabs\Lab7\Output\SpatialErrorModel.mdlResults of Spatia-
lError
```

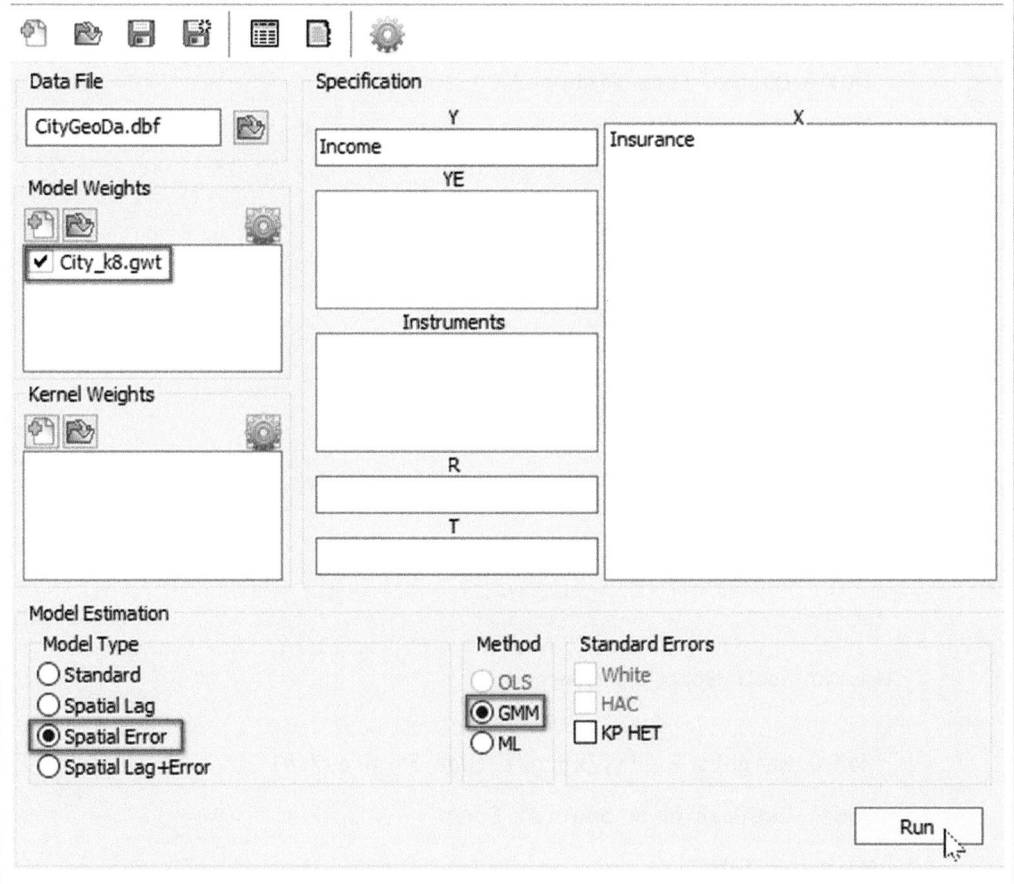

Figure 7.8 Selecting weights and method.

```
Results of SpatialErrorResults.txt are as follows:
```

Exercise 7.2 (*cont.*)

```
REGRESSION

_____

SUMMARY OF OUTPUT: SPATIALLY WEIGHTED LEAST SQUARES (HOM)
_____

Dataset            : CityGeoDa.dbf
Weights matrix     : City_k8.gwt
Dependent Variable :      Income        Number of Observations:      90
Mean dependent var :   16316.7534       Number of Variables   :       2
S.D. dependent var :    4975.6157       Degrees of Freedom    :      88
Pseudo R-squared   :       0.8104          N. of iterations   :       1

------------------------------------------------------------------------
        Variable   Coefficient    Std.Error    z-Statistic   Probability
------------------------------------------------------------------------
        CONSTANT  7693.0776148  1170.7406053    6.5711205     0.0000000
       Insurance    36.5309077     3.0911917   11.8177425     0.0000000
          lambda     0.8333165     0.1027078    8.1134693     0.0000000
------------------------------------------------------------------------

================================ END OF REPORT ========================================
```

Interpreting results: From the interpretation Table 7.2 used in the OLS model in Exercise 7.1, we present only those results relevant to the regression output (Table 7.3).

Table 7.3 Interpretation of basic regression diagnostics for spatial error GMM.

	Value	Significance	Interpretation	Remarks	Evaluation
Model's overall significance/performance					
Pseudo R-squared	0.8104	N/A		Used in spatial econometric models. Pseudo R-squared is not a true R-squared, as it does not reveal the variation explained by the model. It cannot be compared to R-squared.	☑

Exercise 7.2 (*cont.*)

Table 7.3 (*cont.*)

	Value	Significance	Interpretation	Remarks	Evaluation
Coefficients					
z-statistic	See results table	Variables significant at the 0.01 significance level: Insurance, Lamda (λ)	The spatial autoregressive coefficient λ = 0,833 (coefficient of W_Income) is statistically significant, confirming the results of OLS Lagrange multiplier collection tests. The coefficient of Insurance (36.53) is close to the OLS coefficient (39.64). The constant in the spatial error model (7,693.07) is different from the OLS model (6,815.45, see Exercise 7.1).		☑
Overall Conclusion	The model now handles Income and Insurance better, as it accounts for heteroscedasticity. We should also check for spatial autocorrelation of the residuals. **Remark:** Use table SpatialErrorCity_predY_resid.csv and GeoDa capabilities to create Moran's I scatter plot of the residuals.				

Exercise 7.3 OLS with Spatial Regimes

In this exercise, we use the spatial regimes method. We define two regimes. The first one consists of postcodes in the historical and financial business districts of Athens (denoted by 1 in the column Regime; Figure 7.9). The second regime consists of all the remaining postcodes (denoted by 2 in the column Regime). We first run the typical OLS model with spatial regimes where different spatial datasets have different coefficients. Statistically significant different coefficients among regimes for the same variable indicate heterogeneity.

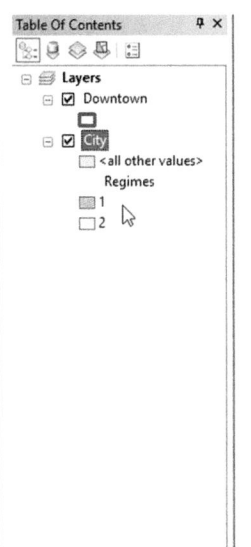
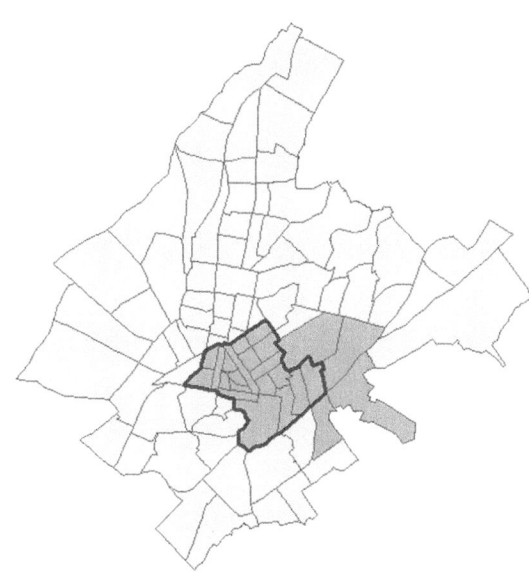

Figure 7.9 Mapping regimes.

We will not refer again to how to set up the variables and the model in GeoDaSpace, as this was explained in Exercise 7.1. The only difference now is that we include a spatial regime, and we also set in the GeoDaSpace Preferences the Error by regimes.

   GeoDa SpaceTools to be used: OLS by regime

ACTION: OLS by regimes

Navigate to the location you have stored the OLS model:

Open GeoDaSpace > Open I:\BookLabs\Lab7\Output\OLSModel.mdl

Main Toolbar > Variables Selection and drag and drop Regimes to R field (see Figure 7.10)

Exercise 7.3 (*cont.*)

```
Main Tool Bar > Show Advanced Settings > TAB = Regimes
```

Error by regimes should be checked (if already checked, do not click it)
Save

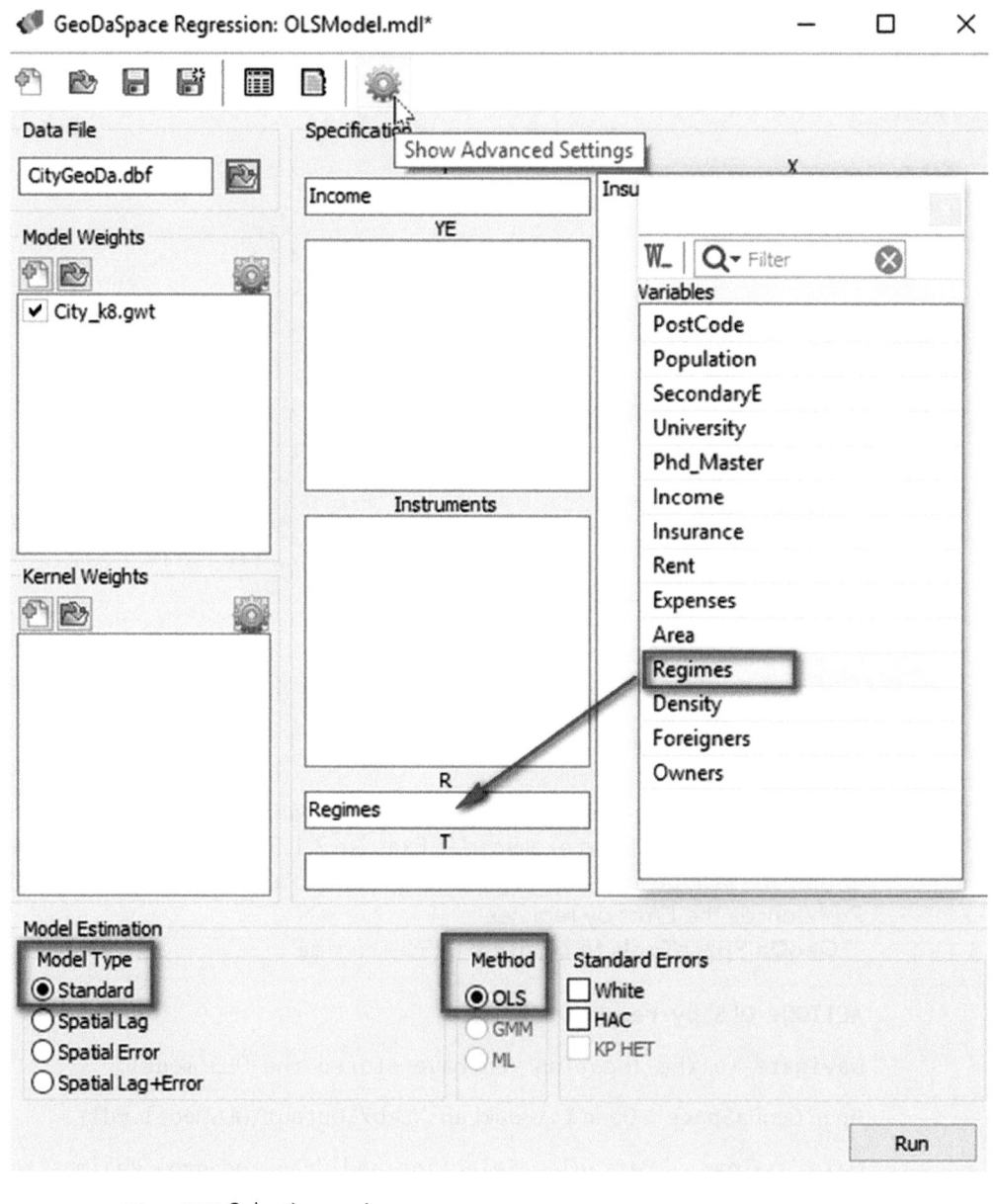

Figure 7.10 Selecting regimes.

Exercise 7.3 *(cont.)*

Model Estimation: Standard

Method: OLS

Run

As soon as you click Run, you will be asked to save the predicted values and residuals in csv format

Save in > I:\BookLabs\Lab7\Output\OLSRegimesCityGeoDa_predY_resid.csv > Save

Save Results File As > I:\BookLabs\Lab7\Output\OLSRegimesResults.txt

Main Toolbar > Save Model As >
I:\BookLabs\Lab7\Output\OLSRegimesModel.mdl

Results of OLSRegimesResults.txt are as follows:
   Regime 1:

REGRESSION
————

SUMMARY OF OUTPUT: ORDINARY LEAST SQUARES ESTIMATION - REGIME 1
——

Dataset	:CityGeoDa.dbf			
Weights matrix	: City_k8.gwt			
Dependent Variable :	1_Income	Number of Observations:		24
Mean dependent var :	22339.3154	Number of Variables :		2
S.D. dependent var :	5248.4513	Degrees of Freedom :		22
R-squared :	0.4322			
Adjusted R-squared :	0.4064			
Sum squared residual:	359717091.199	F-statistic :		16.7482
Sigma-square :	16350776.873	Prob(F-statistic) :		0.0004814
S.E. of regression :	4043.609	Log likelihood :		-232.328
Sigma-square ML :	14988212.133	Akaike info criterion :		468.656
S.E of regression ML:	3871.4612	Schwarz criterion :		471.012

Variable	Coefficient	Std.Error	t-Statistic	Probability
1_CONSTANT	8229.6370715	3545.1511953	2.3213783	0.0299225
1_Insurance	36.9768649	9.0353654	4.0924593	0.0004814

Exercise 7.3 (*cont.*)

Regimes variable: Regimes

REGRESSION DIAGNOSTICS
MULTICOLLINEARITY CONDITION NUMBER   8.472

TEST ON NORMALITY OF ERRORS

TEST	DF	VALUE	PROB
Jarque-Bera	2	1.318	0.5173

DIAGNOSTICS FOR HETEROSCEDASTICITY
RANDOM COEFFICIENTS

TEST	DF	VALUE	PROB
Breusch-Pagan test	1	10.008	0.0016
Koenker-Bassett test	1	7.572	0.0059

DIAGNOSTICS FOR SPATIAL DEPENDENCE

TEST	MI/DF	VALUE	PROB
Lagrange Multiplier (lag)	1	12.860	0.0003
Robust LM (lag)	1	0.098	0.7549
Lagrange Multiplier (error)	1	17.334	0.0000
Robust LM (error)	1	4.572	0.0325
Lagrange Multiplier (SARMA)	2	17.431	0.0002

Interpreting results for Regime 1: Detailed results and interpretation are presented in Table 7.4.

Table 7.4 Interpretation of basic regression diagnostics for spatial error by regimes for regime 1.

	Value	Significance	Interpretation	Remarks	Evaluation
Model Overall Significance/Performance					
R-Squared	0.4322	Not applicable (N/A)	The model explains 43.2% of income variation.	Adjusted R-squared should also be checked to avoid model overfitting. Adjusted R-squared is preferred, especially when their difference is large.	☒

Exercise 7.3 (*cont.*)

Table 7.4 (*cont.*)

	Value	Significance	Interpretation	Remarks	Evaluation
Adjusted R-Squared	0.4064	Not applicable (N/A)	Adjusted R-squared is 0.406, indicating relatively low goodness of fit.	In case of low goodness of fit, we should add extra variables and/or test other regression methods (e.g., spatial lag model).	☒
F-Statistic (Also called Joint F-Statistic	16.748	Significant at the 0.01 significance level	We reject the null hypothesis and accept that the regression model is statistically significant. The trend is a real effect, not due to random sampling of the population.	F-statistic is of minor importance, as usually it is significant.	☑
Akaike Info Criterion (AIC)	468.656	N/A	It is used for comparison with the spatial model. AICc indicates better fit when values decrease.	AICc is a measure of the quality of models for the same set of data. It is used to compare different models and select the most appropriate one.	
Coefficients					
t-statistic	4.092	Variable significant at the 0.01 significance level: Insurance			☑

Exercise 7.3 *(cont.)*

Table 7.4 *(cont.)*

	Value	Significance	Interpretation	Remarks	Evaluation
Spatial Dependence					
Lagrange Multiplier Lag/Robust	12.860/ 0.098	LM Lag significant/ Robust LM Lag not significant	Indication for utilizing the spatial lag model	Used in OLS to guide us in the selection between the spatial lag or the spatial error model. See Figure 7.1.	☑
Lagrange Multiplier Error/ Robust	17.334/ 4.572	LM Error significant/ Robust LM Error significant	Due to spatial error autocorrelation presence, we should consider the spatial error model as an alternative, to account for spatial dependency	In this case, both LM lag and LM error are significant, so we should check their robust forms. Only the robust LM error is significant, so we should test a spatial error model.	☒
Overall Conclusion	Model seems to have low goodness of fit and Lagrange multiplier error is statistically significant.				
Next Step	Check Regime 2 and consider spatial error by regimes.				

```
SUMMARY OF OUTPUT: ORDINARY LEAST SQUARES ESTIMATION - REGIME 2
-----------------------------------------------------------------
Dataset              :CityGeoDa.dbf
Weights matrix       : City_k8.gwt
Dependent Variable   :    2_Income      Number of Observations :       66
Mean dependent var   :  14126.7309      Number of Variables    :        2
S.D. dependent var   :   2426.5459      Degrees of Freedom     :       64
R-squared            :      0.8815
Adjusted R-squared   :      0.8797
Sum squared residual: 45337330.710      F-statistic            :  476.2744
Sigma-square         :    708395.792    Prob(F-statistic)      :  2.386e-31
S.E. of regression   :     841.663      Log likelihood         :  -537.170
Sigma-square ML      :    686929.253    Akaike info criterion  :  1078.339
S.E of regression ML :     828.8120     Schwarz criterion      :  1082.718
```

Exercise 7.3 (*cont.*)

Variable	Coefficient	Std.Error	t-Statistic	Probability
2_CONSTANT	7556.9160121	318.3684549	23.7363844	0.0000000
2_Insurance	34.9396571	1.6009951	21.8237127	0.0000000

Regimes variable: Regimes

REGRESSION DIAGNOSTICS

MULTICOLLINEARITY CONDITION NUMBER   5.979

TEST ON NORMALITY OF ERRORS

TEST	DF	VALUE	PROB
Jarque-Bera	2	18.211	0.0001

DIAGNOSTICS FOR HETEROSCEDASTICITY

RANDOM COEFFICIENTS

TEST	DF	VALUE	PROB
Breusch-Pagan test	1	48.606	0.0000
Koenker-Bassett test	1	21.473	0.0000

DIAGNOSTICS FOR SPATIAL DEPENDENCE

TEST	MI/DF	VALUE	PROB
Lagrange Multiplier (lag)	1	8.469	0.0036
Robust LM (lag)	1	0.709	0.3998
Lagrange Multiplier (error)	1	17.135	0.0000
Robust LM (error)	1	9.375	0.0022
Lagrange Multiplier (SARMA)	2	17.844	0.0001

REGIMES DIAGNOSTICS - CHOW TEST

VARIABLE	DF	VALUE	PROB
CONSTANT	1	0.036	0.8501
Insurance	1	0.049	0.8243
Global test	2	2.683	0.2614

DIAGNOSTICS FOR GLOBAL SPATIAL DEPENDENCE

Residuals are treated as homoscedastic for the purpose of these tests

TEST	MI/DF	VALUE	PROB
Lagrange Multiplier (lag)	1	40.567	0.0000
Robust LM (lag)	1	1.178	0.2777
Lagrange Multiplier (error)	1	64.269	0.0000
Robust LM (error)	1	24.880	0.0000
Lagrange Multiplier (SARMA)	2	65.447	0.0000

============================ END OF REPORT ============================

Exercise 7.3 *(cont.)*

Interpreting results for Regime 2: Detailed results and interpretation are presented in Table 7.5.

Table 7.5 Interpretation of basic regression diagnostics for spatial error by regimes for regime 2.

	Value	Significance	Interpretation	Remarks	Evaluation
Model's Overall Significance/Performance					
R-Squared	0.8815	Not applicable (N/A)	The model explains 88.15% of income variation.	Adjusted R-squared should also be checked to avoid model overfitting. Adjusted R-squared is preferred, especially when their difference is large.	☑
Adjusted R-Squared	0.8797	Not applicable (N/A)	Adjusted R-squared is 0.8797 indicating high goodness of fit.	In cases of low goodness of fit, we should add extra variables and/or test other regression methods (e.g., spatial lag model).	☑
F-Statistic (Also called Joint F-Statistic)	476.274	Significant at the 0.01 significance level	We reject the null hypothesis and accept that the regression model is statistically significant. The trend is a real effect, not due to random sampling of the population.	F-statistic is of minor importance, as usually it is significant.	☑
Akaike Info Criterion (AIC)	1078.339	N/A	It is used for comparison with the spatial model. AICc indicates better fit when values decrease.	AICc is a measure of the quality of models for the same set of data. It is used to compare different models and select the most appropriate one.	
Coefficients					
t-statistic	21.284	Variable significant at the 0.01 significance level: Insurance			☑

Exercise 7.3 (cont.)

Table 7.5 (cont.)

	Value	Significance	Interpretation	Remarks	Evaluation
Spatial Dependence					
Lagrange Multiplier Lag/Robust	8.469/ 0.709	LM Lag significant/ Robust LM Lag not significant	Indication for utilizing the spatial lag model	Used in OLS to guide us in the selection between the spatial lag or the spatial error model. See Figure 7.1.	☑
Lagrange Multiplier Error/ Robust	17.135/ 9.375	LM Error significant/ Robust LM Error significant	Due to spatial error autocorrelation presence, we should consider the spatial error model as an alternative, to account for spatial dependence	In this case, both LM lag and LM error are significant, so we should check their robust forms. Only the Robust LM error is significant, so we should test a spatial error model.	☒
Overall Conclusion	Model seems to have high goodness of fit, and Lagrange multiplier error is statistically significant. Regime 1 has a low goodness of fit, indicating that a potential independent variable is missing.				
Next Step	Check Regime 2 and consider spatial error by regimes.				

In addition to the reports per regime, we also get the Chow test diagnostic for testing the null hypothesis that coefficients are statistical different among regimes, as well as diagnostics for global spatial dependence.

```
REGIMES DIAGNOSTICS - CHOW TEST
              VARIABLE      DF      VALUE          PROB
              CONSTANT       1      0.036         0.8501
             Insurance       1      0.049         0.8243
           Global test       2      2.683         0.2614


DIAGNOSTICS FOR GLOBAL SPATIAL DEPENDENCE
Residuals are treated as homoscedastic for the purpose of these tests
TEST                         MI/DF       VALUE         PROB
Lagrange Multiplier (lag)        1      40.567        0.0000
Robust LM (lag)                  1       1.178        0.2777
Lagrange Multiplier (error)      1      64.269        0.0000
Robust LM (error)                1      24.880        0.0000
Lagrange Multiplier (SARMA)      2      65.447        0.0000
============================ END OF REPORT ============================
```

Exercise 7.3 (*cont.*)

Table 7.6 Regime diagnostics and global spatial dependence

	Value	Significance	Interpretation	Remarks
Regimes Diagnostics				
Chow Test	See regimes diagnostics – Chow test	Global test is not significant at the 0.01 significance level	We cannot reject the null hypothesis (equal coefficients).	Inspecting Insurance coefficients reveal small differences in their coefficients (36.97, 34.93).
Spatial Dependence Global				
Lagrange Multiplier Lag/Robust	See Diagnostics for global spatial dependence results	LM Lag significant/ Robust LM Lag not significant	Due to spatial error autocorrelation presence, we should consider the spatial error model as an alternative, to account for spatial dependency. Results for global spatial dependence are in accordance with results in Exercise 7.1.	Used in OLS to guide us in the selection between the spatial lag or the spatial error model. See Figure 7.1.
Lagrange Multiplier Error/ Robust	See Diagnostics for global spatial dependence results	LM Error significant/ Robust LM Error significant		

Exercise 7.4 Spatial Error by Spatial Regimes

We proceed with the spatial error model with maximum likelihood estimation, considering spatial error by regime.

   GeoDa Tools to be used: Spatial Error by Regimes

ACTION: Spatial Error by Regimes

Open GeoDaSpace > Open I:\BookLabs\Lab7\Output\SpatialErrorMo-del.mdl

Click on Variables Selection and drag and drop Regimes to R field (see Figure 7.11)

Model Estimation: Spatial Error

Method: GMM > Run

As soon as you click Run, you will be asked to save the predicted values and residuals in csv format

Save in >

I:\BookLabs\Lab7\Output\SpatialErrorRegimesCityGeoDa_predY_re-sid.csv

Save Results File As >

I:\BookLabs\Lab7\Output\SpatialErrorRegimes.txt

Main Toolbar > Save Model As >

I:\BookLabs\Lab7\Output\SpatialErrorRegimesModel.mdl

Exercise 7.4 (*cont.*)

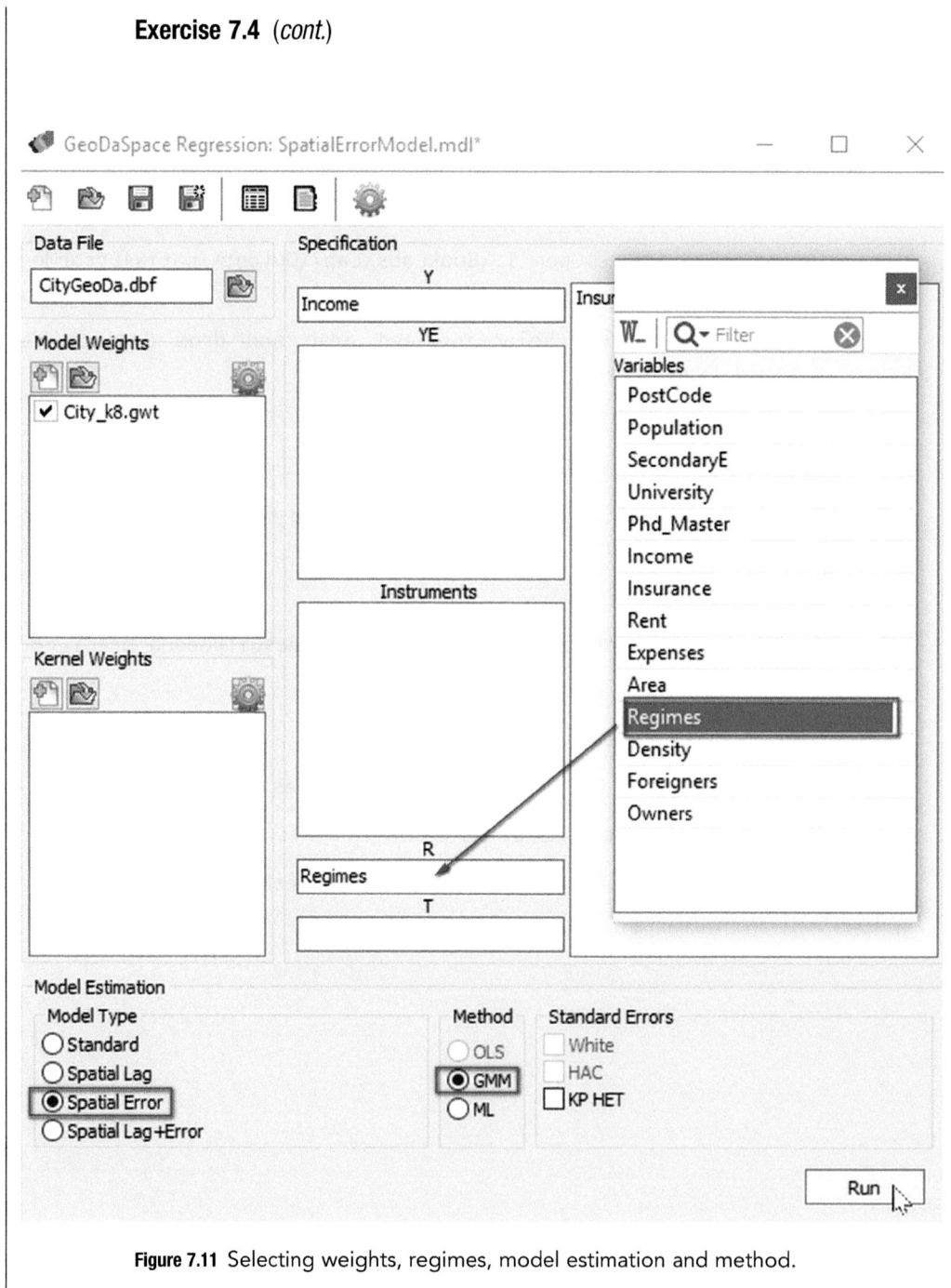

Figure 7.11 Selecting weights, regimes, model estimation and method.

Exercise 7.4 (*cont.*)

Results of O SpatialErrorRegimes.txt are as follows:

REGRESSION
——————————

SUMMARY OF OUTPUT: SPATIALLY WEIGHTED LEAST SQUARES ESTIMATION (HOM) – REGIME 1

Dataset	:CityGeoDa.dbf			
Weights matrix	: City_k8.gwt			
Dependent Variable	: 1_Income	Number of Observations:		24
Mean dependent var	: 22339.3154	Number of Variables :		2
S.D. dependent var	: 5248.4513	Degrees of Freedom :		22
Pseudo R-squared	: 0.4322			

Variable	Coefficient	Std.Error	z-Statistic	Probability
1_CONSTANT	5778.3534159	3146.3243692	1.8365409	0.0662777
1_Insurance	42.8402155	6.9763088	6.1408141	0.0000000
1_lambda	0.7253011	0.1446441	5.0143826	0.0000005

Regimes variable: Regimes
————————————

SUMMARY OF OUTPUT: SPATIALLY WEIGHTED LEAST SQUARES ESTIMATION (HOM) – REGIME 2

Dataset	:CityGeoDa.dbf			
Weights matrix	: City_k8.gwt			
Dependent Variable	: 2_Income	Number of Observations:		66
Mean dependent var	: 14126.7309	Number of Variables :		2
S.D. dependent var	: 2426.5459	Degrees of Freedom :		64
Pseudo R-squared	: 0.8815			

Variable	Coefficient	Std.Error	z-Statistic	Probability
2_CONSTANT	7694.5642575	421.1609982	18.2698880	0.0000000
2_Insurance	34.2548771	2.0069269	17.0683231	0.0000000
2_lambda	0.5390750	0.1196396	4.5058231	0.0000066

Exercise 7.4 (*cont.*)

```
Regimes variable: Regimes

REGIMES DIAGNOSTICS - CHOW TEST
                    VARIABLE      DF       VALUE          PROB
                    CONSTANT      1        0.364          0.5461
                   Insurance     1        1.399          0.2369
                      lambda     1        0.984          0.3212
                  Global test    3        2.572          0.4624
================================ END OF REPORT ================================
```

Interpreting results: The spatial parameter estimate of lambda is significant in both regimes. As the Chow test shows, lamda and coefficients are not statistically significantly different in the two regimes. The overall findings reveal that spatial error by regimes does not yield useful results.

References

Akaike, H. (1974). A new look at the statistical model identification. *IEEE Transactions on Automatic Control*, 19(6), 716–723.

Alpaydin, E. (2009). *Introduction to Machine Learning*. MIT Press.

Alvanides, S., & Openshaw, S. (1999). Zone design for planning and policy analysis. In Stillwell, J., Geertman, S., & Openshaw, S. (eds.). *Geographical Information and Planning* (299–315). Springer.

Ankerst, M., Breunig, M. M., Kriegel, H. P., & Sander, J. (1999). OPTICS: Ordering points to identify the clustering structure. *ACM Sigmod Record*, 28(2), 49–60.

Anselin, L. (1988). *Spatial Econometrics: Methods and Models*. Kluwer Academic. Retrieved from https://link.springer.com/book/10.1007%2F978-94-015-7799-1.

Anselin, L. (1989). *What Is Special about Spatial Data*. Technical Report 8-4. Santa Barbara: University of California, Santa Barbara, National Center for Geographic Information and Analysis.

Anselin, L. (1995). Local indicators of spatial association-LISA. *Geographical Analysis*, 27(2), 93–115.

Anselin, L. (2003). Spatial econometrics. In Baltagi, B. (ed.). *A Companion to Theoretical Econometrics* (310–330). Blackwell.

Anselin, L. (2005). *Exploring Spatial Data with GeoDaTM: A Workbook*. Center for Spatially Integrated Social Science.

Anselin, L. (2010). Thirty years of spatial econometrics. *Papers in Regional Science*, 89(1), 3–25.

Anselin, L. (2014). *Modern Spatial Econometrics in Practice: A Guide to GeoDa, GeoDa-Space and PySAL*. GeoDa Press.

Anselin, L. (2016). *Spatial Autocorrelation*. The Centre of Spatial Data Science. University of Chicago. Retrieved from https://spatial.uchicago.edu/sites/spacial-data.uchicago.edu/files/3_spatial_autocorrelation_r.pdf.

Anselin, L. (2018). *GeoDa: An Introduction to Spatial Data Analysis*. GeoDa. Retreived from https://geodacenter.github.io/workbook/5b_global_adv/lab5b.html#fn1 and https://geodacenter.github.io/workbook/4a_contig_weights/lab4a.html.

Anselin, L., & Kelejian, H. H. (1997). Testing for spatial error autocorrelation in the presence of endogenous regressors. *International Regional Science Review*, 20(1–2), 153–182.

Arthur, D., & Vassilvitskii, S. (2007, January). k-means++: The advantages of careful seeding. In *Proceedings of the Eighteenth Annual ACM-SIAM Symposium on Discrete Algorithms* (1027–1035). Society for Industrial and Applied Mathematics.

Assunção, R. M., Neves, M. C., Câmara, G., & da Costa Freitas, C. (2006). Efficient regionalization techniques for socio-economic geographical units using minimum spanning trees. *International Journal of Geographical Information Science*, 20(7), 797–811.

Bailey, T. C., & Gatrell, A. C. (1995). *Interactive Spatial Data Analysis* (Vol. 413). Longman Scientific & Technical.

Baller, R. D., Anselin, L., Messner, S. F., Deane, G., & Hawkins, D. F. (2001). Structural covariates of US county homicide rates: Incorporating spatial effects. *Criminology*, 39(3), 561–588.

Belsley, D. A., Kuh, E., & Welsch, R. E. (1980). *Regression Diagnostics: Identifying Influential Data and Sources of Collinearity*. John Wiley & Sons. doi: 10.1002/0471725153.

Benjamini, Y. (2010). Discovering the false discovery rate. *Journal of the Royal Statistical Society: Series B (Statistical Methodology)*, 72(4), 405–416.

Benjamini, Y., & Hochberg, Y. (1995). Controlling the false discovery rate: A practical and powerful approach to multiple testing. *Journal of the Royal Statistical Society: Series B (Methodological)*, 57(1), 289–300.

Bivand, R. S., Pebesma, E. J., Gomez-Rubio, V., & Pebesma, E. J. (2008). *Applied Spatial Data Analysis with R* (Vol. 747248717). Springer.

Blaikie, N. (2003). *Analyzing Quantitative Data*. SAGE Publications Ltd. doi: 10.4135/9781849208604.

Bonferroni, C. (1936). Teoria statistica delle classi e calcolo delle probabilita. *Pubblicazioni del R Istituto Superiore di Scienze Economiche e Commericiali di Firenze*, 8, 3–62.

Breusch, T. S., & Pagan, A. R. (1979). A simple test for heteroscedasticity and random coefficient variation. *Econometrica: Journal of the Econometric Society*, 1287–1294.

Briggs, R. (2012). Lecture notes on spatial correlation and regression. Retrieved from www.utdallas.edu/~briggs/.

Brunsdon, C., Fotheringham, A. S., & Charlton, M. E. (1996). Geographically weighted regression: A method for exploring spatial nonstationarity. *Geographical Analysis*, 28(4), 281–298.

Brusco, M. J., & Cradit, J. D. (2001). A variable-selection heuristic for K-means clustering. *Psychometrika*, 66(2), 249–270.

Burt, J. E., Barber, G. M., & Rigby, D. L. (2009). *Elementary Statistics for Geographers*. Guilford Press.

Caldas de Castro, M., & Singer, B. H. (2006). Controlling the false discovery rate: A new application to account for multiple and dependent tests in local statistics of spatial association. *Geographical Analysis*, 38(2), 180–208.

Caliński, T., & Harabasz, J. (1974). A dendrite method for cluster analysis. *Communications in Statistics – Theory and Methods*, 3(1), 1–27.

Campello, R. J., Moulavi, D., Zimek, A., & Sander, J. (2015). Hierarchical density estimates for data clustering, visualization, and outlier detection. *ACM Transactions on Knowledge Discovery from Data (TKDD)*, 10(1), 5.

Cangelosi, R., & Goriely, A. (2007). Component retention in principal component analysis with application to cDNA microarray data. *Biology Direct*, 2(1), 2.

Chambers, J. M., Cleveland, W. S., Kleiner, B., & Tukey, P. A. (1983). Comparing data distributions. In *Graphical Methods for Data Analysis* (47–73). Bell Laboratories.

Charlton, M., Brunsdon, C., Demsar, U., Harris, P., & Fotheringham, S. (2010). Principal components analysis: From global to local. In 13th AGILE International Conference on Geographic Information Science, *Guimaraes, Portugal*.

Clark, P. J., & Evans, F. C. (1954). Distance to nearest neighbor as a measure of spatial relationships in populations. *Ecology*, 35(4), 445–453.

Dall'erba, S. (2009). Exploratory spatial data analysis. In Kitchin, R., & Thrift, N. (eds.). *International Encyclopedia of Human Geography* (Vol. 3) (683–690). Elsevier.

Darmofal, D. (2015). *Spatial Analysis for the Social Sciences* (Analytical Methods for Social Research). Cambridge University Press. doi: 10.1017/CBO9781139051293.

Demšar, U., Harris, P., Brunsdon, C., Fotheringham, A. S., & McLoone, S. (2013). Principal component analysis on spatial data: An overview. *Annals of the Association of American Geographers*, 103(1), 106–128.

de Smith, M. J. (2018). *Statistical Analysis Handbook: A Comprehensive Handbook of Statistical Concepts, Techniques and Software Tools*. Retrieved from www.statsref.com/HTML/index.html.

de Smith, M. J., Goodchild, M. F., & Longley, P. (2018). *Geospatial Analysis: A Comprehensive Guide to Principles, Techniques and Software Tools* (6th edn.). Retrieved from www.spatialanalysisonline.com.

de Vaus, D. (2002). *Analyzing Social Science Data: 50 Key Problems in Data Analysis*. Sage.

Duque, J. C., Ramos, R., & Suriñach, J. (2007). Supervised regionalization methods: A survey. *International Regional Science Review*, 30(3), 195–220.

Ester, M., Kriegel, H. P., Sander, J., & Xu, X. (1996). A density-based algorithm for discovering clusters in large spatial databases with noise. *Proceedings of Knowledge Discovery and Data Mining*, 96(34), 226–231.

ESRI. (2014). How kernel density works. *ESRI*. Retrieved from http://pro.arcgis.com/en/ pro-app/tool-reference/spatial-analyst/how-kernel-density-works.htm.

ESRI. (2015). Incremental spatial autocorrelation. *ESRI*. Retrieved from http://desktop .arcgis.com/en/arcmap/10.3/tools/spatial-statistics-toolbox/incremental-spatial-autocorrelation.htm.

ESRI. (2016a). Interpreting exploratory regression results. *ESRI*. Retrieved from http:// pro.arcgis.com/en/pro-app/tool-reference/spatial-statistics/interpreting-exploratory-regression-results.htm.

ESRI. (2016b). Geographically weighted regression (GWR). *ESRI*. Retrieved from http:// desktop.arcgis.com/en/arcmap/10.3/tools/spatial-statistics-toolbox/geographic ally-weighted-regression.htm.

ESRI. (2017). How emerging hot spot analysis works. *ESRI*. Retrieved from http://pro .arcgis.com/en/pro-app/tool-reference/space-time-pattern-mining/ learnmoreemerging.htm.

ESRI. (2018a). An overview of the mapping clusters toolset. *ESRI*. Retrieved from http:// pro.arcgis.com/en/pro-app/tool-reference/spatial-statistics/an-overview-of-the-map ping-clusters-toolset.htm.

ESRI. (2018b). How similarity search works. *ESRI*. Retrieved from http://pro.arcgis.com/ en/pro-app/tool-reference/spatial-statistics/how-similarity-search-works.htm.

ESRI. (2018c). What is a z-score? What is a p-value?. *ESRI*. Retrieved from https:// pro.arcgis.com/en/pro-app/tool-reference/spatial-statistics/what-is-a-z-score-what-is-a-p-value.htm#ESRI_SECTION1_2C5DFC8106F84F988982CABAEDBF1440.

ESRI. (2019). How grouping analysis works. *ESRI*. Retrieved from http://pro.arcgis.com/en/pro-app/tool-reference/spatial-statistics/how-grouping-analysis-works.htm.

Fischer, M. M., & Getis, A. (eds.). (2010). *Handbook of Applied Spatial Analysis: Software Tools, Methods and Applications*. Springer Science & Business Media.

Fotheringham, S., Brunsdon, C., & Charlton, M. (2000). *Quantitative Geography: Perspectives on Spatial Data Analysis*. Sage.

Fotheringham, S., Brunsdon, C., & Charlton, M. (2002). *Geographically Weighted Regression: The Analysis of Spatially Varying Relationships*. Wiley.

Gangodagamage, C., Zhou, X., & Lin, H. (2008). Autocorrelation, spatial. In Shekhar, S., & Xiong, H. (eds.). *Encyclopedia of GIS*. Springer Science & Business Media.

Gatrell, A. C., Bailey, T. C., Diggle, P. J., & Rowlingson, B. S. (1996). Spatial point pattern analysis and its application in geographical epidemiology. *Transactions of the Institute of British Geographers*, 256–274.

Geary, R. C. (1954). The contiguity ratio and statistical mapping. *The Incorporated Statistician*, 5, 115–145.

Getis, A. (1990). Screening for spatial dependence in regression analysis. *Papers in Regional Science*, 69(1), 69–81.

Getis, A. (1995). Spatial filtering in a regression framework: Examples using data on urban crime, regional inequality, and government expenditures. In Anselin, L., & Florax, R. J. G. M. (eds.). *New Directions in Spatial Econometrics: Advances in Spatial Science*. Springer.

Getis, A., & Griffith, D. A. (2002). Comparative spatial filtering in regression analysis. *Geographical Analysis*, 34(2), 130–140.

Getis, A., & Ord, J. K. (1992). The analysis of spatial association by use of distance statistics. *Geographical Analysis*, 24(3), 189–206.

Gnanadesikan, R., Kettenring, J. R., & Tsao, S. L. (1995). Weighting and selection of variables for cluster analysis. *Journal of Classification*, 12(1), 113–136.

Goodchild, M. (2008). Data analysis, spatial. In Shekhar, S., & Xiong, H., (eds.). *Encyclopedia of GIS*. Springer Science & Business Media.

Goulard, M., Laurent, T., & Thomas-Agnan, C. (2017). About predictions in spatial autoregressive models: Optimal and almost optimal strategies. *Spatial Economic Analysis*, 12(2–3), 304–325.

Grekousis, G. (2013a). Giving fuzziness to spatial clusters: A new index for choosing the optimal number of clusters. *International Journal on Artificial Intelligence Tools*, 22 (03), 1350009.

Grekousis, G. (2018). Further widening or bridging the gap? A cross-regional study of unemployment across the EU Amid Economic Crisis. *Sustainability*, 10(6), 1702.

Grekousis, G. (2019). Artificial neural networks and deep learning in urban geography: A systematic review and meta-analysis. *Computers, Environment and Urban Systems*, 74, 244–256. https://doi.org/10.1016/j.compenvurbsys.2018.10.008.

Grekousis, G., & Fotis, Y. N. (2012). A fuzzy index for detecting spatiotemporal outliers. *Geoinformatica*, 16(3), 597–619.

Grekousis, G., & Gialis, S. (2018). More flexible yet less developed? Spatio-temporal analysis of labor flexibilization and gross domestic product in crisis-hit European Union regions. *Social Indicators Research*, 1–20.

Grekousis, G., & Hatzichristos, T. (2013). Fuzzy clustering analysis in geomarketing research. *Environment and Planning B: Planning and Design*, 40(1), 95–116.

Grekousis, G., & Liu, Y. (2019). Where will the next emergency event occur? Predicting ambulance demand in emergency medical services using artificial intelligence. *Computers, Environment and Urban Systems*, 76, 110–122. https://doi.org/10.1016/j.compenvurbsys.2019.04.006.

Grekousis, G., & Mountrakis, G. (2015). Sustainable development under population pressure: Lessons from developed land consumption in the conterminous US. *PloS One*, 10(3), e0119675.

Grekousis, G., & Photis, Y. N. (2014). Analyzing high-risk emergency areas with GIS and neural networks: The case of Athens, Greece. *The Professional Geographer*, 66(1), 124–137.

Grekousis, G., & Thomas, H. (2012). Comparison of two fuzzy algorithms in geodemographic segmentation analysis: The Fuzzy C-Means and Gustafson–Kessel methods. *Applied Geography*, 34, 125–136.

Grekousis, G., Kavouras, M., & Mountrakis, G. (2015a). Land cover dynamics and accounts for European Union 2001–2011. In *Third International Conference on Remote Sensing and Geoinformation of the Environment (RSCy2015)* (Vol. 9535, p. 953507). International Society for Optics and Photonics.

Grekousis, G., Manetos, P., & Photis, Y. N. (2013). Modeling urban evolution using neural networks, fuzzy logic and GIS: The case of the Athens metropolitan area. *Cities*, 30, 193–203.

Grekousis, G., Mountrakis, G., & Kavouras, M. (2015b). An overview of 21 global and 43 regional land-cover mapping products. *International Journal of Remote Sensing*, 36 (21), 5309–5335.

Grekousis, G., Mountrakis, G., & Kavouras, M. (2016). Linking MODIS-derived forest and cropland land cover 2011 estimations to socioeconomic and environmental indicators for the European Union's 28 countries. *GIScience & Remote Sensing*, 53(1), 122–146.

Griffith, D. A., & Amrhein, C. G. (1997). *Multivariate Statistical Analysis for Geographers*. Prentice Hall.

Guo, D. (2008). Regionalization with dynamically constrained agglomerative clustering and partitioning (REDCAP). *International Journal of Geographical Information Science*, 22(7), 801–823.

Haining, P. R. (2010). The nature of georeferenced data. In Fischer, M., & Getis, A. (eds.). *Handbook of Applied Spatial Analysis*. Springer.

Halkidi, M., Batistakis, Y., & Vazirgiannis, M. (2001). On clustering validation techniques. *Journal of Intelligent Information Systems*, 17(2–3), 107–145.

Hamilton L. (2014). *Introduction to Principal Component Analysis (PCA)*. Retrieved from www.lauradhamilton.com/introduction-to-principal-component-analysis-pca.

Harris, P., Brunsdon, C., & Charlton, M. (2011). Geographically weighted principal components analysis. *International Journal of Geographical Information Science*, 25 (10), 1717–1736.

Illian, J., Penttinen, A., Stoyan, H., & Stoyan, D. (2008). *Statistical Analysis and Modelling of Spatial Point Patterns* (Vol. 70). John Wiley & Sons.

Jackson, J. E. (1991). *A User's Guide to Principal Components*. John Wiley & Sons.

Jain, A. K., & Dubes, R. C. (1988). *Algorithms for Clustering Data*. Prentice-Hall.

Jiang, M. F., Tseng, S. S., & Su, C. M. (2001). Two-phase clustering process for outliers detection. *Pattern Recognition Letters*, 22(6–7), 691–700.

Jolliffe, I. T. (2002). *Principal Component Analysis*. Springer Series in Statistics. Springer.

Kelejian, H. H., & Prucha, I. R. (2010). Specification and estimation of spatial autoregressive models with autoregressive and heteroskedastic disturbances. *Journal of Econometrics*, 157(1), 53–67.

Kim, C. W., Phipps, T. T., & Anselin, L. (2003). Measuring the benefits of air quality improvement: A spatial hedonic approach. *Journal of Environmental Economics and Management*, 45(1), 24–39.

Koenker, R., & Hallock, K. (2001). Quantile regression: An introduction. *Journal of Economic Perspectives*, 15(4), 43–56.

Krivoruchko, K. (2011). *Spatial Statistical Data Analysis for GIS Users*, ESRI Press.

Kulin, H. W., & Kuenne, R. E. (1962). An efficient algorithm for the numerical solution of the generalized Weber problem in spatial economics. *Journal of Regional Science*, 4 (2), 21–33.

Lacey, M. (1997). *Multiple Linear Regression*. Retrieved from www.stat.yale.edu/Courses/1997-98/101/linmult.htm.

Lee, J., & Wong, D. W. (2001). *Statistical Analysis with ArcView GIS*. John Wiley & Sons.

LeSage, J. P. (1998). *Spatial Econometrics*. Report retrieved from http://spatial-econometrics.com/html/wbook.pdf.

Linneman, T. J. (2011). *Social Statistics: The Basics and Beyond*. Routledge.

Longley, P. A., Goodchild, M. F., Maguire, D. J., & Rhind, D. W. (2011). *Geographic Information Science and Systems* (3rd edn.). John Wiley & Sons.

Liu, Y., Wang, R., Grekousis, G., Liu, Y., Yuan, Y., & Li, Z. (2019). Neighbourhood greenness and mental wellbeing in Guangzhou, China: What are the pathways? *Landscape and Urban Planning*, 190, 103603. https://doi.org/10.1016/j.landurbplan.2019.103602.

Lu, Y., & Thill, J. C. (2003). Assessing the cluster correspondence between paired point locations. *Geographical Analysis*, 35(4), 290–309.

MacQueen, J. (1967). Some methods for classification and analysis of multivariate observations. *Proceedings of the Fifth Berkeley Symposium on Mathematical Statistics and Probability*, 1(14), 281–297.

Mitchell, A. (2005). *The ESRI Guide to GIS Analysis, Volume 2*. ESRI Press.

Moran, P. A. P. (1950). Notes on continuous stochastic phenomena. *Biometrika*, 37, 17–23.

Murray, A. T. (2010). Quantitative geography. *Journal of Regional Science*, 50(1), 143–163.

Neyman, J. (1937). X: Outline of a theory of statistical estimation based on the classical theory of probability. *Philosophical Transactions of the Royal Society of London. Series A, Mathematical and Physical Sciences*, 236(767), 333–380.

Newton, R. R., & Rudestam, K. E. (1999). *Your Statistical Consultant: Answers to Your Data Analysis Questions*. Sage.

Ogee, A., et al. (2013). Multiple regression analysis: Use adjusted R-squared and predicted R-squared to include the correct number of variables. *Minitab*. Retrieved from http://blog.minitab.com/blog/adventures-in-statistics-2/multiple-regession-analysis-use-adjusted-r-squared-and-predicted-r-squared-to-include-the-correct-number-of-variables.

Openshaw, S. (1977). A geographical solution to scale and aggregation problems in region-building, partitioning and spatial modelling. *Transactions of the Institute of British Geographers*, 459–472.

Ord, K. (1975). Estimation methods for models of spatial interaction. *Journal of the American Statistical Association*, 70(349), 120–126.

Ord, J. K., & Getis, A. (1995). Local spatial autocorrelation statistics: Distributional issues and an application. *Geographical Analysis*, 27(4), 286–306.

O'Sullivan, D., & Unwin, D. (2003). *Geographic Information Analysis*. John Wiley & Sons.

O'Sullivan, D., & Unwin, D. (2010). *Geographic Information Analysis* (2nd edn.). John Wiley & Sons.

Oyana, T. J., & Margai, F. (2015). *Spatial Analysis: Statistics, Visualization, and Computational Methods*. CRC Press.

Paez, A., & Scott, D. M. (2004). Spatial statistics for urban analysis: A review of techniques with examples. *GeoJournal*, 61(1), 53–67.

Pallant, J. (2013). *SPSS Survival Manual*. McGraw-Hill Education (UK).

Peck, R., Olsen, C., & Devore, J. L. (2012). *Introduction to Statistics and Data Analysis* (4th edn.). Cengage Learning.

Pena, J. M., Lozano, J. A., & Larranaga, P. (1999). An empirical comparison of four initialization methods for the k-means algorithm. *Pattern Recognition Letters*, 20(10), 1027–1040.

Penn State University. (2018). STAT 505: Lesson 11; Principal Components Analysis (pca). Retrieved from https://onlinecourses.science.psu.edu/stat505/node/49/.

Photis, Y. N., & Grekousis, G. (2012). Locational planning for emergency management and response: An artificial intelligence approach. *International Journal of Sustainable Development and Planning*, 7(3), 372–384.

Rawlings, J. O., Pantula, S. G., & Dickey, D. A. (1998). *Applied Regression Analysis: A Research Tool*. Springer Science & Business Media.

Ripley, B. D. (1976). The second-order analysis of stationary point processes. *Journal of Applied Probability*, 13(2), 255–266.

Rogerson, P. A. (2001). *Statistical Methods for Geography*. Sage.

Sander, J., Ester, M., Kriegel, H. P., & Xu, X. (1998). Density-based clustering in spatial databases: The algorithm gdbscan and its applications. *Data Mining and Knowledge Discovery*, 2(2), 169–194. https://doi.org/10.1023/A:1009745219419.

Sankey, T. T. (2008). Statistical descriptions of spatial patterns. In Shekhar, S., & Xiong H., (eds.). *Encyclopedia of GIS* (1135–1141). Springer Science & Business Media.

Schubert, E., Sander, J., Ester, M., Kriegel, H. P., & Xu, X. (2017). DBSCAN revisited, revisited: Why and how you should (still) use DBSCAN. *ACM Transactions on Database Systems (TODS)*, 42(3), 19.

Schwarz, G. (1978). Estimating the dimension of a model. *The Annals of Statistics*, 6(2), 461–464.

Scibila, B. (2017). Regression analysis: How to interpret S, the standard error of the regression. *Minitab*. Retrieved from http://blog.minitab.com/blog/adventures-in-stat istics-2/regression-analysis-how-to-interpret-s-the-standard-error-of-the-regression.

Scott, L. M., & Janikas, M. V. (2010). Spatial statistics in ArcGIS. In Fischer, M. M., & Getis, A. (eds.).*Handbook of Applied Spatial Analysis* (27–41). Springer.

Silverman, B. W. (1986). *Density Estimation for Statistics and Data Analysis*. Chapman and Hall.

Sokal, R. R., & Rohlf, F. J. (1962). The comparison of dendrograms by objective methods. *Taxon*, 33–40.

Stillwell, J., Geertman, S., & Openshaw, S. (eds.). (1999). *Geographical Information and Planning: European Perspectives.* Springer Science & Business Media.

Tabachnick, B. G., Fidell, L. S., & Ullman, J. B. (2012). *Using Multivariate Statistics* (6th edn.). Pearson.

Tobler, W. R. (1979). Cellular geography. In Gale, S., & Olsson, G. (eds.). *Philosophy in Geography* (379–386). Springer.

Thayn, J. B., & Simanis, J. M. (2013). Accounting for spatial autocorrelation in linear regression models using spatial filtering with eigenvectors. *Annals of the Association of American Geographers*, 103(1), 47–66.

Troy, A. (2008). Geodemographic segmentation. In Shekhar, S., & Xiong H. (eds.). *Encyclopedia of GIS* (347–355). Springer.

Tufte, E. R. (2001). *The Visual Display of Quantitative Information* (2nd edn.). Graphics Press.

Wang, F. (2014). *Quantitative Methods and Socio-economic Applications in GIS.* CRC Press.

Wang, R. et al. (2019). Perceptions of built environment and health outcomes for older Chinese in Beijing: A big data approach with street view images and deep learning technique. *Computers, Environment and Urban Systems*, 78. https://doi.org/10.1016/j.compenvurbsys.2019.101386.

Wheeler, D. C., & Páez, A. (2010). *Geographically Weighted Regression.* In Fischer, M., & Getis, A. (eds.). *Handbook of Applied Spatial Analysis* (461–486). Springer.

White, H. (1980). A heteroskedasticity-consistent covariance matrix estimator and a direct test for heteroskedasticity. *Econometrica: Journal of the Econometric Society*, 817–838.

Yrigoyen, C. (2007). *Session 7: Spatial Dependence Models: Estimation and Testing.* Lecture notes. Retrieved from www.uam.es/personal_pdi/economicas/coro/docencia/doctorado/spateconUPC/Slides/Session7_SpatialDepModel_Slides.pdf.

Zhou X., & Lin H. (2008) In Geary's C. Shekhar, S., & Xiong H. (eds.). *Encyclopedia of GIS.* Springer Science & Business Media.

Index